REMOTE
SENSING
of
PROTECTED LANDS

Taylor & Francis Series in
Remote Sensing Applications

Series Editor

Qihao Weng

Indiana State University
Terre Haute, Indiana, USA

Taylor & Francis Series in Remote Sensing Applications
Qihao Weng, Series Editor

REMOTE SENSING

of

PROTECTED LANDS

Edited By

Yeqiao Wang

CRC Press

Taylor & Francis Group

Boca Raton London New York

CRC Press is an imprint of the
Taylor & Francis Group, an **informa** business

CRC Press
Taylor & Francis Group
6000 Broken Sound Parkway NW, Suite 300
Boca Raton, FL 33487-2742

First issued in paperback 2019

ISBN-13: 978-1-4398-4187-7 (hbk)
ISBN-13: 978-0-367-38212-4 (pbk)

Library of Congress Cataloging-in-Publication Data

Remote sensing of protected lands / edited by Yeqiao Wang.
    p. cm.
  Includes bibliographical references and index.
  ISBN 978-1-4398-4187-7 (hardcover : alk. paper)
    1. Remote sensing. 2. Protected areas--Remote sensing. I. Wang, Yeqiao.

G70.4.R4686 2012
333.73'160285--dc23   2011030009

Visit the Taylor & Francis Web site at
http://www.taylorandfrancis.com

and the CRC Press Web site at
http://www.crcpress.com

Contents

Section II Remote Sensing for Inventory, Mapping, and Conservation of Protected Lands and Waters

Series Foreword

Remote Sensing of Protected Lands

Examples of protected lands include national parks, national forests, national seashores, natural preserves and designated areas, wildlife refuges, and last frontiers on the planet Earth. Continued population growth, economic activities, and global climate change have made the protected lands a challenge in resources management and environmental monitoring, since many of them would face the danger of being destroyed, intimidated, or contaminated. Remote sensing is one of the most effective technologies for inventory and monitoring of protected areas, for assessing their conditions, and for resources planning and management. I am pleased that that Professor Yeqiao Wang has the vision, ambition, and energy to edit a special volume in *Remote Sensing of Protected Lands*. Professor Wang is an internationally renowned expert in the field of terrestrial remote sensing. Because of his outstanding achievements, he was awarded the prestigious Presidential Early Career Award for Scientists and Engineers (PECASE) by President Clinton in 2000 and a Chang-Jiang (Yangtze River) endowed professorship by the Ministry of Education of China. It should be noted that Professor Wang has recently published a book on remote sensing of coastal environments in the same series, which is in tremendous demand and has received positive reviews from the readers.

Contributed by a group of leading and well-published scholars in the field, this book is the first to dedicate to the applications of remote sensing science and technology for inventory, monitoring, and managing of protected lands and waters. Professor Wang carefully selected and examined each contribution, and created a well-structured volume to address the use of remote sensing technology in protected lands from four perspectives, namely, changing landscape and change detection; monitoring and mapping methods; inventory and monitoring of frontier lands, and decision support for management of protected lands. This comprehensive approach allows readers to have both a systematic view of the field and a detailed knowledge of a particular topic, seeing both "forest" and "trees."

It has been roughly 4 years since the inception of the *Taylor & Francis Series in Remote Sensing Applications* in 2007. This book marks the eighth volume in the series. The book series is dedicated to recent developments in the theories, methods, and applications of remote sensing. Each book is designed to provide up-to-date developments in a chosen subfield of remote sensing applications. In spite of varying in format, these books contain similar components, such as review of theories and methods, analysis of case studies, and examination of the methods for applying remote sensing techniques to a particular practical area. The series is devised to serve as a guide or references for professionals, researchers, and scientists, as well as textbooks or important supplements for teachers and students. I hope that the publication of this book will promote a better use of remote sensing data, science, and technology, and will facilitate the assessing, monitoring, and managing of protected lands, the last resorts of our Earth. I congratulate Professor Wang on another important milestone in his career.

Qihao Weng, PhD
Hawthorn Woods, Indiana

Preface

About This Book

Humans have created protected areas over the past millennia for a multitude of reasons. Protected lands include areas such as national parks, national forests, national seashores, all levels of natural reserves, wildlife refuges and sanctuaries, and designated areas for conservation of biological diversity and natural and cultural heritage and their significance. Protected lands also include some of the last frontiers that have unique landscape characteristics and ecosystem functions. Along the shoreline and over the ocean and sea, marine-protected areas include part or whole of the enclosed environments of intertidal or subtidal terrain, together with its overlying water and associated flora, fauna, and historical and cultural features. Protected lands and waters serve as the fundamental building blocks of virtually all national and international conservation strategies. Protected areas are increasingly recognized as essential providers of ecosystem services and biological resources. With intensified debates about climate change and challenges in sustainability of natural resources and environments, protected lands become more important in serving as indicators of ecosystem conditions and functions either by the status of themselves and/or by comparisons with adjacent environments.

Traditionally, management of protected lands is operated with field-oriented observation and mapping approaches. Repaid development of remote sensing science and technologies has profoundly influenced and contributed to the practice in understanding and management of protected lands and waters from extended spatial, spectral, temporal, and thematic perspectives for which remote sensing has unique advantages.

This book consists of 23 chapters organized into four sections, including *1. Remote Sensing of Changing Landscape of Protected Lands; 2. Remote Sensing for Inventory; Mapping and Conservation of Protected Lands and Waters; 3. Remote Sensing for Inventory and Monitoring of Frontier Lands;* and *4. Remote Sensing in Decision Support for Management of Protected Lands.* The assembly of chapters with case studies around the world and perspectives from scholars and scientists representing multiple institutions and agencies showcases the state-of-the-art remote sensing applications in understanding the past, present, and future status of protected areas and the environments. As the chapters attest, remote sensing is among the most fascinating frontiers of science and technology. The achievements through applications of such science and technologies, the challenges, and the lessons learned, and recommendations in remote sensing of protected areas deserve special attention. I hope that this book can provide a snapshot about remote sensing applications to address issues that inventory, monitoring, and management of protected lands and waters would need. I also hope that this book can inspire a broader scope of interest in scientific research and management of the natural treasures of protected lands and waters on Earth.

Acknowledgments

My research experience in remote sensing of protected lands and my inspiration for working on this book came from different stages of my professional career. Starting with my graduate advisors and early-career mentors to my supportive colleagues and graduate students, many individuals were involved in providing guidance, support, and assistance throughout my career. Several of the chapters in this book, particularly Chapters 1, 6, 7, 9, 10, 11, 12, 14, and 22, reflect my recent research in remote sensing of protected lands. During these studies I have had the opportunity to work with top-notch scientists, scholars, staff, and enthusiastic graduate students and different agencies for funded scientific research on issues related to inventory, monitoring, management, and conservation of protected lands.

Many individuals helped me tremendously during the progress of this book, and to them I wish to express my sincere gratitude and appreciation. First of all, I thank all 69 contributors. Their insightful expertise and experience on the issues represent an authoritative understanding in inventory, monitoring, and management of protected lands with application of the advancement of remote sensing science and technology. Their dedication and hard work made this book possible. The leading chapter authors made significant contributions in presenting research findings and sharing insights about remote sensing of protected lands and waters. They are, by the order of chapter number and except for myself, John Gross of the National Park Service, Robert Kennedy of Oregon State University, Chengquan Huang of the University of Maryland, Paul Zorn of Parks Canada, Wenfang Tao of Hunan Agricultural University in China, Guofan Shao of Purdue University, Jared Stabach of Colorado State University, Samuel Ayebare of the Wildlife Conservation Society Uganda Program and the University of Rhode Island, Jiawei Xu of the Northeast Normal University in China, Alan Friedlander of the USGS Pacific Islands Ecological Research Center, Yongwei Sheng of the University of California–Los Angles, Kenneth Ranson and Guoqing Sun of the NASA Goddard Space Flight Center, Nancy Sherman of the University of Virginia, Lixin Lu of Colorado State University, Zhong Lu of the USGS, Cuizhen Wang of the University of Missouri, Robert Crabtree and Jennifer Sheldon of Yellowstone Ecological Research Center, Hirofumi Hashimoto of the NASA Ames Research Center, and Wei "Wayne" Ji of the University of Missouri at Kansas City. Many individuals offered constructive comments for the preparation of this book. I wish to express my sincere appreciation to John Gross, Robert Crabtree, Robert Kennedy, Paul Zorn, Zhong Lu, Jennifer Caselle, Alan Friedlander, Wei Ji, Robert Fraser of Natural Resources Canada, Rama Nemani of the NASA Ames Research Center, and Jie Shan of Purdue University for their comments and encouragement.

Irma Shagla, Arlene Kopeloff, and Jennifer Ahringer of Taylor & Francis Group/CRC Press handled the production of this book. Dr. Qihao Weng, the Series Editor of the *Taylor & Francis Series in Remote Sensing Applications*, offered support and encouragement throughout the preparation of this book.

As always, my most special appreciation is dedicated to my wife and daughters for their patience, understanding, and untiring support.

Editor

Dr. Yeqiao Wang is a professor at the Department of Natural Resources Science, University of Rhode Island (URI). He received his BS from the Northeast Normal University in 1982 and his MS in remote sensing and mapping from the Northeast Institute of Geography and Agroecology of the Chinese Academy of Sciences in 1987. He received his MS and PhD in natural resources management and engineering from the University of Connecticut in 1992 and 1995, respectively. From 1995 to 1999, he held the position of assistant professor in GIS and remote sensing in the Department of Geography/Department of Anthropology at the University of Illinois at Chicago. He has been on the faculty of the URI since 1999. Besides his tenured position at URI, he held an adjunct research associate position at the Field Museum of Natural History in Chicago between 1998 and 2003. He has been selected by the Ministry of Education of China and the Li Ka Shing Foundation of Hong Kong as a Yangtze River Scholar Endowed Lecturing Professor at the Northeast Normal University in China.

Dr. Wang received the prestigious Presidential Early Career Award for Scientists and Engineers (PECASE) from President William J. Clinton in 2000. The PECASE Award is the highest honor bestowed by the U.S. government on outstanding scientists and engineers beginning their independent careers. Dr. Wang was a first-place winner of the ESRI Award for Best Scientific Paper in Geographic Information System in 2002 from the American Society for Photogrammetry and Remote Sensing. He received recognition in Outstanding Contributions to Research by the URI in 2003 and received the Research Scientist Excellence Award from the College of the Environment and Life Sciences at URI in 2008.

His specialties are terrestrial remote sensing and modeling in natural resources analysis and mapping. Dr. Wang's particular area of interest is assessment and modeling of the dynamics of landscape and land-cover/land-use change and the effects of such changes on environments, ecosystems, natural resource sustainability, and ecological security. He has published over 50 peer-reviewed journal articles, 70 abstracts and conference papers, and contributed over 20 peer-reviewed book chapters. He also edited *Remote Sensing of Coastal Environments* (CRC Press, 2009). Dr. Wang serves as the editor in chief for *Encyclopedia of Natural Resources* (Taylor & Francis), a three-volume set of *Land*, *Air*, and *Water*. Besides his professional publications in English, he has also authored and edited several science books in Chinese. His research projects have been funded by multiple agencies, such as NASA, USDA, USDI, USAID, among others, which supported his scientific studies in various regions of the United States, East Africa, and Northeast China.

Contributors

Samuel Ayebare
Department of Natural Resources
  Science
University of Rhode Island
Kingston, Rhode Island

and

Wildlife Conservation Society
Albertine Rift Programme
Kampala, Uganda

Justin Braaten
Department of Forest Ecosystems
  and Society
Oregon State University
Corvallis, Oregon

Jennifer E. Caselle
Marine Science Institute
University of California–Santa
  Barbara
Santa Barbara, California

Qingri Chang
College of Resources and
  Environment
Northwest Agricultural and
  Forestry University
Yangling, Shaanxi, People's
  Republic of China

Warren Cohen
Pacific Northwest Research Station
USDA Forest Service
Corvallis, Oregon

Bryan M. Costa
NOAA-Biological Program
Silver Spring, Maryland

Robert L. Crabtree
Yellowstone Ecological Research
  Center
Bozeman, Montana

Lisa Dabek
Field Conservation
  Department
Woodland Park Zoo
Seattle, Washington

Daniel Dzurisin
U.S. Geological Survey
Cascades Volcano Observatory
Vancouver, Washington

Jiang Feng
School of Urban and
  Environmental Sciences
Northeast Normal University
Changchun, Jilin, People's
  Republic of China

Alan M. Friedlander
U.S. Geological Survey—Hawaii
  Cooperative Fishery
  Research Unit
Pacific Islands Ecological
  Research Center
Honolulu, Hawaii

Scott J. Goetz
The Woods Hole Research
  Center
Falmouth, Massachusetts

Samuel N. Goward
Department of Geography
University of Maryland
College Park, Maryland

John E. Gross
National Park Service
Inventory and Monitoring Program
Fort Collins, Colorado

Andrew J. Hansen
Ecology Department
Montana State University
Bozeman, Montana

Hirofumi Hashimoto
Division of Science and
    Environmental Policy
California State University–
    Monterey Bay
Seaside, California

and

NASA Ames Research Center
Moffett Field, California

Hong S. He
School of Natural Resources
University of Missouri
Columbia, Missouri

Samuel H. Hiatt
Division of Science and
    Environmental Policy
California State University–
    Monterey Bay
Seaside, California

and

NASA Ames Research Center
Moffett Field, California

Joanne Howl
Sigma Space Corporation
Lanham, Maryland

Chengquan Huang
Department of Geography
University of Maryland
College Park, Maryland

Rigel Jensen
Australian Wildlife Conservancy
Subiaco, Western Australia,
    Australia

Wei "Wayne" Ji
Department of Geosciences
University of Missouri–Kansas City
Kansas City, Missouri

Hyung-Sup Jung
Department of Geoinformatics
University of Seoul
Seoul, Republic of Korea

Robert E. Kennedy
Department of Forest Ecosystems
    and Society
Oregon State University
Corvallis, Oregon

Viatcheslav I. Kharuk
V. N. Sukachev Institute of Forest
Russia Academy of Science
Academgorodok, Krasnoyarsk,
    Russia

Junli Li
Department of Geography
University of California–
    Los Angeles
Los Angeles, California

Yu Liang
Institute of Applied Ecology
Chinese Academy of Sciences
Shenyang, Liaoning, People's
    Republic of China

T. V. Loboda
Department of Geography
University of Maryland
College Park, Maryland

Lixin Lu
Department of Atmospheric Science
Colorado State University
Fort Collins, Colorado

Zhong Lu
U.S. Geological Survey
EROS Center and Cascades Volcano
    Observatory
Vancouver, Washington

Forrest M. Melton
Division of Science and
    Environmental Policy
California State University–
    Monterey Bay
Seaside, California

and

NASA Ames Research Center
Moffett Field, California

Andrew R. Michaelis
Division of Science and
    Environmental Policy
California State University–
    Monterey Bay
Seaside, California

and

NASA Ames Research Center
Moffett Field, California

Cristina Milesi
Division of Science and
    Environmental Policy
California State University–
    Monterey Bay
Seaside, California

and

NASA Ames Research
    Center
Moffett Field, California

David Moyer
Wildlife Conservation Society
Albertine Rift Programme
Kampala, Uganda

Peder Nelson
Department of Forest Ecosystems
    and Society
Oregon State University
Corvallis, Oregon

Ramakrishna R. Nemani
NASA Ames Research Center
Moffett Field, California

Sally O'Grady
Parks Canada, Ecological Integrity
    Branch
Gatineau, Quebec, Canada

Nathan B. Piekielek
Ecology Department
Montana State University
Bozeman, Montana

Andrew J. Plumptre
Wildlife Conservation Society
Iringa, Tanzania

Gabriel Porolak
Tree Kanbaroo Conservation
    Program
Lae, Morobe Province,
    Papua New Guinea

Jiaguo Qi
Center for Global Change
    and Earth Observations
    and Department of
    Geography
Michigan State University
East Lansing, Michigan

Kenneth J. Ranson
NASA Goddard Space Flight
    Center
Greenbelt, Maryland

Karen Schleeweis
Department of Geography
University of Maryland
College Park, Maryland

Guofan Shao
Department of Forestry and
    Natural Resources
Purdue University
West Lafayette, Indiana

Rajeev Sharma
Parks Canada, Ecological Integrity
    Branch
Gatineau, Quebec, Canada

Jennifer W. Sheldon
Yellowstone Ecological Research
    Center
Bozeman, Montana

Yongwei Sheng
Department of Geography
University of California–
    Los Angeles
Los Angeles, California

N. J. Sherman
Department of Environmental
    Sciences
University of Virginia
Charlottesville, Virginia

H. H. Shugart
Center for Regional Environmental
    Studies
University of Virginia
Charlottesville, Virginia

Jared A. Stabach
Natural Resource Ecology
    Laboratory
Colorado State University
Fort Collins, Colorado

Guoqing Sun
Department of Geography
University of Maryland
College Park, Maryland

Wenfang Tao
Institute of Tobacco
Hunan Agricultural University
Changsha, Hunan, China

David M. Theobald
Natural Resource Ecology
    Laboratory and Human
    Dimensions of Natural
    Resources
Colorado State University
Fort Collins, Colorado

Nancy Thomas
Department of Geography
University of Maryland
College Park, Maryland

Darien Ure
Parks Canada
Ecological Integrity Branch
Gatineau, Quebec, Canada

Petr Votava
Division of Science
    and Environmental
    Policy
California State University
    Monterey Bay
Seaside, California

and

NASA Ames Research Center
Moffett Field, California

Cuizhen Wang
Department of Geography
University of Missouri
Columbia, Missouri

Weile Wang
Division of Science and
  Environmental Policy
California State University–
  Monterey Bay
Seaside, California

and

NASA Ames Research Center
Moffett Field, California

Yeqiao Wang
Department of Natural Resource
  Science
University of Rhode Island
Kingston, Rhode Island

Lisa M. Wedding
NOAA–Biological Program
Silver Spring, Maryland

Zhengfang Wu
School of Urban and
  Environmental Sciences
Northeast Normal University
Changchun, Jilin, People's Republic
  of China

Jiawei Xu
School of Urban and
  Environmental Sciences
Northeast Normal University
Changchun, Jilin, People's Republic
  of China

Jian Yang
Institute of Applied Ecology
Chinese Academy of Sciences
Shengyang, China

Zhiqiang Yang
Department of Forest Ecosystems
  and Society
Oregon State University
Corvallis, Oregon

Xing Yuan
School of Urban and
  Environmental Sciences
Northeast Normal University
Changchun, Jilin, People's Republic
  of China

Hongyan Zhang
School of Urban and
  Environmental Sciences
Northeast Normal University
Changchun, Jilin, People's Republic
  of China

Jiquan Zhang
School of Urban and
  Environmental Sciences
Northeast Normal University
Changchun, Jilin, People's Republic
  of China

Yuyu Zhou
Department of Natural Resources
  Science
University of Rhode Island
Kingston, Rhode Island

Paul Zorn
Parks Canada, Ecological Integrity
  Branch
Gatineau, Quebec, Canada

1

Remote Sensing of Protected Lands: An Overview

Yeqiao Wang

CONTENTS

1.1 Introduction

The World Commission on Protected Areas (WCPA) adopted a definition that describes a protected area as clearly defined geographical space, recognized, dedicated, and managed, through legal or other effective means, to achieve the long-term conservation of nature with associated ecosystem services and cultural values (Dudley 2008). In general, protected lands include areas such as national parks, national forests, national seashores, all levels of natural reserves, wildlife refuges and sanctuaries, and designated areas for conservation of native biological diversity and natural and cultural heritage and significance. Protected lands also include some of the last frontiers that have unique landscape characteristics and ecosystem functions. Along the shoreline and over the ocean and sea, the International Union for the Conservation of Nature (IUCN) has defined marine-protected areas (MPAs) as any area of intertidal or subtidal terrain, together with its overlying water and associated flora, fauna, and historical and cultural features, which has been reserved by law or other effective means to protect part or the entire enclosed environment (Kelleher 1999). As reported by the World Database on

Protected Areas (IUCN and UNEP-WCMC 2010), as of 2009, worldwide approximately 13% of the lands are designated as protected areas and about 0.8% waters along the shoreline and over the ocean are set as MPAs. In the United States, 14.81% of the terrestrial lands have been set as protected, and along the shoreline and over the ocean 24.75% of the terrestrial waters up to 12 nautical miles are set as MPAs. Protected lands and waters serve as the fundamental building blocks of virtually all national and international conservation strategies, supported by governments and international institutions. Those provide the core of efforts to protect the world's threatened species and are increasingly recognized as essential providers of ecosystem services and biological resources; key components in climate change mitigation strategies; and in some cases also vehicles for protecting threatened human communities or sites of great cultural and spiritual value (Dudley 2008).

Humans have created protected areas over past millennia for a multitude of reasons (Crabtree and Sheldon 2011). The establishment of Yellowstone National Park in 1872 by the United States Congress ushered in the modern era of governmental protection of natural areas that catalyzed a global movement (Heinen and Hite 2007; IUCN 2008). However, even with implementations of a tremendous variety of monitoring programs and conservation planning efforts and achievements, species' population declines, biodiversity loss, extinctions, system degradation, pathogen spread, and state change events are occurring at unprecedented rates (Hoffmann et al. 2010; Pereira et al. 2010). The effects are augmented by continued changes in land use, invasive spread, alongside the direct, indirect, and interactive effects of climate change and disruption (Crabtree and Sheldon 2011). Protected lands become more important in serving as indicators of ecosystem conditions and functions either by the status of themselves and/or by comparisons with unprotected adjacent areas. Protected lands are prized highly by society, and yet differently with diversified representative characteristics. Therefore, inventory, monitoring, understanding, and management of protected lands present challenges.

Field-oriented observation and mapping approach have been implemented in the past for management of protected lands. Repaid development of remote sensing science and technologies has profoundly changed and contributed to the practice in land management. Remote sensing has been broadly applied for inventory and monitoring of protected lands. Remote sensing imageries and derivatives of thematic maps have been considered among essentials for conservation planning and informed decision-making for managing protected lands.

Remote sensing is defined as the art, science, and technology of obtaining reliable information about physical objects and the environment, through the process of recording, measuring, and interpreting imagery and digital representations of energy patterns derived from satellite and airborne sensor systems. Remote sensing can provide comprehensive geospatial information to map and study protected lands in different spatial scales (e.g., high spatial resolution, large area coverage, etc.), different temporal frequencies (daily, weekly, monthly, annual observations, etc.), different spectral properties (visible light, near infrared, microwave, etc.), and with spatial contexts (e.g., immediate adjacent areas of protected lands vs. broader background of land bases). Remote sensing in combination with field-based studies has created new and exciting opportunities to meet monitoring needs for studying protected areas (Fancy et al. 2009).

A variety of ecological applications require data from broad spatial extents that cannot be collected using field-based methods alone. Space-based sensors and computational models are valuable tools for environmental monitoring, reporting, and forecasting at large spatial and temporal scales (Theobald 2001; Hansen and Rotella 2002; Jantz et al. 2003; Parmenter et al. 2003; Wessels et al. 2004; DeFries et al. 2005). Remote sensing data and analytical techniques address needs such as identifying and detailing biophysical characteristics of habitats, predicting the distribution of species and spatial variability in species richness, and detecting natural and human-caused changes at different scales (DeFries et al. 1999; Kerr and Ostrovsky 2003). Remote-sensing-based biodiversity indices have been designed and applied to detect biodiversity hotspots with high conservation value (Gould 2000).

One particular advantage that remote sensing can provide for inventory and monitoring of protected lands is the information for understanding the past and current status, the changes that occurred under different impacting factors and management practices, the trends of changes in comparison with adjacent areas, and implications of changes on ecosystem functions (e.g., Hansen and DeFries 2007a,b). This is reflected as a component of the land-use and land-cover change (LULCC) studies. LULCC is an interdisciplinary scientific theme that includes performing repeated inventories of landscape change from space; developing scientific understanding and models necessary to simulate the processes taking place; evaluating consequences of observed and predicted changes; and further understanding consequences on environmental goods and services and management of natural resources. Information about LULCC is critical to improve our understanding of human interaction with the environment and to provide a scientific foundation for sustainability, vulnerability, and resilience of land systems and their use.

Remote sensing has been identified as one of the primary data sources to produce land-cover maps that indicate landscape patterns and human development processes (Turner 1990; Coppin and Bauer 1996; Rogan et al. 2002; Wilson and Sader 2002; Griffith et al. 2003; Turner et al. 2003; Wang et al. 2003). Issues and methodologies in change detection have long been discussed in literatures (e.g., Lambin and Strahler 1993; Mas 1999; Hayes and Sader 2001; Wilkinson et al. 2008). The articles in a recent special issue of *Remote Sensing of Environment* (Gross et al. 2009) addressed issues of the conceptual approach to remote sensing monitoring of the landscape context of protected areas (Goetz et al. 2009; Jones et al. 2009; Svancara et al. 2009; Townsend et al. 2009) for biodiversity conservations (Wiens et al. 2009), for change in detection tools and considerations (Kennedy et al. 2009), for land-cover change analysis through case studies (Dennison et al. 2009; Fraser et al. 2009; Huang et al. 2009; Reed et al. 2009; Wang et al. 2009), for monitoring ecosystem functions and environmental change (Crabtree et al. 2009; Nagler et al. 2009), for monitoring active fire and burnt areas (Ressl et al. 2009), and modeling tools for forecasting ecosystem dynamics (Nemani et al. 2009).

Remote sensing has unique advantages in monitoring frontier lands, which are always in remote and difficult-to-reach locations and huge in area coverage. Different types of remote sensing data have been applied in study of frontier lands, for example, using hyperspectral and radar data for monitoring forests in the Amazon (Sheng and Alsdorf 2005; Arima et al. 2008; Mena 2008; Walsh et al. 2008; Wang et al. 2005; Wang and Qi 2008) and in Siberia (Sun et al. 2002; Bergen et al. 2008; Kharuk et al. 2009), and for detecting hydrologic change in the lake-rich Arctic region (Stow et al. 2004; Sheng et al. 2008).

Besides broadly applied Landsat and Système Probatoire d'Observation de la Terre (SPOT) types of optical or passive multispectral remote sensing that make use of visible, near-infrared, and short-wave infrared sensors to form images of the landscape and seascape, development in active remote sensing such as interferometric synthetic aperture radar (InSAR) brings new types of data and enhanced capacities in the study of protected lands. The SAR backscattering signal is composed of intensity and phase components. The intensity component of the signal is sensitive to terrain slope, surface roughness, and dielectric constant. Active radar sensors are well known for their all-weather imaging capabilities, which are effective for mapping habitats over cloud-prone regions. Studies have demonstrated that SAR intensity images can map and monitor forested and nonforested wetlands (e.g., Ramsey 1995). SAR intensity data have been used to monitor floods and dry conditions, temporal variations in the hydrological conditions of wetlands, such as for delineation of inundated area and vegetation (Hess et al. 1995; Costa et al. 2002; Costa 2004), for monitoring wetlands (Kasischke and Bourgeau-Chavez 1997), for monitoring regional inundation pattern and hydroperiod (Bourgeau-Chavez et al. 2005), and for mapping height and biomass of mangrove forests (Simard et al. 2006).

When the phase components of two SAR images of the same area acquired from similar vantage points at different times are combined through InSAR processing, an interferogram can be constructed to depict range changes between the radar and the ground, and can be further processed with a digital elevation model (DEM) to produce an image with centimeter to subcentimeter vertical precision under favorable conditions (Massonnet and Feigl 1998). InSAR has been extensively utilized to study ground surface deformation associated with volcanic activities, earthquakes, landslides, water-level change, and land subsidence, which happen frequently in protected lands (Lu et al. 2011). InSAR has been used to measure the change in water-level on the Amazon floodplain (Alsdorf et al. 2000). L-band Japanese Earth Resource Satellite 1 (JERS-1) images have been used to map changes in water level over the Everglades in Florida (Wdowinski et al. 2004). InSAR has been used for three-dimensional habitat mapping and morphological change analysis (Lu and Kwoun 2008; Lu et al. 2010).

Field survey and *in situ* observations are essential to identify protected habitats through remote sensing. Almost every remote sensing exercise requires field survey to define habitats, calibrate remote sensing imagery, and evaluate the accuracy of remote sensing outputs (Wang et al. 2007). With GPS-guided positioning and field survey becoming a routine operation, challenges remain for the incorporation of *in situ* measurements with remote sensing observations for quantitative analyses of protected habitats.

MPAs are among critical components of protected lands and waters. Important factors that affect the way plants and animals respond to MPAs include distribution of habitat types, level of connectivity to nearby fished habitats, wave exposure, depth distribution, prior level of resource extraction, regulations, and level of compliance to regulations (Friedlander et al. 2011). Conservation benefits are evident through increased habitat heterogeneity at the seascape level, increased abundance of threatened species and habitats, and maintenance of a full range of genotypes (Edgar et al. 2007). Remote sensing data that quantify spatial patterns in habitat type, oceanographic conditions, and benthic complexity can be integrated with *in situ* ecological data for design, evaluation, and monitoring of MPA networks to design, assess, and monitor MPAs (Wedding and Friedlander 2008; Wedding et al. 2008). Combining remote sensing products with *in situ* ecological and physical data can support the development of a statistically robust monitoring program of living marine resources within and adjacent to MPAs (Friedlander et al. 2007).

With solid progress in remote sensing technology and increasingly broader scope of applications, this book provides conceptual framework and case studies to showcase state-of-the-art remote sensing for inventory, monitoring, mapping, reporting, and management of protected lands and waters. The chapters in this book are organized into four sections and described as follows.

1.2 Remote Sensing of Changing Landscape of Protected Lands

Section I of this book consists of seven chapters that discuss the general topic of remote sensing of the changing landscape of protected lands. Representative case studies include remote sensing for inventory and monitoring of the U.S. national parks; monitoring landscape dynamics of national parks in the western United States; forest dynamics within and around Olympic National Park (ONP) assessed using time-series Landsat observations; using earth observation to monitor species-specific habitat change in the Greater

Kejimkujik National Park region of Canada; land-cover change and conservation of protected lands in urban and suburban settings; land-cover change and conservation of the protected ancient city park in Xi'an, Northwestern China; and remote sensing for accurate assessment of the user's perspective.

In Chapter 2, Gross et al. introduce the framework and background of the inventory and monitoring (I&M) networks and remote sensing applications of U.S. national parks. The PALMS (Park AnaLysis and Monitoring Support) project focused on four sets of national parks to develop and demonstrate the approach. These parks included Sequoia-Kings Canyon and Yosemite National Parks (Sierra Nevada I&M Network), Yellowstone and Grand Teton National Parks (Greater Yellowstone I&M Network), Rocky Mountain National Park (Rocky Mountain I&M Network), and a combination of Delaware Water Gap National Recreation Area and Upper Delaware Scenic and Recreational River (Eastern Rivers and Mountains I&M Network). The PALMS project started with the development of a suite of indicators including measurements of weather and climate, stream health (water), land cover and land use, disturbances, primary production, and monitoring area. The project team recognized the challenges in identifying suitable area of analysis (AOA) because the extent of the most appropriate AOA varies with the specific issue, process, or species that is of most interest. To address the challenge, the PALMS project team developed a framework for delineating the ecosystem surrounding a protected area that is likely to strongly influence ecological function and biodiversity within the protected area. The so-called protected area-centered ecosystem (PACE) was defined as the logical place to focus monitoring, research, and collaborative management in order to maintain the function and condition of the protected area. The approach helps facilitate research and conservation across the parks and important surrounding lands. Detailed project methods and findings and insights about lessons learned and recommendations presented in Chapter 2 deserve special attention.

In Chapter 3, Kennedy et al. introduce a study about monitoring landscape dynamics of national parks in the western United States by Landsat-based approaches. The study is established by the approaches that translate ecologically based views of change into the spectral domain when the entire archive of Thematic Mapper (TM) spectral imagery is considered. A spectral index is used as a proxy for ecological attributes, and that index can be tracked as a time-series trajectory. The newly developed algorithms, collectively known as "LandTrendr" (Kennedy et al. 2010), use simple statistical fitting rules to identify periods of consistent progression in the spectral trajectory (segments) and turning points (vertices) that separate those periods. The change-monitoring methods capture a wide range of processes affecting vegetation inside and outside of western national parks, such as decline/mortality processes, growth/recovery processes, and combinations of different processes, among others. The study concluded that even though national parks are protected from many forms of direct human intervention, their landscapes are anything but static. Vegetation is changing continuously in response to both endogenous and exogenous pressures, and monitoring these dynamic landscapes in coming decades will require tools that capture a wide range of processes over large areas. With temporal segmentation of the spectral signals in the archive, the authors argue that satellite data are moving toward a characterization of dynamic landscapes that is more consistent with an ecological view of change than has been possible in the past.

In Chapter 4, Huang et al. evaluate forest dynamics within and around the ONP using time-series Landsat observations. The ONP was established in the 1930s to protect the diversity of its ecosystems. It has some of the best examples of intact temperate rainforest in the Pacific Northwest and is home to endemic species. In 1976, it was designated by the United Nations Educational, Scientific and Cultural Organization (UNESCO) as an International Biosphere Reserve, and was declared a World Heritage Site in 1981. The old growth forests that were once widely distributed across the Olympic Peninsula provide critical habitat for the northern spotted owl and other endangered species (McNulty 2009). Due to extensive logging in the early-twentieth century, however, much of the unprotected old growth forests outside the ONP and the Olympic National Forest were lost, and some of the private lands in the peninsular are among the most heavily logged in the United States. Viable protection of the unique biodiversity of the Olympic Peninsula requires continuous monitoring of forest dynamics across the peninsula. The time-series approach described in this study represents a new approach for monitoring protected forest lands and their surrounding areas. This approach allows reconstruction of disturbance history over the last two decades by taking advantage of the temporal depth of the Landsat archive, and can be used to provide continuous monitoring as new satellite images are acquired.

In Chapter 5, Zorn et al. highlight the use of earth observation and remote sensing for monitoring wildlife habitat change in Kejimkujik National Park and National Historic Site in southern Nova Scotia of the Canadian Atlantic Coastal Uplands Natural Region. The study addresses the major goals of protecting and maintaining ecological integrity through two key measures of "effective habitat amount" and "effective habitat connectivity." The study employed the Automated Multi-temporal Updating by Signature Extension (AMUSE) protocols to generate multitemporal land-cover maps from 1985, 1990, 1995, 2000, and 2005 Landsat remote sensing data. It created a mask of potentially changed pixels using change vector analysis (CVA) between two consecutive dates, followed by thresholding to identify a change/no-change mask. Landscape ecology methods were applied to measure the effect of land-cover change from the perspective of a specific species or species group following the concept of ecologically scaled landscape indices. The results are gradient maps that show areas of high and low effective habitat amount and connectivity throughout the park and greater park ecosystem for each species. Thresholds were then applied to identify species landscapes that are estimated to contain sufficient habitat. Statistics on the total amount and

connectivity of effective habitat for Kejimkujik and the greater park ecosystem were then extracted for each five-year time step and compared. The
visual output from these analyses provides powerful communication
products on how land use can affect habitat patterns around the park. These
communication products can support the park in engaging partner agencies
and stakeholders in the development of collaborative conservation strategies
that are vital in allowing Kejimkujik to manage its long-term ecological
integrity goals.

In Chapter 6, Wang presents a perspective of land-cover change and conservation of protected lands in urban and suburban settings. A recent study
reported that 1.4 million hectares of open space were lost to urban sprawl in
the United States between 1990 and 2000 (McDonald et al. 2010). Urban
sprawl fragments habitats, isolates populations, and is among the most
important human activity stressors reshaping natural processes and most
threatening to the sustainability of ecological communities. This is particularly true for protected areas that are situated in urban and suburban environments. The chapter takes case studies from the Chicago Wilderness
project and selected national parks in the northeastern United States to
reveal significant amounts of human-induced urban land-cover change and
the effects on protected lands. The chapter cites President John F. Kennedy's
1961 message to the congress on housing that emphasized the need for preserving open space in metropolitan areas. President Kennedy said that
"Land is the most precious resource of the metropolitan area. The present
patterns of hazard suburban development are contributing to a tragic waste
in the use of a vital resource now being consumed at an alarming rate"
(NIPC 1962: 358). With the rate of land-cover change in the past three decades
revealed by remote sensing data and the forecast pattern of population and
employment growth, the region's protected natural areas will come under
tremendous pressures. Simulation of the region's landscape demonstrates a
trend of dramatic increase of interface between natural areas and urban
land. If the rate of consumption of open land was alarming, as acknowledged by President Kennedy five decades ago, the current rate of land-cover
change and the future trends of the change could approach catastrophic
rates. In the epilogue of his 2005 book, *Cities in the Wilderness: A New Vision of
Land Use in America*, Bruce Babbitt, the former U.S. Secretary of the Interior,
argues that the fundamental issue facing public lands is that we are yet to
reach consensus as to their ultimate placement on the "use spectrum" from
cities to wilderness. Public lands should now be administered primarily,
although not exclusively, to maintain and restore their natural values. He
continues that a basic measure of good land use is sustainability, a word that
has come to signify living in respectful relationship with the land, passing it
on unimpaired, and even renewed and restored, to future generations.
Development should enlarge the possibilities for human progress, creativity,
and quality of life, which it cannot accomplish by continually eroding the
beauty and productivity of the natural world (Babbitt 2005). Informed man-

agement and conservation of protected lands in urban and suburban settings are steps toward the ultimate goals as Babbitt envisioned. The chapter suggests that there is no single solution to the problems associated with suburban sprawl. The remote-sensing-derived information represents an effort to delineate and clarify understanding of the problem. Information is vital to sound decision-making at all levels. Remote sensing and the related information technology have been playing and will continue to play critical roles in natural resource monitoring and management.

In Chapter 7, Tao et al. present a case study about land-cover change and conservation of the protected ancient city park in Xi'an, Northwestern China. Xi'an is among the oldest cities in the world with over 3100 years of history. It served as the capital city for 13 dynasties of ancient China and it was the beginning terminus of the "silk road" connecting the east and the west. The city has preserved the largest ancient city wall and over a hundred square kilometers of protected significant sites from early dynasties of the Zhou (1046–256 BC), Qin (221–206 BC), Han (206 BC to AD 220), and Tang (AD 618–907). The process of urbanization has profoundly changed the spatial patterns of the landscape under China's recent economic development policy. Protection and conservation of historical landmarks of the ancient civilization and the development of a new city in this world famous ancient capital represent a significant challenge. The study used Landsat remote sensing data and derivative maps to reveal spatial distribution and patterns of urban development. The study made attempts to analyze driving factors and the influence of policies for urban development and conservation of protected areas.

In Chapter 8, Shao presents a study about the accuracy of remote sensing-derived habitat maps for application usage. The author points out that any derivatives of remotely sensed data can contain uncontrollable error and uncertainty when technology is used carelessly or incorrectly. This is a serious issue because remotely sensed derivatives are not always the final products from the user's perspective. The maps, for example, are always used or referenced for various decision-making exercises. Chapter 8 provides an overview about errors involved in remotely sensed map products and possible measures that users can take to reduce the error. The chapter introduces an error matrix algorithm (EMA) that can be used for calibrating various remote-sensing-derived maps and for improving map accuracy.

1.3 Remote Sensing for Inventory, Mapping, and Conservation Planning of Protected Lands and Waters

Section II of this book consists of six chapters that examine remote sensing for inventory, mapping, and conservation planning of protected lands and waters. Assembled case studies include remote sensing of dominant forest types for Matschie's tree kangaroo conservation and planning in Papua New Guinea, remote sensing for biodiversity conservation of the Albertine Rift in eastern Africa, remote sensing assessment of natural resource and ecological security of the Changbai Mountain region in Northeast Asia, estimating effects of past volcanic eruptions on forests of the Changbai Mountain with *in situ* and remote sensing observations, integration of remote sensing and *in situ* ecology for the design and evaluation of MPAs through examples of tropical and temperate ecosystems, and remote sensing assessment of wildfire impacts and simulation modeling of short-term post-fire vegetation recovery within Dixie National Forest in Utah in the United States.

In Chapter 9, Stabach et al. report a study that used Landsat 7 Enhanced Thematic Mapper plus (ETM+) data and GPS collars to provide detailed information for Matschie's tree kangaroo conservation in Papua New Guinea. Matschie's tree kangaroos (*Dendrolagus matschiei*) are arboreal marsupials endemic to the Huon Peninsula in Papua New Guinea. Due primarily to increased hunting pressure and loss of habitat from agricultural expansion, *D. matschiei* are currently listed as endangered by the IUCN. The study concluded that *Dacrydium nidulum*-dominant forests are the most widespread forest throughout the study area and are also where tree kangaroos are located. However, additional research has shown that these are not the only forest type that is used by the species. Clustered and independent movement locations indicate that animals do not utilize their habitat uniformly. These data provide vital information toward a better understanding of the habitat, the requirements of the animals, and the long-term conservation of Matschie's tree kangaroo habitat in Papua New Guinea.

In Chapter 10, Ayebare et al. describe how aerial remote sensing from the Wildlife Conservation Society's Flight Program is being used to support biodiversity conservation in Madagascar, eastern and southern Africa, with a focus on the Albertine Rift. The Albertine Rift stretches from the northern end of Lake Albert to the southern end of Lake Tanganyika, with the five countries of Uganda, Rwanda, Democratic Republic of Congo, Burundi, and Tanzania bordering this region. The Albertine Rift ecoregion is identified as one of the richest sites in Africa for biodiversity and a priority for conservation actions. Many sites in the Albertine Rift are protected as national parks, wildlife reserves, or forest reserves. Major threats to biodiversity conservation in the Albertine Rift are increasing human population, civil strife, and industrialization. These have led to large-scale LULCC in the region. There has been a tremendous increase in the numbers of people leading to increased pressures on the natural resource base around protected areas. With the conversion of buffer zones to agricultural activities and subsequent competition between people and wild animals for the same resources, there has been an increase in human–wildlife conflict. The aerial imageries have been used to map threats to biodiversity, to develop land-use plans for protected area management, and to measure vegetation cover and dynamics.

In Chapter 11, Wang et al. introduce a study that used remote sensing to assess natural resource and ecological security of the Changbai Mountain region in Northeast Asia. Inventory and monitoring of protected lands and the supporting ecosystems become critically important under the facts of intensified human-induced LULCC, fragmentation of habitats, and effects of climate changes. This is particularly true for the regions that contain protected lands with significant biological diversity and yet fall in the transboundary-sensitive international hotspots of different political regimes and social administration systems. Ecological security is an essential cornerstone for the sustainability of any human and nature systems. It depends on the balance between human demands and actions in consumption and alteration of resource base and the sustainability, vulnerability, and resilience of environmental systems that provide ecosystem services.

In a large spatial context, the Changbai Mountain range extends along the border between Northeast China and North Korea. Toward the northeast, it connects the Sikhote-Alin Mountains in the Russian Far East. The highest section and the most representative of this ecoregion is the Changbai Mountain, where the Changbai Mountain Nature Reserve (CMNR) is situated. The CMNR was established in 1961 and admitted into the UNESCO's Man and Biosphere Program in 1979. The unique and distinctive vertical zonal pattern of vegetation and the ecosystems showcase a condensed configuration and composition of temperate and boreal forests found across Northeast Asia. Socioeconomic development, aggressive logging, intensified urban and agricultural land use, demographic change, and pollution through air and water systems accelerate the degradation of natural resources of this region of international interest. Chapter 11 makes efforts to articulate remote sensing assessment of natural resources and ecological security through perspectives of LULCC, volcanic risks, postimpact vegetation succession, geospatial modeling, and regional inventory and monitoring. Remote sensing data and products, such as Moderate Resolution Imaging Spectroradiometer (MODIS), Landsat, IKONOS, Shuttle Radar Topography Mission (SRTM), and InSAR, are used and presented through case studies.

Xu et al., in Chapter 12, present a study about the integration of remote sensing and *in situ* observations for examining effects of past volcanic eruptions on forests of the CMNR. In particular, the study evaluated the vegetation patterns documented by remote sensing and integrated with samples of carbonized wood to reveal the pre-eruption vegetation structure and posteruption succession process in this important protected area. Samples of carbonized woods that are buried in volcanic ash provide evidence of pre-eruption forest structure, species composition, spatial distributions, temporal variations, and the impacted areas. The study concluded that the distinctive quality of vegetation in the CMNR is determined by the influence of volcanic eruptions on the ecosystems. The traces of volcanic eruptions exist in the forest ecosystems in the CMNR after hundreds of years of devastated volcanic eruptions.

In Chapter 13, Friedlander et al. introduce, through examples from tropical and temperate ecosystems, the integration of remote sensing and *in situ* ecology for the design and evaluation of MPAs. The authors point out that individual MPAs need to be networked in order to provide large-scale ecosystem benefits and to have the greatest chance of protecting all species, life stages, and ecological linkages if they encompass representative portions of all ecologically relevant habitat types in a replicated manner. This chapter highlights several examples of marine biogeographic assessments that have been implemented using a combination of remotely sensed and *in situ* datasets. These case studies serve as models to illustrate the integration of remotely sensed data to better understand marine ecosystems at multiple spatial scales and support management actions in both temperate and tropical MPAs. The chapter focuses on high-resolution sensors (0.01–10 m) capable of mapping physical and biological features of benthic habitat. The studies include monitoring of coral reef in the Hawaii Archipelago and near-shore protected areas in California and New England. The chapter concludes with constraints and considerations in the use of remote sensing in tropical and temperate marine environments and the use of new technologies, data fusion, and decision support tools that serve to synthesize remotely sensed data into a format amenable for marine spatial planning.

In Chapter 14, Wang et al. introduce a study in simulation modeling of short-term post-fire vegetation recovery within Dixie National Forest in Utah. Wildfires are a growing natural hazard in the United States and around the world. Although as a natural process a wildfire can be beneficial to ecosystem functions, direct fire impacts and secondary effects such as erosion, landslides, introduction of invasive species, and changes in water quality and hydrology are often disastrous. Pattern, severity, and timing of wildfires affect significantly the succession process of vegetation. Understanding the effects of fire impacts and the pathway of vegetation recovery is critical for community actions on resource management planning, land-use decision, treatment procedure, and habitat restoration and for studies of ecological and economic complexities in association with wildfires. The study developed an approach that integrates LANDFIRE data products, differenced normalized burn ratio (DNBR) data, and LANDISv4.0a modeling in the simulation exercise. A combination of DNBR and LANDFIRE data provided information on burn severity, prefire existing vegetation types, vegetation composition and structure, and the biophysical gradients that affect the distribution of ecosystem components. Such critical information established the initial states for LANDIS parameterization and simulation modeling. The simulation was at a 30-m cell size, a 1-year time interval, and a 10-year duration. The results indicated that this approach provided the necessary datasets and modeling mechanism for updating vegetation maps affected by wildfires.

1.4 Remote Sensing of Frontier Lands

Section III of this book consists of six chapters that discuss remote sensing for inventory and monitoring of frontier lands. Representative case studies include satellite-observed endorheic lake dynamics across the Tibetan Plateau between circa 1976 and 2000; multisensor remote sensing of forest dynamics in Central Siberia; remote sensing and modeling for assessment of complex Amur tiger and Far Eastern leopard habitats in the Russian Far East; incorporating remotely sensed land surface properties to regional climate modeling; and InSAR for monitoring of natural hazards in protected lands and for characterizing biophysical properties in protected tropical forests in Southeast Asia.

The Tibetan Plateau, known as the "Roof of the World," is considered to be one of the last frontier lands of the world. With a pronounced temperature rise, the plateau is one of the areas of the world most vulnerable to climate change. Temperatures in the Tibetan Plateau have been accelerating since the 1950s, and the accelerated warming is expected to drive an array of complex physical and ecological changes in the region. Tibetan lakes in the endorheic basins serve as a sensitive indicator to regional climate and water cycle variability. They are dynamic in their inundation area in response to climate and hydrological conditions. However, these lake dynamics at regional scales are not well understood due to the inaccessibility and the inhospitable environment of this remote plateau, making satellite remote sensing the only feasible tool to detect lake dynamics across the plateau. In Chapter 15, Sheng and Li report a study that monitors Tibetan lake change using nearly a hundred Landsat scenes acquired around 1976 and 2000. The lake change analysis reveals that Tibetan lakes in general had expanded by ~9.5% and their dynamics varies from region to region with strong spatial patterns. The chapter introduces Tibetan lake dynamics in the context of climate change, briefs lake dynamics mapping methods from available satellite images, examines lake changes between ~25 years in endorheic basins across this broad plateau, and analyzes the spatial patterns of the observed lake dynamics.

The forested regions of Russian Siberia are vast and contain about a quarter of the world's forests that have not experienced harvesting. However, many Siberian forests are facing twin pressures of rapidly changing climate and increasing timber harvest activity. Monitoring the dynamics and mapping the structural parameters of the forest are important for understanding the causes and consequences of changes observed in these areas. Because of the inaccessibility and large extent of these forests, remote sensing data can play an important role for observing forest state and change. In Chapter 16, Ranson et al. introduce a case study that engaged multisensor remote sensing data to monitor forest disturbances and to map above-ground biomass from Central Siberia. Radar images from the Shuttle Imaging Radar-C (SIR-C)/XSAR mission are used to estimate forest biomass in the Sayan Mountains. Radar images from JERS-1, European Remote Sensing Satellite 1 (ERS-1), and Canada's RADARSAT-1, and data from ETM+ on-board Landsat 7 are used to characterize forest disturbances from logging, fire, and insect damages.

Listed as endangered and critically endangered by the IUCN, respectively, fewer than 400 adult and subadult Amur (Siberian) tigers (*Panthera tigris altaica*) and only between 14 and 20 adult Amur (Far Eastern) leopards (*Panthera pardus orientalis*) persist in the wild in a topographically complex and biologically diverse landscape in the Russian Far East and in the Changbai Mountains in Northeast China near the Russia–China border. These endangered species face threats mostly from loss of habitat due to forest fire and logging to illegal poaching and competing for prey with man. Remote sensing is a valuable tool for characterizing the vast habitat of wide-ranging endangered species such as the Amur tiger and the critically endangered Amur leopard. In Chapter 17, Sherman et al. describe a study that uses remote sensing for mapping, monitoring, and modeling tiger and leopard habitats related to vegetation and terrain. The study helps target resources toward locations that are most important for ensuring a future in the wild for these species as well as helps to identify new areas that would be appropriate to connect good habitat, including that in established reserves.

In Chapter 18, Lu introduces research about incorporating remotely sensed land surface properties to regional climate modeling. The chapter first discusses the Regional Atmospheric Modeling System (RAMS) and builds the assimilation RAMS (ASSM-RAMS) to ingest the leaf area index (LAI) derived from the normalized difference vegetation index (NDVI), and examine the sensitivity of simulated regional climate to multiyear satellite observations of vegetation changes (NDVI). Second, recognizing the large sensitivity of RAMS to seasonal variation in vegetation phenology, the chapter introduces the adoption of the CENTURY ecosystem model to simulate vegetation growth, and links it with RAMS to form a regional climate modeling system (RAMS-CENTURY) that includes a two-way feedback between the atmosphere and biosphere. Third, the chapter compares the results from both ASSM-RAMS and the RAMS-CENTURY modeling system to comprehensively understand the impact of heterogeneous vegetation distribution on the seasonal climate prediction. The chapter points out that, in its current form, RAMS land-surface hydrological processes (e.g., evaporation and transpiration), energy exchanges (e.g., latent heat and sensible heat fluxes), momentum exchanges (e.g., roughness length), and biophysical parameters (e.g., vegetation albedo, transmissivity, and stomatal conductance) are parameterized to have a strong dependence on the value of LAI. Consequently, an inadequate and unrealistic description of the vegetation distribution and its evolution in the current RAMS land-surface models is considered a major deficiency. Using NDVI datasets to derive LAI and the CENTURY model to simulate LAI can provide a more realistic vegetation distribution, both of which have the potential to improve the regional climate model simulations. The model domain used in this chapter comprises a coarse grid covering the entire conterminous United States at a 200-km grid spacing and a finer nested grid covering Kansas, Nebraska, South Dakota, Wyoming, and Colorado at a 50-km grid spacing. The finer grid covers an area of 1500 km in the east–west direction and 1300 km in the north–south direction, which includes Rocky Mountain National Park, Yellowstone National Park, Arapaho and Roosevelt National Forests, and Pawnee National Grassland, which are designated as protected natural areas.

Very often natural hazards occur in remotely located protected lands or impact on such large areas that *in situ* measurements are not feasible, accurate, or timely enough for monitoring or early warning. InSAR technology provides an all-weather imaging capability for measuring ground-surface deformation with centimeter-to-subcentimeter precision and inferring changes in landscape characteristics over a large region. With its global coverage and all-weather imaging capability, InSAR is an important remote sensing technique for measuring ground-surface deformation of various natural hazards and the associated landscape changes. The spatial distribution of surface deformation data, derived from InSAR imagery, enables the construction of detailed numerical models to enhance the study of physical processes of natural hazards. In Chapter 19, Lu et al. introduce the basics of InSAR for deformation mapping and landscape change detection, discuss state-of-the-art technical issues in InSAR processing and interpretation, and showcase the application of InSAR to the study of volcano, earthquake, landslide, and glacier movement and the mapping of high-resolution digital elevation model and fire progression with InSAR imagery over protected lands.

In tropical forests in Southeast Asia, national parks have been established as a means of reducing deforestation in ecologically and economically favorable ways. Accurate estimates of forest biophysical properties in these national parks are important in assessing their current status and in supporting sustainable management of these protected lands. In Chapter 20, Wang and Qi introduce a study that used Landsat ETM+ and JERS-1 SAR images to estimate green and woody structures in a protected forest in northern Thailand. Forest fractional cover and LAI are estimated from the ETM+ image using a linear unmixing model. A microwave/optical synergistic model is modified to improve the estimation of woody biomass by removing leaf contribution from radar backscatter in tropical forests. Aside from seasonal variations in different forest types, forest degradation by human disturbances is observed, which results in reduced green and woody biomass in forests closer to human settlements. The study demonstrates that synergistic use of optical and SAR images could extract quantitative biophysical information of protected forests in tropical mountains.

1.5 Remote Sensing in Decision Support for Management of Protected Lands

Section IV of this book consists of three chapters that discuss remote sensing in decision support for management of protected lands. The chapters address the topics of integration of focal species populations and remote sensing for monitoring and modeling environmental change in protected areas; monitoring and forecasting ecosystem dynamics over protected lands using the Terrestrial Observation and Prediction System (TOPS); and geospatial decision models for management of protected wetlands through a thorough review and case studies.

In Chapter 21, Crabtree and Sheldon share their insights and research by integration of focal species populations and remote sensing for monitoring and modeling environmental change in protected areas. As the authors point out, the intrinsic value of protected areas is in part set by the unique and rare species and processes they encompass. For these reasons, the chapter focuses on the role of long-term legacy species datasets and their value, in a protected area setting, in helping elucidate the metrics of a resilient, intact, and sustainable ecosystem. Monitoring, when coupled with modeling, can provide a powerful basis for guiding management actions, while simultaneously advancing science through hypothesis testing and predictions. Modeling of ecosystem indicators informed by monitoring information such as that provided from remote sensing is essential for efficient, transparent, repeatable, and defensible decision-making in ecological systems. These decisions should include outcome-based activities driving toward goals such as population recovery, critical habitat restoration, biodiversity increase, and improved ecosystem services. Narrative, conceptual, visual, ecological, or statistical models provide a common language for scientists and practitioners, permitting hypothesis testing about the mechanisms or drivers underlying the observed variation in data. The chapter describes a series of linked decision support tools, collectively called EAGLES (Ecosystem Assessment, Geospatial Analysis, and Landscape Evaluation System), as the framework of modeling, combining architecture for integrating species legacy data to provide insights into cause and consequence. This monitoring and modeling approach provides a common platform, shared data, and data analysis protocols across jurisdictional and ecological boundaries, needed to craft science-based management strategies to adapt to current and future impacts, especially those imposed by climate.

In Chapter 22, Hashimoto et al. introduce a representative study in monitoring and forecasting ecosystem dynamics along the Appalachian Trail (A.T.) using TOPS. The A.T. traverses most of the high elevation ridges of the eastern United States, extending about 3676 km across 14 states, from Springer Mountain in Northern Georgia to Mount Katahdin in central Maine. The A.T. and its corridor intersects 8 national forests and 6 national park units; crosses more than 70 state park, forest, and game management units; and passes through 287 local jurisdictions. The gradients in elevation, latitude, and moisture of the A.T. sustain a rich biological assemblage of temperate zone forest species. The A.T. and its surrounding protected lands harbor forests with some of the greatest biological diversity in the United States including rare, threatened, and endangered flora and fauna, and are the headwaters of important water resources for millions of people. As the north–south alignment of the A.T. represents a mega-transect of the forests and alpine areas of the eastern United States, it offers a setting for collecting scientifically valid and relevant data on the health of the ecosystems and the species that inhabit them. The high-elevation setting of the A.T. and its protected corridor can serve as a mega-transect for understanding undesirable changes in the natural resources of the eastern United States, from development encroachment to recreational misuse, acid precipitation, invasions of exotic species, and climate change (Dufour and Crisfield 2008). Better information and information assessment for the public about the effect of environmental stress of the A.T. and citizen engagement are considered one of the essentials. With such a spatial extent and dynamics of landscape features, timely monitoring of the ecological conditions of protected lands along the A.T. presents a challenge.

Chapter 22 describes the TOPS framework structure and components and demonstrates the use of TOPS for analyzing current vegetation status, assessing the impact of recent climate change on vegetation, and projecting future vegetation conditions using climate scenarios downscaled for the protected lands and ecosystems along the A.T. The ultimate goal of this study is to provide seamless data and derived information to establish a decision support system for monitoring, reporting, and forecasting ecological conditions of the A.T region (Wang et al. 2010).

Finally, in Chapter 23, Ji presents reviews and case studies of geospatial decision models in management of protected wetlands. Recognizing the lack of systematic review of geospatial decision models evolving and increasingly being applied to address various wetland management questions, this chapter intends to fill this information gap by analyzing common characteristics of geospatial decision models, reviewing the history of methodology development and widely used models, and describing a case project to demonstrate technical details in the development of a geospatial decision model.

1.6 Concluding Remarks

Gross et al. (Chapter 2) point out that many remote-sensing-based monitoring projects, especially those that involve many sites and collaborators, will likely face challenges similar to those described in the chapter. Some of these are common to most large problems, while others are more specific to working with complex technologies and management agencies. The lessons learned and recommendations put forward by the authors of Chapter 2 include: *"allocate sufficient time to develop a genuine science-management partnership," "communicate results in a management-relevant context," "confirm or embellish existing frameworks and processes," "plan for persistence and change,"* and *"build on existing, widely used data analysis tools and software frameworks, even if they seem inefficient."* The detailed discussions in the chapter provide insights that should benefit data and model developers, users and resource managers, and policy and decision-makers.

Zorn et al. (Chapter 5) point out that ecological monitoring is critical to the success of protected areas and remote sensing is critical to the success of ecological monitoring. Both of these statements are true, in part, because of the uncertainty inherent in protected area management. A dominant paradigm in protected area management is the "uncertainty principle" (Hockings et al. 2000) which simply states that decisions can never be made with complete certainty regarding the outcome of management actions. This uncertainty is due to the complex and dynamic nature of ecosystems that protected areas are established to protect. For this reason, adaptive management is a key strategy in the conservation of protected areas (Hockings et al. 2000). Results of decisions are monitored to ensure that they support goals and objectives. If monitoring suggests that actions (or inactions) are not effective then they are adapted in light of the new information and adjustments are made. This cycle of decision-making–monitoring–adaption is repeated in an iterative manner until protected area goals are met. For protected areas, adaptive management cannot be successful without effective ecological monitoring. When supplemented by *in situ* monitoring, remote sensing can be sensitive to many relevant ecological changes due to climate change and land use. The authors declare that remote sensing information possesses the characteristics to provide managers with scientifically credible, timely information and powerful images for communicating key management issues.

Crabtree and Sheldon (Chapter 21) point out that science and decision-making should go hand in hand, because they both measure success by their ability to predict the consequences of actions. Although monitoring is the critical first step needed for science-based decisions, it is often treated as an end unto itself. How do we explain variation across space and over time? To what is it attributable? What are the anticipated long- and short-term outcomes from our management activities? At its best, ecological modeling provides answers to these questions by (1) exploratory and synthetic analysis of possible causes using explanatory variables, (2) investigating possible causes while testing hypotheses (diagnostic models), and (3) predicting consequences, for example, under future scenarios (prognostic models). Arguably, ecological models are as powerful as the quality and relevance of the causal and explanatory variables ("covariates") we are able to include. Lack of these explanatory covariates in predictive modeling is equivalent to conducting science without explanatory alternate hypotheses. Ecosystem indicators, whether process based (e.g., productivity), pattern based (e.g., land-use activities), or component based (species populations), vary in space and time; yet a major limiting factor in the comprehensive ecological models is the lack of explanatory geospatial data. Although these geospatial data may exist, impediments to access by ecologists in protected area contexts may include issues as varied as data formats, technical barriers to data integration, validation issues, documentation issues, uniform standards and protocols accompanying the remote sensing data, increasing specialization of disciplines, cost, lack of requisite technical expertise, and time for dealing with all of the above complexities.

These issues conspire against the ready, standardized integration of remote sensing into ecological research for protected area management. Nonetheless, remote sensing science is a universal tool for managers and researchers across many domains (Kennedy et al. 2009). Remote sensing data and data products coupled with user-friendly data exploration, data management, analyses, and modeling tools, in an accessible common platform, will allow both scientists and practitioners a better understanding of how environmental impacts affect species populations and the ecosystem goods and services that sustain them.

Remote sensing is among the most fascinating frontiers of science and technology that are constantly improving. Protected lands are by no means uniform entities and they have a wide range of management aims and are governed by many different stakeholders. As the chapters in this book attest, advances in remote sensing have enabled us to gather and share information about the protected lands at unprecedented rates and scales. There are many new and exciting applications of remotely sensed data that will contribute to better informing management of protected lands. The achievements through applications of such science and technologies, the challenges, the lessons learned, and recommendations in remote sensing of protected areas deserve special attention.

It is a challenge to put together a book that covers important topics and latest developments under the general term of remote sensing of protected lands. I hope that this book can provide a snapshot about remote sensing applications to address issues of inventory, monitoring, and management of protected lands. I also hope that this book can inspire a broader scope of interests in scientific research and management of the natural treasures of protected lands and waters on Earth.

Acknowledgments

I would like to express my sincere gratitude and appreciation to the several individuals who helped tremendously during the development of this book.

The names of those individuals are identified in the Acknowledgment of this book. I thank all 69 contributors of this book, whose dedication and hard work made this book the best of its kind. In particular, the lead authors of each chapter worked very hard to present their research findings and to share their insights about the practice in remote sensing of protected lands and waters.

References

Alsdorf, D., J. Melack, T. Dunne, L. Mertes, L. Hess, and L. Smith, 2000. Interferometric radar measurements of water level changes on the Amazon floodplain. *Nature* 404: 174–177.

Arima, E.Y., R.T. Walker, M. Sales, C. Souza, Jr., and S.G. Perz. 2008. The fragmentation of space in the Amazon basin: Emergent road networks. *Photogrammetric Engineering & Remote Sensing* 74(6): 699–709.

Babbitt, B. 2005. *Cities in the Wilderness: A New Vision of Land Use in America*. Washington, DC: Island Press, 200pp.

Bergen, K.M., T. Zhao, V. Kharuk, Y. Blam, D.G. Brown, L.K. Peterson, and N. Miller. 2008. Changing regimes: Forested land cover dynamics in Central Siberia 1974–2001. *Photogrammetric Engineering & Remote Sensing* 74(6): 787–798.

Bourgeau-Chavez, L.L., K.B. Smith, S.M. Brunzell, E.S. Kasischke, E.A. Romanowicz, and C.J. Richardson. 2005. Remote monitoring of regional inundation patterns and hydroperiod in the greater Everglades using synthetic aperture radar. *Wetlands* 25(1): 176–191.

Coppin, P.R., and M.E. Bauer. 1996. Digital change detection in forest ecosystem with remotely sensed imagery. *Remote Sensing Review* 13: 207–234.

Costa, M.P.F. 2004. Use of SAR satellites for mapping zonation of vegetation communities in the Amazon floodplain. *International Journal of Remote Sensing* 25(10): 1817–1835.

Costa, M.P.F., O. Niemann, E. Novo, and F. Ahern. 2002. Biophysical properties and mapping of aquatic vegetation during the hydrological cycle of the Amazon floodplain using JERS-1 and RADARSAT. *International Journal of Remote Sensing* 23(7): 1401–1426.

Crabtree, R., C.S. Potter, R.S. Mullen et al. 2009. Synthesis of ground and remote sensing data for monitoring ecosystem functions in the Colorado River Delta, Mexico. *Remote Sensing of Environment* 113(7): 1486–1496.

Crabtree, R., and J. Sheldon, 2011. Chapter 21—Monitoring and modeling environmental change in protected areas: Integration of focal species populations and remote sensing. In Y. Wang (ed.), *Remote Sensing of Protected Lands*. Boca Raton, FL: CRC Press.

DeFries, R., C.B. Field, I. Fung, G.J. Collatz, and L. Bounoual. 1999. Combining satellite data and biogeochemical models to estimate global effects of human-induced land cover change on carbon emissions and primary productivity. *Global Biogeochemical Cycles* 13: 803–815.

DeFries, R., A.J. Hansen, A.C. Newton, M. Hansen, and J. Townshend. 2005. Isolation of protected areas in tropical forests over the last twenty years. *Ecological Applications* 15: 19–26.

Dennison, P.E., P.L. Nagler, K.R. Hultine, E.P. Glenn, and J.R. Ehleringer. 2009. Remote monitoring of tamarisk defoliation and evapotranspiration following salt cedar leaf beetle attack. *Remote Sensing of Environment* 113(7): 1462–1472.

Dudley, N., ed. 2008. *Guidelines for Applying Protected Area Management Categories.* Gland, Switzerland: IUCN. x + 86pp.

Dufour, C., and E. Crisfield, eds. 2008. *The Appalachian Trail MEGA-Transect.* Harpers Ferry, WV: Appalachian Trail Conservancy.

Edgar, G.J., G.R. Russ, and R.C. Babcock. 2007. Marine protected areas. In S. Connell and B.M. Gillards (eds), *Marine Ecology.* South Melbourne, VIC, Australia: Oxford University Press, pp. 533–555.

Fancy, S.G., J.E. Gross, and S.L. Carter. 2009. Monitoring the condition of natural resources in U.S. National Parks. *Environmental Monitoring and Assessment* 151(1–4): 161–174.

Fraser, R., I. Olthof, and D. Pouliot. 2009. Monitoring land cover change and ecological integrity in Canada's national parks. *Remote Sensing of Environment* 113(7): 1397–1409.

Friedlander, A.M., E.K. Brown, and M.E. Monaco. 2007. Coupling ecology and GIS to evaluate efficacy of marine protected areas in Hawaii. *Ecological Applications* 17: 715–730.

Friedlander, A.M., L.M. Wedding, J.E. Caselle, and B.M. Costa. 2011. Chapter 13—Integration of remote sensing and *in situ* ecology for the design and evaluation of marine protected areas: Examples from tropical and temperate ecosystems. In Y. Wang (ed.), *Remote Sensing of Protected Lands.* Boca Raton, FL: CRC Press.

Goetz, S.J., P. Jantz, and C.A. Jantz. 2009. Connectivity of core habitat in the northeastern United States: Parks and protected areas in a landscape context. *Remote Sensing of Environment* 113(7): 1421–1429.

Gould, W. 2000. Remote sensing of vegetation, plant species richness, and regional biodiversity hotspots. *Ecological Applications* 10: 1861–1870.

Griffith, J.A., S.V. Stehman, T.L. Sohl, and T.R. Loveland. 2003. Detecting trends in landscape pattern metrics over a 20-year period using a sampling-based monitoring programme. *International Journal of Remote Sensing* 24: 175–181.

Gross, J., G. Goetz, and J. Cihlar. 2009. Application of remote sensing to parks and protected area monitoring: Introduction to the special issue. *Remote Sensing of Environment* 113(7): 1343–1345.

Hansen, A.J., and J.J. Rotella. 2002. Biophysical factors, land use, and species viability in and around nature reserves. *Conservation Biology* 16: 1–12.

Hansen, A.J., and R. DeFries. 2007a. Land use change around nature reserves: Implications for sustaining biodiversity. *Ecological Applications* 17(4): 972–973.

Hansen, A.J., and R. DeFries. 2007b. Ecological mechanisms linking protected areas to surrounding lands. *Ecological Applications* 17(4): 974–988.

Hayes, D.J., and S.A. Sader. 2001. Comparison of change-detection techniques for monitoring tropical forest clearing and vegetation regrowth in a time series. *Photogrammetric Engineering & Remote Sensing* 67(9): 1067–1075.

Heinen, J. (Lead Author) and K. Hite (Topic Editor). 2007. Protected natural areas. In Cutler J. Cleveland (ed.), *Encyclopedia of Earth.* Washington, DC: Environmental Information Coalition, National Council for Science and the Environment. [First published in the *Encyclopedia of Earth* November 29, 2007; Last revised November 29, 2007; Retrieved February 5, 2011 <http://www.eoearth.org/article/Protected_natural_areas>.]

Hess, L.L., J.M. Melack, S. Filoso, and Y. Wang. 1995. Delineation of inundated area and vegetation along the Amazon floodplain with the SIR-C synthetic aperture radar. *IEEE Transactions on Geoscience and Remote Sensing* 33(4): 896–904.

Hockings, M., S. Stolton, and N. Dudley. 2000. *Evaluating Effectiveness: A Framework for Assessing the Management of Protected Areas*, World Commission on Protected Areas. Best Practice Protected Area Guidelines Series No. 6. Gland, Switzerland and Cambridge, UK: IUCN.

Hoffmann, M., C. Hilton-Taylor, A. Angulo et al. 2010. The impact of conservation on the status of the world's vertebrates. *Science* 330(6010): 1503–1509.

Huang, C., S.N. Goward, K. Schleeweis, N. Thomas, J.G. Masek, and Z. Zhu. 2009. Dynamics of national forests assessed using the Landsat record: Case studies in eastern United States. *Remote Sensing of Environment* 113(7): 1430–1442.

IUCN. 2008. Shaping a sustainable future. In *The IUCN Programme 2009–2012*. Gland, Switzerland: IUCN.

IUCN and UNEP-WCMC. 2010. *The World Database on Protected Areas (WDPA): January 2010*. Cambridge, UK: UNEP-WCMC.

Jantz, C.A., S.J. Goetz, and M.A. Shelley. 2003. Using the SLEUTH urban growth model to simulate the land use impacts of policy scenarios in the Baltimore-Washington metropolitan region. *Environment and Planning* 31(2): 251–271.

Jones, D.A., A.J. Hansen, K. Bly et al. 2009. Monitoring land use and cover around parks: A conceptual approach. *Remote Sensing of Environment* 113(7): 1346–1356.

Kasischke, E.S., and L.L. Bourgeau-Chavez. 1997. Monitoring south Florida wetlands using ERS-1 SAR imagery. *Photogrammetric Engineering and Remote Sensing* 63: 281–291.

Kelleher, G. 1999. *Guidelines for Marine Protected Areas*. Gland, Switzerland and Cambridge, UK: IUCN.

Kennedy, R.E., P.A. Townsend, J.E. Gross, W.B. Cohen, P. Bolstad, Y.Q. Wang, and P. Adams. 2009. Remote sensing change detection and natural resource monitoring for managing natural landscapes. *Remote Sensing of Environment* 113(7): 1382–1396.

Kennedy, R.E., Z. Yang, and W.B. Cohen. 2010. Detecting trends in forest disturbance and recovery using yearly Landsat time series: 1. LandTrendr—Temporal segmentation algorithms. *Remote Sensing of Environment* 114(12): 2897–2910.

Kerr, J.T., and M. Ostrovsky. 2003. From space to species: Ecological applications for remote sensing. *TRENDS in Ecology and Evolution* 18(6): 299–306.

Kharuk, V. I., K.J. Ranson, and S.T. Im. 2009. Siberian silkmoth outbreak pattern analysis based on SPOT VEGETATION data. *International Journal of Remote Sensing* 30(9): 2377–2388.

Lambin, E.F., and A.H. Strahler. 1993. Change-vector analysis in multitemporal space: A tool to detect and categorize land-cover change processes using high-temporal resolution satellite data. *Remote Sensing Environment* 48: 231–244.

Lu, Z., D. Dzurisin, J. Biggs, C. Wicks, Jr., and S. McNutt. 2010. Ground surface deformation patterns, magma supply, and magma storage at Okmok volcano, Alaska, from InSAR analysis: I. Inter-eruption deformation, 1997–2008. *Journal of Geophysical Research* 115: B00B03.

Lu, Z., D. Dzurisin, and H.S. Jung. 2011. Chapter 19—Monitoring natural hazards in protected lands using interferometric synthetic aperture radar (InSAR). In Y. Wang (ed.), *Remote Sensing of Protected Lands*. Boca Raton, FL: CRC Press.

Lu, Z., and O. Kwoun. 2008. RADARSAT-1 and ERS interferometric analysis over southeastern coastal Louisiana: Implication for mapping water-level changes beneath swamp forests. *IEEE Transactions on Geoscience and Remote Sensing* 46: 2167–2184.

Mas, J.F. 1999. Monitoring land-cover changes: A comparison of change detection techniques. *International Journal of Remote Sensing* 20(1): 139–152.

Massonnet, D., and K. Feigl. 1998. Radar interferometry and its application to changes in the Earth's surface. *Reviews of Geophysics* 36: 441–500.

McDonald, R., R. Forman, and P. Kareiva. 2010. Open space loss and land inequality in United States' cities, 1990–2000. *PLoS ONE* 5(3): e9509.

McNulty, T. 2009. *Olympic National Park, A Natural History.* Seattle: University of Washington Press, 338pp.

Mena, C.F. 2008. Trajectories of land-use and land-cover in the Northern Ecuadorian Amazon: Temporal composition, spatial configuration, and probability of change. *Photogrammetric Engineering & Remote Sensing* 74(6): 737–751.

Nagler, P.L., E.P. Glenn, and O. Hinojosa-Huerta. 2009. Synthesis of ground and remote sensing methods for monitoring ecosystem functions in the Colorado River delta, Mexico. *Remote Sensing of Environment* 113(7): 1473–1485.

Nemani, R.R., H. Hashimoto, P. Votava et al. 2009. Monitoring and forecasting ecosystem dynamics using the using the Terrestrial Observation and Prediction System (TOPS). *Remote Sensing of Environment* 113(7): 1497–1509.

NIPC. 1962. *Open Space Policies for Northeastern Illinois: A Statement Adopted by the Northeastern Illinois Metropolitan Area Planning Commission.* Chicago: NIPC, p. 358.

Parmenter, A.P., A. Hansen, R. Kennedy, W. Cohen, U. Langner, R. Lawrence, B. Maxwell, A. Gallant, and R. Aspinall. 2003. Land use and land cover in the Greater Yellowstone Ecosystem: 1975–95. *Ecological Applications* 13(3): 687–703.

Pereira, H.M., P.W. Leadley, V. Proença et al. 2010. Scenarios for Global biodiversity in the 21st century. *Science* 330(6010): 1496–1501.

Ramsey, III, E.W. 1995. Monitoring flooding in coastal wetlands by using radar imagery and ground-based measurements. *International Journal of Remote Sensing* 16(13): 2495–2502.

Reed, R., M. Budde, P. Spencer, and A.F. Miller. 2009. Integration of MODIS-derived metrics to assess interannual variability in snowpack, lake ice, and NDVI in southwest Alaska. *Remote Sensing of Environment* 113(7): 1443–1452.

Ressl, R., G. Lopez, I. Cruz, S. Ressl, M.I. Schmidt, and R. Jiménez. 2009. Operational active fire mapping and burnt area identification applicable to Mexican nature protection areas using MODIS-DB data. *Remote Sensing of Environment* 113(6): 1113–1126.

Rogan, J., J. Franklin, and D.A. Roberts. 2002. A comparison of methods for monitoring multitemporal vegetation change using thematic mapper imagery. *Remote Sensing of Environment* 80: 143–156.

Sheng, Y., and D. Alsdorf. 2005. Automated ortho-rectification of Amazon basin-wide SAR mosaics using SRTM DEM data. *IEEE Transactions on Geoscience and Remote Sensing* 43(8): 1929–1940.

Sheng, Y., C.A. Shah, and L.C. Smith. 2008. Automated image registration for hydrologic change detection in the lake-rich arctic. *IEEE Geoscience and Remote Sensing Letters* 5(3): 414–418.

Simard, M., K. Zhang, V.H. Rivera-Monroy, M.S. Ross, P.L. Ruiz, E. Castañeda-Moya, R.R. Twilley, and E. Rodriguez. 2006. Mapping height and biomass of mangrove forests in Everglades National Park with SRTM elevation data. *Photogrammetric Engineering & Remote Sensing* 72(3): 299–311.

Stow, D.A., A. Hope, D. McGuire et al. 2004. Remote sensing of vegetation and land-cover change in Arctic tundra ecosystems. *Remote Sensing of Environment* 89(3): 281–308.

Sun, G., K.J. Ranson, and V.I. Kharuk. 2002. Radiometric slope correction for forest biomass estimation from SAR data in Western Sayani mountains, Siberia. *Remote Sensing of Environment* 79: 279–287.

Svancara, L., J.M. Scott, T.R. Loveland, and A.B. Pidgorma. 2009. Assessing the landscape context and conversion risk of protected areas using remote-sensing derived data. *Remote Sensing of Environment* 113(7): 1357–1369.

Theobald, D.M. 2001. Land use dynamics beyond the American urban fringe. *Geographical Review* 91: 544–564.

Townsend, P.A., T.R. Lookingbill, and C.C. Kingdon. 2009. Spatial pattern analysis for monitoring protected areas. *Remote Sensing of Environment* 113(7): 1410–1420.

Turner, M.G. 1990. Spatial and temporal analysis of landscape patterns. *Landscape Ecology* 4: 21–30.

Turner, W., S. Spector, N. Gardiner, M. Fladeland, E. Sterling, and M. Steininger. 2003. Remote sensing for biodiversity science and conservation. *TRENDS in Ecology and Evolution* 18(3): 306–314.

Walsh, S.J., J.P. Messina, and D.G. Brown. 2008. Mapping & modeling land use/land cover dynamics in frontier settings. *Photogrammetric Engineering & Remote Sensing* 74(6): 677–679.

Wang, C., and J. Qi. 2008. Biophysical estimation in tropical forests using JERS-1 SAR and VNIR imagery: II—Aboveground woody biomass. *International Journal of Remote Sensing* 29(23): 6827–6849.

Wang, C., J. Qi, and M. Cochrane. 2005. Assessment of tropical forest degradation with canopy fractional cover from Landsat ETM+ and IKONOS imagery. *Earth Interactions* 9(22): 1–18.

Wang, Y., G. Bonynge, J. Nugranad, M. Traber, A. Ngusaru, J. Tobey, L. Hale, R. Bowen, and V. Makota. 2003. Remote sensing of mangrove change along the Tanzania coast. *Marine Geodesy* 26(1–2): 35–48.

Wang, Y., R. Nemani, F. Dieffenbach et al. 2010. Development of a decision support system for monitoring, reporting and forecasting ecological conditions of the Appalachian trail. In *Proceedings of the 2010 IEEE International Geoscience and Remote Sensing Symposium, IEEE Xplore* entry: 978-1-4244-9566-5, pp. 2095–2098.

Wang, Y., M. Traber, B. Milestead, and S. Stevens. 2007. Terrestrial and submerged aquatic vegetation mapping in Fire Island National seashore using high spatial resolution remote sensing data. *Marine Geodesy* 30(1): 77–95.

Wang, Y.Q., B.R. Mitchell, J. Nugranad-Marzilli, G. Bonynge, Y. Zhou, and G.W. Shriver. 2009. Remote sensing of land-cover change and landscape context of the national parks: A case study of the Northeast Temperate Network. *Remote Sensing of Environment* 113(7): 1453–1461.

Wdowinski, S., F. Amelung, F. Miralles-Wilhelm, T. Dixon, and R. Carande. 2004. Space-based measurements of sheet-flow characteristics in the Everglades wetland, Florida. *Geophysical Research Letters* 31: L15503.

Wedding, L., and A.M. Friedlander. 2008. Determining the influence of seascape structure on coral reef fishes in Hawaii using a geospatial approach. *Marine Geodesy* 31: 246–266.

Wedding, L., A.M. Friedlander, M. McGranaghan, R. Yost, and M.E. Monaco. 2008. Using bathymetric Lidar to define nearshore benthic habitat complexity: Implications for management of reef fish assemblages in Hawaii. *Remote Sensing of Environment* 112: 4159 – 4165.

Wessels, K.J., R.S. DeFries, J. Dempewolf, L.O. Anderson, A.J. Hansen, S.L. Powell, and E.F. Moran. 2004. Mapping regional land cover with MODIS data for biological conservation: Examples from the Greater Yellowstone Ecosystem, USA and Pará State, Brazil. *Remote Sensing of Environment* 92: 67–83.

Wiens, J.A., R.D. Sutter, M. Anderson, J. Blanchard, A. Barnett, and N. Aguilar-Amuchastegui. 2009. Selecting and conserving lands for biodiversity: The role of remote sensing. *Remote Sensing of Environment* 113(7): 1370–1381.

Wilkinson, D.W., R.C. Parker, and D.L. Evans. 2008. Change detection techniques for use in a statewide forest inventory program. *Photogrammetric Engineering & Remote Sensing* 74(7): 893–901.

Wilson, E.H., and S.A. Sader. 2002. Detection of forest type using multiple dates of Landsat TM imagery. *Remote Sensing of Environment* 80: 385–396.

Section I

Remote Sensing of Changing Landscape of Protected Lands

2

Remote Sensing for Inventory and Monitoring of U.S. National Parks

John E. Gross, Andrew J. Hansen, Scott J. Goetz, David M. Theobald,
Forrest S. Melton, Nathan B. Piekielek, and Ramakrishna R. Nemani

CONTENTS

2.1 Introduction

U.S. National Park Service (NPS) units ("parks") are important components in a system of reserves that protect biodiversity and other natural and cultural resources. To meet the NPS mission to manage resources so that they are left ". . . unimpaired for the enjoyment of future generations" (16 USC 1) it is essential to know what resources occur in parks and to monitor the status and trends in the condition of key resource indicators. The NPS Inventory and Monitoring (I&M) Program (see Table 2.1 for a list of acronyms) was designed to provide the infrastructure and staff to identify critical environmental indicators ("vital signs") and to implement long-term monitoring of natural resources in more than 270 parks that contain significant natural resources (Fancy et al. 2009). These parks are organized into 32 ecoregional networks (Figure 2.1). Each of the 32 I&M networks consists of core professional staff (program manager, data manager, ecologists, field technicians, etc.), and each I&M network supports monitoring in parks within the network.

The overall purpose of I&M is to provide sound scientific information that enhances management of natural resources. To do so, I&M collects, organizes, and makes available natural resource data and contributes to the

TABLE 2.1

Acronyms Used in This Chapter

Acronym	Meaning
AOA	Area of analysis
DEWA	Delaware Water Gap Recreation Area
GIS	Geographical Information System
GPP	Gross primary productivity
I&M	Inventory and Monitoring Program
LAI	Leaf area index
MODIS	Moderate Resolution Imaging Spectroradiometer
N	Naturalness
NARSEC	North American Network for Remote Sensing Park Ecological Condition
NASA	U.S. National Aeronautics and Space Administration
NDVI	Normalized difference vegetation index
*n*EPT	Number of *Ephemeroptera, Plecoptera,* and *Tricoptera* species
NPP	Net primary productivity
NPS	National Park Service
PACE	Protected-area-centered ecosystem
PALMS	Park AnaLysis and Monitoring Support
SOP	Standard operating procedure
TOPS	Terrestrial Observation and Prediction System
UPDE	Upper Delaware Scenic and Recreational River

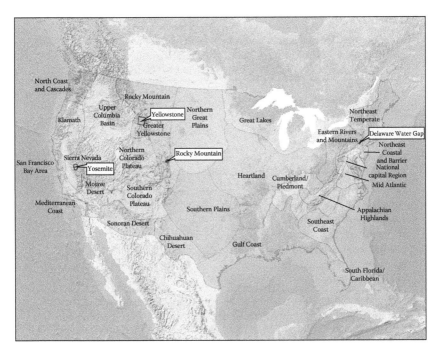

FIGURE 2.1
Map of the U.S. National Park Service ecoregional inventory and monitoring (I&M) networks for the continental United States (Alaska and Pacific Islands not shown). Each network consists of staff and infrastructure to support long-term ecosystem monitoring for natural resource parks within the network. Focal parks (names in boxes) are served by the Sierra Nevada, Greater Yellowstone, Rocky Mountain, and Eastern Rivers and Mountains Networks.

service's knowledge by adding value to data though analysis, synthesis, and modeling. I&M initiated 12 basic natural resource inventories to collect the information needed as a foundation for monitoring and to determine the current status of park resources (Table 2.2). Most inventories are now complete, except for the more expensive and time-consuming vegetation and geological resource inventories.

The NPS I&M instituted systems-based "vital signs" monitoring to provide sound scientific information on trends in the condition of park natural resources. "Vital signs" are a subset of physical, chemical, and biological elements and processes of park ecosystems that are selected to represent the overall health or condition of park resources, known or hypothesized effects of stressors, or elements that have important human values (Fancy et al. 2009). I&M networks worked extensively with park personnel and other experts to identify the highest-priority vital signs—a lengthy process that involved more than 1000 people. Landscape dynamics, along with climate and invasive species, was ranked as one of the highest priorities for long-term monitoring across all the 32 I&M networks. Despite the high ratings,

TABLE 2.2

Baseline Inventories Undertaken by the National Park Service Inventory
and Monitoring Program for Parks with Significant Natural Resources,
and the Percent Completed by July 2011

Inventory	Percent Completed
Geologic resources inventory	32[a]
Vegetation inventory	38[b]
Soil resources inventory	77
Baseline water quality data	100
Base cartography data	100
Species lists	99
Air quality data	100
Air quality-related values	100
Climate inventory	100
Natural resource bibliography	100
Species occurrence and distribution	100
Water body location	100

[a] A total of 184 parks (68% of total) have completed digital maps; 88 (32% of total) of these have a final comprehensive report.

[b] An additional 154 parks (57% of total) have vegetation inventories in progress.

few I&M networks have successfully developed landscape monitoring protocols and implemented landscape monitoring.

The slow development of landscape-scale monitoring reflects the complex decisions needed to identify a small set of indicators that are reasonably comprehensive, informative, relevant, and affordable. To facilitate progress in developing operational landscape monitoring, NPS, Parks Canada Agency, U.S. National Aeronautics and Space Administration (NASA), and other agencies cosponsored workshops to share experiences and knowledge (NARSEC 2005, 2007; Gross et al. 2009). A clear need identified at these workshops was for organized teams of experts to focus on developing general methods, at relevant scales, which could be widely applied to distribute and share the costs of development. It is simply too difficult and expensive for individual parks or I&M networks, on their own, to undertake development of a full suite of landscape dynamics monitoring protocols. To address needs for broad-scale data on landscape attributes across the entire system of parks, the I&M Program Office developed the NPScape project. NPScape provides landscape-level data, methods, tools, and evaluations for a limited set of attributes derived from data on land cover, population and housing, roads, and land ownership (NPScape 2010). Data and results from NPScape are provided for all the more-than-270 I&M park units. A central goal of NPScape was to reduce per-park cost by identifying and documenting a small set of highly relevant landscape-scale measurements that could be derived from national-scale data, and then centralizing data acquisition, processing, analysis, and reporting. NPScape is founded on the principle of economy of scale, and the huge variations in park geographical location, ecological context, and size make it impossible for NPScape to address many questions that require park-specific data or other local data. Although the needs for landscape-scale monitoring in and around Canadian parks differ somewhat from those in the United States, Parks Canada Agency found they were in a similar position. In response to these needs, Parks Canada Agency and the Canadian Space Agency cofunded multiyear studies to develop and enhance operational use of remotely sensed data for park monitoring.

Although NPScape and other national programs will meet many NPS needs for broad-scale, relatively coarse resolution indicators, there remained a need for complementary monitoring protocols that operate at finer resolutions, that can address park-specific contexts, but that are still broadly relevant and easily adopted by and incorporated into the NPS I&M. The goal of this chapter is to describe a project that focused on addressing this need and to facilitate further progress in using remotely sensed data to support the management of protected areas. We describe a multiyear project that worked with geographically dispersed parks from a variety of settings. This chapter is effectively a case study, illustrating approaches and results that will help implement routine use of remotely sensed data for monitoring in and around parks. We describe the rationale, design, and products of a project to enhance use of NASA data and technology by the NPS I&M. While we focused on the needs of the NPS I&M, the issues, approaches, and results are broadly applicable to monitoring many types of protected areas. Interested readers can refer to other sources for detailed reviews of the NPS I&M Program (Fancy et al. 2009; http://science.nature.nps.gov/im/index.cfm), conceptual frameworks that support landscape-scale monitoring (Hansen and DeFries 2007; Jones et al. 2009), and technical considerations that must be addressed when designing remote-sensing-based monitoring indicators (Phinn et al. 2003; Kennedy et al. 2009; papers in Gross et al. 2009).

2.2 PALMS: Park AnaLysis and Monitoring Support

The overall goal of the PALMS (Park AnaLysis and Monitoring Support) project was to enhance the quality of natural resource management in parks by better integrating the routine acquisition and analysis of NASA Earth System Science products and other data sources into the NPS I&M. NASA supported the project via a program that specifically targets applied science (versus basic research). Each participating I&M network and the national I&M office supported the project by allocating time of personnel with expertise that would contribute to the project. This included time of Geographic Information System (GIS)/data specialists and ecologists with local knowledge of focal parks. I&M networks also served as liaisons with the (much larger) park staff, thereby ensuring participation of decision-makers and others when appropriate. We felt the explicit contribution of NPS resources to the project was important to encourage shared ownership of results and to share risks that might result from inadequate engagement.

Specific objectives of the PALMS project were to

1. (a) Identify NASA and other products useful as indicators for NPS I&M monitoring, and (b) delineate the boundaries of the surrounding protected-area-centered ecosystems (PACE) appropriate for monitoring.
2. Add value to these datasets for understanding change through analysis and forecasting.
3. Deliver these products and a means to integrate them into the NPS I&M decision support framework.

The project focused on four sets of national parks to develop and demonstrate the approach (Figure 2.1): Sequoia-Kings Canyon and Yosemite National Parks (Sierra Nevada I&M Network), Yellowstone and Grand Teton National Parks (Greater Yellowstone I&M Network), Rocky Mountain National Park (Rocky Mountain I&M Network), and a combination of Delaware Water Gap National Recreation Area (DEWA) and Upper Delaware Scenic and Recreational River (UPDE) (Eastern Rivers and Mountains I&M Network). Selection of focal parks was based almost entirely on the familiarity of the principal investigators with these parks, and access to data and resources that supported the goals of the project. Other parks and networks were keen to participate in this project, but we lacked the capacity to expand the study and include additional parks. An expanded, follow-on project is pending.

The PALMS project was designed from the outset to be highly collaborative. All the investigators were experienced, had worked with the NPS, and had some idea of the type and extent of communication that would be required. The explicit contribution of staff time from each participating I&M network clearly promoted this approach. Nonetheless, a surprisingly substantial and sustained effort by all project staff was required to keep park and network collaborators informed and engaged throughout the project. Park personnel, especially those in supervisory positions, tend to have many fixed-time commitments that made scheduling complicated. When working with parks, the time required to schedule meetings or provide products and obtain reviews can be considerable.

2.3 PALMS Ecological Indicators

Every monitoring project must balance the desire to deliver the most comprehensive, useful, and interesting information with constraints imposed by technical feasibility, cost, staff expertise, and available resources (Phinn et al. 2003; Jones et al. 2009; Kennedy et al. 2009). All NPS I&M networks undertook a multiyear effort to identify high-priority vital signs before we initiated this project, and "landscape dynamics" was consistently ranked among the highest of all monitoring needs. Beyond identifying the need for landscape-scale monitoring, few networks had identified any specific variables for monitoring. Furthermore, networks clearly understood the importance of landscape changes outside park boundaries (GAO 1994; Parks and Harcourt 2002; Hansen and DeFries 2007), but all networks were struggling to define the boundaries of scientifically credible and defensible areas for monitoring landscape-scale changes outside park boundaries.

Our first step was to identify candidate indicators for further development by consulting I&M network monitoring plans and related documents—that is, glean what we could from existing information (Jean et al. 2005; Britten et al. 2007; Marshall and Piekielek 2007; Mutch et al. 2008). I&M monitoring plans described park resources and threats to resources, existing and planned monitoring, and related information that could help identify suitable indicators. We held a series of meetings with park and network staff to discuss and refine definitions of indicators, and we also relied on our collective experiences and expertise. The process of identifying and refining indicators was iterative, and the final resolution of some indicators took more than two years of discussion and development. All forms of inputs proved to be valuable contributions to the final selection and development of the indicators.

The complete set of PALMS indicators and their geospatial attributes is summarized in Table 2.3. The suite of PALMS indicators includes measurements of weather and climate, stream health (water), land cover and land use, disturbances, primary production, and monitoring area. In the following sections, we briefly summarize features of exemplar indicators that are novel to this project or that are otherwise of particular interest. More complete descriptions of methodology and results are available in the following descriptions of PALMS products and other publications (Goetz and Fiske 2008; Jantz and Goetz 2008; Goetz et al. 2009; Nemani et al. 2009; Theobald et al. 2009; Bierwagen et al. 2010; Jantz et al. 2010; Theobald 2010; Hansen et al. 2011).

2.3.1 Protected-Area-Centered Ecosystems

Identifying a suitable area of analysis (AOA) is challenging because the extent of the most appropriate AOA varies with the specific issue, process, or species that is of most interest. Ideally, a long-term monitoring program would simply define an AOA that encompassed the broadest-scale issue anticipated. This is an impractical solution for most parks because the cost of imagery acquisition, processing, and analysis is directly related to the size of the AOA. There is thus a strong incentive to constrain many analyses to the smallest area necessary. Following Hansen and DeFries (2007), we developed

TABLE 2.3

Indicators Selected for Development by the Park AnaLysis and Monitoring Support Project and Some of Their Attributes

Level	Category	Indicator	Extent	Resolution
Air and climate	Weather and climate	Phenology (normalized difference vegetation analysis—NDVI; annual anomaly)	CONUS	1 km (all); 8 and 16 days
		Climate gridded daily (1980–2010)	DEWA, ROMO, YELL, YOSE	1 km
		Climate scenarios (monthly)	YOSE, DEWA, GYE, CONUS	12 km
Water	Stream health	Bioitic index of biological integrity; sensitive taxa	DEWA	1:24 K; 1:100 K
Landscape dynamics	Land cover	Ecosystem-type composition; summary by spatial scale	DEWA, ROMO, YELL, YOSE	30 m
		Bird hotspots and key habitat types	GYE	1 km
		Impervious cover change	DEWA	30 m
		Housing density class (1940–2100, decadal)	CONUS	100 m
		Landscape connectivity of forests	Eastern United States	270 m
		Pattern of natural landscapes	CONUS	270 m
		Past to future modeling	DEWA	30 m
	Extreme disturbance events	Fire effects via changes in phenology and related measures	DEWA, ROMO, YELL, YOSE	1 km; monthly anomalies/ annual summaries
	Primary production	Gross and net primary productivity (via simulation model results)	DEWA, ROMO, YELL, YOSE	1 km daily and/ or monthly summaries; annual trends
	Monitoring area	Greater park ecosystem boundaries	DEWA, ROMO, YELL, YOSE	30 m
	Land use	Land use	CONUS	90 m

CONUS, Continental United States (lower 48 states); DEWA, Delaware Water Gap National Recreation Area (including Upper Delaware Scenic and Recreational River); GYE, Greater Yellowstone Ecosystem; ROMO, Rocky Mountain National Park; YELL, Yellowstone National Park; YOSE, Yosemite National Park.

a framework for delineating the ecosystem surrounding a protected area that is likely to strongly influence ecological function and biodiversity within the protected area. Termed "protected area centered ecosystems" (PACEs; Hansen et al., 2011), this area becomes the logical place to focus on monitoring, research, and collaborative management to maintain protected area function and condition. The PACE framework is founded on five ecological mechanisms (processes) by which human activities impact on ecosystem functioning (Table 2.4). The PACE served two very important purposes. First, it defined a spatial context for conducting analyses and reporting results. Second, it is, by itself, an indicator of landscape condition, because the shape, composition, and extent of the PACE respond to and reflect human impacts in the area around a park.

To illustrate the approach in a variety of geographic and land-use settings, we defined PACEs for the NPS units included in the PALMS project and in two additional regions (the Pacific Northwest and the Appalachian Highlands). The resulting PACEs were on average 6.7 times larger than the

TABLE 2.4

Mechanism, Rationale, and Criteria Used to Define the Protected-Area-Centered Ecosystem (PACE)

Mechanism[a]	Rationale	PACE Criterion
Change in effective size of reserve	Fewer species are supported in small areas; species can be lost as habitats are isolated.	Specific habitat areas in the PACE are proportional to those in the park, up to the area specified by the species–area relationship.
Changes in ecological flows into and out of reserve	Water, sediments, nutrients, hydrological patterns may be altered by upstream land uses.	Watershed boundaries around park; or subbasins or subwatersheds that intersect park boundaries.
	Atmospheric transport of dust and pollutants affect parks; upwind land use can affect local climates.	Airsheds based on sources of pollutants or climate.
	Disturbances that originate outside parks can move into parks; conditions in initiation and run-on zones affect likelihood of disturbance and provide key habitats.	Perimeter around park based on historic disturbance rates, size, and shape.
Loss of crucial habitats	Includes seasonal habitats or ranges, movement paths, source populations, and parts of large home ranges that are outside of parks and that may be altered or destroyed.	Key habitats for migration, seasonal use, or otherwise crucial for park organisms (requires local knowledge).
Edge effect due to human activity	Human activities in areas adjacent to parks can directly or indirectly disturb or kill wildlife. Examples include hunting, poaching, pets (dogs, cats), introduction of exotic species, effects of noise and light, etc.	Create 25 km buffer around park and select human-dominated areas; create 5 km buffer around crucial habitat polygons.

Source: Adapted from Hansen, A. J., and R. DeFries. *Ecological Applications*, 17:974–988, 2007; Hansen, A. J. et al., *BioScience*, submitted.

[a] Mechanisms describe the ways in which human activities around parks may impact on ecosystem processes and biodiversity.

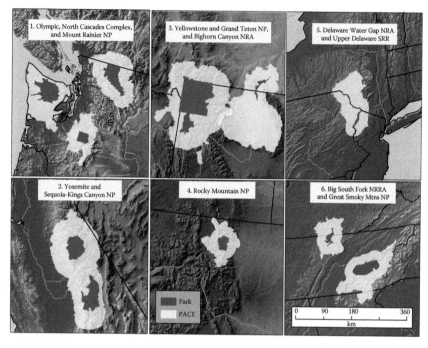

FIGURE 2.2
(**See color insert.**) Maps of protected-area-centered ecosystems (PACEs) delineated in this study for 13 U.S. National Park Service units. PACEs were defined by the criteria in Table 2.4.

parks for those in upper watersheds and 44.6 times larger for those in mid-watersheds (Figure 2.2). PACEs in the eastern US were dominated by private lands with high rates of land development, suggesting that they offer the greatest challenge for management. Our NPS collaborators generally embraced this approach for delineating the area to be monitored around national parks and suggested that the approach helps facilitate research and conservation across the parks and important surrounding lands.

2.3.2 Stream Biota

Stream macroinvertebrate diversity is a commonly used indicator of aquatic health, reflecting overall ecological integrity within a watershed (Van Sickle et al. 2006). Urbanization and associated impervious surface cover have adverse effects on aquatic systems, including greater variability in stream flow (flashiness), lower base flows, and increased bank and stream bed erosion (Schueler et al. 2009). These effects can be mitigated by near-stream vegetation buffers and other actions that reduce the force of overland flows, absorb excess nutrients, maintain stream bank integrity, and provide shade that reduces warming of stream water (Snyder et al. 2003; Goetz 2006). We mapped and modeled these processes in watersheds that encompass the

UPDE and the DEWA. Of primary interest to the Eastern Rivers and Mountains Network, and more generally to the NPS I&M effort, is information on stream biota and how these are likely to be impacted on by expanding urbanization, including low-density residential development. We addressed this need by adapting statistical models of the relationship of stream health indicators developed in data-rich watersheds of the mid-Atlantic region (Goetz and Fiske 2008). These models were based on relating *in situ* observations from the Maryland Biological Stream Survey (Roth et al. 2004; Kazyak et al. 2005) to land-cover variables, translated into relatively simple procedures that can be conducted in a GIS environment (Goetz and Fiske 2010). The procedures allow prediction of the richness and abundance of stream macroinvertebrates as well as integrated indices of stream biological integrity. As the models use land-cover metrics to predict the variation of stream biotic metrics, they can be used across small watersheds as indicators of stream impairment and thus to focus on monitoring, restoration, and protection management objectives. An example prediction of the diversity of *Ephemeroptera*, *Plecoptera*, and *Tricoptera* species (*n*EPT; genera are mayfly, stone fly, and caddis fly, respectively), which are known to be sensitive to stream pollution and sedimentation, is shown in Figure 2.3.

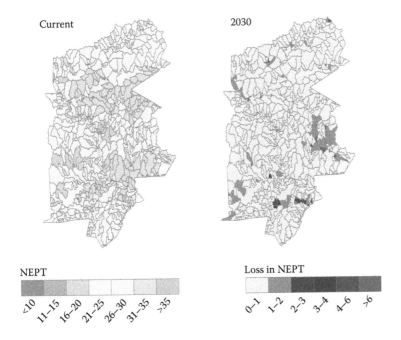

FIGURE 2.3
Maps of the Upper Delaware river basin showing the predicted number of sensitive stream taxa (abundance of EPT species; see text) for the present (left) and as predicted to change by 2030 using future land cover based on simulations of continued urbanization trends for watersheds (right).

Future predictions of urbanization under different land management scenarios, where they exist, can also be used to assess the potential impact of impervious cover in new residential and commercial developments on stream biota. As part of PALMS, we developed such predictions (Jantz et al. 2010) and used them to predict the status of future stream biotic condition, as expressed by the nEPT (Figure 2.3). The results clearly show the potential for reducing the impacts of impervious areas through mitigation measures such as maintaining riparian buffers and overall natural vegetation cover within a watershed.

Watershed biotic diversity maps of this sort, based on land-cover variables, provide a baseline against which *in situ* stream measurements can be compared and assessed as NPS monitoring programs develop. Moreover, the predictions are useful to I&M and park staff as they evaluate the sampling design for long-term monitoring of stream health and assess the risk of future residential and commercial development on aquatic biota.

2.3.3 Connectivity

Habitat fragmentation poses one of the foremost threats to biodiversity in parks and other protected areas in the United States (Hilty et al. 2006). Fragmentation is generally caused by loss of habitat, and results in the isolation of parks. Isolated parks are unable to support levels of biodiversity that existed prior to landscape changes (Newmark 1986; Parks and Harcourt 2002), and the ability of animals to move between large tracts of natural habitat is necessary to sustain the full range of biota and ecological processes in parks.

Connectivity of landscapes for the conterminous United States was estimated using a GIS-based least-cost–distance method that provides two novel aspects. First, this approach does not require patches to be first identified, as do patch-matrix approaches. Rather, the method considers the landscape as a gradient (Kupfer et al. 2006; McGarigal et al. 2009), which better reflects the gradual transitions that commonly occur between many land-cover types. Second, the method provides a quantitative estimate of the importance of each linkage or movement pathway. The application of these quantitative estimates can assist selection and prioritization of local and on-the-ground efforts.

Two products were generated from our approach. First, four cost–distance maps, each reflecting the weighted distance from the left, right, top, and bottom of the map extent, are averaged together to compute an overall landscape connectivity surface, similar to traditional least-cost "corridor" maps generated from the average value from two cost–distance maps (Beier et al. 2006; Theobald 2006). These maps are useful to understand general patterns of natural landscapes, where additional information about the landscape configuration is added. Permeability to movement was estimated by the "naturalness," N, which ranged from 1.0 (natural) to 0.0 (intensely human-modified) at 270 m resolution as a function of land-cover types, housing density, presence of roads, and effects of highway traffic (Theobald 2010). Resistance values (or cost weights) were calculated as $1/N$. Second, pathways that flow across the surface maps are found, similar to the flow of water across the terrain forming dendritic networks (Figure 2.4). Rather than forming a hydrologic network, a network of potential movement pathways is formed. The flow accumulation is weighted by N at each pixel, so that movement pathways incorporate both the pattern of movement as well as the importance of that movement. This flow-accumulated value is a computationally efficient approximation of the "betweenness" centrality measure (Borgatti 2005).

Identification of corridors by this approach has proved useful to parks. A key attribute of the method is the clarity of the result. "Dendritic" corridors identified by the method were easily interpreted, and they were very intuitive to park staff and partners. The approach permits calculations over very large grids ($>10^8$ cells); hence, we were able to provide results that identified important corridors at local to continental scales. These results are valuable

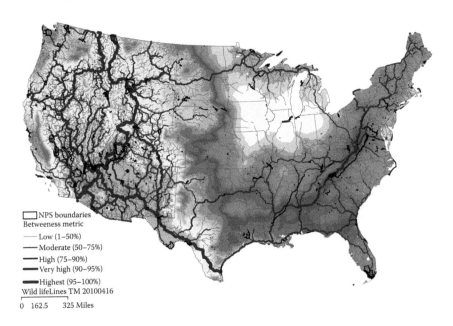

FIGURE 2.4
(See color insert.) Map showing connectivity of natural landscapes in the United States. The thickness of red lines indicates magnitude of cumulative movement, assuming that animals avoid human-modified areas. The surface underneath the pathways depicts the averaged cost–distance surfaces, or the overall landscape connectivity surface. Colors range from green through yellow and purple to white, where green is greatest connectivity (lowest travel cost) and white indicated lowest connectivity (highest travel cost). National Park Service units are outlined in black.

to parks because they clearly communicated the important role of parks as links in pathways that provide for very broad-scale movements (Figure 2.4). Results for DEWA, in particular, aligned very well with landscape-scale connectivity assessments (Goetz et al. 2009) and ongoing local conservation efforts. Local-scale analyses had identified high-priority areas for conservation at a fine scale, but they had not identified or realized how these local corridors would likely contribute to regional-scale conservation.

2.3.4 Land Surface Phenology

Variability and trends in the timing of seasonal biological events (phenology) are thought to be responsive indicators of global change (Schwartz 2003; Morisette et al. 2009). The onset and length of the growing season through their impact on primary production or simply plant growth are excellent indicators of ecosystem function with broad consequences for biodiversity. For example, spatial and temporal patterns of grassland and shrubland productivity in and near Yellowstone National Park are of particular interest due to their importance to migratory elk and bison (NRC 2002; White et al. 2010). These seasonally migratory ungulates have historically crossed public–private land boundaries in search of high-quality forage and to avoid deep snow during winter months, sometimes creating conflicts between landowners and wildlife managers. Working with the PALMS team, Yellowstone National Park staff identified forage phenology as a high-priority indicator, with a desire to better understand how land use and climate patterns influence forage availability and thus the spatial distribution of ungulates.

PALMS developed multiple phenology indicators based on NASA Moderate Resolution Imaging Spectroradiometer (MODIS) 250 m normalized difference vegetation index (NDVI) data products (Justice et al. 1998; Huete et al. 1999; Huete et al. 2002). For a pilot study centered on the Yellowstone Northern Range, we created annual NDVI curves and calculated phenology metrics based on properties of those curves for each eight-day interval for 2003–2009. These phenology metrics included measures of date of spring green-up, length of the growing season, and peak annual NDVI (White et al. 2009; deBeurs and Henebry 2010). Collectively, these metrics describe annual characteristics of grassland growth across space and interannual patterns of growth through time. We separated habitats that provide ample grassland cover for ungulate foraging and incorporated these into an annual, three-dimensional animation of greenness to help park staff visualize patterns of forage productivity at the landscape scale in and adjacent to their park. Further investigation of the spatial and temporal dimensions of grassland productivity demonstrates the degree to which productivity is influenced by climate and land use. With interacting effects, land use and climate change have the potential to significantly alter spatial and temporal patterns of grassland productivity in the Greater Yellowstone ecosystem in ways which will increase the likelihood of future conflicts between private landowners and wildlife managers. One intended use of phenology measurements is to use phenology as a leading indicator of animal movements, thereby enabling Yellowstone National Park managers to anticipate animal space use and plan strategies to mitigate conflicts with private landowners in areas surrounding the park.

2.3.5 Primary Production

Gross primary production (GPP) is the rate at which plants and other producers in an ecosystem capture and store energy as biomass via photosynthesis. Some fraction of this energy is used to maintain existing tissues or is lost through plant respiration, and net primary production (NPP) is the remaining amount that is "fixed" or stored by an ecosystem. As indicators of ecosystem productivity, GPP and NPP provide an integrative measure of ecosystem condition that incorporates seasonal climatic influences and satellite measures of vegetation condition, as well as information on topography, soils, and water availability.

To characterize ecosystem productivity for each of the partner I&M networks, we followed the general approach used by the MODIS MOD17A2 algorithms (Running et al. 2000) and applied a simplified version of the BIOME-BioGeochemical Cycles (BGC) ecosystem model (Thornton et al. 2002; Thornton et al. 2005) within the Terrestrial Observation and Prediction System (TOPS) (Nemani et al. 2009). TOPS is a modeling and climate and satellite data assimilation framework maintained by NASA Ames for use in ecological forecasting and ecosystem modeling research and applications. Relative to standard MODIS productivity products, TOPS uses gridded climate data at a much finer spatial resolution (1 km) to account for the steep heterogeneous terrain in many of our partner I&M networks and parks. TOPS uses satellite-derived estimates of leaf area to estimate various water (evaporation, transpiration, stream flows, and soil water), carbon (net photosynthesis, plant growth), and nutrient (uptake and mineralization) flux processes on a daily time step. BIOME-BGC requires as inputs spatially continuous data layers to describe the land cover, soil texture and depth, daily meteorology, and elevation across the land surface. To evaluate spatial and temporal patterns in GPP, daily maps were produced for the PACE surrounding each of the focal parks for the period from 2001 to 2010. Feedback from collaborators indicated that daily and monthly GPP maps were useful, but difficult to translate into summary products. We thus compiled the GPP data into seasonal and annual summaries of cumulative GPP (Figure 2.5) by park, PACE, and major ecosystem type, and evaluated the data to characterize baseline conditions for future monitoring and identify any emerging trends over the past decade. A standard operating procedure (SOP) was prepared for the productivity products (Melton et al. 2010), and the summary products were distributed via a dynamic web interface (Figure 2.6).

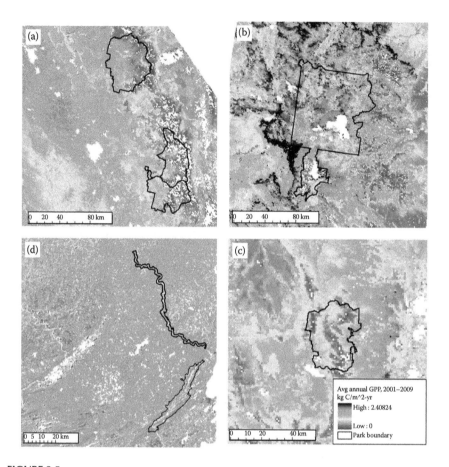

FIGURE 2.5
Daily estimated gross primary production (GPP) was summarized to convey spatial and temporal patterns in productivity in the parks and surrounding ecosystems. Maps of average annual total GPP for 2001–2009 are shown for these National Park Service units: (a) Yosemite/Sequoia-Kings Canyon, (b) Yellowstone/Grand Teton, (c) Rocky Mountain, (d) Delaware Water Gap/Upper Delaware.

Patterns in GPP varied by park, region, and ecosystem type. For example, in the Sierra Nevada parks, the indicator captured the significant interannual variability in productivity driven by year-to-year variations in the timing of snow accumulation and melt. In contrast, parks in the Eastern Rivers and Mountains I&M network showed sustained declines in GPP over the past decade, which may be due in part to increasing tree mortality resulting from infestations of the hemlock wooly adelgid (*Adelges tsugae*) throughout the region. While a 10-year data record is too short to identify long-term trends, the indicator was shown to capture the impact of climate variation and disturbance events on ecosystem condition.

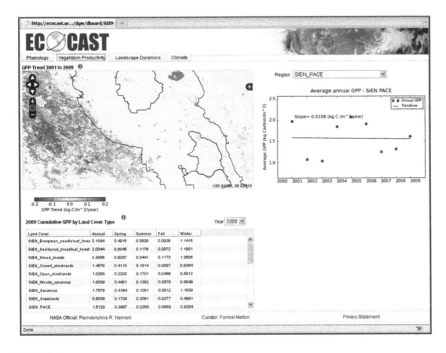

FIGURE 2.6
The Ecocast dashboard was used to display summaries of results for indicators directly estimated from remotely sensed data (e.g., phenology, snow cover), and results from simulations that estimated many other ecosystem variables (e.g., gross primary production—GPP—illustrated here; see text for explanation). The Ecocast summaries included maps showing regions with emerging trends or anomalies, and graphs and charts summarizing patterns by park or protected-area-centered ecosystem (PACE), time period, and/or ecosystem type.

2.4 Effectively Delivering Results to the NPS

There was an unusually high rate of turnover in cooperating NPS staff during our project, which led us to reconsider our plans for transferring PALMS products and knowledge to NPS. We had planned to place a high priority on training individual staff who would serve as NPS experts on PALMS products and methods. This strategy involved considerable investment in individuals, and that investment would be lost if they left their NPS positions. We consulted NPS collaborators and concluded that the most effective means for transferring project results included NPS-hosted websites, a set of site-specific project completion calls, park-specific reports, datasets, detailed methods (SOPs, see below), and peer-reviewed publications. This variety of products clearly reflected the general desires articulated by park managers in an earlier survey (Hubbard 2006). The suite of products and close-out activities we employed are, in our experience, rather unusual, and we believe this can serve as a good model for many projects that seek transfer of knowledge and technologies to specific partner programs or agencies. The following sections describe our strategy in more detail.

2.4.1 Documentation of Indicators and Methods

Our project partners felt that one- to two-page "resource briefs" on individual indicators would effectively communicate results to decision-makers and serve as quick introductions to the indicators for ecologists and other resource professionals. Each brief included a short description of the indicator issue, why it was useful, and a very short (one to two paragraphs) summary of results. Results were always illustrated with one or more maps, tables, and/or graphs. For each park, the briefs were combined into a single package (document) that included an abstract, table of contents, one-page overview of the project, and a table (similar to Table 2.3) with information on all the indicators for that park. The set of briefs did not include details on methods, and they included only the highlights of results. Recipients found the set of briefs to be much more accessible than a technical report or peer-reviewed publication.

A fundamental goal for PALMS was to develop indicators and methods that would be adopted by the NPS I&M. A major impediment to adopting an indicator or new method is the cost of development of an approved protocol. All NPS I&M networks are required to develop a detailed peer-reviewed protocol that meets published guidelines for each indicator they monitor (Oakley et al. 2003; Fancy et al. 2009). These guidelines were established to ensure that I&M procedures are completely documented and remain consistent through time and across changes in personnel. The work required to write a complete protocol is usually well beyond the scope of an externally funded research or development project, but projects may be able to draft parts of protocols and greatly reduce the time and cost required to complete a protocol.

Protocols compliant with NPS I&M standards consist of a narrative describing the goals and overall approach of the protocol, and a set of SOPs that describe, in detail, the specific procedures for a discrete task or operation. The PALMS team focused on writing SOPs for the core procedures for calculating each project indicator. These SOPs are highly detailed documents that permit I&M staff to repeat analyses or conduct the same analysis on new datasets. SOPs contain more details than the methods section in a typical peer-reviewed paper. For protocols that rely on GIS software and remotely sensed data, SOPs are usually illustrated with screen shots of key steps and, when appropriate, include step-by-step instructions for computer procedures.

To facilitate replication of GIS-based PALMS analyses, we developed ArcGIS (ESRI 2009) tools with Arc ModelBuilder. These tools automated complex or repetitive tasks, and served to reduce the level of software-specific expertise needed to reproduce our results or to repeat analyses with other datasets for different locations or time frames.

2.4.2 Websites

The range of products from PALMS is probably typical of a large, complex, multiagency monitoring development project. The large number of products, diverse array of product formats, extended period for delivery, and large volumes of data motivated the use of a tiered website to communicate and deliver products to project and park participants. We developed a public website on an NPS server for posting SOPs, reports, links to related sites, and links to data or information for acquiring large datasets. Because NPS was the target "client," the use of an NPS server (rather than one hosted elsewhere) helped ensure delivery of all relevant products and methods to NPS and it increased the likelihood that products would be properly cataloged, archived, and remain accessible to NPS staff in the long term. These websites will be removed as the required quality checks are completed and the products are fully integrated into and retrievable from the NPS information system.

Sustained interactions with park-based personnel required the addition of site-specific web pages that supported the completion calls (see Section 2.4.4) and facilitated review and discussion of products as they were being developed.

2.4.3 Dynamic Web Interfaces and Data Services

Satellite data analysis and ecosystem modeling are specialized fields, and park managers may be unfamiliar with satellite-derived indices and model parameters (e.g., NDVI, leaf area index, GPP, NPP), presenting a barrier to their adoption and use in park monitoring. To address this challenge, we developed a dynamic web interface based on the TOPS Ecocast framework (Figure 2.6) to present visual examples to NPS collaborators and to demonstrate how indicators derived from satellites and ecosystem models could be applied to characterize spatial and temporal patterns in park ecosystem conditions. This interface utilized open-source tools and software libraries to provide an interface that was driven by an OPeNDAP data server (Opensource Project for a Network Data Access Protocol; http://www.opendap. org/) and included dynamic web maps to characterize spatial patterns, and graphs and charts to summarize temporal patterns in the satellite- and model-derived indicators. This interface was effective in providing concrete examples of the use of the satellite and model data to dynamically summarize park ecosystem conditions.

To be sustainable as long-term I&M indicators, the source data used to calculate indicators must be readily accessible via data services and tools that are compatible with NPS information systems. Source data are needed to allow I&M networks to update indicators, and to develop customized analyses and summaries that can address park-specific issues. The PALMS project used ArcGIS (ESRI 2009) compatible geodatabases to store and distribute data for many of the PALMS indicators. We also tested the use of open-source software for data distribution, including an OPeNDAP server, which was used to distribute satellite and ecosystem model results from TOPS (http://ecocast.arc.nasa.gov/opendap). OPeNDAP is optimized to distribute large archives of raster data, such as those provided by TOPS, and it provides functionality for both temporal and spatial subsetting of geospatial data archives. ArcGIS tools are widely used within the NPS and this approach had the advantage of providing data in a form that can be directly imported into I&M geodatabases and incorporated into NPS projects. The NPS is currently enhancing the I&M data system with tools that automate data retrieval via web services, data transformation and analysis, and visualization of results. Data services used by the PALMS project are fully compliant with and support this growing I&M data infrastructure.

2.4.4 Project Completion Calls

Several factors posed significant challenges to using traditional project meetings for presenting our results. The integrated nature of the project meant that all investigators contributed important results for all parks, but project and park personnel were located at more than a dozen sites across the United States. The number and complexity of project indicators (Table 2.3) ensured that any presentation of all site-specific results would be an overwhelming volume of information. Furthermore, we were convinced that deep local knowledge was required to fully interpret our results and ensure that they addressed issues that were relevant and important to managers. Full interpretation required a series of conversations.

To meet these challenges, we scheduled a series of project "completion calls" lasting about 2 h. Each site participated in at least three webinars, and we scheduled additional webinars on specific results, methods, or topics as required. Call participants included project staff, principal NPS collaborators from each park, and interested management staff. These calls seemed to be effective by delivering results in measured "doses" and facilitating discussions of the results and outcomes. They also permitted time for park staff to review and discuss results between calls, and for additional interaction that might be needed to clarify, refine, or revise our work. We conducted a total of 14 such sessions over the final year of the project.

2.5 Lessons Learned and Recommendations

Many remote-sensing-based monitoring projects, especially those that involve many sites and collaborators, will likely face challenges similar to those we experienced. Some of these are common to most large problems, whereas others are more specific to working with complex technologies and management agencies such as the NPS. Here, we summarize a few important lessons, emphasizing things that worked well for us.

2.5.1 Allocate Sufficient Time to Develop a Genuine Science–Management Partnership

To effect a genuine collaboration between scientists, resource specialists, and managers takes more time, potentially *much* more time, to design, develop, implement, conduct, communicate, report, and deliver products than is typical for research projects. Remote-sensing projects tend to involve complex technology, sophisticated methods, and sometimes obscure measurements. "Black box" calculations that managers do not understand are unlikely to sway opinion or usefully contribute to important decisions unless they are skillfully explained by scientists. Time is required to develop a common language and explain how results were obtained and what they mean.

The transition of methods and results from research to operations requires a long-term commitment from all parties. Efforts to apply research data products for operational decision support often discover that more research is needed. Methods that apply at one site may not work well elsewhere, or it may be necessary to develop additional ecological or physical relationships to convert results of spectral analysis into units that are meaningful to managers. There is rarely a finite hand-off or delivery of research results accompanied by a seamless integration into a management decision support framework. In most cases, only through the long-term development of scientific understanding and collaboration with managers will decision-making be positively influenced by research results.

2.5.2 Communicate Results in a Management-Relevant Context

Uptake of results occurs most readily when they are available to the right people, at the right time, and in the right format. In parks, budget exercises, annual work plan, and field activities are typically conducted at the same time each year. Monitoring data need to be available when results can feed into decisions. Results should be expressed using formats and language familiar to managers and connections should be made between results and attributes that affect decisions. For example, it may be possible to correlate soil moisture and plant stress (as estimated from a simulation model driven by MODIS products) into a coarse measure of fire risk. Patterns of soil moisture may be of little interest to managers, but fire risk is almost always of great interest. Critical evaluation by end users will likely be required to ensure that the data products for decision support are available at the appropriate spatial and temporal scales.

2.5.3 Conform or Embellish Existing Frameworks and Processes

For PALMS, this included using the existing I&M network and program structure as a primary means to communicate across parks and project staff. We built on existing guidelines and formats for publications, method documents, and fact sheets, and we linked our efforts to specific personnel and positions within I&M. NPS collaborators were familiar with these products, and we minimized the costs associated with designing these products. We largely followed existing practices and produced reports and results for specific audiences.

2.5.4 Plan for Persistence and Change

NPS I&M is charged with conducting long-term monitoring. Protocols or products that do not persist through time will not meet program goals. The PALMS team's strategy to produce versioned SOPs was very well aligned with I&M protocol development needs. Production of detailed methods ensures persistence of standard methodology, and versioning provides a clear means to update individual procedures or an entire protocol with changes in technology or understanding.

2.5.5 Build on Existing, Widely Used Data Analysis Tools and Software Frameworks, Even If They Seem Inefficient

The use of existing tools and frameworks permits rapid development and reduces development costs. It increases client "buy-in" because the efficacy of application components is known, and it ensures usability. If possible, personnel should be exchanged to gain cross-enterprise experience in the tools and day-to-day processes used for data management and decision-making. NPS and the NASA-Ames groups employed software development teams with complementary skills; each group was familiar with the technologies, programming languages, and infrastructure that they could support after the initial development project ended. Communication between the groups was important to identify technologies that would most likely be adopted.

2.5.6 Practice Rigorous Scope Control to Maximize the Chance of Success

Operational use of remotely sensed data and technology requires robust, repeatable, credible, and defensible methods. Through discussion with collaborators, we continually refined the scope of work and avoided "mission creep" by focusing on specific functions, variables, and reporting products.

2.6 Summary

The PALMS framework, approach, and methods were developed specifically to meet the needs of NPS I&M, but the resources and impacts the indicators address are common to protected areas worldwide. Very few North American parks—and probably no NPS units—are sufficiently remote and large enough to sustain the biodiversity once native to the park, or to be unaffected by activities outside park boundaries (GAO 1994; Carroll et al. 2004). Human development is increasing more rapidly near the boundaries of protected areas than elsewhere in the United States (Radeloff et al. 2010) and other continents (Wittemyer et al. 2008). Furthermore, climate changes are projected to result in huge shifts in ranges of species and habitats (Iverson et al. 2008; Belant et al. 2010; Cole 2010; Gonzalez et al. 2010). These threats emphasize the need for integrated assessments of the condition of landscapes around protected areas at a range of spatial scales.

PALMS is unusual among monitoring projects for the breadth of attributes addressed by the suite of indicators, and the use of various models to assimilate data. The suite of indicators developed by PALMS can provide a rich picture of landscape context and the condition of attributes that conserve or threaten biodiversity in and around parks. Other reviews have illustrated the value of remotely sensed data to monitor traits not addressed by PALMS, but also important to supporting biodiversity (Turner et al. 2003; Bergen et al. 2009) and the broader goals of protected area monitoring (Kerr and Ostrovsky 2003; Gross et al. 2006; Gross et al. 2009; this volume). The potential to increase the use of remote sensing for operational monitoring is great, especially when the value of remotely sensed data is enhanced through multifactor analyses and modeling.

Here, we illustrated just a few of the indicators developed by PALMS, and we focused on the approaches that worked for us. I&M networks have worked with partners to explore a variety of useful methods for monitoring landscapes that are well suited to specific situations. These include multiscale monitoring of land-cover change (Wang et al. 2009), graph-based analyses of connectivity (Goetz et al. 2009; Townsend et al. 2009), phenology and snow cover (Reed et al. 2009), and forest monitoring (Kennedy et al. 2011, this volume).

Remotely sensed data will become increasingly important to NPS I&M as technologies improve, costs decline, and analyses become more integrative and sophisticated. As the chapters in this book attest, there are many new and exciting applications of remotely sensed data that will contribute to better informing management of protected areas.

Acknowledgments

We thank I&M and park staff who contributed to this project—especially Ben Bobowski, Mike Britten, Kristina Callahan, Jeff Connor, Robert Daley, Richard Evans, Don Hamilton, Andi Heard, Cathie Jean, Bill Kuhn, Matthew

Marshall, Leslie Morlock, Linda Mutch, Tom Olliff, Roy Renkin, Ann Rodman, Billy Schweiger, and Judy Visty. We thank TOPS team members Sam Hiatt and Andrew Michaelis for their insight and expertise. This project was supported by NASA Earth Science Directorate, Decision Support through Earth Science Research Results (DECISIONS), and the NPS I&M. We thank Woody Turner and Gary Geller for their support and help throughout the project and Bill Monahan for comments on the manuscript.

References

Beier, P., K. Penrod, C. S. W. Luke, and C. Cabanero. 2006. South coast missing linkages: Restoring connectivity to wildlands in the largest metropolitan area in the United States. In K. R. Crooks and M. A. Sanjayan (eds), *Connectivity Conservation*. Cambridge, UK: Cambridge University Press, pp. 555–586.

Belant, J. L., E. A. Beever, J. E. Gross, and J. J. Lawler. 2010. Ecological responses to contemporary climate change within species, communities, and ecosystems. *Conservation Biology* 24:7–9.

Bergen, K. M., S. J. Goetz, R. O. Dubayah et al. 2009. Remote sensing of vegetation 3-D structure for biodiversity and habitat: Review and implications for lidar and radar spaceborne missions. *Journal of Geophysical Research-Biogeosciences* 114 G00E06, doi: 10.1029/2008JG000883.

Bierwagen, B. G., D. M. Theobald, C. R. Pyke, A. Choate, P. Groth, J. V. Thomas, and P. Morefield. 2010. National housing and impervious surface scenarios for integrated climate impact assessments. *Proceedings of the National Academy of Sciences*.

Borgatti, S. P. 2005. Centrality and network flow. *Social Networks* 27:55–71.

Britten, M., E. W. Schweiger, B. Frakes, D. Manier, and D. Pillmore. 2007. Rocky Mountain Network vital signs monitoring plan. Natural Resource Report ROMN/NRR—2007/010. Fort Collins, CO: National Park Service.

Carroll, C., R. E. Noss, P. C. Paquet, and N. H. Schumaker. 2004. Extinction debt of protected areas in developing landscapes. *Conservation Biology* 18:1110–1120.

Cole, K. L. 2010. Vegetation response to early Holocene warming as an analog for current and future changes. *Conservation Biology* 24:29–37.

deBeurs, K. M., and G. M. Henebry. 2010. Spatio-temporal statistical methods for modelling land surface phenology. In L. L. Hudson, and M. R. Keatley (eds), *Phenological Research*. Dordrecht, Netherlands: Springer Science and Business Media B.V., pp. 177–208.

ESRI. 2009. ArcGIS version 9.3. Redlands, CA: ESRI, Inc.

Fancy, S. G., J. E. Gross, and S. L. Carter. 2009. Monitoring the condition of natural resources in US national parks. *Environmental Monitoring and Assessment* 151:161–174.

GAO (Government Accounting Office). 1994. Activities outside park borders have caused damage to resources and will likely cause more. GAO/T-RCED-94-59, pp. 1–34.

Goetz, S. J. 2006. Remote sensing of riparian buffers: Past progress and future prospects. *Journal of the American Water Resources Association* 42:133–143.

Goetz, S. J., and G. Fiske. 2008. Linking the diversity and abundance of stream biota to landscapes in the mid-Atlantic USA. *Remote Sensing of Environment* 112:4075–4085.

Goetz, S. J., and G. Fiske. 2010. *PALMS SOP—Estimating Impervious Cover Change*. Fort Collins, CO: National Park Service.

Goetz, S. J., P. Jantz, and C. A. Jantz. 2009. Connectivity of core habitat in the northeastern United States: Parks and protected areas in a landscape context. *Remote Sensing of Environment* 113:1421–1429.

Gonzalez, P., R. P. Neilson, J. M. Lenihan, and R. J. Drapek. 2010. Global patterns in the vulnerability of ecosystems to vegetation shifts due to climate change. *Global Ecology and Biogeography* 19:755–768.

Gross, J. E., S. J. Goetz, and J. Cihlar. 2009. Application of remote sensing to parks and protected area monitoring: Introduction to the special issue. *Remote Sensing of Environment* 113:1343–1345.

Gross, J. E., R. R. Nemani, W. Turner, and F. Melton. 2006. Remote sensing for the national parks. *Park Science* 24:30–36.

Hansen, A. J., and R. DeFries. 2007. Ecological mechanisms linking protected areas to surrounding lands. *Ecological Applications* 17:974–988.

Hansen, A. J., C. Davis, N. B. Piekielek et al. 2011. Delineating the ecosystems containing protected areas for monitoring and management. *BioScience*. 61: 263–273.

Hilty, J. A., W. Z. Lidicker, Jr., and A. M. Merenlender. 2006. *Corridor Ecology*. Washington, DC: Island Press.

Hubbard, A. 2006. Survey results of management needs: Recap of what IMR managers want from Inventory and Monitoring Networks. Available at: http://www.southwestlearning.org/topics/sciencemanagement/workshops/chico Accessed 7 July 2011.

Huete, A., K. Didan, T. Miura, E. P. Rodriquez, X. Gao, and L. G. Ferreira. 2002. Overview of the radiometric and biophysical performance of the MODIS vegetation indices. *Remote Sensing of Environment* 83:195–213.

Huete, A., C. Justice, and W. van Leeuwen. 1999. MODIS vegetation index (MOD 13) algorithm theoretical basis document (ATBD-MOD-13 version 3). Available at http://modis.gsfc.nasa.gov/data/atbd/atbd_mod13.pdf Accessed 7 July 2011.

Iverson, L. R., A. M. Prasad, S. N. Matthews, and M. Peters. 2008. Estimating potential habitat for 134 eastern US tree species under six climate scenarios. *Forest Ecology and Management* 254:390–406.

Jantz, C. A., S. J. Goetz, D. Donato, and P. Claggett. 2010. Designing and implementing a regional urban modeling system using the SLEUTH cellular urban model. *Computers, Environment and Urban Systems* 34:1–16.

Jantz, P., and S. J. Goetz. 2008. Using widely available geospatial data sets to assess the influence of roads and buffers on habitat core areas and connectivity. *Natural Areas Journal* 28:261–274.

Jean, C., A. M. Schrag, R. E. Bennetts, R. Daley, E. A. Crowe, and S. O'Ney. 2005. Vital signs monitoring plan for the Greater Yellowstone Network. Unpublished report. Fort Collins, CO: National Park Service Inventory and Monitoring Program, 107pp.

Jones, D. A., A. J. Hansen, K. Bly et al. 2009. Monitoring land use and cover around parks: A conceptual approach. *Remote Sensing of Environment* 113:1346–1356.

Justice, C. O., E. Vermote, J. R. G. Townshend et al. 1998. The Moderate Resolution Imaging Spectroradiometer (MODIS): Land remote sensing for global change research. *IEEE Transactions on Geoscience and Remote Sensing* 36:1228–1249.

Kazyak, P. F., J. V. Kilian, S. A. Stranko et al. 2005. *Maryland Biological Stream Survey 2000–2004. Volume 9: Stream and Riverine Biodiversity.* Annapolis, MD: Maryland Department of Natural Resources.

Kennedy, R. E., P. A. Townsend, J. E. Gross et al. 2009. Remote sensing change detection tools for natural resource managers: Understanding concepts and tradeoffs in the design of landscape monitoring projects. *Remote Sensing of Environment* 113:1382–1396.

Kennedy, R. E., Z. Yang, J. Braaten, P. Nelson, and W. B. Cohen. 2011. Chapter 3: Monitoring of landscape dynamics of national parks in the western United States. In Y. Wang (ed.), *Remote Sensing of Protected Lands.* Boca Raton, FL: CRC Press.

Kerr, J. T., and M. Ostrovsky. 2003. From space to species: Ecological applications for remote sensing. *Trends in Ecology and Evolution* 18:299–305.

Kupfer, J. A., G. P. Malanson, and S. B. Franklin. 2006. Not seeing the ocean for the islands: The mediating influence of matrix-based processes on forest fragmentation effects. *Global Ecology and Biogeography* 15:8–20.

Marshall M.R., and N. B. Piekielek. 2007. Eastern Rivers and Mountains Network ecological monitoring plan. Natural Resource Report NPS/ERMN/NRR—2007/017. Fort Collins, CO: National Park Service.

McGarigal, K., S. Tagil, and S. A. Cushman. 2009. Surface metrics: An alternative to patch metrics for the quantification of landscape structure. *Landscape Ecology* 24:433–450.

Melton, F., S. Hiatt , G. Zhang, and R. Nemani. 2010. *PALMS SOP—Estimating Landscape Indicators of Phenology from Satellite Observations: Start of Season.* Fort Collins, CO: National Park Service.

Morisette, J. T., A. D. Richardson, A. K. Knapp et al. 2009. Tracking the rhythm of the seasons in the face of global change: Phenological research in the 21st century. *Frontiers in Ecology and the Environment* 7:253–260.

Mutch, L. S., M. G. Rose, A. M. Heard, R. R. Cook, and G. L. Entsminger. 2008. Sierra Nevada Network vital signs monitoring plan. Natural Resources Report NPS/SIEN/NRR—2008/072. Fort Collins, CO: National Park Service, 130pp.

NARSEC (North American Network for Remote Sensing Park Ecological Condition). 2005. http://science.nature.nps.gov/im/monitor/meetings/StPetersburg_05_rs_pa/rs_pa_workshop.cfm. Accessed October 25, 2010.

NARSEC (North American Network for Remote Sensing Park Ecological Condition). 2007. http://science.nature.nps.gov/im/monitor/meetings/NARSEC_2007/index.cfm Accessed October 25, 2010.

Nemani, R., H. Hashimoto, P. Votava et al. 2009. Monitoring and forecasting ecosystem dynamics using the Terrestrial Observation and Prediction System (TOPS). *Remote Sensing of Environment* 113:1497–1509.

Newmark, W. D. 1986. Species area relationship and its determinants for mammals in western North American National Parks. *Biological Journal of the Linnean Society* 28:83–98.

NPScape. 2010. http://science.nature.nps.gov/im/monitor/npscape/index.cfm. Accessed July 7, 2011.

NRC (National Research Council). 2002. *Ecological Dynamics on Yellowstone's Northern Range*. Washington, D.C.: National Academy Press.

Oakley, K. L., L. P. Thomas, and S. G. Fancy. 2003. Guidelines for long-term monitoring protocols. *Wildlife Society Bulletin* 31:1000–1003.

Parks, S. A., and A. H. Harcourt. 2002. Reserve size, local human density, and mammalian extinctions in US protected areas. *Conservation Biology* 16:800–808.

Phinn, S. R., D. A. Stow, J. Franklin, L. A. K. Mertes, and J. Michaelsen. 2003. Remotely sensed data for ecosystem analyses: Combining hierarchy theory and scene models. *Environmental Management* 31:429–441.

Radeloff, V. C., S. I. Stewart, T. J. Hawbaker et al. 2010. Housing growth in and near United States protected areas limits their conservation value. *Proceedings of the National Academy of Sciences of the United States of America* 107:940–945.

Reed, B., M. Budde, P. Spencer, and A. E. Miller. 2009. Integration of MODIS-derived metrics to assess interannual variability in snowpack, lake ice, and NDVI in southwest Alaska. *Remote Sensing of Environment* 113:1443–1452.

Roth, N. E., M. T. Southerland, G. M. Rogers, and J. H. Vølstad. 2004. *Maryland Biological Stream Survey 2000–2004. Volume 3: Ecological Assessment of Watersheds Sampled in 2002*. Annapolis, MD: Maryland Department of Natural Resources, 318pp.

Running, S. W., P. E. Thornton, R. R. Nemani, and J. M. Glassy. 2000. Global terrestrial gross and net primary production from the Earth Observing System. In O. Sala, R. Jackson, H. Mooney, and R. W. Howarth (eds), *Methods in Ecosystem Science*, New York: Springer-Verlag, pp. 44–57.

Schueler, T. R., L. Fraley-McNeal, and K. Cappiella, K. 2009. Is impervious cover still important? Review of recent research. *Journal of Hydrologic Engineering* 14:309–315.

Schwartz, M. D. 2003. *Phenology: An Integrative Environmental Science*. Dordrecht, Netherlands: Kluwer Academic Publishing.

Snyder, C. D., J. A. Young, R. Villella, and D. P. Lemarie. 2003. Influences of upland and riparian land use patterns on stream biotic integrity. *Landscape Ecology* 18:647–664.

Theobald, D. M. 2006. Exploring the functional connectivity of landscapes using landscape networks. In K. R. Crooks and M. Sanjayan (eds), *Connectivity Conservation*. Cambridge, UK: Cambridge University Press, pp. 416–444.

Theobald, D. M. 2010. Estimating natural landscape changes from 1992 to 2030 in the conterminous US. *Landscape Ecology* 25:999–1011.

Theobald, D. M., S. J. Goetz, J. B. Norman, and P. Jantz. 2009. Watersheds at risk to increased impervious surface cover in the conterminous United States. *Journal of Hydrologic Engineering* 14:362–368.

Thornton, P. E., B. E. Law, H. L. Gholz et al. 2002. Modeling and measuring the effects of disturbance history and climate on carbon and water budgets in evergreen needleleaf forests. *Agricultural and Forest Meteorology* 113:185–222.

Thornton, P. E., S. W. Running, and E. R. Hunt. 2005. Biome-BGC: Terrestrial ecosystem process model and users guide, version 4.1.1. Model product. Oak Ridge, TN: Oak Ridge National Laboratory Distributed Active Archive Center. Available from: http://daac.ornl.gov; doi: 10.3334/ORNLDAAC/805.

Townsend, P. A., T. R. Lookingbill, C. C. Kingdon, and R. H. Gardner. 2009. Spatial pattern analysis for monitoring protected areas. *Remote Sensing of Environment* 113:1410–1420.

Turner, W., S. Spector, N. Gardiner, M. Fladeland, E. Sterling, and M. Steininger. 2003. Remote sensing for biodiversity science and conservation. *Trends in Ecology and Evolution* 18:306–314.

Van Sickle, J., J. L. Stoddard, S. G. Paulsen, and A. R. Olsen. 2006. Using relative risk to compare the effects of aquatic stressors at a regional scale. *Environmental Management* 38:1020–1030.

Wang, Y. Q., B. R. Mitchell, J. Nugranad-Marzilli, G. Bonynge, Y. Y. Zhou, and G. Shriver. 2009. Remote sensing of land-cover change and landscape context of the national parks: A case study of the Northeast Temperate Network. *Remote Sensing of Environment* 113:1453–1461.

White, M. A., K. M. deBeurs, K. Didan, et al. 2009. Intercomparison, interpretation, and assessment of spring phenology in North America estimated from remote sensing for 1982–2006. *Global Change Biology* 15:2335–2359.

White, P.J., K.M. Proffit, L.D. Mech, S.B. Evans, J.A. Cunningham, and K.L. Hamlin. 2010. Migration of northern Yellowstone elk: Implications of spatial structuring. *Journal of Mammalogy* 91:827–837.

Wittemyer, G., P. Elsen, W. T. Bean, A. C. O. Burton, and J. S. Brashares. 2008. Accelerated human population growth at protected area edges. *Science* 321:123–126.

3

Monitoring Landscape Dynamics of National Parks in the Western United States

Robert E. Kennedy, Zhiqiang Yang, Justin Braaten, Peder Nelson, and Warren Cohen

CONTENTS

3.1 Introduction

Many emblematic national parks in the western United States face pressures from climate change, human development, altered fire regimes, and pollution (Lovejoy, 2006; Pressey et al., 2007). These stressors could affect vegetation and habitat quality and arrangement, which in turn affect cycles of nutrients, water, and wildlife, as well as the scenic and natural character of the parks themselves. Effective management under such stressors requires effective monitoring of landscape dynamics in and around the parks, but the large-sized, diverse ecosystems represented, and the wilderness character of many national parks in the western United States pose challenges for such monitoring.

Within the National Park Service Inventory and Monitoring Program (http://science.nature.nps.gov/im/), many networks of parks have determined that remote sensing may aid monitoring of landscape dynamics in and around parks. Remote sensing measurements are attractive because they provide consistent measurements across large areas regardless of on-the-ground accessibility; they are relatively low cost for the area covered; and some forms of remote sensing data can provide a link to conditions several decades before current monitoring programs were begun.

   Despite the potential utility of remote sensing for monitoring, key methodological challenges must be considered in the design of a remote sensing-
based monitoring project (Kennedy et al., 2009). Evaluating the arc of a
remote-sensing-based project requires balancing cost and benefit for different remote sensing technologies and analytical techniques, as well as availability of sufficient independent reference data to train and validate those
data and algorithms. A key component of the potential benefit of a given
technology is the extent to which the changes of interest on the ground affect
the spectral signal (diagnostic contrasts in reflectance of electromagnetic
energy from different parts of the spectrum) on which the detection of
change is based. For a remote sensing monitoring project to be successful,
the expected changes must be reliably captured with the remote sensing
technologies that are affordable.
   This fact underscores three major challenges to using remote sensing
data to detect the effects of stressors in national parks. First, the potential
impacts of many stressors are diverse in terms of vegetative community
affected, spatial and temporal scale of impact, and the magnitude of the
effect. Ideally, monitoring must capture all of these potential factors everywhere, all the time, but to date much of the focus in remote sensing change
detection is on single drivers in one focal ecosystem at a time. Second, while
some stressors will result in abrupt, high-magnitude changes easily
captured with traditional remote sensing change detection approaches,
many stressors on vegetation are expected to cause subtle or slow changes
in vegetation over time. These changes have typically been undetected in
many studies to date because subtle changes on the ground cause lower
magnitude change in the remotely sensed signal. Such signals become
increasingly difficult to separate from background spectral change caused
by either systemic noise or by real but nontarget changes on the surface
(Figure 3.1). For example, decreased vigor in a forest caused by pollution
may result in a subtle change in signal that is mimicked by satellite sensor
deterioration over time, or by the signal change that occurs every year as
leaves senesce. Separating durable from ephemeral effects is critical to
avoid false alarms.
   The final challenge in applying remote sensing to landscape monitoring in
national parks is effective attribution of the cause of the change. In much of
the remote sensing change detection literature, the cause of change is simply
implied because only one or two agents were detectable. When detection of
a diverse range of impacts in diverse ecosystems is sought, the agent of
change is critical to inform scientists and managers.
   This chapter describes our experiences building remote-sensing-based
monitoring programs to address these challenges in a range of different
national parks across the western United States. Working in ecosystems
ranging from southwestern Alaska to the desert Southwest, we have
encountered similar challenges and developed approaches designed to
address these challenges in a consistent manner across ecosystems.

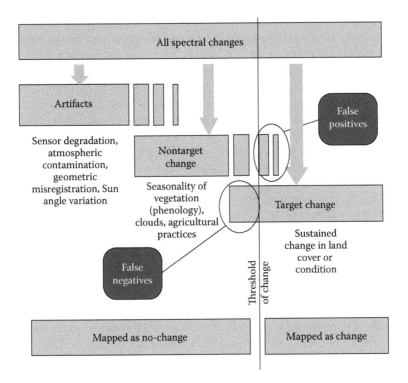

FIGURE 3.1
Challenges in detecting real change. Of all spectral changes that can be observed by comparing remotely sensed images, only a small portion may be associated with changes of interest on the ground. Typically, the choice of a threshold separating change from no-change results in both false-positive and -negative errors. Strategies to reduce cumulative error must improve separation of nontarget and target change.

3.2 Landsat-Based Approaches

The diversity of remote sensing tools available to observe conditions in national parks of the United States continually increases, but several key considerations have led to the selection of Landsat Thematic Mapper (TM) imagery as a key component of monitoring in all our efforts (Kennedy et al., 2007). TM imagery represents a useful balance between spatial extent and grain that has repeatedly shown to be useful in natural resource applications. The data are free (Woodcock et al., 2008), meaning that resources can focus on extracting information. Most importantly, however, the data are consistent and have a long-enough observation record to place current changes in the context of past events (Cohen and Goward 2004). Therefore, we have focused on methods to extract the most possible information on landscape change from the archive of TM imagery.

Historically, remote sensing algorithms approached change as an anomalous event on an otherwise static landscape (Coppin and Bauer 1996). This was partially driven by data availability and processing cost: Because the affordable approach to detecting change involved comparison of one image before and one after presumed change, it was difficult to unambiguously determine whether any observed spectral change was the result of a discrete event (such as a fire or land clearing action), an ephemeral event (phenological change), or an ongoing process (such as recovery from a prior event).

This contrasts with the ecological view of change that must underlie many monitoring programs in national parks. Rather than viewing change as an anomalous event, we recognize that processes of change are occurring over much of any landscape at any point in time, but that different types of change can be differentiated by the speed and direction of change in a core ecological attribute such as vegetative biomass or cover. As a first-order simplification, the changes can be conceptualized as vectors with different slopes and durations (Figure 3.2). Within the time period of any single vector, the process causing change is considered to be relatively consistent. A critical point in the time series occurs when the vector of change alters course, often indicating a change in the process dominating that area. Also, the slope and sequence of vectors often provide insight into the types of processes occurring. For example, changes in the state of the landscape can be distinguished from cyclical changes, and both can be separated from long-duration processes that slowly change the condition of an area.

The ecologically based view of change can be translated into the spectral domain when the entire archive of TM spectral imagery is considered. A spectral index is used as a proxy for ecological attributes, and that index can be tracked as a time-series trajectory. We have developed algorithms collectively known as "LandTrendr" (Kennedy et al., 2010) that use simple statistical fitting rules to identify periods of consistent progression in the spectral trajectory (segments) and turning points (vertices) that separate those periods. This process of "temporal segmentation" is carried out for

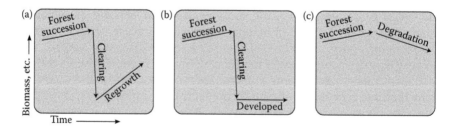

FIGURE 3.2
In an ecological view, change in core attributes (such as biomass, cover, etc.) is occurring constantly and can be conceptualized as vectors of change. When these change direction, an important change in process is implied, ranging from cyclical (a) to transitional (b) to chronic (c).

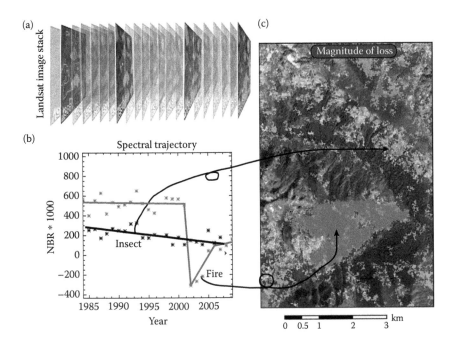

FIGURE 3.3
(See color insert.) Temporal segmentation of the spectral trajectory using LandTrendr for data recorded by two pixels in a stack of Landsat images for North Cascades National Park (a). The record of spectral data, here represented by the normalized burn ratio (NBR), bounces up and down from year to year (asterisks), but a set of mathematical algorithms can detect longer-term trends and abrupt changes in those trends (b). The magnitude of change, as captured by the difference of endpoints of fitted segments, can be mapped to show patterns of disturbance on a landscape (c).

every pixel in a sequentially ordered stack of TM images (Figure 3.3). Although TM image data are free and computer resources are inexpensive, data preparation and processing remain important and time-consuming components of the process, and are detailed in the study by Kennedy et al., (2010).

Temporal segmentation allows direct mapping of the processes underlying the ecologically based view of change. After temporal segmentation in each pixel, summary attributes of each segment can be mapped to highlight attributes of the processes of interest. For example, fire in a forested area will result in an abrupt drop in spectral indices associated with vegetative cover (such as the normalized difference vegetation index [NDVI (Tucker 1979)] or the normalized burn ratio [NBR (Key and Benson 2005)]). The segment associated with the fire event can be selected based on its direction of change, and the magnitude of the change calculated by differencing the endpoints of the segment.

In addition to its conceptual link with ecological processes, the temporal segmentation approach offers two practical benefits relevant for monitoring in national parks. First, the method is capable of capturing both trends and events occurring in the same location over time. For example, an insect outbreak in a forest may cause low-level mortality that accumulates for years (a trend), which then may be followed by a fire (an abrupt event) (Figure 3.3). Methods that can detect events may be confused by the presence of a trend before or after the event; similarly, detection of trends may be confounded by the presence of abrupt events. More importantly, progressions of events to trends (or vice versa) could not be captured by methods that target only one of the two types. The second practical benefit of the temporal segmentation approach is its increased sensitivity to subtle change. Even with rigorous data preparation, the comparison of any pair of remotely sensed images will reveal residual differences in atmospheric conditions, vegetative phenological state, and illumination (Sun angle) between the two images. By considering a sequence of images over time, these factors become noise that is implicitly accounted for in the statistical fitting approach. For example, a pixel with 15 years of relatively stable conditions on the ground provides 15 examples of what stability "looks like" spectrally; when a small change in the 16th year extends beyond that range, it can more reliably be considered a change than if that same change were observed without reference to what happened before it.

3.3 Monitoring Change in Western National Parks

Our change-monitoring methods capture a wide range of processes affecting vegetation inside and outside of western national parks. Below, we describe our experiences mapping such processes, focusing on mortality processes, on growth processes, on sequences of mortality and growth processes, and finally on tools to attribute the agent of change to those processes.

3.3.1 Decline/Mortality Processes

Decline and mortality processes are those that cause vegetation to lose biomass over time, either through reduction in vigor or outright death. These processes involve change (or disturbance) types such as fire, landslides, or human-mediated harvest in forests, as well as insect- or drought-related canopy thinning or mortality. Although these processes all cause declines in spectral indices such as the NBR or NDVI, the magnitudes of change and the duration over which that change occurs vary. Many fires and human-mediated vegetation clearing processes (forest harvest, development) result in segments showing abrupt, steep declines in spectral indices such as the NBR or NDVI, whereas insect-caused mortality often creates segments showing long, subtle declines over time.

Although vegetation clearing for economic extraction is not found in protected areas, it can be prevalent outside the boundaries of those areas. Parks in the coastal states of the western United States are often embedded in a matrix of forest-dominated public lands where economic extraction of timber is common, making abrupt, high-magnitude change a common feature outside of many parks we have examined.

Unlike forest harvest, fire is an abrupt, often high-magnitude process that occurs both outside and inside national parks. Although most common in drier parks such as Zion or Yosemite National Parks (Figure 3.4), fire can occur even in the moist Olympic National Park. Fire can also be used as a management tool to reduce fuel loads that might otherwise lead to catastrophic fire.

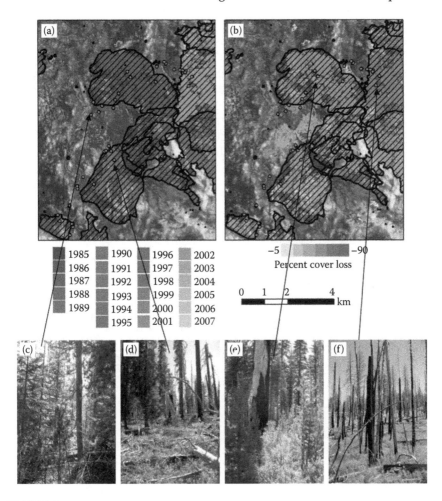

FIGURE 3.4
(**See color insert.**) Fire plays an important role in many national parks of the western United States, including Yosemite National Park shown here. Maps of year of fire disturbance (a) and estimated percent vegetation cover loss (b) show a mosaic of timing and intensity of fires. Field-visited locations show the range of both fire severity and post-fire recovery, including unburned (c), moderate severity burn (d), older burn with substantial recovery (e), and recently burned with substantial cover loss (f).

Prescribed burning in Sequoia and Kings Canyon National Parks is captured not by its immediate effects on spectral data, but by the multiyear cumulative loss of vegetation following the prescribed burn (Figure 3.5). Such delayed-effect processes can only be captured and interpreted if the entire time series is considered as it is for temporal segmentation within LandTrendr; mapping-prescribed fire effects only based on the year immediately following burning would lead to missed or underestimated disturbance magnitudes.

Long-term decline of forests caused by insect pests can lead to chronic deterioration of spectral vegetation indices. We found evidence for such long-term decay in forests in or near every national park we studied (see Figure 3.6 for an example shown from Lake Clark National Park in southwest Alaska). Note that even Olympic National Park has experienced long-term spectral decay caused by an insect, in this case the introduced balsam woolly adelgid, which has been killing subalpine fir communities in the drier northeast quadrant of the park.

Some insect outbreaks cause more abrupt spectral response, such as the well-known bark beetle outbreak occurring on the western edge of Rocky Mountain National Park (Figure 3.7). In practice, we have found the spectral

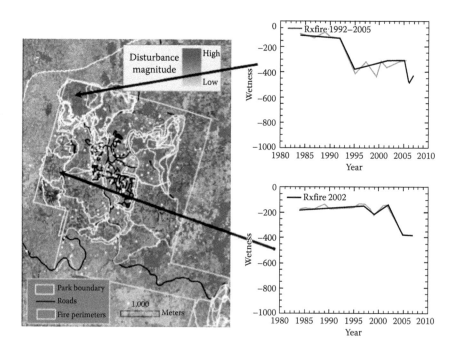

FIGURE 3.5
Prescribed fire in Grant Grove, Kings Canyon National Park. Prescribed burns typically are intended to minimize impact on dominant canopy trees, making detection with simple change detection difficult. By examining the full trajectory, the effects of delayed mortality and branch fall can be detected to infer timing of the original event.

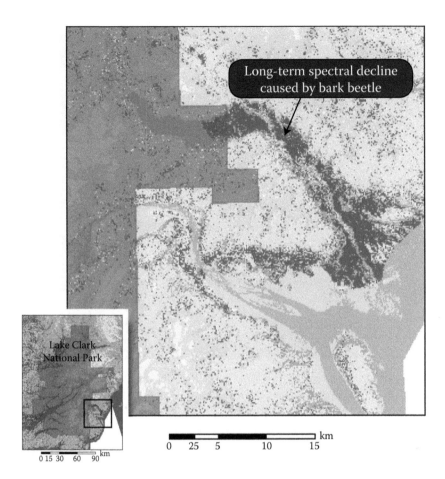

FIGURE 3.6
Insect-related mortality in forests of Lake Clark National Park. LandTrendr based maps of mortality processes show large swaths of forest with long-term decline in spectral indices related to vegetation vigor; these correspond to known bark-beetle outbreaks in the last two decades.

effects of insect-related mortality to be consistent in direction of spectral change (decreasing spectral indices) but highly variable in duration (both abrupt and longer-duration segments). This underscores the need to detect insect effects with algorithms that can capture both the abrupt change and the longer-term trends.

3.3.2 Growth/Recovery Processes

We define growth and postdisturbance recovery processes as those that cause an increase in biomass over time. From a strictly biological perspective, vegetative growth occurs everywhere that vegetation occurs. To be detected using remote sensing, that growth must result in a change in

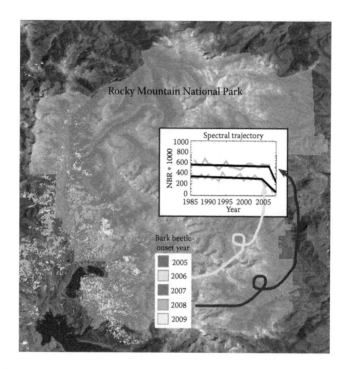

FIGURE 3.7
The spread of the bark-beetle mortality signal over several years for Rocky Mountain National Park, inferred from timing in the change of direction of spectral indices fit using LandTrendr.

spectral properties. Except in agricultural systems, these processes evolve over years rather than abruptly within a single year. Although traditional two-date change methods can identify such change, the spectral signal of increased growth is identical to that caused by comparing vegetation at peak biomass to vegetation at off-peak biomass, making separation of false signals problematic. When an upward trend in a spectral vegetation index is observed consistently over multiple years, however, the odds of that signal being caused by random phenological differences are greatly reduced. Thus, trend-based detection approaches can separate growth from background noise at a much lower threshold of change than traditional two-date change detection approaches. Moreover, the temporal segmentation approach allows calculation of the slope of the increase in the spectral index over time, providing insight into the rate of growth.

Vegetative recovery after disturbance is one of the most common vegetative growth signals observed in national parks in the western United States. Parks with frequent fire, such as Kings Canyon National Park, show a rich array of recovery rates after recent and even decades-old fires. Rates of spectral change vary considerably within a given recovery patch, which is visually consistent with ecological expectation.

When vegetative growth occurs in areas where vegetation was previously absent or especially sparse, it can signal changes in important drivers. For example, in Yosemite National Park, lodgepole pine forest is known to be encroaching on high-elevation meadows. Because lodgepole trees cast shadows and because their needle clumps are themselves darker than the soil and bright herbaceous meadows into which they are encroaching, the encroachment process can be captured (Figure 3.8) as a long-term decrease in overall reflectance (here reflected in the tasseled-cap brightness index; Crist and Cicone 1984). A similar signal is found in meadows of North Cascades National Park (Figure 3.9). In Lake Clark National Park in southwest Alaska, alder shrubs have been establishing and densifying further upslope than in the past as climate warms, and the apparent signal of this change is captured by long-term increase in vegetation.

FIGURE 3.8
Encroachment of lodgepole pine into meadows of Yosemite National Park, here detected by noting long-term decline in brightness as dark, shade-casting trees fill brighter, herbaceous-dominated background (a). Maps suggested rings of lodgepole encroachment around meadows (b), confirmed by field visits (c).

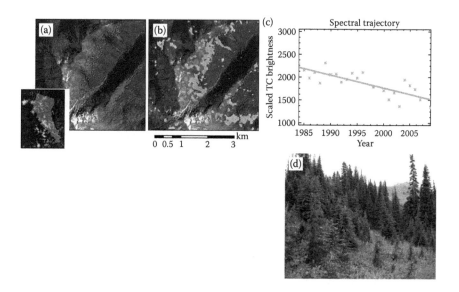

FIGURE 3.9
Encroachment of trees into meadows also occurs in North Cascades National Park (a, b, and d), again captured as a long-term decline in brightness (c).

3.3.3 Combinations of Different Processes

Although mortality and growth processes are important to capture in isolation, the richer ecological story of a park requires understanding how these processes follow each other over time at the same location or how they are distributed at a given time across a park. The information content needed to find such patterns is also captured in the temporal segmentation.

One approach to capturing such change is to map the snapshot of the instantaneous rate of change at a given point in the temporal record. Pixels experiencing growth can be differentiated from those experiencing mortality, and the spatial arrangement of the two processes can reveal ecologically meaningful patterns. For example, the instantaneous rate of change of pixels several years after a fire in Yosemite National Park shows that residual post-fire mortality continues to occur around the margins of the fire-affected area, while the central core of the fire shows robust regrowth (Figure 3.10). These differences are likely to have ecological consequences that persist long after the fire event itself.

At any given point in time, the landscape is a mosaic of growth and mortality processes. The sum of those processes across the geographic footprint of a park in a given year can provide an indication of the park landscape's net balance; by examining that balance over time, we can begin describing the overall momentum of the park (Figure 3.11).

The temporal sequence of growth and mortality may also tell a richer ecological story than simple univariate change mapping. For example, pixels

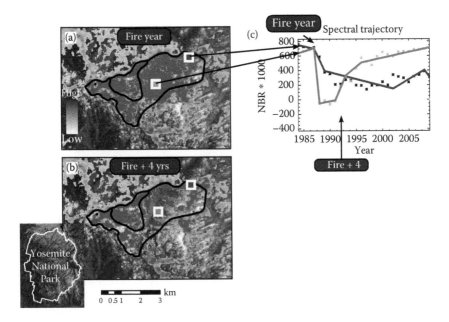

FIGURE 3.10
Signals suggesting postfire mortality in Yosemite National Park. By observing instantaneous direction of change from year to year in LandTrendr-fitted image stacks, mortality processes can be mapped for the 1988 Walker fire (a). Four years after the fire, the signal of mortality persisted around the margins of the fire (b). In the center of the fire, initial fire magnitude was high, but the spectral signal shows that vegetation recolonized quickly, while the margins of the fire experienced lower magnitude change in the initial fire event, but long postfire mortality signals (c).

that experience only long, slow decay or mortality are different from those that experience the same decay followed by a fire, and these are different again from those pixels that showed signs of growth preceding the same fire. These distinctions can be captured as different "change classes" that are distinct in the sequence of segments resulting from temporal segmentation: pixels with a growth segment followed by an abrupt mortality segment can be classified as different from a slow mortality segment followed by an abrupt mortality segment. When the entire landscape is classified by the temporal sequence of segments, a rich mosaic of change processes can be seen (Figure 3.12). Such a mosaic provides nonspecialists with a quick overview of the dynamism of landscapes both inside and outside parks and protected areas.

3.3.4 Attribution

Detection of different change processes is necessary but not sufficient for monitoring needs in national parks in the western United States. To be truly useful for park managers and researchers, we must provide a sense of what

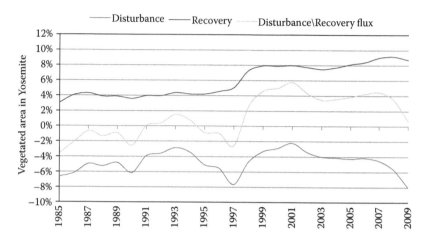

FIGURE 3.11

The balance between mortality (disturbance) and growth (recovery) processes in Yosemite National Park can be graphically depicted by summarizing the instantaneous rates of change of fitted spectral data. These summaries provide a sense for overall momentum of the park, here suggesting that the balance in Yosemite has been toward growth for more than a decade.

caused the change. For example, both fire and mechanical clearing are mortality processes, but have distinctly different drivers whose management responses would be entirely different. Similarly, a low-intensity fire and an insect-caused mortality process may result in a similar diminishment of spectral signal, but would have substantially different effects on the fate of carbon and habitat because the former removes biomass quickly through combustion whereas the latter transfers it from living to dead pools that remain in place. Labeling the agent of change is a process we refer to as "change attribution."

Our strategy for attribution of change is based on four principles. First, attribution requires that analysis moves from the pixel to the patch level. This is needed both because agents of change often have definitions that are only meaningful at the patch level, and because agents of change are rarely distinguishable based on spectral value alone. Second, attribution of change should leverage the types and direction of change occurring before and after the target change to aid in inference of agent. A clearing event that is followed by a flat, nonrecovering segment may indicate permanent change in land cover associated with development, whereas a similar event that experiences rapid vegetative recovery may indicate a cyclical disturbance event such as an avalanche or fire. Third, attribution will be most successful when it is anchored on human interpretation. Spatial, temporal, and ecological contexts are critical to distinguish among change agents, and are extremely difficult to define *a priori* for rule-based algorithms. In contrast, humans naturally consider context in everyday life and easily apply such strengths to assessing what could have caused a given change. Finally, algorithm-based

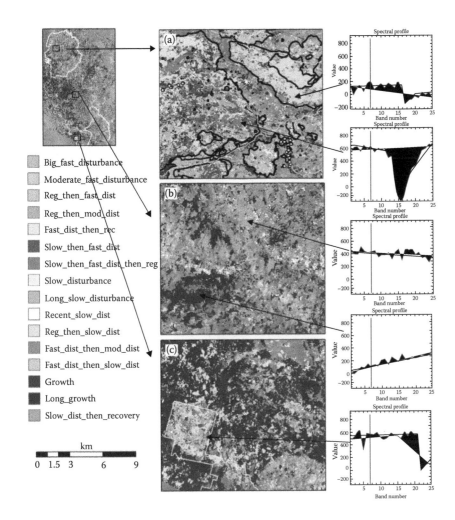

FIGURE 3.12
(**See color insert.**) By grouping pixels with similar sequences of mortality and growth segments, the rich dynamics in and around national parks can be revealed. Here, a suite of change classes show variability in pre- and postfire processes (a), growth of forest after harvest as well as slow decay caused by insects (b), and prescribed burning in a matrix of rapidly growing forest (c).

attribution is mathematically a classification exercise in a multivariate space, where the classes are rarely linear or well behaved and are often incompletely distinguished.

These four principles drive the attribution workflow, results of which are shown for Mount Rainier National Park (Figure 3.13). First, groups of pixels in maps of change must be grouped together into patches. For abrupt disturbance events, the year of onset is a logical grouping variable, but for slowly spreading, long-duration processes such as growth or chronic insect-related

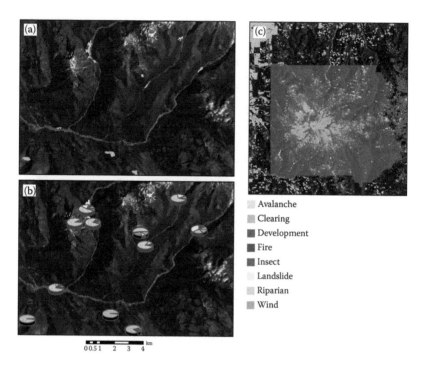

FIGURE 3.13
The spectral, spatial, and temporal dimensions of change processes can provide clues to the cause of change. For Mount Rainier National Park, expert interpreters assigned labels to agent of change (a) for selected polygons derived from disturbance maps created by LandTrendr. A statistical approach is trained using those interpretations, and then is applied back to the dataset to vote on likely cause (b). These rules can also be applied to map change agent for all polygons (c).

mortality, the duration of the process must also be considered as a grouping variable. Once patches have been defined, a wide range of spatial data layers must be summarized and extracted by patch (mean and variances of attributes within each patch, as well as statistics describing the patch shape and landscape location), and those layers must not only include information on the magnitude, timing, and duration of the target change segment but also include basic information about the type of spectral change before and after the target segment. A subset of patches is then selected using statistical sampling rules, and these are then evaluated by a trained interpreter to assign an agent of change. Any relevant external datasets can be used to aid in this process, including ground-based data, high spatial resolution photos, existing spatial data layers derived from other sources, and personal knowledge of the park. Additionally, a software tool we have developed (TimeSync; Cohen et al., 2010) to aid in corroboration of temporal segmentation can provide a rich view of the temporal evolution of the spectral trajectories for pixels within the patches. Once these attribution calls have been assigned, they are considered dependent variables to be predicted using the patch-level data summaries of spectral change, landscape position, patch size, and so on. We use the statistical classification tool known as "random forest" (Breiman 2001) because it makes no assumptions about data distribution, because it is relatively robust to noisy data, and because it allows for a voting process by which each patch can have a range of possibly likely attribution calls. This provides a means of evaluating where training data have been ambiguous and need to be improved.

3.4 Conclusions

Even though national parks are protected from many forms of direct human intervention, their landscapes are anything but static. Vegetation is changing continuously in response to both endogenous and exogenous pressures, and monitoring these dynamic landscapes in coming decades will require tools that capture a wide range of processes over large areas. Landsat TM imagery has long been used for natural resource monitoring because of its useful balance between spatial grain and large extent, and the advent of free imagery from the U.S. Geological Survey archive has added temporal richness and consistency as attractions of the data. With temporal segmentation of the spectral signals in the archive, we argue that satellite data are moving toward a characterization of dynamic landscapes that is more consistent with an ecological view of change than has been possible in the past.

References

Breiman, L. 2001. Random forests. *Machine Learning, 45*, 5–32.
Cohen, W.B., and Goward, S.N. 2004. Landsat's role in ecological applications of remote sensing. *BioScience, 54*, 535–545.
Cohen, W.B., Zhiqiang, Y., and Kennedy, R.E. 2010. Detecting trends in forest disturbance and recovery using yearly Landsat time series: 2. TimeSync—Tools for calibration and validation. *Remote Sensing of Environment, 114*, 2911–2924.
Coppin, P.R., and Bauer, M.E. 1996. Digital change detection in forested ecosystems with remote sensing imagery. *Remote Sensing Reviews, 13*, 207–234.
Crist, E.P., and Cicone, R.C. 1984. A physically-based transformation of thematic mapper data—The TM tasseled cap. *IEEE Transactions on Geoscience and Remote Sensing, GE 22*, 256–263.
Kennedy, R.E., Cohen, W.B., Kirschbaum, A.A., and Haunreiter, E. 2007. Protocol for Landsat-based monitoring of landscape dynamics at North Coast and Cascades Network Parks. In *U.S. Geological Survey Techniques and Methods*. Reston, VA: USGS Biological Resources Division. Available at: http://pubs.usgs.gov/tm/2007/tm2g1/.

Kennedy, R.E., Townsend, P.A., Gross, J.E., Cohen, W.B., Bolstad, P., Wang, Y.Q., and Adams, P.A. 2009. Remote sensing change detection tools for natural resource managers: Understanding concepts and tradeoffs in the design of landscape monitoring projects. *Remote Sensing of Environment, 113*, 1382–1396.

Kennedy, R.E., Yang, Z., and Cohen, W.B. 2010. Detecting trends in forest disturbance and recovery using yearly Landsat time series: 1. LandTrendr—Temporal segmentation algorithms. *Remote Sensing of Environment, 114*, 2897–2910.

Key, C.H., and Benson, N.C. 2005. Landscape assessment: Remote sensing of severity, the normalized burn ratio. In D.C. Lutes (ed.), *FIREMON: Fire Effects Monitoring and Inventory System*. Ogden, UT: USDA Forest Service, Rocky Mountain Research Station.

Lovejoy, T.E. 2006. Protected areas: A prism for a changing world. *Trends in Ecology & Evolution, 21*, 329–333.

Pressey, R.L., Cabeza, M., Watts, M.E., Cowling, R.M., and Wilson, K.A. 2007. Conservation planning in a changing world. *Trends in Ecology & Evolution, 22*, 583–592.

Tucker, C.J. 1979. Red and photographic infrared linear combinations for monitoring vegetation. *Remote Sensing of Environment, 8*, 127–150.

Woodcock, C.F., Allen, A.A., Anderson, M., Belward, A.S., Bindschadler, R., Cohen, W.B., Gao, F. et al. 2008. Free access to Landsat imagery. *Science, 320*, 1011.

4

Forest Dynamics within and around Olympic National Park Assessed Using Time-Series Landsat Observations

Chengquan Huang, Karen Schleeweis, Nancy Thomas, and Samuel N. Goward

CONTENTS

4.1 Introduction

Olympic National Park (ONP) was established in the 1930s to protect the diversity of its ecosystems. It has some of the best examples of intact temperate rainforest in the Pacific Northwest and is home to 24 endemic species, including 8 plant species and 16 animal species (National Park Service 2007). In 1976, it was designated by the United Nations Educational, Scientific, and Cultural Organization (UNESCO) as an International Biosphere Reserve, and was declared a World Heritage Site in 1981. The old growth forests that were once widely distributed across the Olympic Peninsula provide critical habitat for the northern spotted owl and other endangered species (McNulty 2009). Due to extensive logging in the early-twentieth century, however, much of the unprotected old growth forests outside the ONP and the Olympic National Forest (ONF) were lost, and some of the private lands in the peninsula are among the most heavily logged in the United States. Viable protection of the unique biodiversity of the Olympic Peninsula requires continuous monitoring of forest dynamics across the peninsula.

The main purpose of this chapter is to evaluate forest dynamics within and around the ONP using time-series Landsat observations. With the successful launch of the first Landsat in 1972, a series of six Landsat systems have produced one of the longest imagery records of the earth's surface (Goward and Williams 1997). The subhectare spatial resolutions of this record make it suitable for characterizing many of the changes arising from natural or anthropogenic disturbances (Townshend and Justice 1988). Landsat images have been used to evaluate forest change in numerous studies. Comprehensive reviews of different change detection techniques using satellite imagery have been provided in a number of publications (e.g., Singh 1989; Coppin et al. 2004; Lu et al. 2004). Many studies were concerned with changes between two dates evaluated using bitemporal change detection algorithms (e.g., Malila 1980; Nelson 1983; Genc and Smith 2003), while bitemporal and multitemporal change detection techniques have also been used to analyze changes between multiple dates (e.g., Cohen et al. 1998; Miller et al. 1998; Franklin et al. 2002; Yang and Lo 2002; Healey et al. 2005). Use of multiple image acquisitions in change analysis over a relatively long period (e.g., a decade or longer) is necessary not only for understanding the temporal variability of changes but also for capturing transient changes, especially in regions where trees grow rapidly. Significant portions of those transient changes likely will not be captured if change analysis is performed using sparse image acquisitions (Lunetta et al. 2004; Masek et al. 2008). To reduce such omission errors, new change detection techniques have been developed for mapping forest change using temporally dense Landsat observations, among which include the vegetation change tracker (VCT) (Huang et al. 2010a) and LandTrendr (Kennedy et al. 2010). In this chapter, the VCT is used to map forest change over the Olympic Peninsula.

4.2 Data and Methods

4.2.1 Study Area

The Olympic Peninsula is located in the northwest of Washington State in the United States, with surrounding water in the east, north, and west (Figure 4.1). At the center of the peninsula is the ONP, which also includes a large stretch of the Pacific coast. In addition to extensive forest cover, the park

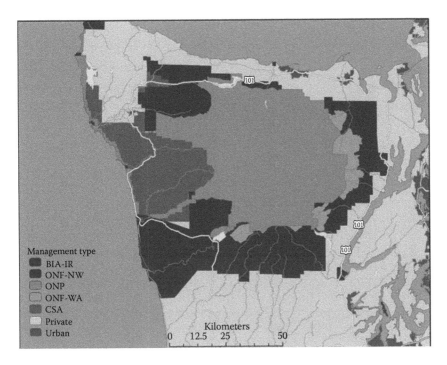

FIGURE 4.1
Olympic National Park (ONP) and other land management areas of the Olympic Peninsula, including areas of the Indian Reservation of the Bureau of Indian Affairs (BIA-IR), wilderness area of the Olympic National Forest (ONF-WA), nonwilderness area of the Olympic National Forest (ONF-NW), spotted owl conservation support area (CSA), private land, and urban areas.

also contains glacier-clad peaks interspersed with alpine meadows. The majority of the park is surrounded by wilderness area and other areas (i.e., nonwilderness area) of the ONF (ONF-WA and ONF-NW, respectively), and the spotted owl conservation support area (CSA). The rest of its border is shared with an Indian Reservation of the Bureau of Indian Affairs (BIA-IR) and some private lands.

4.2.2 Landsat Imagery

The majority of the Olympic Peninsula is covered by two Landsat path/row tiles as defined by the World Reference System (WRS). A third WRS tile is necessary to cover the northwest tip of the peninsula. Due to time constraints, however, that tile was not included in this study. For each of the two WRS tiles, we assembled Landsat time series stacks (LTSSs) consisting of images acquired between 1984 and 2007 (Huang et al. 2009a). The acquisition dates of the images used in this study are listed in Table 4.1. Although our goal was to obtain one cloud-free image acquired during the leaf-on growing season for each year, no image was acquired for some of the years due to lack

TABLE 4.1

Acquisition Dates of Landsat Images Used in This Study

Path 47/Row 27		Path 48/Row 27	
Year	Day of Year	Year	Day of Year
1984	208	1984	199
1985	194	1985	169
1986	213	1986	220
1987	168	1987	239
1987	232	1987	271
1987	264	1988	242
1988	203	1988	274
1989	205	1990	263
1989	253	1991	186
1990	192	1992	253
1990	224	1993	271
1990	256	1995	245
1991	211	1999	264
1991	227	2000	235
1991	259	2001	253
1993	216	2002	224
1994	171	2003	203
1994	203	2004	222
1995	174	2005	208
1995	254	2006	179
1996	225	2007	214
1997	211	2007	262
1997	227		
1997	243		
1998	214		
1998	246		
1999	193		
1999	201		
1999	209		
2000	212		
2000	236		
2001	222		
2002	225		
2003	244		
2004	167		
2004	199		
2005	217		
2006	204		
2007	191		

of data in the U.S. Geological Survey (USGS) Landsat archive. For the years where no cloud-free images were available, several partly cloudy images were acquired, which were later mosaicked to reduce cloud cover.

The LTSS consisted of mostly Thematic Mapper (TM) images. From 1999 to 2003, we considered both TM and Enhanced Thematic Mapper plus (ETM+) images. To avoid processing complications arising from dealing with data gaps caused by the scan line corrector failure that occurred in May 2003, no ETM+ images acquired after that date were used in this study. Note that we did not draw a distinction between TM and ETM+ images in this study because the two types of images had very similar spatial and spectral characteristics and there was essentially no difference between them in terms of the preprocessing and change mapping algorithms described in the subsequent sections.

4.2.3 Image Corrections

All Landsat images used in this study were downloaded from the USGS archive through the GLOVIS web site (www.glovis.gov). These images had been corrected by the USGS using the standard terrain correction method (level 1T). Geolocation accuracies were generally found within one pixel. No further geometric correction was performed in this study. Radiometrically, the images had been calibrated using the best available calibration methods (Chander et al. 2009). In this study, the raw digital numbers (DNs) of the reflective bands were further converted to top-of-atmosphere (TOA) reflectance following Markham and Barker (1986) and the *Landsat 7 Science Data User's Handbook* (Landsat Project Science Office 2000). To further reduce atmospheric effect, the images were corrected using an atmospheric correction algorithm adapted from the Moderate Resolution Imaging Spectroradiometer (MODIS) 6S radiative transfer approach (Vermote et al. 2002). Validation of the derived Landsat surface reflectance using simultaneously acquired MODIS daily reflectance products revealed that the discrepancies between the two products were generally within the uncertainty of the MODIS products themselves—the greater of 0.5% absolute reflectance or 5% of the retrieved reflectance value (Masek et al. 2006).

For the thermal band, the raw DN was converted to TOA (apparent) temperature using the standard approach provided by Markham and Barker (1986) for TM images and by the Landsat Project Science Office (2000) for ETM+ images.

4.2.4 Forest Change Mapping

The LTSSs were used to map forest disturbances using the VCT. This method differs from most existing land-cover change detection methods in that it automatically detects and tracks changes by analyzing all images of an LTSS at the same time, which allows it to take advantage of the rich temporal information of the LTSS in characterizing forest, nonforest, and disturbance. It is very efficient. For each LTSS, it took less than 4 h for the current version of this algorithm to produce the disturbance products described in Section 4.3.1, which, based on our experiences in forest cover change analysis using bitemporal change detection techniques, could take an experienced image analyst at least 10 days or much longer. Here, we provide a brief description of the VCT algorithm. Detailed description of all components of this algorithm has been provided in previous publications (Huang et al. 2010a; Huang, 2011).

The VCT algorithm consists of two major steps. During the first step, each image of an LTSS is analyzed independently to create a mask and to calculate an integrated forest z-score index (IFZ). This step is called single-image masking and normalization. Once this step is complete for all images of an LTSS, the derived forest index images are stacked to form an IFZ time series for each pixel, which is then analyzed to detect and track forest changes.

4.2.4.1 Single-Image Masking and Normalization

The major goal of this step is to use the spectral signature of known forest pixels within each image to normalize that image. Suppose the mean and standard deviation of the band i spectral values of known forest pixels within an image are $\bar{b}_i$ and SD_i, respectively, then for any pixel (b_{pi}) in that image, a forest z-score (FZ) value for that band can calculated as follows:

$$FZ_i = \frac{b_{pi} - \bar{b}_i}{SD_i}$$

For multispectral satellite images, the IFZ value of each pixel is defined by integrating FZ_i over the spectral bands as follows:

$$IFZ = \sqrt{\frac{1}{NB} \sum_{i=1}^{NB} (FZ_i)^2}$$

where NB is number of bands used. For Landsat TM and ETM+ images, bands 3, 5, and 7 are used to calculate the IFZ. Bands 1 and 2 are not used because they are highly correlated with band 3. The near infrared band (band 4) is excluded because, though forest canopy typically has high reflectance values in this band, nonforest surfaces can have high- or low-reflectance values in this band depending on the nonforest cover type. As a result, forest disturbances do not necessarily lead to spectral changes in a particular direction, and spectral changes in this band do not necessarily indicate real disturbances.

Notice that if the spectral signature of forest pixels has a normal distribution, FZ_i can be directly related to the probability of a pixel being a forest pixel using the standardized normal distribution table published in statistical textbooks (e.g., Davis 1986). As the root sum square of FZ_i, IFZ can be interpreted similarly. Although the forest pixels within a Landsat image may not have a rigorous normal distribution in all bands, an approximate probability interpretation of FZ_i and IFZ makes it possible to use probability-based threshold values later on during time series analysis that might be applicable to images acquired in different dates over different places. An intuitive interpretation of IFZ is that it is an inverse measure of the likelihood of a pixel being a forest pixel. Pixels having low IFZ value near 0 are close to the spectral center of forest samples and therefore have high probability of being forest pixels, whereas those having high IFZ values are likely nonforest pixels.

For each satellite image, confident forest samples are delineated using a dark object approach (Huang et al. 2008). This approach is based on a well-known observation that, due to substantial shadows cast within tree canopy, forest is generally darker than most other vegetated surfaces in the visible and shortwave infrared bands (Colwell 1974; Goward et al. 1994; Huemmrich and Goward 1997). In a histogram created using a local image window consisting of substantial forest pixels, those forest pixels are located toward the lower end of the histogram and often form a peak called the *forest peak*. Automatic delineation of forest pixels is achieved by locating the forest peak and then thresholding the local image window using threshold values defined by the forest peak. A detailed description of this dark object approach for delineating confident forest pixels has been provided by Huang et al. (2008).

Although every effort was made to select cloud-free images in assembling the LTSS, due to data availability constraints some images contained small portions of cloud cover. To minimize the risk of clouds over forest being mapped as disturbances, a cloud and shadow masking algorithm was applied to each image to mask out clouds and shadow. This algorithm uses the forest samples delineated using the above-described method as reference and makes use of spectral, thermal, spatial, and elevation information. A detailed description of this algorithm has been provided by Huang et al. (2010b).

4.2.4.2 Tracking Forest Change Using the IFZ

Because IFZ measures the likelihood of a pixel being a forest pixel, its change over time can be used to track forest change. For pixels masked as cloud or shadow in the single-image masking and normalization step (see Section 4.2.4.1), their IFZ values were calculated through linear interpolation using good observations (i.e., not contaminated by cloud or shadow) that were acquired in the years immediately before and after the concerned acquisition year. Figure 4.2 shows the typical temporal profiles of the IFZ for major forest cover change processes. For a persisting forest pixel that did not experience disturbance during the entire observing period of an LTSS, the IFZ value

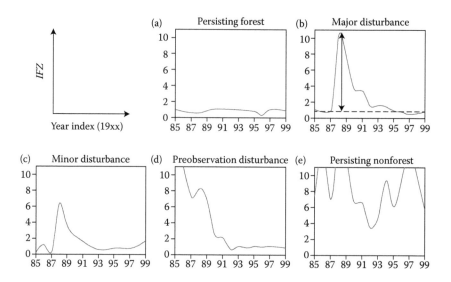

FIGURE 4.2
Typical IFZ temporal profiles of forest cover change processes. The location of the double arrow in (b) indicates the *disturbance year* and its length indicates *disturbance magnitude*. The dashed line in (b) indicates the mean IFZ value of forest observations.

stays low and is stable throughout the monitoring period (Figure 4.2a). The occurrence of a major disturbance will result in a sharp increase in the IFZ value in the image acquired immediately after the disturbance (Figure 4.2b). The acquisition year of that image is defined by the VCT algorithm as the *disturbance year* for that disturbance, although the actual disturbance year can be that year or any year between that year and the available previous image acquisition. A less dramatic change will result in a moderate increase in the IFZ value (Figure 4.2c). For recovery from a disturbance that occurred before the acquisition of the first image in the LTSS or conversion from non-forest to forest, the IFZ will start with high values but will go down gradually as the trees grow (Figure 4.2d). Finally, the IFZ is generally at an all-time high for persisting nonforest land. For cropland the IFZ may also fluctuate greatly as surface conditions change from one year to another due to harvesting and crop rotation (Figure 4.2e).

The very distinctive IFZ temporal profiles for different forest cover change processes allow identification of those change processes using the following simple rules:

- Pixels having low IFZ values through the entire observing period are classified as persisting forest.
- Pixels having low IFZ values for at least two consecutive observations but not for the entire observing period are classified as disturbed forest pixels. The disturbance year is determined as the

acquisition year when the IFZ value increases sharply from a low level (Figure 4.2b).

- Pixels having high IFZ values or having undulating IFZ values throughout the entire observing period are classified as persisting nonforest.

It should be noted that while only the IFZ is used here to track forest changes, the band-specific FZ values could be used to determine change types such as logging or fire. These issues are the subject of ongoing investigations.

4.2.5 Validation of the Disturbance Products

Validation of land-cover products derived using satellite observations typically relied on independent reference data collected through ground-based field work, visual interpretation of high-resolution images, or both (Congalton 1991; Stehman and Czaplewski 1998). For the disturbance products derived in this study, however, such a validation approach has many practical difficulties. In order to validate the disturbance year mapped by the VCT, for example, one would have to conduct field work or acquire high-resolution images in each of the acquisition years of an LTSS. This would be impossible even when resource availability is not a constraint, because (1) most of the required historical high-resolution images do not exist, and (2) due to the natural growth of vegetation, one cannot reliably determine the ground conditions immediately before and after the occurrence of a disturbance through a field trip conducted at the time of disturbance mapping, which could be many years after the occurrence of that disturbance. To avoid these problems, we designed a hybrid approach to validate the VCT disturbance product.

In the hybrid validation approach, we determined the disturbance year through visual inspection of all Landsat images of each LTSS. For most major disturbances, the disturbance year determined this way should be reliable because (1) most forests are spectrally distinctive in Landsat images and can be easily separated from nonforest by experienced image analysts; (2) major disturbances often yield spectral change signals in the Landsat images that are significant enough to recognize; and (3) visual interpretation is one of the most reliable approaches for analyzing satellite images because of the ability of human eyes to combine spectral, spatial (including texture and contextual information), and temporal information in image analysis. In order to avoid any possible confusions, for each reference point we also checked the high-resolution images available at GoogleEarth to assist visual interpretation of the Landsat images. Although for most locations the high-resolution images were available for one or a limited number of dates, which typically did not coincide with the disturbance year of a concerned disturbance, the surface conditions determined based on the high-resolution images at a particular time can be used as a reference by an image analyst in interpreting Landsat images acquired in other years. This method has been used to validate VCT disturbance products over six locations selected across the United States (Huang et al. 2009b; Huang et al. 2010a; Thomas et al. 2010).

Since only a small portion of the land area of the Olympic Peninsula was in the path 48/row 27 tile, only the disturbance product for the path 47/row 27 tile was validated using this method. The validation points were selected using a stratified random sampling approach. The disturbance year map was used to define the strata, where each class was a stratum. The number of points selected from each stratum was proportionate to its areal proportion within the concerned image tile. For rare classes where the number of points calculated this way was too low, it was increased to allow reliable calculation of class-specific accuracy values. Points located along polygon edges as determined through visual inspection were moved such that they were at least two pixels away from any edge to avoid the difficulty in determining the reference label for such points that may arise due to residual misregistration errors among the Landsat images. This sampling method resulted in a total of over 800 validation points over the path 47/row 27 site. For each point, the above-described hybrid approach was used to determine the appropriate land-cover type or disturbance year. The inclusion probability of each selected point was tracked appropriately and was used in calculating the accuracy estimates according to Stehman et al. (2003). As the edge pixels accounted for only small portions of the total pixels within each LTSS, the potential biases in the derived accuracy estimates as a result of not including the edges in the accuracy assessment should be small. It is also worth noting that the residual misregistration errors among the Landsat images should not necessarily lead to the forest edges being more likely to be mapped as disturbances. Such edges could be mapped as disturbances only in the unlikely event when the misregistration errors had a temporal trend such that they caused the IFZ profile for a persisting forest (Figure 4.2a) to look like that for a disturbance (Figure 4.2b–d).

4.3 Results

4.3.1 Accuracy of the Disturbance Products

A disturbance year map for the Olympic Peninsula was produced by applying the VCT to the images listed in Table 4.1 (Figure 4.3). This map shows where and when disturbances occurred. Specific disturbance year values were assigned to disturbances that occurred after 1984 or the first Landsat acquisition in an LTSS. A pre-1985 disturbance class was assigned to pixels that were not forested in 1984 but became forested in later years. A pixel of this class indicates either an afforestation process if that pixel was never forested before 1984, or a recovery following a disturbance that occurred in 1984 or earlier if it was forested some time before 1984. Pixels that maintained

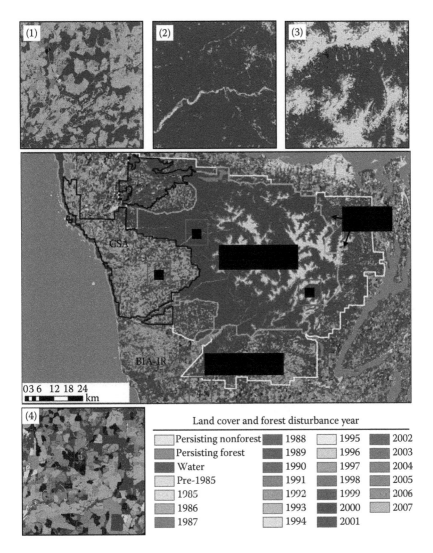

FIGURE 4.3
An overview of the disturbance map derived using the Landsat time series stack (LTSS)–vegetation change tracker (VCT) approach over the Olympic Peninsula with four full-resolution examples for (1) extensive recovery from previous harvest in the spotted owl conservation support area (CSA), (2) river channel shifts within Olympic National Park (ONP), (3) possible damages due to avalanche or mudslide, and (4) extensive logging on private land. The area covered by each full-resolution map is 12 km by 12 km.

the same land-cover type between 1984 and 2007 were classified as one of the three static classes: persisting forest, persisting nonforest, and water.

Table 4.2 shows the confusion matrix derived using the validation method described in Section 4.2.5 for the disturbance map for path 47/row 27. The overall user's and producer's accuracies were calculated according to

TABLE 4.2

Confusion Matrix for the Vegetation Change Tracker (VCT)-Derived Disturbance Year Map for World Reference System Path 47/Row 27

VCT	Reference																									Grand total	User's accuracies
	1	2	14	1985	1986	1987	1988	1989	1990	1991	1993	1994	1995	1996	1997	1998	1999	2000	2001	2002	2003	2004	2005	2006	2007		
1	33.75	0.55	0.55	0.28																						35.1	96.1
2	0.26	28.05	0.41	0.06			0.06														0.04					28.9	97.1
14	0.20	1.33	7.95	0.09										0.05		0.04	0.04			0.04					0.05	9.7	81.9
1985		0.04	0.09	1.02	0.09																					1.2	82.3
1986				0.20	1.75																					2.0	89.6
1987						1.51					0.06															1.6	96.4
1988							1.41																			1.4	100.0
1989								1.23	0.11																	1.3	92.0
1990								0.05	1.25																	1.3	92.5
1991									0.04	1.06																1.1	96.1
1993											1.56															1.6	100.0
1994												0.87	0.04													0.9	95.8
1995													1.03	0.22	0.05											1.3	79.2
1996													0.05	1.03	0.09											1.2	87.7
1997															1.05	0.09		0.04					0.05	0.04	0.04	1.4	76.9
1998									0.05							0.98										1.0	95.4

1999	0.04							0.81	0.12														1.0		83.3
2000	0.04								0.97														1.0		96.0
2001								0.77	0.12														1.0		79.2
2002				0.08						0.77													0.8		100.0
2003											0.89												0.9		100.0
2004												1.02											1.0		100.0
2005													1.04	0.10	0.10								1.2		84.1
2006			0.04											0.83	0.04								1.0		87.0
2007															1.14								1.1		100.0
Grand total	34.2	30.1	9.3	1.0	2.2	1.5	1.5	1.3	1.4	1.1	1.7	0.9	1.1	1.3	1.2	1.2	0.9	1.1	0.8	0.9	0.9	1.1	1.1	1.0	1.4	100.0
Producer's accuracies	98.6	93.2	85.2	100.0	80.3	100.0	96.3	95.8	89.0	100.0	91.9	100.0	91.8	79.1	87.9	84.8	95.2	88.9	95.0	83.0	96.0	100.0	95.5	85.5	83.4	Overall: 93.8

Note: Agreements are highlighted in bold. The overall accuracy is 93.8%. Class code is defined as follows: 1, Persisting nonforest; 2, Persisting forest; 14, Pre-1985 disturbance; 1985–2007, Disturbance year.

Stehman and Czaplewski (1998) and Stehman et al. (2003). When all land-cover and disturbance classes listed in Table 4.2 were considered, the disturbance map had an overall accuracy of 93.8% at the per-pixel level. Most change classes had user's and producer's accuracies over 80%, and the average user's and producer's accuracies were 91.5% and 91.8%, respectively.

4.3.2 Forest Disturbances in Different Management Areas

The VCT disturbance products allowed calculation of forest disturbance rates for different management areas. Figure 4.4 shows the percentage of the total area of each management type that was classified as persisting forest and two disturbance categories: pre-1985 disturbance and post-1984 disturbance (i.e., disturbances that occurred in or after 1985). The total percentage of the three categories within each management area gives the total percentage of forest land within that management area, where forest land is defined to include both undisturbed and disturbed forest pixels. The percentage of nonforest (including water) in each management area is the difference between 100% and the total percentage of forest land within that management area.

Figure 4.4 shows that forest land accounted for over 75% of the total area in all six management types. The ONP and the ONF-WA had smaller portions of forest land than most of the other management types, but they have the highest percentage of persisting forest and low percentage of disturbed forests. Nearly 70% of the areas within the ONF-NW were persisting forest, over 15% pre-1985 disturbance, and over 10% post-1984 disturbance. The other three management types had more disturbed forest pixels than

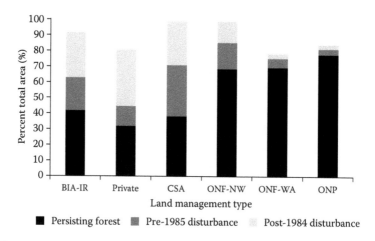

FIGURE 4.4
Percentage of persisting forest, pre-1985 disturbance, and post-1984 disturbance as mapped by the Landsat time series stack (LTSS)–vegetation change tracker (VCT) for different land management areas.

persisting forest pixels, suggesting that these areas had more active for-estry activities than the ONP, ONF-WA, and ONF-NW. Much of the distur-bances in those three management types and the ONF-NW, however, were pre-1985 disturbances, indicating widespread recovery from disturbances that occurred in or before 1984 over the Olympic Peninsula (Figures 4.3 and 4.4), which likely was the result of the 1984 Washington Wilderness Act and other conservation laws that were passed around the 1980s (National Park Service 2007; McNulty 2009).

The dense temporal intervals among the LTSS acquisitions allowed calcu-lation of the percentage of forest land being disturbed on an annual basis. For each of the few time periods where the temporal interval between two consecutive Landsat acquisitions was more than one year (see Table 4.1), the annual disturbance rate for that period was derived by dividing the disturbance rate calculated using the VCT disturbance product for that period by the length of that period. Figure 4.5 shows the annual forest dis-turbance rates over the six management areas from 1985 to 2007. It reveals that both the ONP, over 90% of which has been designated as wilderness area (National Park Service 2007), and the ONF-WA maintained low distur-bance rates (mostly <0.5%) over the entire observing period. Almost all the disturbances mapped in these two management areas were due to natural events, including river channel shift (Figure 4.3, full resolution example 2), avalanche or mudslide (Figure 4.3, full resolution example 3), or damages caused by snow storm, insect, disease, or fire. In contrast to these two man-agement areas, the forests on private land were heavily logged (Figure 4.3,

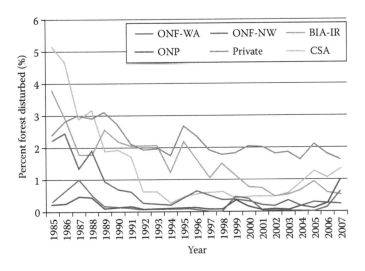

FIGURE 4.5
Percent forest area disturbed annually as mapped by the Landsat time series stack (LTSS)–vegetation change tracker (VCT) for different land management areas.

full resolution example 4), and had relatively high disturbance rates (2% or higher) from 1985 to 2007.

The ONF-NW had disturbance rates of over 2% in 1985 and 1986, but the rate decreased gradually until 1992, and remained relatively low since then. Similarly, the spotted owl CSA had decreasing disturbance rates between 1985 and 1994, but the percentage of forest area disturbed in each year was substantially high compared to that of ONF-NW, and the disturbance rate had an increasing trend after 2003. The BIA-IR area had disturbance rates comparable to those of private lands before the mid-1990s, but since then the rate decreased gradually and remained below 1% after 2000. In addition to the 1984 Washington Wilderness Act mentioned earlier, the Northwest Forest Plan that was signed into law in 1994 likely also contributed to the decreasing trend of disturbance rates seen in ONF-NW, CSA, and BIA-IR areas. However, these legislations did not lead to substantial reduction in disturbance rates on private land.

4.4 Conclusions

This chapter describes an approach for monitoring forest dynamics using LTSS and VCT. This approach was used to evaluate forest dynamics from 1984 to 2007 within the ONP and other management areas surrounding it, including the ONF-WA, ONF-NW, private land, spotted owl CSA on private land, and BIA-IR area. An accuracy assessment revealed that the disturbance map produced using the LTSS–VCT approach had an overall accuracy of 93.8%, and its average user's and producer's accuracies were 91.5% and 91.8%, respectively. As expected, this map revealed that both the ONP and ONF-WA had very low disturbance rates throughout the observing period. While most of those disturbances were caused by natural events, local experts and *in situ* observations or high-resolution image will need to be consulted to determine the exact causes of the mapped disturbances.

The disturbance map also revealed that reforestation or afforestation occurred in most areas disturbed in or before 1984, and that the ONF-NW and CSA experienced significant decreases in forest disturbance rates before the mid-1990s, while the BIA-IR area had decreasing disturbance rates until 2000. The private forest lands, however, maintained high disturbance rates (about 2%) throughout the observing period. Linkages were found between changes in forest disturbance rates and conservation legislations, including the 1984 Washington Wilderness Act and the 1994 Northwest Forest Plan. However, more studies are needed to quantify the roles other conservation regulations may have on observed changes in forest disturbance rates (e.g., Kay et al. 2007; Healey et al. 2008).

The time series approach described in this study represents a new approach for monitoring protected forest lands and their surrounding areas. This approach allows reconstruction of disturbance history over the last two decades by taking advantage of the temporal depth of the Landsat archive, and can be used to provide continuous monitoring as new satellite images are acquired. Now that the entire USGS Landsat archive can be accessed with no data cost to users, the Landsat images needed for assembling the required LTSS should exist for most places in the United States (Goward et al. 2006). Therefore, this time-series approach can be used to monitor protected areas across the United States and those in other countries where Landsat images needed for developing LTSS exist.

Acknowledgments

This study was made possible through funding support from the U.S. National Aeronautics and Space Administration's (NASA) Terrestrial Ecology, Carbon Cycle Science, and Applied Sciences Programs, the U.S. Geological Survey, and the LANDFIRE project, which was sponsored by the inter-governmental Wildland Fire Leadership Council of the United States. The datasets used in this study were assembled through the North American Forest Dynamics (NAFD) project, which contributes to the North American Carbon Program.

References

Chander, G., Markham, B.L., and Helder, D.L. 2009. Summary of current radiometric calibration coefficients for Landsat MSS, TM, ETM+, and EO-1 ALI sensors. *Remote Sensing of Environment, 113*, 893–903.

Cohen, W.B., Fiorella, M., Bray, J., Helmer, E., and Anderson, K. 1998. An efficient and accurate method for mapping forest clearcuts in the Pacific Northwest using Landsat imagery. *Photogrammetric Engineering & Remote Sensing, 64,* 293–300.

Colwell, J.E. 1974. Vegetation canopy reflectance. *Remote Sensing of Environment, 3,* 174–183.

Congalton, R. 1991. A review of assessing the accuracy of classifications of remotely sensed data. *Remote Sensing of Environment, 37,* 35–46.

Coppin, P., Lambin, E., Jonckheere, I., Nackaerts, K., and Muys, B. 2004. Digital change detection methods in ecosystem monitoring: A review. *International Journal of Remote Sensing, 25,* 1565–1596.

Davis, J.C. 1986. *Statistics and Data Analysis in Geology.* New York: John Wiley & Sons, Inc., 646pp.

Franklin, S.E., Lavigne, M.B., Wulder, M.A., and Stenhouse, G.B. 2002. Change detection and landscape structure mapping using remote sensing. *Forestry Chronicle, 78,* 618–625.

Genc, L., and Smith, S. 2003. Forest cover change assessment for North Central Florida using Landsat thematic mapper data. *Surveying and Land Information Science, 63,* 149–154.

Goward, S., Irons, J., Franks, S., Arvidson, T., Williams, D., and Faundeen, J. 2006. Historical record of landsat global coverage: Mission operations, NSLRSDA, and international cooperator stations. *Photogrammetric Engineering and Remote Sensing, 72,* 1155–1169.

Goward, S.N., Huemmrich, K.F., and Waring, R.H. 1994. Visible-near infrared spectral reflectance of landscape components in western Oregon. *Remote Sensing of Environment, 47,* 190–203.

Goward, S.N., and Williams, D.L. 1997. Landsat and earth systems science: Develoment of terrestrial monitoring. *Photogrammetric Engineering & Remote Sensing, 63,* 887–900.

Healey, S.P., Cohen, W.B., Spies, T.A., Moeur, M., Pflugmacher, D., Whitley, M.G., and Lefsky, M. 2008. The relative impact of harvest and fire upon landscape-level dynamics of older forests: Lessons from the Northwest Forest Plan. *Ecosystems, 11,* 1106–1119.

Healey, S.P., Cohen, W.B., Zhiqiang, Y., and Krankina, O.N. 2005. Comparison of tasseled cap-based Landsat data structures for use in forest disturbance detection. *Remote Sensing of Environment, 97,* 301.

Huang, C. 2011. Forest change analysis using time series Landsat observations. In: Q. Weng (ed.), *Advances in Environmental Remote Sensing: Sensors, Algorithms, and Applications.* New York: CRC Press, pp. 339–365.

Huang, C., Goward, S.N., Masek, J.G., Gao, F., Vermote, E.F., Thomas, N., Schleeweis et al. 2009a. Development of time series stacks of Landsat images for reconstructing forest disturbance history. *International Journal of Digital Earth, 2,* 195–218.

Huang, C., Goward, S.N., Masek, J.G., Thomas, N., Zhu, Z., and Vogelmann, J.E. 2010a. An automated approach for reconstructing recent forest disturbance history using dense Landsat time series stacks. *Remote Sensing of Environment, 114,* 183–198.

Huang, C., Goward, S.N., Schleeweis, K., Thomas, N., Masek, J.G., and Zhu, Z. 2009b. Dynamics of national forests assessed using the Landsat record: Case studies in eastern U.S. *Remote Sensing of Environment, 113,* 1430–1442.

Huang, C., Song, K., Kim, S., Townshend, J.R.G., Davis, P., Masek, J., and Goward, S.N. 2008. Use of a dark object concept and support vector machines to automate forest cover change analysis. *Remote Sensing of Environment, 112,* 970–985.

Huang, C., Thomas, N., Goward, S.N., Masek, J., Zhu, Z., Townshend, J.R.G., and Vogelmann, J.E. 2010b. Automated masking of cloud and cloud shadow for forest change analysis. *International Journal of Remote Sensing, 31,* 5449–5464.

Huemmrich, K.F., and Goward, S.N. 1997. Vegetation canopy PAR absorptance and NDVI: An assessment for ten tree species with the SAIL model. *Remote Sensing of Environment, 61,* 254–269.

Kay, W.M., Donoghue, E.M., Charnley, S., and Moseley, C. 2007. Northwest Forest Plan, The first ten years 1994–2003, Socioeconomic monitoring of the Mount Hood National Forest and three local communities. General Technical Report General Technical Report, PNW-GTR-701. Portland, OR: U.S. Department of Agriculture, Pacific Northwest Research Station.

Kennedy, R.E., Yang, Z.G., and Cohen, W.B. 2010. Detecting trends in forest disturbance and recovery using yearly Landsat time series: 1. LandTrendr—Temporal segmentation algorithms. *Remote Sensing of Environment, 114,* 2897–2910.

Landsat Project Science Office. 2000. *Landsat 7 Science Data User's Handbook.* Greenbelt, MD: National Aeronautics and Space Administration.

Lu, D., Mausel, P., Brondízio, E., and Moran, E. 2004. Change detection techniques. *International Journal of Remote Sensing, 25,* 2365–2407.

Lunetta, R.S., Johnson, D.M., Lyon, J.G., and Crotwell, J. 2004. Impacts of imagery temporal frequency on land-cover change detection monitoring. *Remote Sensing of Environment, 89,* 444–454.

Malila, W.A. 1980. Change vector analysis: An approach for detecting forest changes with Landsat. In: *Proceedings of the 6th Annual Symposium on Machine Processing of Remotely Sensed Data.* West Lafeyette, IN: Purdue University, pp. 326–335.

Markham, B.L., and Barker, J.L. 1986. Landsat MSS and TM post-calibration dynamic ranges, exoatmospheric reflectances and at-satellite temperatures. *EOSAT Landsat Technical Notes, 1,* 3–8.

Masek, J.G., Huang, C., Cohen, W., Kutler, J., Hall, F., and Wolfe, R.E. 2008. Mapping North American forest disturbance from a decadal Landsat record: Methodology and initial results. *Remote Sensing of Environment, 112,* 2914–2926.

Masek, J.G., Vermote, E.F., Saleous, N.E., Wolfe, R., Hall, F.G., Huemmrich, K.F., Feng, G., Kutler, J., and Teng-Kui, L. 2006. A Landsat surface reflectance dataset for North America, 1990–2000. *IEEE Geoscience and Remote Sensing Letters, 3,* 68–72.

McNulty, T. 2009. *Olympic National Park, A Natural History.* Seattle: University of Washington Press, 338pp.

Miller, A.B., Bryant, E.S., and Birnie, R.W. 1998. An analysis of land cover changes in the Northern Forest of New England using multitemporal Landsat MSS data. *International Journal of Remote Sensing, 19,* 245–265.

National Park Service. 2007. *Final general management plan, environmental impact statement, Olympic National Park, Washington.* Washington, DC: National Park Service, U.S. Department of the Interior, 473pp.

Nelson, R.F. 1983. Detecting forest canopy change due to insect activity using Landsat MSS. *Photogrammetric Engineering and Remote Sensing, 49,* 1303–1314.

Singh, A. 1989. Digital change detection techniques using remotely-sensed data. *International Journal of Remote Sensing, 10,* 989–1003.

Stehman, S.V., and Czaplewski, R.L. 1998. Design and analysis for thematic map accuracy assessment: Fundamental principles. *Remote Sensing of Environment, 64,* 331–344.

Stehman, S.V., Wickham, J.D., Smith, J.H., and Yang, L. 2003. Thematic accuracy of the 1992 national land-cover data for the eastern United States: Statistical methodology and regional results. *Remote Sensing of Environment, 86,* 500–516.

Thomas, N., Huang, C., Goward, S.N., Powell, S., Rishmawi, K., Schleeweis, K., and Hinds, A. 2010. Validation of North American forest disturbance dynamics derived from Landsat time series stacks. *Remote Sensing of Environment, 115,* 19–32.

Townshend, J.R.G., and Justice, C.O. 1988. Selecting the spatial resolution of satellite sensors required for global monitoring of land transformations. *International Journal of Remote Sensing*, 9, 187–236.

Vermote, E.F., El Saleous, N.Z., and Justice, C.O. 2002. Atmospheric correction of MODIS data in the visible to middle infrared: First results. *Remote Sensing of Environment*, 83, 97–111.

Yang, X., and Lo, C.P. 2002. Using a time series of satellite imagery to detect land use and land cover changes in the Atlanta, Georgia metropolitan area. *International Journal of Remote Sensing*, 23, 1775–1798.

5

Using Earth Observation to Monitor Species-Specific Habitat Change in the Greater Kejimkujik National Park Region of Canada

Paul Zorn, Darien Ure, Rajeev Sharma, and Sally O'Grady

CONTENTS

5.1 Introduction

Ecological monitoring is critical to the success of protected areas and remote sensing is critical to the success of ecological monitoring. Both of these statements are true, in part, because of the uncertainty inherent in protected area management. A dominant paradigm in protected area management is the "uncertainty principle" (Hockings et al. 2000) which simply states that decisions can never be made with complete certainty regarding the outcome of management actions. This uncertainty is due to the complex and dynamic nature of ecosystems that protected areas are established to protect. For this reason, adaptive management is a key strategy in the conservation of protected areas (Hockings et al. 2000). Adaptive management is an iterative approach to management where decisions are treated as hypotheses that are to be tested through monitoring (Holling 1978). Results of decisions are monitored to ensure that they support goals and objectives. If monitoring suggests that actions (or inactions) are not effective then they are adapted in light of the new information and adjustments are made. This cycle of decision-making–monitoring–adaption is repeated in an iterative manner until protected area goals are met. For protected areas, adaptive management cannot be successful without effective ecological monitoring.

Effective ecological monitoring programs all possess common characteristics (Table 5.1). They must represent major park ecosystems and be sensitive to changes of management concern. They must have the ability to detect change at spatial and temporal scales appropriate for protected area boundaries and management cycles. Individual monitoring measures, taken as a set, should be integrated such that they give managers comprehensive information regarding ecosystem dynamics. Effective monitoring programs must also be operational, cost effective, and be consistently implemented to give unbiased results. Lastly, monitoring programs must give managers useful and engaging information to communicate to stakeholders and visitors in order to change human behavior and land-use patterns in and around protected areas.

Earth observation/remote sensing (EO/RS) information possesses these characteristics and both methods are vital tools in the protected area manager's toolbox for the development of effective ecological monitoring programs. EO/RS allow for the assessment of ecological change at the scale of entire protected areas. When supplemented by *in situ* monitoring, EO/RS

TABLE 5.1

Characteristics of an Effective Ecological Monitoring Program

Sensitivity	Ecological monitoring programs must comprise measures that are sensitive to ecological processes or stressors of management concern.
Representivity	Due to the complexity of ecosystems, monitoring measures must be accurate surrogates for protected area ecosystem dynamics and adequately represent key ecological features.
Spatial scale	Measures must be monitored at a spatial scale (extent and resolution) commensurate with the ecological process of interest.
Temporal scale	Measures must be monitored at a temporal scale (frequency and length of time series) commensurate with the ecological process of interest.
Integration	Different monitoring measures will possess varying sensitivities to different ecological processes or stressors. Ideally, the interpretation of individual measures can be integrated to give a holistic assessment of protected area ecosystems.
Operations	The implementation of methods for monitoring measures must be operational given available human resources and expertise. This includes monitoring remote areas with a study design that will not bias results.
Cost	Monitoring measures must not be cost prohibitive over the long term given uncertain finances.
Consistency	Methods must be able to be implemented consistently and repeatable given staff turnover and different users over time.
Communication	Ideally, monitoring measures should generate information with a high communicative value for a range of different target audiences.

can be sensitive to many relevant ecological changes due to climate change and land use. EO/RS can provide managers with scientifically credible, timely information and provide powerful images for communicating key management issues.

The purpose of this chapter is to provide an example of the application of EO/RS to the monitoring of protected areas. Specifically, this chapter highlights the use of EO/RS for monitoring wildlife habitat change in Kejimkujik National Park and National Historic Site (Kejimkujik). A major goal of national parks in Canada is the protection and maintenance of ecological integrity. Ecological integrity refers to "with respect to a park, . . . a condition that is determined to be characteristic of its natural region and likely to persist, including abiotic components and the composition and abundance of native species and biological communities, rates of change and supporting processes" (Government of Canada 2000 Section 2(1), p. 1). For Kejimkujik, a measure of ecological integrity is the provision of habitat for wildlife that represent major park ecosystems.

In order to protect viable populations of wildlife species, the metapopulation theory suggests that two characteristics are critical: habitat amount and habitat connectivity (Hanski 1999). Habitat amount is a surrogate for resource availability for wildlife and habitat connectivity is a surrogate for resource accessibility. Generally speaking, landscapes with greater amounts of accessible resources can support larger wildlife populations and larger populations tend to be more viable and resilient to ecological change (Hanski 1999). For this reason, EO/RS methods are used at Kejimkujik to monitor two key attributes of ecological integrity: "effective habitat amount" and "effective habitat connectivity."

5.2 Study Area

Kejimkujik was established in 1974 to protect a representative example of the Canadian Atlantic Coastal Uplands Natural Region. It is situated in southern Nova Scotia (Figure 5.1) and is representative of the Acadian forest region which is a transition zone between southern deciduous forests and northern coniferous forest and is one of the most richly diverse temperate forests in the world (Simpson 2008). Rowe (1972) divided the Acadian forest into seven subsections based on recurring tree communities of which the Atlantic Coastal Uplands Region covers half of Nova Scotia and includes all of Kejimkujik. The Atlantic Coastal Uplands' forest mosaic is less boreal than many other areas of the Acadian forest, characterized by elevated levels of pine, Eastern hemlock (*Tsuga Canadensis*), Red maple (*Acer rubrum*), and Red oak (*Quercus rubra*) (Basquill et al. 2001). The forests of Kejimkujik contain mature to old-growth stands of Eastern hemlock, Red spruce (*Picea rubens*)

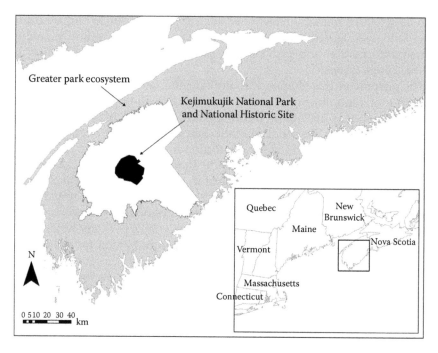

FIGURE 5.1
Regional context of Kejimkujik National Park and National Historic Site. Study area boundary represents the greater park ecosystem area.

and Sugar maple (*Acer saccharum*). Red maple floodplains flank river systems and wetter areas. Black spruce (*Picea mariana*) and mixed-wood forests are also locally important forested wetland ecosystems. Large White pine (*Pinus strobes*) and Red pine (*Pinus resinosa*) occur in older mixed-wood stands, with American beech (*Fagus grandifolia*) and Ironwood (*Ostrya virginiana*) in mid-seral stands. Much of the forest landscape surrounding Kejimkujik is subject to active timber management, including clear cutting, making forestry practices one of the major agents of land-cover change around the park.

5.3 Methods

Different species respond to changes in land cover differently. Depending on their landscape requirements, a particular pattern of land-cover change may represent a significant effect on habitat for one species and no effect on habitat for another species. For this reason, Kejimkujik selected a set of focal species that vary in their habitat associations and dispersal abilities as surrogates for assessing the effects of forest land-cover change in and around the park from 1985 to 2005. Five species or species groups were chosen: (1) American marten (*Martes americana*), (2) Northern (*Glaucomys sabrinus*) and Southern (*Glaucomys volans*) flying squirrel, (3) Northern goshawk (*Accipiter gentilis*), (4) Coniferous forest birds (i.e., Blackburnian [*Dendroica fusca*], Black-throated green [*Dendroica virens*] warblers), and (5) Deciduous forest birds (i.e., Ovenbird [*Seiurus aurocapillus*], Black-throated blue warbler [*Dendroica caerulescens*]). These species and species groups are all associated with mature forests and are particularly sensitive to the effects of regional habitat loss and fragmentation (Chapin et al. 1996; Betts and Forbes 2005; Beazley et al. 2005; Betts et al. 2006; Ritchie et al. 2009).

To determine the landscape-scale effects of land-cover change on the habitat of these species it was necessary to link a standardized classified land-cover image to a suitable habitat for each species. Once species-specific habitat maps were generated, changes with respect to "effective habitat amount" and "effective habitat connectivity" were assessed for Kejimkujik's greater park ecosystem for a 20-year period between 1985 and 2005. The first set of methods involves EO/RS image processing procedures to create a time series of land-cover maps for the study area. The second set of methods includes a series of Geographic Information System (GIS)-based landscape ecology procedures to create trends in habitat amount and connectivity for each of the five species groups.

5.3.1 Earth Observation/Remote Sensing Methods

The main remote sensing data sources for this study were multitemporal, multispectral digital data, acquired from the thematic mapper/enhanced thematic mapper plus (ETM+), onboard Landsat 5 and Landsat 7, respectively, and included rows 08/29 and 09/29 (Table 5.2). These scenes were procured (free of charge) from the U.S. Geological Survey Landsat archives (available at: http://landsat.usgs.gov/USGSLandsatGlobalArchive.php). Images representing 1985, 1990, 1995, 2000, and 2005 were used in the analysis.

Multitemporal land-cover maps were generated following Automated Multitemporal Updating by Signature Extension (AMUSE) protocols, developed by the Canada Centre for Remote Sensing (Fraser et al. 2008). The approach broadly consists of: (i) geometric registration of multidate Landsat images, (ii)

TABLE 5.2

Landsat Data Acquired for Project

Year	P08 R29	P09 R29	Sensor
1985	September 1, 1986	June 23, 1986	Landsat 5 TM
1990	July 4, 1988	June 18, 1991	Landsat 5 TM
1995	August 9, 1995	July 31, 1995	Landsat 5 TM
2000	July 13, 2000	June 18, 2000	Landsat 7 ETM+
2005	July 3, 2005	August 27, 2005	Landsat 5 TM

radiometric normalization of 1985, 1990, 1995, and 2005 images with respect to the 2000 reference imagery, (iii) generation of master baseline land-cover classification, (iv) identifying change pixels using change vector analysis, (v) update land cover for other dates using constrained signature extension, and (vi) validation of baseline land cover and change. The image analysis procedure is described in the draft AMUSE manual (Fraser et al. 2008).

The baseline land-cover map was generated using an unsupervised clustering approach that combines features of the enhancement classification (ECM) (Beaubien et al. 1999) and classification by progressive generalization (CPG) (Cihlar et al. 1998) methods. The ECM procedure is intended to capture most of the information visible in an enhanced three-channel image by converting the image into a classification. CPG can be used to merge a large number of unsupervised clusters (e.g., 150) to a smaller number (e.g., 60) before labeling based on spectral and spatial distance among clusters. The spectral clusters were assigned class labels (Table 5.3) using ground truth data collected during field work from another study conducted in 2005, provincial forest cover maps, and local knowledge.

Classifications were produced for 1985, 1990, 1995, and 2005 via signature extension and thresholds from change vector analysis were used to forward-update the 1990, 1995, and 2000 classifications. In order to constrain the changes to real changes in the land cover and not seasonal variation due to crop rotation, agricultural masks were created to map areas of extensive agriculture and wetlands and water masks were also employed to prevent back-and-forth transitions of wetlands between dates that were likely due to yearly differences in precipitation.

TABLE 5.3

Land-Cover Classification Scheme

Code	Class Label (Cover Type)
1	Evergreen forest (>75% cover)
2	Evergreen open canopy (~50% cover)
3	Mixed coniferous (~50 cover)
4	Mixed deciduous (~50% cover)
5	Deciduous forest (>75% cover)
6	Deciduous open (~50% cover)
15	Low regenerating to young mixed cover
16	Deciduous shrubland (>75% cover)
20	Wetlands
34	Annual row-crop forbs and grasses—low biomass
40	Mostly disturbed bare areas (e.g., cutovers)
41	Low vegetation cover
42	Urban and built-up
43	Water bodies
46	Clouds

A mask of potentially changed pixels was first created using change vector analysis (CVA) between two consecutive dates, followed by thresholding to identify a change/no-change mask. CVA is a robust method for detecting radiometric change in multispectral imagery (Johnson and Kasischke 1998). CVA images were generated by calculating the multispectral Euclidean distance for Landsat channels 3, 4, and 5 on a per-pixel basis. This distance was computed as follows: CVA = ((Red_Date1 – Red_Date2)^2) + ((NIR_Date1 – NIR_Date2)^2) + ((SWIR_Date1 – SWIR_Date2)^2)^0.5.

In a CVA image, gray values range between 0 and 255. The lower the value, the lower the spectral change in a category between the two dates and vice versa. Thresholds were determined interactively by examining known land-cover changes in pseudo color composite images (Red: Band 4, Green: Band 5, Blue: Band 3) from two dates against change masks created over a range of thresholds. Land-cover maps were then updated for pixels in the change mask only, by spectral signature extension, whereby changed pixels were matched to the most similar labeled cluster from a baseline land-cover map (Fraser et al. 2008).

Using CVA, a change magnitude image was produced between each consecutive pair of image dates (i.e., 1985–1990, 1990–1995, 1995–2000, and 2000–2005). A liberal threshold, ranging from 60- to 70-pixel value, was used for identifying change areas for each of the change magnitude image. All the gray values above this threshold represent change.

5.3.2 Landscape Ecology Methods

The intent of these landscape ecology methods is to measure the effect of land-cover change from the perspective of a specific species or species group following the concept of ecologically scaled landscape indices (ESLIs) (Vos et al. 2001). ESLIs address the issue that responses to land-cover change tends to be species specific. Even for species that are associated with the same major park ecosystem (e.g., forest) there will likely be major differences in how they respond to spatial–temporal changes in forest cover. These differences may be influenced by a species' habitat associations with different ecosites, whether a species is a habitat generalist or specialist, the movement abilities and behavior of a species, and the scale at which a species perceives and utilizes its environment (to name only a few). Given these differences, a monitoring protocol that looks only at land-cover change and/ or fragmentation metrics associated with discrete land-cover units may not be sensitive for specific species characteristics that represent the major park ecosystem of interest. In short, the degree to which land-cover change matters with respect to ecological integrity will be determined by an interaction between the pattern of landscape change and the specific species occupying that landscape. A certain pattern of land-cover change in or around a park may functionally disconnect a landscape for some species but not for others. To address these differences any measure of landscape change must be appropriately "scaled" to accommodate species-specific characteristics, hence ESLIs.

There are essentially two main consequences of habitat loss and fragmentation: a decrease in total habitat area and an increase in the isolation of remaining habitat patches (Hanski and Gilpin 1991; Opdam 1990). This follows the basic metapopulation theory that the persistence of a species' metapopulation is determined by the relationship between patch colonization and extinction rates within a region (Levins 1970). Colonization rates are influenced by patch isolation such that as isolation (interpatch distance) increases, successful immigration by individuals becomes less likely, thereby reducing colonization. Extinction rates, in addition to being influenced by patch isolation, are also impacted by habitat amount. As habitat decreases patches tend to become smaller with lower carrying capacities that support fewer individuals (Hanski 1994). As population size decreases extinction probabilities increase. While this simplifies the dynamics of habitat change effects, decades of research in metapopulation theory, island biogeography, and landscape ecology support the notion that the key factors influencing a species' sensitivity to habitat loss and fragmentation are area-dependent and isolation-dependent factors (Real and Brown 1991).

Following this argument, these methods focus on two monitoring measures: (1) effective habitat amount and (2) effective habitat isolation. The term "effective" is used here to refer to the species-specific relationship between land-cover change and habitat. So, for example, a change in forest cover may be detected from remote sensing but the extent to which this affects the habitat of forest-dwelling species may be something quite different from the pattern of forest cover itself depending on what other information is used to delineate "effective habitat" for the species. In order to determine what information is needed to operationally delineate effective habitat a species profile must be created. Species profiles provide the rules by which land-cover information becomes translated into species-specific habitat units. Species profiles contain three elements: (1) species-specific habitat associations based on land-cover type and spatial context (e.g., mature conifer forest with >75% cover, forest patches must be greater than 25 ha, 150 m edge effect for forest patches adjacent to nonforest land-cover types), (2) a species-specific cost surface that estimates landscape-scale movement barriers that may inhibit dispersal between habitat patches (e.g., road density), and (3) an explicit spatial scale that estimates home range size and is used in a moving window analysis. The specific combination of land-cover type and spatial context that these profiles represent is unique to each species. Table 5.4 contains the species profiles used for the five species monitored at Kejimkujik.

These sets of rules were applied using ArcGIS 9.3 and FRAGSTATS 3.3 (McGarigal and Marks 1995) to create an "effective habitat patch" map for each species. Second, a moving window analysis was applied to each home range or dispersal area of the species. Within the window, the total

TABLE 5.4

Species Profiles Used for Habitat Fragmentation Analysis

Species	Patch Type	Cost Surface	Dispersal Range	Threshold	References
Northern and southern flying squirrel	Coniferous, deciduous, and mixed forest polygons; >40% suitable habitat within home range	Moderate weighting for regenerating forest (×5); high weighting for open nonforest areas >30 m (×10)	1.414 km moving window (based on 2 km² dispersal area)	>40% suitable habitat across park and GPE	Lavers 2004; A. Lavers, pers. comm.; M. Smith, pers. comm.
American marten	Coniferous and mixed forest polygons; 100 m buffer around nonforest (excluding lakes and wetlands); >70% suitable habitat within home range	Moderate weighting for road density (×2); high weighting on >200 m nonforest cover (×10); complete barrier for 1–2 km nonforest gaps (×2000)	2 km radius moving window (based on 4 km² home range)	>20% suitable habitat across park and GPE	P. Austin-Smith, pers. comm.
Northern goshawk	Coniferous, deciduous, and mixed forest polygons; 170 ha post-fledgling area with >70% suitable habitat and >3 nesting patches of continuous forest >24 ha; >105 ha area of nonforest within 170 ha post-fledgling area are unsuitable; 170 ha area post-fledgling area must be >150 m from permanent roads; >55% suitable habitat within home range	N/A	2.38 radius moving window (based on 570 ha home range)	>22% minimum habitat across park and GPE; ideally >40% suitable habitat	P. Bush, pers. comm., P. Austin-Smith, pers. comm.

(Continued)

TABLE 5.4 (continued)

Species	Patch Type	Cost Surface	Dispersal Range	Threshold	References
Deciduous forest birds	Deciduous and mixed forest polygons; 150 m buffer around nonforest (excluding lakes and wetlands); 28.2% suitable habitat within home range	High weighting for nonforest gaps >30 m between patches (×2)	2 km radius moving window (based on dispersal scale)	>28.2% suitable habitat across park and GPE	Betts et al. 2006; C. Staicer, pers. comm.
Coniferous forest birds	Coniferous forest polygons; 150 m buffer around nonforest (excluding lakes and wetlands); 29.6% suitable habitat within home range	High weighting for nonforest gaps >30 m between patches (×2)	2 km radius moving window (based on dispersal scale)	>29.6% suitable habitat across park and GPE	Betts et al. 2006; C. Staicer, pers. comm.

Note: GPE, greater park ecosystem.

amount and connectivity of effective habitat were calculated to determine the suitability of species landscapes throughout the study area. Amount is measured as the total area of effective habitat patches summed per window. Connectivity is measured as the mean cost-weighted distance between patches where cost is determined by a species-specific cost surface (e.g., road density, nonforest gap crossings) developed in GIS. The results are gradient maps that show areas of high and low effective habitat amount and connectivity throughout the park and greater park ecosystem for each species. Thresholds were then applied to identify species landscapes that are estimated to contain sufficient habitat (i.e., a certain proportion of a moving window must contain effective habitat to be suitable; the average cost-weighted distance between habitat patches must be less than the width of the moving window to be connected). Statistics on the total amount and connectivity of effective habitat (scaled to species home range) for Kejimkujik and the greater park ecosystem were then extracted for each five-year time step and compared.

5.4 Results

Figure 5.2a,b, and c show trends in effective habitat amount for northern and southern flying squirrel, northern goshawk, and softwood forest birds, respectively (for brevity, maps for other species are omitted). Generally speaking, trends in species-specific habitat amount within the park are relatively stable compared to the larger, surrounding region where timber harvesting is permitted. Table 5.5 summarizes the trends for all species selected for analysis. Estimated habitat amount for American marten has increased in Kejimkujik by 10.5% compared to a 78.3% reduction in the greater park ecosystem between 1985 and 2005. Northern and southern flying squirrel habitat shows a slight 3.9% decline in the park compared to 48.2% in the greater park ecosystem. Estimated northern goshawk habitat declined by 6.2% in the park compared to 67.3%. Hardwood forest bird habitat declined by 5.3% in the park versus a 57.3% decline in the surrounding region and softwood forest bird habitat was stable in the park (0.6% change) compared to a 47.7% decline in the greater park ecosystem.

Table 5.5 also summarizes trends in effect habitat connectivity for the same set of species. In this table, values for "Connectivity" represent the proportion of species-specific habitat that is considered connected from the cost-weighted distance between patches. For American marten, 78.3% of the study area was considered connected in 1985 compared to only 39.4% of the area in 2005 (a decline of 49.7%) (Figure 5.3). This pattern is in contrast to trends in the park where estimated habitat connectivity for American marten has been stable

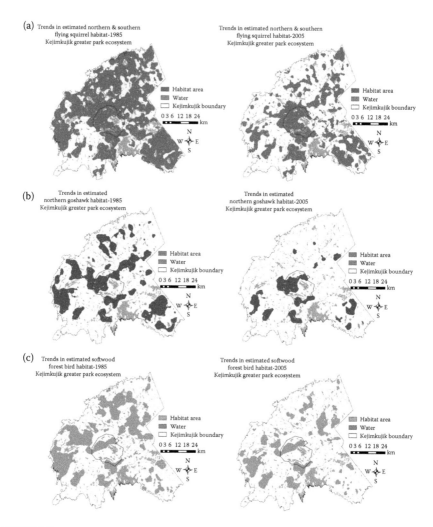

FIGURE 5.2
(See color insert.) Estimated habitat amount from 1985 to 2005 for (a) northern and southern flying squirrel, (b) northern goshawk, and (c) softwood forest birds.

across the same time period. For all other species, trends in estimated habitat connectivity have been stable in both the greater park ecosystem and the park despite changes in estimated habitat amount. This is because these species are good dispersers and relatively unaffected by land-cover patterns in the non-habitat matrix. While habitat amount reduces the ability of the regional landscape to sustain viable populations of species, habitat remnants are likely to still provide "stepping stones" that allow individuals to move throughout the landscape.

TABLE 5.5

Change in Habitat Amount in Kejimkujik and in the Greater Park Ecosystem Study Area (GPE)

Species	Year	Amount (GPE) (km²)	Percentage Change	Amount (Park) (km²)	Percentage Change	Connectivity (GPE)	Percentage Change	Connectivity (Park)	Percentage Change
American marten	1985	266.4	n/a	39.5	n/a	78.3	n/a	83.3	n/a
	1990	196.9	−26.1	40.7	3.1	74.5	−4.9	84.7	1.7
	1995	152.9	−42.6	40.0	1.5	65.9	−15.8	83.8	0.6
	2000	55.6	−79.1	38.4	−2.6	47.3	−39.6	84.3	1.2
	2005	57.8	−78.3	43.6	10.5	39.4	−49.7	85.5	2.6
Northern and southern flying squirrel	1985	3481.5	n/a	311.8	n/a	100.0	n/a	100.0	n/a
	1990	3144.6	−9.7	310.0	−0.6	100.0	0.0	100.0	0.0
	1995	2880.4	−17.3	309.4	−0.8	100.0	0.0	100.0	0.0
	2000	2269.8	−34.8	302.4	−3.0	100.0	0.0	100.0	0.0
	2005	1804.8	−48.2	299.6	−3.9	100.0	0.0	100.0	0.0
Northern goshawk	1985	1380.2	n/a	177.3	n/a	100.0	n/a	100.0	n/a
	1990	1251.6	−9.3	184.1	3.9	100.0	0.0	100.0	0.0
	1995	951.4	−31.1	169.6	−4.3	100.0	0.0	100.0	0.0
	2000	674.1	−51.2	175.2	−1.2	100.0	0.0	100.0	0.0
	2005	451.9	−67.3	166.3	−6.2	100.0	0.0	100.0	0.0
Hardwood birds	1985	364.8	n/a	93.4	n/a	100.0	n/a	100.0	n/a
	1990	295.7	−18.9	87.3	−6.5	100.0	0.0	100.0	0.0
	1995	287.2	−21.3	90.5	−3.1	100.0	0.0	100.0	0.0
	2000	220.2	−39.6	88.5	−5.3	100.0	0.0	100.0	0.0
	2005	155.9	−57.3	88.4	−5.3	100.0	0.0	100.0	0.0
Softwood birds	1985	1582.2	n/a	113.5	n/a	99.0	n/a	93.1	n/a
	1990	1324.1	−16.3	110.9	−2.3	98.7	−0.3	93.0	−0.1
	1995	1159.5	−26.7	110.4	−2.7	98.3	−0.7	92.9	−0.2
	2000	922.0	−41.7	109.1	−3.9	97.5	−1.5	93.0	−0.1
	2005	827.0	−47.7	114.1	0.6	96.8	−2.3	92.6	−0.5

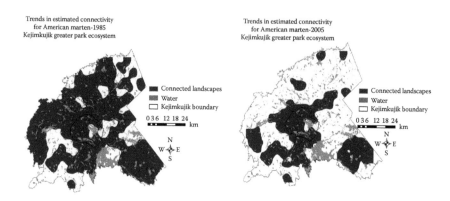

FIGURE 5.3
Estimated trends in landscape connectivity for American marten from 1985 to 2005.

5.5 Conclusion

A main objective of Kejimkujik, and protected areas in general, is the provision of suitable habitat to wildlife. From the metapopulation theory, defining "suitable" habitat requires estimates of habitat amount and connectivity. For this reason, Kejimkujik monitors spatial–temporal trends in estimated habitat amount and connectivity as part of its long-term ecological integrity monitoring program. EO/RS methods allow the park to monitor the potential effects of surrounding land-cover patterns on ecological integrity through habitat loss of resident wildlife populations and increasing isolation of the park from the surrounding region.

The use of EO/RS data for this purpose allows the park to monitor these important patterns in a manner that satisfies criteria for an effective monitoring program (Table 5.1). EO/RS data are sensitive to land uses that result in land-cover change that, in term, results in changes in species-specific habitat (Figures 5.2 and 5.3). Landsat TM allows the park to assess these changes at sufficient spatial and temporal scales that provide managers with an opportunity to evaluate Kejimkujik's role in a broader context. Since Landsat data are now freely available these assessments can be made into the future in an affordable and consistent manner. The visual output from these analyses provides powerful communication products on how land use can affect habitat patterns around the park. These communication products can support the park in engaging partner agencies and stakeholders in the development of collaborative conservation strategies that are vital in allowing Kejimkujik to manage its long-term ecological integrity goals.

References

Basquill, S.P., Woodley, S.J., and A.B. Pardy. 2001. The history and ecology of fire in Kejimkujik National Park. Technical Reports in Ecosystem Science, Report No. 029. Halifax, NS: Parks Canada.

Beaubien, J., Cihlar, J., Simard, G., and R. Latifovic. 1999. Land cover from multiple Thematic mapper scenes using a new enhancement-classification methodology. *Journal of Geophysical Research* 104: 27909–27920.

Beazley, K., Smandych, L., Snaith, T., MacKinnon, D., Austin-Smith Jr., P., and P. Duinker. 2005. Biodiversity considerations in conservation system planning: Map-based approach for Nova Scotia, Canada. *Ecological Applications* 15(6): 2192–2208.

Betts, M.G. and G.J. Forbes (eds). Greater Fundy Ecosystem Research Group. 2005. Forest management guidelines to protect native biodiversity in the Greater Fundy Ecosystem. New Brunswick: New Brunswick Co-operative Fish and Wildlife Research Unit, University of New Brunswick.

Betts, M.G., Forbes, G.J., Diamond, A.W., and P.D. Taylor. 2006. Independent effects of fragmentation on forest songbirds: An organism-based approach. *Ecological Applications*, 16(3): 1076–1089.

Chapin, T.G., Harrison, D.J., and D.D. Katnik. 1996. Influence of landscape pattern on habitat use by American marten in an industrial forest. *Conservation Biology* 12(6): 1327–1337.

Cihlar, J., Xiao, Q., Chen, J., Beaubien, J., Fung, K., and R. Latifovic. 1998. Classification by progressive generalization: A new automated methodology for remote sensing multichannel data. *International Journal of Remote Sensing* 19: 3141–3168.

Fraser, R., Olthof, I., and D. Pouliot. 2008. A manual for performing land cover change detection in national parks using Landsat TM and ETM+ satellite data, Automated Multi-temporal Updating by Signature Extension (AMUSE) (Draft report). Ottawa, ON: Canada Centre For Remote Sensing, Natural Resources.

Government of Canada. 2000. Canada National Parks Act. Queen's Writer, Ottawa, ON.

Hanski, I. 1994. A practical model of metapopulation dynamics. *Journal of Animal Ecology* 63: 151–162.

Hanski, I. 1999. *Metapopulation Ecology.* Oxford Series in Ecology and Evolution. Oxford: Oxford University Press.

Hanski, I., and M. Gilpin. 1991. Metapopulation dynamics: Brief history and conceptual domain. *Biological Journal of the Linnean Society* 42: 89–103.

Hockings, M., S. Stolton, and N. Dudley. 2000. *Evaluating Effectiveness: A Framework for Assessing the Management of Protected Areas.* World Commission on Protected Areas. Best Practice Protected Area Guidelines Series No. 6. Gland, Switzerland and Cambridge, UK: IUCN.

Holling, C. S. (ed.). 1978. *Adaptive Environmental Assessment and Management.* Chichester: Wiley.

Johnson, R.D. and E.S. Kasischke. 1998. Change vector analysis: A technique for the multispectral monitoring of land cover and condition. *International Journal of Remote Sensing* 19: 411–426.

Levins, R. 1970. Extinction. In M. Gerstenhaber (ed.), *Some Mathematical Questions in Biology. Volume 2: Lectures on Mathematics in Life Sciences*. Providence, RI: American Mathematical Society, pp. 77–107.

McGarigal, K., and B. J. Marks. 1995. FRAGSTATS: Spatial pattern analysis program for quantifying landscape structure. USDA Forest Service General Technical Report PNW-351. Portland, OR: U.S. Department of Agriculture, Forest Service, Pacific Northwest Research Station.

Opdam, P. 1990. Dispersal in fragmented populations: The key to survival. In R.G.H. Bruce and D.C. Howard (eds), *Special Dispersal in Agricultural Habitats*. London: Belhaven Press, pp. 3–17.

Real, L.A., and J.H. Brown (eds). 1991. *Foundations of Ecology: Classic Papers with Commentaries*. Chicago, IL: University of Chicago Press.

Ritchie, L.E., Betts, M.G., Forbes, G., and K. Vernes. 2009. Effects of landscape composition and configuration on northern flying squirrels in a forest mosaic. *Forest Ecology and Management* 257: 1920–1929.

Rowe, J.S. 1972. *Forest Regions of Canada*, Canadian Forest Service Publication No. 1300. Ottawa, ON: Department of Environment, Canadian Forest Service.

Simpson, J. 2008. *Restoring the Acadian Forest: A Guide to Forest Stewardship for Woodlot Owners in the Maritimes*. Res Tellurius: Canning, Nova Scotia.

Vos, C.C., J.Verboom, P.F.M. Opdam, and C.J.F. Ter Braak. 2001. Towards ecologically scaled landscape indices. *American Naturalist* 157: 24–41.

6

Land-Cover Change and Conservation of Protected Lands in Urban and Suburban Settings

Yeqiao Wang

CONTENTS

6.1 Introduction

Protected lands such as parks, recreational areas, greenways, trails are among important components of conservation efforts in metropolitan regions (e.g., Trzyna 2001; Robinson et al. 2005) and suburban settings (e.g., Radeloff et al. 2005). Conservation and management decisions require information from all aspects of human dimensions and ecosystem concerns (e.g., Gobster and Westphal 2004). During the past few decades, suburban sprawl has dominated the growth of nearly all American metropolitan areas (Johnson 2001). Suburban sprawl, understood as a pattern of low-density development reinforced by a strict separation of land uses, is one of the main forces driving land-cover change. Suburban sprawl often encroaches both on agricultural land and on natural areas for residential and retail development. A recent study reported that 1.4 million hectares of open space was lost to urban sprawl in the United States from 1990 to 2000 (McDonald et al. 2010). Housing growth and its environmental effects pose major conservation challenges (Radeloff et al. 2005). Urban sprawl fragments habitats, isolates populations, and is among the most important human activity stressors reshaping natural processes and most threatening to the sustainability of ecological communities. This is particularly true for protected areas that are situated in urban and suburban environments.

Remote sensing is a proven technology that is effective for mapping and characterizing cultural and natural resources. The multispectral capabilities of remote sensing allow observation and measurement of biophysical characteristics of the landscape. The multitemporal nature of remote sensing allows tracking of changes in landscapes over time. This chapter takes case studies from the Chicago Wilderness project and selected national parks to showcase remote sensing of protected lands in urban and suburban settings, in particular due to human-induced land-use and land-cover change.

6.2 Chicago Wilderness: A Case Study in the Metropolitan Area

6.2.1 About Chicago Wilderness

Within the metropolis survive some of the world's best remaining examples of eastern tallgrass prairie, oak savanna, open oak woodland, and prairie wetland. These natural areas have survived due to the deliberate protection in the past few decades. The "Plan of Chicago," developed in 1909, endorsed a system of forest preserves within the region and recommended the early acquisition of all suitable tracts (CPC 1961). After a century of development, the preserved areas that were originally designed as the outer shelter of the city have now been made inner city parks. The conditions of these communities depend on proper management of more extensive, restorable lands that surround and connect the patches of high-quality remnants.

Chicago Wilderness is more than 81,000 ha of protected areas in national parks, state parks, forest preserve, and conservation districts, as well as in private institutions and corporations in the urban and suburban matrix. It extends in a crescent around Lake Michigan, from southeastern Wisconsin, through Illinois, into northwestern Indiana. These protected lands contain important natural communities or they serve as buffers, protecting and supporting natural areas (Chicago Wilderness 1999). Under the name of Chicago Wilderness, an alliance of diverse and determined institutions, including local, state, and federal government agencies, private landowners, research institutions, and conservation organizations, joined to protect, restore, and manage these globally prominent natural communities.

Like many metropolitan areas in the United States, the Chicago region has experienced dramatic land-cover change during the last 30 years (Wang and Moskovits 2001). Information critical to the success of conservation efforts in the region includes (1) a regional vegetation map in sufficient detail to allow quantitative goal setting for the region's biodiversity recovery plan; (2) quantified fragmentation status of the natural communities; and (3) patterns of land-cover change and their effects on the vitality of communities under threat. In such a vast and complex landscape, biodiversity conservation and urban planning require remote sensing technology that can span spatial and temporal scales to assist in resource evaluation and management.

6.2.2 Data and Information Extraction

The study employed multispectral data from the Landsat Thematic Mapper (TM) and associated ground references to produce the vegetation map. With Landsat data acquired in 1972, 1985, and 1997, the study obtained land-cover maps of the region at roughly equivalent intervals over the past decades.

The generalized land-cover categories in the analysis include urban and built-up areas, agriculture, natural areas, and unassociated vegetation. Natural areas refer to five generalized categories of *upland forest/woodland, savanna/oak-woodland, floodplain forest, prairie,* and *wetland*. The unassociated vegetation has been defined as an individual type of land cover. This category includes both woody and grassy vegetation types and is considered a part of the cultural community. The unassociated woody growth is a mix of shrubs and trees which owe their existence to recent human land-use practices. It is so named because its constituent species do not naturally occur together, either historically, or as associates in long-term self-perpetuating communities. However, all the native constituent species do occur in other natural community types. Most unassociated woody growth develops as woody plants colonize Eurasian meadows, abandoned farm fields, prairies, sedge meadows, or cut-over forest, woodland, and savanna. Other than by comparison with the original natural community that the unassociated woody growth ultimately replaced, there is no standard by which to assess unassociated woody growth. The diversity of herbaceous flora tends to be exceedingly low in the unassociated woody growth, as there are no processes that promote survival of such flora. Without a stabilizing herbaceous layer, the presence of the unassociated woody growth can promote soil erosion and lead to degraded water quality.

The unassociated grassy areas are relatively open fields that have developed after weedy alien and native herbaceous species colonize barren soils or decimated natural areas. Most unassociated grassy areas occur as recolonization of plants in the years following the cessation of agricultural cultivation or grazing. Natural prairie grasses and other grassy categories are the dominant features. If not maintained as open sites, unassociated grassy areas are succeeded by unassociated woody growth. Prairie restoration sites recovering from agricultural practice are considered as unassociated grassy areas.

From the TM-derived regional vegetation map, locations of the protected natural sites were identified and extracted (Figure 6.1). The protected sites included national parks, state parks, preserved areas of the forest preserve and conservation districts of the counties, and the sites designated by the Illinois Natural Area Inventory. Area extents of the five natural categories and the unassociated vegetation within the protected sites were calculated. Figure 6.2 illustrates the protected natural areas on the generalized 1972,

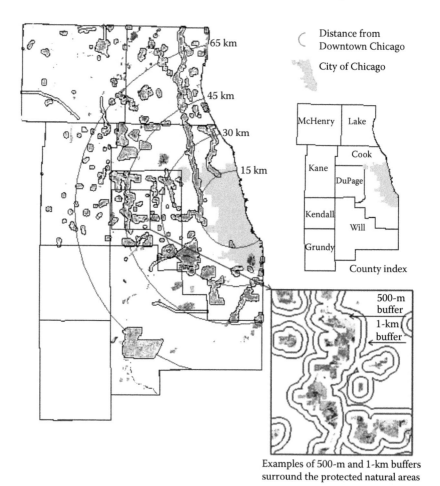

Examples of 500-m and 1-km buffers surround the protected natural areas

FIGURE 6.1

Protected natural areas in the northeastern Illinois counties of the Chicago metropolitan region. The map inset shows the 500-m and 1-km buffer zones around the protected natural areas.

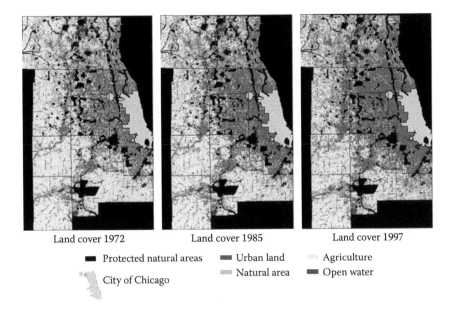

Land cover 1972   Land cover 1985   Land cover 1997

■ Protected natural areas   ■ Urban land   ▒ Agriculture

▒ City of Chicago   ▒ Natural area   ■ Open water

FIGURE 6.2
Land-cover maps of the Chicago metropolitan region in 1972, 1985, and 1997 depicting the land-cover change outside the protected natural areas.

1985, and 1997 land-cover maps. Table 6.1 summarizes the land-cover types and the change that occurred outside the protected areas.

To quantify the effects of urban sprawl on the protected natural areas, a 500-m and a 1-km buffer surrounding the protected sites were created. The 500-m buffer would cover the immediate neighboring areas of the protected sites. The 1-km buffer would cover the site's more general land-cover environment. The land-cover types within the buffer zones were extracted and compared. Changes in area and percentage were calculated. Table 6.2 quantifies the land-cover composition of the region bordering protected natural areas.

TABLE 6.1

Land-Cover and Area Change That Occurred Outside the Protected Natural Areas (in ha)

Category	1972	1985	1997	1972–1985	1985–1997	1972–1997
Urban land	205,217	244,330	317,841	+ 39,113 (19%)	+73,511 (30%)	+112,624 (55%)
Natural area	182,566	160,597	136,877	−21,969 (−12%)	−23,720 (−15%)	−45,689 (−25%)
Agriculture	440,969	332,588	248,151	−108,381 (−25%)	−84,436 (−25%)	−192,817 (−44%)
Unassociated vegetation	53,783	148,019	185,543	+ 94,236 (175%)	+37,523 (25%)	+131,760 (245%)

TABLE 6.2

Change of Neighboring Land-Cover Types of the Protected Natural Areas (in ha)

Land Cover	1972	1985	1997	1972–1985 Change	1985–1997 Change	1972–1997 Change
500-m Buffer area						
Natural area	31,027	27,901	23,530	−3126 (−10%)	−4371 (−16%)	−7497 (−24%)
Unassociated vegetation	7891	20,042	22,712	12,151 (154%)	2670 (13%)	14,821 (188%)
Agriculture	32,831	19,660	13,040	−13,171 (−40%)	−6620 (−34%)	−19,791 (−60%)
Urban land	24,210	28,830	37,845	4620 (19%)	9015 (31%)	13,635 (56%)
1-km Buffer area						
Natural area	58,844	52,659	44,860	−6185 (−11%)	−7799 (−15%)	−13,984 (−24%)
Unassociated vegetation	15,457	39,531	44,094	24,074 (156%)	4563 (12%)	28,637 (185%)
Agriculture	71,126	44,392	30,157	−26,734 (−38%)	−14,235 (−32%)	−40,999 (−58%)
Urban land	49,734	59,522	78,185	9788 (20%)	18,663 (31%)	28,451 (57%)

Besides protected sites, the region has unprotected natural areas. Over the past few years, local preservation agencies have been steadily acquiring land for a variety of purposes. The regional biodiversity recovery plan recognizes that high priority should be given to identifying and preserving important but unprotected natural communities, especially those threatened by development. Protection of areas that can function as large blocks of natural habitat through restoration and management is also among the region's priorities (Chicago Wilderness 1999). The same plan recommends that these identified remaining unprotected natural areas be protected by the expansion of public preserves, by the public acquisition of large new sites, or by actions of qualified private owners. Therefore, identification of these areas and evaluation of their conditions are critically important. These areas are commonly mixed with residential and other types of urban land use, such as commercial, industrial, institutional, transportation facilities, as well as agricultural land use. The unprotected natural areas and their associated land use were extracted from the TM-derived vegetation map coupled with the land-use map of the Northeastern Illinois Planning Commission (NIPC). Table 6.3 summarizes the unprotected natural areas and illustrates how these areas are being used. The information depicts the expansion potential of the region's natural preserves.

6.2.3 Results

Satellite images provided striking visual comparisons of land-use and land-cover patterns. They also provided banks of geographically referenced data that made quantitative tracking of trends possible. The data on habitat degradation and fragmentation provided one of the basic datasets as

TABLE 6.3

Unprotected Natural Areas and Their Associated Land Use (in ha)

Natural Area	Land Use	Area Extent	Total
Woodland	Residential	53,716	78,978
	Commercial/Industrial	1393	
	Office/Institutional	1927	
	Transportation	340	
	Agricultural	3375	
	Recreational	18,227	
Savanna	Residential	12,948	22,646
	Commercial/Industrial	257	
	Office/Institutional	811	
	Transportation	73	
	Agricultural	1322	
	Recreational	7235	
Prairie	Residential	1305	14,973
	Commercial/Industrial	364	
	Office/Institutional	426	
	Transportation	309	
	Agricultural	7930	
	Recreational	4639	
Wetland	Residential	3742	27,359
	Commercial/Industrial	1052	
	Office/Institutional	492	
	Transportation	728	
	Agricultural	3180	
	Recreational	18,165	
Unassociated woody vegetation	Residential	20,085	46,408
	Commercial/Industrial	1311	
	Office/Institutional	2014	
	Transportation	564	
	Agricultural	11,326	
	Recreational	11,108	
Unassociated grassy vegetation	Residential	33,888	130,933
	Commercial/Industrial	3843	
	Office/Institutional	6867	
	Transportation	8236	
	Agricultural	47,723	
	Recreational	30,376	

biological foundation of quantitative goals for regional restoration (Wang and Moskovits 2001).

Most of the region's high-quality natural remnants are already under protection. Among the protected natural areas, the five categories of natural areas cover 36,396 ha, accounting for 44% of the region's natural areas. In addition, 24,771 ha of unassociated vegetation existed within the protected areas. The presence of unassociated vegetation within the protected area indicated both the degradation of natural areas and restoration of natural areas from other types of land use.

Comparison of the 1972, 1985, and 1997 land covers outside the protected area demonstrated that urban land expansion dominated the regional land-cover change (Table 6.1). Between 1972 and 1985, urban land increased by 19%. This trend accelerated to 30% between 1985 and 1997. Within the 25 years from 1972 to 1997, urban land increased 55%. Most of the suburban expansion occurred by consumption of agricultural land in the outer ring and the collar counties. Agricultural land declined 25% for each of the time periods from 1972 to 1985 and from 1985 to 1997. Most of the lost agricultural land was converted into urban land use. Overall, agricultural land decreased 44% from 1972 to 1997. Natural areas also declined 12% from 1972 to 1985, and 15% from 1985 to 1997. Within the 25 years, 25% of the natural areas outside the protected sites were converted into other types of land use.

Most of the urban land expansion occurred between 30 and 65 km away from downtown Chicago. This belt is the home of many small- and medium-sized natural preserves (Figure 6.1). Dramatic rates of residential and commercial developments are evident in this belt. Examples include the residential and commercial developments in the northwest Cook County, the southern part of Lake and McHenry Counties, most part of DuPage County, the eastern Kane County, and the northern Will County (Figure 6.2).

Data in Table 6.1 reveal a vast increase of the unassociated vegetation areas. Most of the unassociated woody growth resulted from woody plants colonizing Eurasian meadows, abandoned farm fields, prairies, sedge meadows, or cut-over forest, woodland, and savannas. The unassociated grassy areas resulted from recolonization of plants in the years following the cessation of agricultural cultivation or grazing.

6.2.4 Discussion

The region's natural communities within the protected area are severely threatened by the effects of urban sprawl. The land-cover types immediately adjacent to protected natural areas, as within the 500-m buffer zone, changed dramatically during the 25 years. This is an irreversible conversion of the environment neighboring on the region's natural preserves. On the other hand, the agricultural areas directly connected to the natural preserves declined 40% between 1972 and 1985 and then an additional 34% between

1985 and 1997. The areas of unassociated vegetation increased dramatically during the time period. This increase altered the environment surrounding the natural preserves as well. The areas of unassociated vegetation are more likely to be developed for future urban land use.

A similar pattern of change was observed in the 1-km buffer zone surrounding the protected sites (Table 6.2). The altered land-cover composition and the increased proportion of urban land indicated that the region's protected natural areas were exposed to manmade environment more frequently and were facing greater threats from the neighboring urban landscape.

A direct result of the suburban sprawl is the fragmentation of natural areas. Fragmentation particularly threatens the once widespread prairie, savanna, woodland, and upland forest communities. The effects of fragmentation include not only the partitioning of sites but also what happens in the remaining small and isolated patches. Roads and areas of human occupation divide up the natural areas and affect the community in a number of ways, including altering gene flow, increasing predation, and increasing opportunities for invasive species. Fragmentation is a particular problem for animal species, most notably grassland and forest birds that can only breed successfully in large, contiguous habitat blocks. The site of Springbrook Natural Preserve, shown in Figure 6.3/Site 3, illustrated a typical example of this impact. DuPage County Forest Preserve District purchased this 688 ha piece of land in 1973 and started a prairie restoration project on the abandoned farmlands. At that time, the site was in the midst of agricultural fields. Twelve years later, in 1985, the preserve was still surrounded primarily by agricultural fields (Figure 6.3a) supportive of prairie animals and birds. But another 12 years later, in 1997, the preserve had become an island in a sea of newly developed residential subdivisions and commercial service areas (Figure 6.3b). New infrastructures, such as the I-355 tollway (Figure 6.3/Site 1) and the DuPage County airport (Figure 6.3/Site 2), were built between 1985 and 1997. Nature preserves in this subset area have been almost completely isolated by the dramatically increased suburban residential developments. Many other natural preserves in the region face similar threats from suburban sprawl.

Increase of management practices and development of biodiversity planning for the remaining unprotected natural areas have been recognized as an important component of the regional biodiversity conservation (Chicago Wilderness 1999). For some community types, such as woodlands, savannas, and wetlands, a substantial portion of these areas lie on privately owned lands. Most of the unprotected natural areas are associated with other types of urban land use. Table 6.3 lists the areas' extent and their links. The data indicated that most of the region's remaining unprotected woodland, savanna, wetland, and prairie areas are associated with residential land use. Some patches are associated with recreational and conservation parks (arboretums, botanical gardens, and golf courses), or with office and institutional

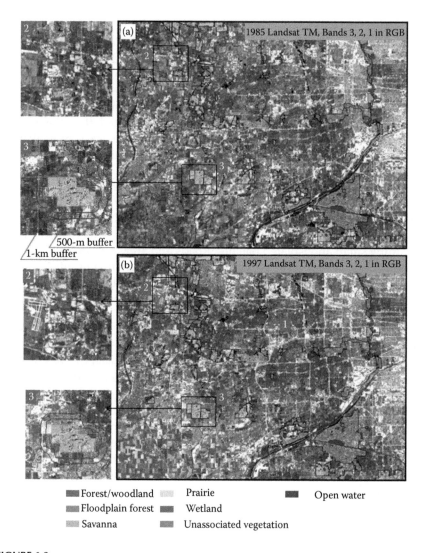

Forest/woodland   Prairie   Open water
Floodplain forest   Wetland
Savanna   Unassociated vegetation

FIGURE 6.3
(See color insert.) Comparative display illustrating land-cover change between 1985 and 1997 in the west suburb of Chicago. Site 1 depicts the newly built I-355 tollway which is identifiable in the 1997 image but does not exist in the 1985 image. Site 2 illustrates the area around the West Chicago Natural Preserve and the nearby constructed DuPage County airport. Site 3 indicates the impact of the newly developed residential area on the Springbrook Natural Preserve. Most protected natural areas in this subset are isolated and fragmented by urban land.

land use (college campus and office complexes), or with commercial and industrial land use (industrial parks), or with transportation facilities. These areas are unlikely to change their ownership. Strategies need to be developed to work with various landowners to protect and manage these remaining natural communities.

The value of unassociated vegetation should not be ignored. Table 6.3 shows that a large amount of unassociated woody and grassy vegetation exists outside of the protected areas. Some of these areas have potentials for restoration to natural area. These areas may also provide easement buffers or function as corridors to connect the protected natural sites.

The results indicate that wholesale conversion of land was not the only threat to the natural communities, but that fragmentation, isolation, and the quality of the matrix were equally important and must be considered. With a clear understanding of past patterns of change, effective plans and decisions could be made for the conservation of biodiversity in the region.

6.3 National Parks in Urban and Suburban Areas

Information about the magnitude and pattern of land-cover change helps understand the landscape context of national historical parks (NHPs) and national historical sites (NHSs) and offers a better understanding of how park ecosystems fit into the broader landscape. In particular, the inventory and monitoring networks of the National Park Service (NPS) identified landscape dynamics and land-use and land-cover change as one of the vital signs to implement long-term monitoring of natural resources for the parks that contain significant natural resources (Fancy et al. 2009; Gross et al. 2011— Chapter 2 of this book).

Minute Man NHP was established in 1959 to consolidate, preserve, and selectively restore and interpret portions of the Lexington-Concord Battle Road in Concord, Lincoln and Lexington in Massachusetts. The 3 km² park is the place where the opening battle of the American Revolution was initiated. The primary resource management objectives of the park are to preserve cultural resources and reestablish the historic landscape. Cultural resources at Minute Man consist of buildings, monuments, and archaeological sites, and the historic landscape consists of fields, forests, and wetlands. Within a suburban setting, invasive exotic plants are among natural and cultural resource management concerns due to their impacts on natural communities and the cultural landscape.

Morristown NHP consists of about 7 km² distributed across four geographically separate units in New Jersey. The NHP was established in 1933 as the first national historical park to preserve the lands and resources associated with the winter encampments used by the Continental Army during the Revolutionary War. Vegetation in the park is dominated by a mix of mowed fields, orchards, planted gardens, and forest stands. Changing land-use patterns have dramatically altered the character of the area from farmed or hardwood forested areas intersected by streams to low-density residential development, expanding networks of roads, and commercial and recreational development.

Roosevelt–Vanderbilt NHS consists of three sites including the Eleanor Roosevelt Mansion, the home of Franklin D. Roosevelt, and the Vanderbilt Mansion in Hyde Park, New York. The sites, totaling about 3 km², are located within 5 km of each other in the Eastern Great Lakes and Hudson Lowlands ecoregion. Both the home of Franklin D. Roosevelt and the Vanderbilt Mansion border on the Hudson River. The presence of the river brings a marine influence far inland, resulting in unique plant communities and animal species otherwise uncommon to the region. Significant threats from exotic species are affecting the natural and cultural landscapes.

Saratoga NHP was established in 1938 to commemorate the first significant American military victory during the Revolutionary War. The Battles of Saratoga are considered by some to rank among the 15 most decisive battles in world history. The park consists of three separate units of a battlefield and other monumental sites. Forests dominate the Saratoga landscape. Grasslands, brush/shrub areas, and wetlands make up the rest of the park landscape. The Hudson River floodplain and associated streams and wet meadows support unique habitats within and around the park.

Those parks are examples of many that are relatively small in area and exposed to historically or newly developed urban and suburban environment. Conservation and protection of the parks' resources and ecosystems are dependent not only on the management of protected parks but also on the land-use practices in adjacent neighboring communities.

Remote sensing can be used to provide the baseline data. A recent study employed Landsat remote sensing data for three time periods of the early 1970s, mid-1980s, and 2002 to quantify land-cover change within and adjacent to the above parks using selected buffer areas and to reveal patterns of land-cover change of the parks and the neighboring areas.

6.3.1 Data and Information Extraction

The study selected Landsat data that represented the best match in time frame and, if possible, selected to the anniversary of image acquisition in order to reduce seasonal effects. The study employed supervised, unsupervised, and stratified classifications for obtaining the land-cover data. Multiple training signatures were selected to represent the spectral variations for land-cover types.

The NPS vegetation mapping project utilizes the national vegetation classification standard as the classification scheme. The maps consisted of vector Geographic Information System (GIS) data that defined boundaries of vegetation types based on plot information and manual delineations from aerial photos. Roosevelt–Vanderbilt NHS and Saratoga NHP were among the park units that had available NPS vegetation mapping project data at the time of image classifications. For these two parks a stratified classification technique was used in land-cover classification so that the NPS vegetation mapping data can be referenced (Figure 6.4).

FIGURE 6.4
Vegetation maps of the three sites of the Roosevelt–Vanderbilt National Historic Site (NHS) displayed on top of the Landsat thematic mapper image. The Landsat imagery data reveal adjacent landscape context of this NHS.

Stratification involves segmentation of an image into focused areas and categories based on existing GIS land-cover data in order to improve a classification (Wang et al. 2007). NPS vegetation mapping data were referenced as the baseline to mask corresponding pixels for each vegetation/land-cover category from the subset of Landsat images, so that the follow-up classifications were focused on each of the vegetation/land-cover types identified by the vegetation mapping projects. The stratified classification allowed identification and reduction of conflicts between data products as well as used the NPS vegetation mapping data to monitor landscape context (Wang et al. 2009).

After standard process in accuracy assessment, the post-classification comparison method was adopted to obtain changes in areas of land-cover type. Post-classification comparison involves independently produced spectral classification data from each end of the time interval of interest, followed by comparison of data to detect changes in cover type (Coppin and Bauer 1996;

Mas 1999; Coppin et al. 2004). The principal advantage of post-classification comparison is that the images are separately classified, thereby minimizing the problem of radiometric calibration between dates (Song et al. 2001) and reducing the amount of data preprocessing. By choosing an appropriate classification scheme, post-classification comparison can be made insensitive to a variety of transient changes. With appropriately developed land-cover maps in separate time periods, the class changes during the time interval and transition rate between classes can be calculated (Hall et al. 1991).

Spatial buffers were established to extract land-cover and landscape context information in different buffer zones. The buffering analysis provided land-cover and landscape context information of within-the-park boundaries and the zones within 500 m, 1 km, and 5 km of the park boundaries.

6.3.2 Results

The primary objectives of resource management at Minute Man NHP are to preserve the cultural resources and reestablish the historical landscape. The data show an increasing trend of urban area both within the park boundary and the adjacent buffer zones, in particular during the time period between 1974 and 1987. Decline of forests is also evident, in particular between 1974 and 1987 (Table 6.4). Changing land-use pattern around Morristown NHP has dramatically altered the character of the area from farmed or hardwood forested areas intersected by streams to low-density residential development, expanding networks of roads, and commercial and recreational development. The data show that although urban land did not change much within the park boundary, the increase was significant in all the buffer zones, particularly in the time period between 1976 and 1988 (Table 6.4). Major development adjacent to the park occurred between 1975 and 1988. As the available land became saturated, the change in urban area slowed down between 1988 and 2002 within all the buffer zones. Urban area increased from 40 ha in 1973 to 100 ha in 2002 within Roosevelt–Vanderbilt NHS. More significant increases of urban land occurred outside the park in all buffer zones. Although deciduous forests increased, coniferous and mixed forests showed a declining pattern, and overall forested area also declined. For Saratoga NHP, the sequence of the park's land acquisition and land-use history has produced a mosaic of old field, shrub, and forest communities. As with many of the other parks, significant urban development occurred in all buffer zones, and forest area declined.

6.3.3 Discussion

Landsat images represent snapshots of the landscape at the time of image acquisition. Many factors on the ground may affect the quality of the image and the landscape features recorded. The complexity of the landscape of the study areas and image processing and classification processes may affect the

TABLE 6.4

Comparison of Urban and Forest Lands within Park Boundaries and Buffer Zones Adjacent to Parks

NHP	Land Cover	Within Park Boundary (in ha)			500-m Buffer Adjacent Park (in ha)		
		1970s	1980s	2000s	1970s	1980s	2000s
MIMA	Urban	6	45	61	106	324	420
	Forest	345	266	243	876	624	524
MORR	Urban	0	8	6	70	272	280
	Forest	671	630	659	1095	823	883
ROVA	Urban	40	81	100	74	197	235
	Forest	409	384	324	664	589	507
SARA	Urban	1	5	5	32	66	80
	Forest	623	717	583	566	401	323
		1-km Buffer Adjacent Park (in ha)			5-km Buffer Adjacent Park (in ha)		
		1970s	1980s	2000s	1970s	1980s	2000s
MIMA	Urban	264	724	913	3021	9155	11,466
	Forest	1846	1302	1124	28,535	19,296	17,935
MORR	Urban	191	735	763	951	5127	5704
	Forest	2323	1602	1728	15,321	9108	9432
ROVA	Urban	209	413	490	1093	2239	2528
	Forest	1628	1507	1290	11,676	10,914	9429
SARA	Urban	55	104	129	246	434	547
	Forest	1341	952	783	13,864	10,451	8768

accuracy of the land-cover maps. Therefore, the land-cover maps generated by such data and technology may differ from the understanding of the landscape by resource managers, because the snapshot view may sometimes capture atypical transient conditions.

The supervised and stratified classifications are appropriate approaches for extracting land-cover information from Landsat sensor data. NPS has made a huge investment in vegetation mapping projects across the parks. The resulting products provide valuable information for resource management and reliable guidance for extended classification of Landsat remote sensing data. With the availability of NPS vegetation mapping project data, future landscape characterization should take advantage of this important investment. The stratified classification, that is, the local-to-global approach, is a practical tool for involving NPS vegetation mapping project data into characterization of landscape dynamics.

The post-classification change detection is a common and justifiable approach for this type and scope of study. Such an approach is particularly merited when multiple Landsat sensor data types (Multispectral Scanner,

Thematic Mapper, and Enhanced Thematic Mapper plus) are used. It also reduces the significant amount of workload for preprocessing atmospheric corrections of Landsat data, which would tax the image processing capacity of park managers. However, it is needed to acknowledge the risk of error propagation among individually classified land-cover maps in different time periods, in particular for the categories that have low agreement with reference data in the accuracy assessment. Detected changes could result from erroneous pixels due to misclassification, misinterpretation, or misregistration on original remote sensing images and the derivatives of thematic information. Consistency is another challenge. Since the multidate thematic data could be generated by different individuals, differences between classification or interpretation procedures may eventually cause a false change detection result. These factors must be considered when conducting a postclassification change analysis.

6.4 Concluding Remarks

In his 1961 message to the congress on housing, President John F. Kennedy emphasized the need for preserving open space in metropolitan areas. He said, "Land is the most precious resource of the metropolitan area. The present patterns of hazard suburban development are contributing to a tragic waste in the use of a vital resource now being consumed at an alarming rate" (NIPC 1962: 358). As reported by the Northeastern Illinois Planning Committee (NIPC 1999), the population and employment in the metropolitan Chicago region grew by 4% and 21%, respectively, between 1970 and 1990. Forecasts indicate that by the year 2020 the population of the region will grow by 25% while the employment will increase by 37% from 1990. This forecast illustrates that the rate of growth in DuPage and northeast Cook Counties will slow and the growth rate in Lake, McHenry, Kane, and Will Counties will increase substantially. With the current rate of land-cover change and the forecasted pattern of population and employment growth, the region's protected natural areas will come under tremendous pressures. Simulation of the region's landscape demonstrates a trend of dramatic increase of interface between natural areas and urban land, as well as between natural areas and the unassociated vegetation. If the rate of consumption of open land was alarming four decades ago, as acknowledged by President Kennedy, the current rate of land-cover change and the future trends of the change could approach to the catastrophic.

Not everyone agrees that the present pattern of metropolitan development is a problem. Some feel that decentralization is just part of the natural, efficient evolution of cities, as residents and business express their preferences in a free market (Gordon and Richardson 1997). Regardless of whether one believes that the current patterns of metropolitan decentralization and growth are desirable, inevitable, or problematic, there is little doubt that there are a variety of negative effects, particularly on urban biology and biodiversity. These effects have given rise to a range of efforts to change the development patterns, or at least to mitigate the negative consequences (Wiewel and Schaffer 2001).

As pointed out by Illinois's Critical Trends Assessment Project (CTAP 1994), humans already have become so ecologically dominant in the region that it is impossible to draw a clear line separating natural systems from the social, economic, technological, and political systems that influence them. As natural systems are more and more affected by human activity, government incentive programs will become more important. One unambiguous trend over the past years has been the increased number of public programs to moderate the human environmental impact (CTAP 1994). That project also states that habitat fragmentation and other physical changes have surpassed conventional pollution as threats to ecosystem functioning in Illinois. The status of Chicago region's natural preserves echoes trends in the state. CTAP says that the State "may be said to be moving away from complex natural systems toward simpler ones, from stable systems toward unstable ones, from native species toward non-native ones, from integrated systems toward fragmented ones, from preserved systems to restored or created ones" (1994, p. 89).

The suggestions from the Openlands Project (1999) include the establishment of programs and provision for greater funding to protect resource-rich lands and to guide land acquisition decisions. It also suggests the establishment of a state office of planning and land conservation to develop and implement state land-use goals and strategies. Municipalities should update comprehensive plans and zoning and subdivision ordinances to require open space and resource protection as a first step in site design.

In the epilogue of his 2005 book, Bruce Babbitt, the former U.S. Secretary of the Interior, argues that the fundamental issue facing public lands is that we have yet to reach consensus as to their ultimate placement on the use spectrum from cities to wilderness. Public lands should now be administered primarily, although not exclusively, to maintain and restore their natural values. He continues that a basic measure of good land use is sustainability, a word that has come to signify living in a respectful relationship with the land, passing it on unimpaired, and even renewed and restored, to future generations. Development should enlarge the possibilities for human progress, creativity, and quality of life, which it cannot accomplish by continually eroding the beauty and productivity of the natural world (Babbitt 2005). Informed management and conservation of protected lands in urban and suburban settings are one step toward the ultimate goals as Babbitt envisioned.

There is no single solution to the problems associated with suburban sprawl. The data presented for the Chicago Wilderness case study represent an effort to delineate and clarify understanding of the problem. Information is vital to sound decision-making at all levels. Remote sensing and the related information technology have been playing and will continue to play critical roles in the development of the understanding of natural resource monitoring and management.

The case studies of the NHP/NHS identified several potential points for improvement or future studies. Although in general the change analyses are informational, for smaller parks especially those situated within urban centers and suburban areas, finer spatial resolution satellite and/or aerial remote sensing data may provide specific spatial details to meet the needs of small park resource management. Certain buffer size surrounding parks should be considered for future aerial data acquisition. The buffer areas adjacent to the parks can make the mapping results and assessments more relevant to the landscape context and land-use patterns of the surrounding communities, as well as help to identify where the major stressors are located, and assist resource managers and the nearby communities in management planning.

Acknowledgments

The original research for the Chicago Wilderness case study was supported by U.S. National Aeronautics and Space Administration (NASA) (Grant No. NAG5-8829) and the land-cover change detection project (NASA Grant No. GP37J) and upon which the regional vegetation map and land-cover change data were developed. The GP37J project was a collaborated effort of an enormous group of ecologists and land managers in the Chicago metropolitan region and the Chicago Wilderness. Laura Barghusen and Ruth Harari-Kremer performed postclassification refinement of the Chicago Wilderness land-cover data. Kiven Zhang and Jason Yang helped in GIS data preparation. Other individuals contributed significantly to the efforts, in particular Debra Moskovits, Stephen Packard, Wayne Lampa, Lisle Burns, and Wayne Schennum, for the expertise and insights about regional ecosystems and issues in conservation. The original research for the case studies of the national parks was funded by the Northeast Temperate Network of the NPS Inventory & Monitoring program.

References

Chicago Wilderness. 1999. *Biodiversity Recovery Plan: A Final Draft for Public Review*. Chicago Wilderness, Chicago, 190p, http://www.chicagowilderness.org/pdf/biodiversity_recovery_plan.pdf.

Coppin, P., I, Jonckheere, K. Nackaerts, B. Muys, and E. Lambin. 2004. Digital change detection methods in ecosystem monitoring: A review. *International Journal of Remote Sensing*, 25(9), 1565–1596.

Coppin, P.R., and M.E. Bauer. 1996. Digital change detection in forest ecosystem with remotely sensed imagery. *Remote Sensing Review* 13: 207–234.

CPC. 1961. *The Chicago Plan Commission: A Historical Sketch, 1909–1960*. Chicago: Department of City Planning, 41pp.

CTAP. 1994. *The Changing Illinois Environment: Critical Trends. Summary Report of the Critical Trends Assessment Project*. Chicago: Illinois Department of Energy and Natural Resources and the Nature of Illinois Foundation, p. 89.

Babbitt, B., 2005. *Cities in the Wilderness: A New Vision of Land Use in America*. Washington D.C.: Island Press, 200pp.

Fancy, S.G., J.E. Gross, and S.L. Carter. 2009. Monitoring the condition of natural resources in US national parks. *Environmental Monitoring and Assessment*, 151: 161–174.

Gobster, P.H., and L.M. Westphal. 2004. The human dimensions of urban greenways: Planning for recreation and related experiences. *Landscape and Urban Planning*, 68(2004): 147–165.

Gordon, P., and H.W. Richardson. 1997. Are compact cities a desirable planning goal? *Journal of the American Planning Association*, 63: 95–106.

Gross, J.E., A.J. Hansen, S.J. Goetz, D.M. Theobald, F.M. Melton, N.B. Piekielek, and R. Nemani. 2011. Chapter 2—Remote sensing for inventory and monitoring of the U.S. national parks. In Y. Wang (ed.), *Remote Sensing of Protected Lands*. Boca Raton, FL: CRC Press.

Hall, F.G., D.B. Botkin, D.E. Strebel, K.D. Woods, and S.J. Goetz. 1991. Large-scale patterns of forest succession as determined by remote sensing. *Ecology*, 72, 628–640.

Johnson, E.W. 2001. *Chicago Metropolis 2020: The Chicago Plan for the Twenty-First Century*. Chicago: University of Chicago Press, 188pp.

Mas, J.F. 1999. Monitoring land-cover changes: A comparison of change detection techniques. *International Journal of Remote Sensing*, 20(1): 139–152.

McDonald, R., R. Forman, and P. Kareiva. 2010. Open space loss and land inequality in United States' cities, 1990–2000. *PLoS One*, 5(3): e9509.

NIPC. 1962. *Open Space Policies for Northeastern Illinois: A Statement Adopted by the Northeastern Illinois Metropolitan Area Planning Commission*. Chicago: NIPC, p. 358.

NIPC. 1999. *Population, Household and Employment Forecasts for Northeastern Illinois: 1990 to 2020*. Chicago: NIPC, 14pp.

Openlands Project. 1999. *Pressure: Land Consumption in Chicago Region, 1998–2028*. Openlands Project, Chicago, Illinois. 31 pp.

Radeloff, V.C., R.B. Hammer, and S.I. Stewart. 2005. Rural and suburban sprawl in the U.S. midwest from 1940 to 2000 and its relation to forest fragmentation. *Conservation Biology*, 19(3): 793–805.

Robinson, L., J.P. Newell, and J.M. Marzluff. 2005. Twenty-five years of sprawl in the Seattle region: Growth management responses and implications for conservation. *Landscape and Urban Planning*, 71: 51–72.

Song, C., C.E. Woodcock, K.C. Seto, M.P. Lenney, and S.A. Macomber. 2001. Classification and change detection using Landsat TM data: When and how to correct atmospheric effects? *Remote Sensing of Environment*, 75, 230–244.

Trzyna, T. 2001. California's urban protected areas: Progress despite daunting pressures. *Parks: The International Journal for Protected Area Managers*, 11(3): 4–15.

Wang, Y., and D.K. Moskovits. 2001. Tracking fragmentation of natural communities and changes in land cover: Application of Landsat data for conservation in an urban landscape (Chicago Wilderness). *Conservation Biology*, 15(4): 835–843.

Wang, Y., M. Traber, B. Milestead, and S. Stevens. 2007. Terrestrial and submerged aquatic vegetation mapping in Fire Island National seashore using high spatial resolution remote sensing data. *Marine Geodesy*, 30(1): 77–95.

Wang, Y.Q., B.R. Mitchell, J. Nugranad-Marzilli, G. Bonynge, Y. Zhou, and G.W. Shriver. 2009. Remote sensing of land-cover change and landscape context of the national parks: A case study of the Northeast Temperate Network. *Remote Sensing of Environment* 113(7): 1453–1461.

Wiewel, W., and K. Schaffer. 2001. Learning to think as a region: Connecting suburban sprawl and city poverty. *European Planning Studies*, 9(5): 593–611.

7

Land-Cover Change and Conservation of the Protected Ancient City Park in Xi'an, Northwestern China

Wenfang Tao, Qingri Chang, and Yeqiao Wang

CONTENTS

7.1 Introduction

Xi'an is among the oldest cities with over 3100 years of history and has served as the capital city for 13 dynasties of ancient China. At the geographic center of China, Xi'an deserves the title of a natural history museum. The city has preserved the largest ancient city wall and over 100 km² of protected significant sites from early dynasties of Zhou (1046–256 BC), Qin (221–206 BC), Han (206 BC to AD 220), and Tang (AD 618–907). As part of the economic revival of interior regions under China's recent "Western Development" policy, Xi'an has reemerged as one of China's major cities with a population of over 7 million. The urbanization process has profoundly changed the spatial patterns of the landscape. Protection and conservation of historical landmarks of ancient civilizations and the development of a new city in this world famous ancient capital represent a significant challenge.

Urban development is usually associated with conversion of rural land into residential and commercial land use. Urban sprawl is a very significant environmental concern in most parts of the world, especially in densely populated countries such as China. With 30% of its 1.3 billion inhabitants residing in urban areas and a projection that more than 50% of the population will be in urban areas by 2030 (United Nations 2001), the urban landscape in China has been in the past, and will be, expanding at a rapid rate (Yu and Ng 2007). With such growth, cities exert a heavy pressure on lands and resources at their periphery (Leao et al. 2004). As the extent of built-up land increases, further development generally raises concerns about impacts of land-use and land-cover change (LULCC) on urban and rural environmental conditions and on quality of life. Spatial distributions and LULCC patterns often affect socioeconomic status (Douglass 2000), environmental conditions (Wang and Moskovits 2001; Gillies et al. 2003), and regional climatic variations (Arnfield 2003; Kalnay and Cai 2003; Voogt and Oke 2003). Monitoring of spatial–temporal changes in large urban and suburban settings has been the research focus in the past years and is becoming increasingly important (Small 2001). The ability to monitor urban LULCC is necessary to provide detailed information for understanding of anthropogenic impacts on the landscape and environments (Carlson 2003).

As the largest city in Northwestern China, Xi'an has witnessed rapid growth both in extent and population in the last two decades. Xi'an benefited economically from China's "Western Development" policy, which aims at the development of the western regions of China and at bridging the growing economic disparity between the coastal and interior regions. According to the Eleventh National Five-Year Plan, the Guan-Zhong region is one of the three major economic zones in western China along with Chengdu-Chongqing and the northern Gulf Economic Zone. Xi'an plays a significant role for economic and social development in the Guan-Zhong economic zone.

Because of the need for information to assess LULCC, different methodologies and data resources have been continuously explored. Techniques that have been explored for characterization and quantification of urban areas include both ground-based measurements and remotely sensed observations (Bauer et al. 2007). Remote sensing is particularly effective in monitoring spatial distribution and patterns of urban LULCC (Sudhira et al. 2004; Yang and Liu 2005). Remote sensing has been well documented for characterizing landscape (Jensen 1996), for studying urban environments (Seto and Liu 2003; Yang and Lo 2003; Helmer and Ruefenacht 2005), for urban growth modeling (Herold and Goldstein 2003; Seto and Kaufmann 2003; Wilson et al. 2003), and patterns of urban sprawl (Wang and Moskovits 2001; Sutton 2003). Data from multiple resources such as population density and distributions have been integrated with remotely sensed data in a variety of studies (Hutchinson 1982; Mesev 1998; Radeloff et al. 2000).

Change analysis has been supported through image-to-image or post-classification comparisons (Gomarasca et al. 1993; Green et al. 1994; Yeh and Li 2001). Approaches include spectral mixture analysis (Wu and Murray 2003; Wu 2004; Lu and Weng 2006), regression tree modeling (Yang et al. 2003a; Yang et al. 2003b; Xian and Crane 2005), subpixel classification (Civco et al. 2002; Wang and Zhang 2004), and neural network classifications (Civco 1993).

These studies demonstrated that the Landsat type of remote sensing data can be used for mapping and monitoring urban built-up areas and their changes.

Socioeconomic information, for example, household size as an indicator of occupants of a single dwelling house, is an important factor for studying urban environment and related LULCC (Liu et al. 2003). With economic development and consumption of land resources, changing household size is considered one of the important driving factors affecting biodiversity (Pimm et al. 1995; Cincotta et al. 2000). Changing living patterns, such as decreasing fertility rate, increasing per capita income, aging population, declining occupation of dwelling house by multigeneration family, may contribute to the reduction in household size. Changing pattern of household size may lead to higher per capita resource consumption. Reduction in average household size was an important factor causing the increase of household number and thus generally increasing the amount of land and materials needed for housing construction (Liu et al. 2003).

In this chapter, we present a case study that employed remote sensing data to reveal spatial pattern of urban development in the past decades using Landsat Thematic Mapper (TM) data and the derivatives of land-cover maps. We made comparisons of urban expansions over time, evaluated factors behind the urban land-cover change, and studied the effects on the protected ancient city park in Xi'an.

7.2 Methods

7.2.1 Study Area

Xi'an has an area of about 828 km² and is the capital and the largest city of the Shaanxi Province in Northwestern China. Xi'an was not only the cultural and political center of the early Chinese empires for almost 1,000 years, but also the beginning terminus of the Silk Road connecting the East and the West. It has been one of the most popular tourist destinations, famous for the astounding tomb of the Qin Emperor and the Army of the Terracotta Warriors. Xi'an is situated in the Guanzhong Plain, with borders on the Loess Plateau to the north and the Qinling Mountains to the south. It is in the temperate zone with a monsoon climate. The annual average temperature is between 9°C and 14°C, and the annual precipitation is about 580 mm. The study area included six administrative districts of Weiyang, Lianhu, Xincheng, Beilin, Baqiao, and Yanta. In this study, we emphasized on assessment of urban built-up area and the land-cover changes (Figure 7.1).

7.2.2 Data

We used three sets of Landsat TM images acquired on August 23, 1988, May 7, 1995, and July 24, 2006. The TM data possess a 30-m spatial resolution

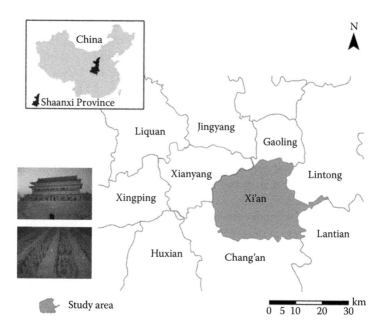

FIGURE 7.1
Location of the study area.

and multiple spectral bands. The three images were georeferenced to the Universal Transverse Mercator map coordinates. We also collected population data within the area as a basic attribute. We referenced data from online open resources such as those with authoritative statistical and spatial data for the area of interest.

7.2.3 Land-Cover Mapping and Change Analysis

A variety of methods in remote sensing change detection has been developed and evaluated (Rogan and Roberts 2002; Woodcock and Ozdogan 2004; Healey et al. 2005). We followed the general procedures of (1) image preprocessing, (2) image classification, (3) accuracy assessment, (4) postclassification processing, and (5) related analysis.

In the preprocessing stage we used a 1:700,000 administrative map of the Shaanxi Province as a reference for extracting the areas of the administrative districts involved in this study from Landsat TM imageries (Figure 7.2). We performed supervised classification on the subset images. We selected multiple training signatures to represent the spectral variation for the identified representative land-cover types. The study area contains major rivers and channels passing through the city center and the nearby sites. As commonly understood, water areas may share similar spectral features with certain

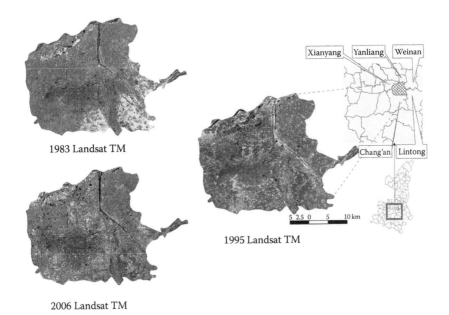

1983 Landsat TM

1995 Landsat TM

2006 Landsat TM

FIGURE 7.2
Landsat imageries of the study area in 1988, 1995, and 2006.

urban pavements and shadows in Landsat data. To improve the classification accuracy we removed the water areas from the TM image using delineated area of interest. We performed supervised classification on the water-removed subset images. We then mosaicked classified water areas back and recoded the land-cover classes into four general categories of urban, water, forest, and agriculture.

We repeated the procedures for classification of the 1988 and 1995 TM images to obtain generalized land-cover maps (Figure 7.3). Mixed pixels of land-cover categories can have a direct influence on statistical variation of each land-cover class. The problem of mixed pixels is difficult to overcome due to the spatial resolution and the accuracy of classification. We made efforts to reduce the logical errors by using a rule-based model.

Accuracy assessment is critical for maps generated by remote sensing. Methodologies in accuracy assessment and related concerns have been addressed in numerous studies (Lunetta and Lyon 2004; Congalton and Green 2008). We used random sampling to select reference pixels for evaluation of the classification accuracy. We compared random reference points selected from classified land-cover types with available referencing data, such as open-source data and the fine resolution imagery. An evaluation of the error matrix included using simple desecrate descriptive statistics of overall accuracy and producer's, and user's accuracies, and statistical-based measures of agreement by κ coefficient.

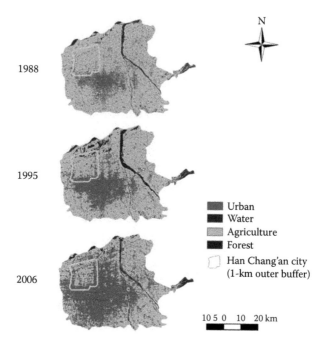

FIGURE 7.3
Land-cover maps of the study area derived from classification of Landsat imageries.

Change detection is the process of identifying differences in the state of an object or phenomenon by observing it from data at different times (Singh 1989). In this study, we used the postclassification comparison method to obtain changes in areas of land-cover types.

7.3 Results

The overall accuracy and the user's and producer's accuracies report the contribution of classification errors from each land-cover type for each time period (Table 7.1). The classification of urban and water areas achieved lower accuracies than that of agriculture and forest categories. The land-cover map of 1995 achieved the highest overall accuracy of 95.70%, followed by the 1988 map with an accuracy of 94.92%, and the 2006 map with an accuracy of 92.19%.

As shown in Table 7.2, the land cover experienced significant changes in the last two decades. Overall, this region showed a large increase in urban area and decrease in agricultural area between 1988, 1995, and 2006. Agricultural area was the most abundant land-cover type in all three time periods with about 666 km² in 1988, 565 km² in 1995, and 420 km² in 2006. Meanwhile, the urban land cover increased 205% between 1988 and 2006. During this

TABLE 7.1

Accuracy Assessments for the Land-Cover/Land-Use Data Developed from Classification of Landsat Data for Xi'an in 1988, 1995, and 2006

	Reference Data					
	Urban	Water	Agri	Forest	Total	User's Accuracy
1988						
Urban	**32**	2	0	0	34	94.12%
Water	2	**14**	0	0	16	87.50%
Agriculture	5	1	**188**	2	196	95.92%
Forest	0	0	1	9	10	90.00%
Total	39	17	189	11	256	
Producer's accuracy	82.05%	82.35%	99.47%	81.82%		Overall 94.92%
1995						
Urban	**39**	5	1	0	44	88.64%
Water	3	**8**	1	0	12	66.67%
Agriculture	1	1	**193**	0	195	98.97%
Forest	0	0	0	5	5	100.00%
Total	43	14	195	5	256	
Producer's accuracy	90.70%	57.14%	99.48%	100.00%		Overall 95.70%
2006						
Urban	**80**	7	1	0	88	90.91%
Water	1	**15**	0	0	16	93.75%
Agriculture	7	4	**131**	0	142	92.25%
Forest	0	0	0	10	10	100.00%
Total	88	23	132	10	256	
Producer's accuracy	90.91%	57.69%	99.24%	100.00%		Overall 92.19%

time period the region lost about 246 km² (40%) of agricultural and 4.7 km² (36%) of forest areas to urban land cover. The postclassification comparison identified other land-cover types converted to urban land use (Tables 7.2 and 7.3). Quantitative analyses revealed that about 108 km² of agricultural land were converted into urban land use between 1988 and 1995 resulting in about 94% increase in urban areas. Analyses of land-cover changes between

TABLE 7.2

Land-Cover Types and Changes between 1988 and 2006 (in km²)

	Land Cover			Percentage Change in Land Cover		
Category	1988	1995	2006	1988–1995	1995–2006	1988–2006
Urban	122.55	206.59	373.46	+68.65	+80.77	+204.74
Agriculture	666.21	564.77	419.96	−15.23	−25.64	−39.96
Water	25.39	45.69	26.20	+79.95	−42.66	+3.19
Forest	13.12	10.98	8.40	−16.31	−23.50	−35.98

TABLE 7.3

Land-Cover Changes from Other Land-Cover Types to Urban between 1988 and 2006 (in km^2)

	Agriculture to Urban	Water to Urban	Forest to Urban
1988–1995	109.38	1.09	0.30
1995–2006	173.23	10.19	0.04

1995 and 2006 indicates that about 173 km^2 of agricultural area were lost to urban land use. These significant changes resulted in an increase of about 90% in urban land cover.

The Han Chang'an city is situated 5 km northwest of Xi'an. The area not only has the largest relics of any urban area but also is the well-preserved ancient capital. The Han Chang'an city is one of China's key national cultural relics and heritage sites for conservation and protection by the order of the State Council in 1961. The city has a circumference of 25.7 km, covering an area of 36 km^2 in the Weiyang District. Urban development in the region has affected both protected relics and immediate adjacent areas.

An increase in new construction sites inside the Han Chang'an city is observable by a comparison of the 1995 and 2006 maps. Land-cover types have changed in all directions outside the Han Chang'an city. The maps illustrate that, in 1988, the southern area of the Han Chang'an city was mainly the urban land cover connecting to downtown Xi'an. The other areas were mainly agricultural lands. In 1995, as Xi'an expanded northward, urban development along the transportation lines made Xi'an and Han Chang'an cities more integrated. In the 2006 map, almost all agricultural lands were changed to urban land cover around the Han Chang'an city.

7.4 Discussion and Conclusions

Population growth, policy incentives, and economic development all contribute to urban sprawl. The variability in the demographic change, especially the growth of nonagricultural employees, the socioeconomic development, and the geophysical conditions, also had effects in urban expansion (Liu 2002; Zhang 2002; Zhao et al. 2002; Ding 2003). Economic reforms have led to massive rural-to-urban migrations (Smith 2002) and generated a bilevel labor market with disparate wages between urban and rural workers (Wang et al. 2000; Meng and Zhang 2001). The influx of population from rural areas has led to housing demand in the cities.

The change of household size is another factor driving housing demands and development. In Xi'an, the annual growth rate of households was 3.52%, which was substantially higher than the population growth rate of 1.83% between 1978 and 2006 (Figure 7.4). There were about 1.043 million families

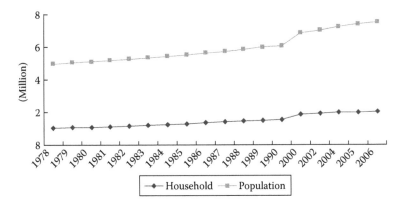

FIGURE 7.4
Household and population in Xi'an from 1978 to 2006.

with about 4.78 people per household in 1978. The number of families increased to about 2.07 million with about 3.64 people per household in 2006. The average household size in 2006 was about 1.14 persons less than that in 1978. The decreasing average household size is among important factors that caused increase in number of houses and drove the demands on housing and urban land-cover change.

The conflict between urban development and protection of heritage sites represents a challenge in decision-making in the planning process. Although the Han Chang'an city was included in the national heritage conservation order since 1961, movements for the reconstruction or renovation of old cities in the 1980s started the real estate development. It inevitably brought in massive urbanization in the area.

Xi'an has a long history of land-use planning. Protection of relics was given top priority in urban development, which brought restrictions to the overall city development plans.

Compared with other regions in southeastern China, the urbanization process in this region is still relatively slow. Monitoring and understanding the patterns of urban development and effects on the environment in such a major metropolitan area should be continued in order to balance its economic development and environmental conservation for sustainable development in the ecologically fragile western section of China and for the conservation of protected lands.

Acknowledgments

This study was supported by the Chinese Scholarship Council from which the principal author pursued her graduate study at the University of Rhode

Island as part of her PhD program of study at the Northwest A&F University. The authors wish to express their thanks to the individuals who provided help and assistance during the research work, including Dr. Peter August, Roland Duhaime, Yuyu Zhou, Eivy Y. Monroy, John Clark, and Sam Sayebare from the University of Rhode Island in the United States; Kun Deng from Northwest Agricultural and Forestry University; and Jiheng Zhou and Yiyang Zhang from Hunan Agricultural University in China.

References

Arnfield, A.J. 2003. Two decades of urban climate research: A review of turbulence exchanges of energy and water, and the urban heart island. *International Journal of Climatology*, 231, 1.

Bauer, M.E., B.C. Loffelholz, and B. Wilson. 2007. Estimating and mapping impervious surface area by regression analysis of landsat imagery. In Qihao Weng et al. (eds), *Remote Sensing of Impervious Surfaces*. Indiana State University, Terre Haute, IN: CRC Press, pp. 3–19.

Carlson, T.N. 2003. Applications of remote sensing to urban problems. *Remote Sensing of Environment*, 86, 273.

Cincotta, R.P., J. Wisnewski, and R. Engelman. 2000. Human population in the biodiversity hotspots. *Nature*, 404, 990–992.

Civco, D.L. 1993. Artificial neural networks for land-cover classification and mapping. *International Journal of Geographical Information Science*, 7(2), 173–186.

Civco, D.L., J.D. Hurd, E.H. Wilson, C.L. Arnold, and M.P. Prisloe, Jr. 2002. Quantifying and describing landscapes in the northeast United States. *Photogrammetric Engineering and Remote Sensing*, 68(10), 1083–1090.

Congalton, R.G., and K. Green. 2008. *Assessing the Accuracy of Remotely Sensed Data: Principles and Practices*, 2nd edition. Boca Raton, FL: CRC Press.

Ding, C. 2003. Land policy reform in China: Assessment and prospects. *Land Use Policy*, 20, 109–120.

Douglass, M. 2000. Mega-urban regions and world city formations: Globalization, the economic crisis and urban policy issues in Pacific Asia. *Urban Studies*, 37, 2315.

Gillies, R.R., J.B. Box, J. Symanzik, and E.J. Rodemaker. 2003. Effects of urbanization on the aquatic fauna of the Line Creek watershed, Atlanta—A satellite perspective. *Remote Sensing of Environment*, 86, 411–422.

Gomarasca, M.A., P.A. Brivio, F. Pagnoni, and A. Galli. 1993. One century of land use changes in the metropolitan area of Milan (Italy). *International Journal of Remote Sensing*, 14(2), 211–223.

Green, K., D. Kempka, and L. Lackey. 1994. Using remote sensing to detect and monitor land-cover and land-use change. *Photogrammetric Engineering & Remote Sensing*, 60, 331–337.

Healey, S.P., W.B. Cohen, Y. Zhiqiang, and O.N. Krankina. 2005. Comparison of tasseled cap-based Landsat data structures for use in forest disturbance detection. *Remote Sensing of Environment*, 97: 301–310.

Helmer, E.H., and B. Ruefenacht. 2005. Cloud-free satellite image mosaics with regression trees and histogram matching. *Photogrammetric Engineering and Remote Sensing*, 71(9): 1079–1089.

Herold, M., and N.C. Goldstein. 2003. The spatiotemporal form of urban growth: Measurement, analysis and modeling. *Remote Sensing of Environment*, 86, 286–302.

Hutchinson, C.F. 1982. Techniques for combining Landsat and ancillary data for digital classification improvement. *Photogrammetric Engineering & Remote Sensing*, 8(1), 123–130.

Jensen, J.R. 1996. *Introductory Digital Image Processing: A Remote Sensing Perspective.* Upper Saddle River, NJ: Prentice Hall.

Kalnay, E., and M. Cai. 2003. Impact of urbanization and land-use change on climate. *Nature*, 423, 528.

Leao, S., I. Bishop, and D. Evans. 2004. Simulating urban growth in developing nation's region using a CA-based model. *Journal of Urban Planning and Development*, 130(3), 145–158.

Liu, J., G.C. Dally, P.R. Ehrllch, and G.W. Luck. 2003. Effects of household dynamics on resource consumption and biodiversity. *Nature*, 421(30), 459–558.

Liu, S.H. 2002. Spatial pattern and dynamic mechanisms of urban land growth in China: Case studies of Beijing and Shanghai. *International Institute for Applied Systems Analysis (IIASA) International Report IR-02-05.* Luxemburg: IIASA, 38pp.

Lu, D., and Q. Weng. 2006. Use of impervious surface in urban land-use classification. *Remote Sensing of Environment*, 102(1–2), 146–160.

Lunetta, R.S., and J.G. Lyon (eds). 2004. *Remote Sensing and GIS Accuracy Assessment.* Boca Raton, FL: CRC Press.

Meng, X., and J.S. Zhang. 2001. The two-tier labor market in urban China: Occupational segregation and wage differentials between urban residents and rural migrants in Shanghai. *Journal of Comparative Economics*, 29(3), 485–504.

Mesev, V. 1998. Remote sensing of urban system: Hierarchical integration with GIS. *Computers Environment and Urban Systems*, 21(3/4), 175–187.

Pimm, S.L., G.J. Russell, J.L. Gittleman, and T.M. Brooks. 1995. The future of biodiversity. *Science*, 269, 347–350.

Radeloff, V.C., A.E. Hagen, P.R. Voss, P.R., D.R. Field, and D.V. Mladenhoff. 2000. Exploring the spatial relationships between census and land-cover data. *Society & Natural Resources*, 13(6), 599–609.

Rogan, J., and D.A. Roberts. 2002. A comparison of methods for monitoring multitemporal vegetation change using Thematic Mapper imagery. *Remote Sensing of Environment*, 80, 143–156.

Seto, K.C., and R.K. Kaufmann. 2003. Modeling the drivers of urban land use change in the Pearl River delta, China: Integrating remote sensing with socioeconomic data. *Land Economics*, 79(1), 106–121.

Seto, K.C., and W.G. Liu. 2003. Comparing ARTMAP neural network with maximum-likelihood for detecting urban change: the effect of class resolution. *Photogrammetric Engineering and Remote Sensing*, 69(9), 981–990.

Singh, A. 1989. Digital change detection techniques suing remotely sensed data. *International Journal of Remote Sensing*, 10(6), 989–1003.

Small, C. 2001. Estimation of urban vegetation abundance by spectral mixture analysis. *International Journal of Remote Sensing*, 22, 1305–1334.

Smith, C.J. 2002. The transformative impact of capital and labor mobility on the Chinese city. *Urban Geography*, 21(8), 670–770.

Sudhira, H.S., T.V. Ramachandra, and K.S. Jagadish. 2004. Urban sprawl: Metrics, dynamics and modelling using GIS. *International Journal of Applied Earth Observation and Geoinformation*, 5, 29–39.

Sutton, P. 2003. A scale-adjusted measure of "urban sprawl" using nighttime satellite imagery. *Remote Sensing of Environment*, 88(3), 370–384.

United Nations. 2001. *World Population Prospects: The 2000 Revision*. New York: United Nations.

Voogt, J.A., and T.R. Oke. 2003. Thermal remote Sensing of urban climates. *Remote Sensing of Environment*, 86, 370–384.

Wang, T. H., A. Maruyama, and M. Kikuchi. 2000. Rural–urban migration and labor markets in China: A case study in a northeastern province. *Development Economies*, 38(1), 80–104.

Wang, Y. and D. Moskovits. 2001. Tracking fragmentation of natural communities and changes in land cover: Applications of Landsat data for conservation in an urban landscape (Chicago Wilderness). *Conservation Biology*, 15(4), 835–843.

Wang, Y., and X. Zhang. 2004. A SPLIT model for extraction of subpixel impervious surface information. *Photogrammetric Engineering & Remote Sensing*, 70(7), 821–828.

Wilson, E.H., J.D. Hurd, D.L. Civco, M.P. Prisloe, and C. Arnold. 2003. Development of a geospatial model to quantify, describe and map urban growth. *Remote Sensing of Environment*, 86(3), 275–285.

Woodcock, C.E., and M. Ozdogan. 2004. Trends in land cover mapping and monitoring. In G. Gutman (ed.), *Land Change Science*. New York: Springer, pp. 367–377.

Wu, C. 2004. Normalized spectral mixture analysis of monitoring urban composition using ETM+ imagery. *Remote Sensing of Environment*, 93(4), 480–492.

Wu, C., and A.T. Murray. 2003. Estimating impervious surface distribution by spectral mixture analysis. *Remote Sensing of Environment*, 84(4), 493–505.

Xian, G., and M. Crane. 2005. Assessments of urban growth in Tampa Bay watershed using remote sensing data. *Remote Sensing of Environment*, 97(2), 203–215.

Yang, L., C. Huang, and C.G. Homer. 2003a. An approach for mapping large-area impervious surfaces: Synergistic use of Landsat-7 ETM+ and high spatial resolution imagery. *Canadian Journal of Remote Sensing*, 29(2), 230–234.

Yang, L., G. Xian, J.M. Klaver, and D. Brian. 2003b. Urban land-cover change detection through sun-pixel imperiousness mapping using remotely sensed data. *Photogrammetric Engineering & Remote Sensing*, 69(9), 1003–1010.

Yang, X., and Z. Liu. 2005. Use of satellite derived landscape imperviousness index to characterize urban spatial growth. *Computers Environment and Urban Systems*, 29, 524–540.

Yang, X., and C.P. Lo. 2003. Modelling urban growth and landscape changes in the Atlanta metropolitan area. *International Journal of Geographical Information Science*, 17(5), 463–488.

Yeh, A.G.O., and X. Li. 2001. Urban growth management in Pearl River delta—An integrated remote sensing and GIS approach. *ITC Journal I*, 77–85.

Yu, X.J., and C.N. Ng. 2007. Spatial and temporal dynamics of urban sprawl along two urban-rural transects: A case study of Guangzhou, China. *Landscape and Urban Planning*, 79, 96–109.

Zhang, T. 2002. Land market forces and government's role in sprawl, the case of China. *Cities*, 17, 123–135.
Zhao, X.B., Z.G. Chen, and D.A. Xue. 2002. Globalization and trend of major cities development in China [in Chinese]. *Urban Planning Overseas*, 5, 7–14.

8

Accurately Assessing Habitat with Remote Sensing: User's Perspective

Guofan Shao

CONTENTS

8.1 Errors and Uncertainties in Habitat Assessment: An Introduction

Habitat loss and degradation are among primary threats to biodiversity conservation (Fischer and Lindenmayer 2007). Habitat degradation makes the habitat less suitable for supporting a single species or a group of species. Habitat fragmentation is a type of human-induced habitat degradation that contributes to the declines in certain biodiversity. Habitat loss and fragmentation are two distinct but closely related processes. Remote sensing technology has been broadly used for assessing habitat loss and fragmentation in protected areas. Remote sensing has unique capabilities to acquire accurate, timely, and broad-scale habitat information that cannot be obtained easily with ground surveys (Duro et al. 2007; Lengyel et al. 2008; Newton et al. 2009). One of the simplest remote sensing approaches in habitat loss detection is to compare temporal changes in the extent/area of specific habitat or a variety of habitats in a landscape or region. The detection of habitat fragmentation is relatively more complicated and requires spatially explicit habitat maps. In either case, remote sensing is almost the only practical approach to quantitatively evaluate and monitor habitat dynamics in protected areas, particularly in relatively large, remote geographical regions.

Any derivatives from remotely sensed data can contain uncontrollable errors and uncertainties if remote sensing technology is used carelessly or incorrectly. This is a serious issue because remotely sensed derivatives, such as maps, are not always the final products from the user's perspective and they can be used for various decision-making processes. Using maps with errors could mislead in decision-making, causing economic and ecological losses. The sad truth is that the majority of mapping activities in reality fail to provide any assessment of uncertainties or errors in the mapping products (Newton et al. 2009), and error-implicit maps normally are used in the same way as error-explicit ones in real-world applications. For example, Nagendra (2008) used 34 peer-reviewed papers to evaluate the impact of protected areas on land-cover clearing in 22 countries, but the remotely sensed land-cover maps in half of the papers were not assessed by the original producers. Most of the remaining half of the papers contained either incomplete or low classification accuracy values.

Continuous efforts have been made to improve the accuracy of habitat detection with advanced remote sensing technology, including high spatial resolution, high spectral resolution, and active remote sensing. Sound use of remote sensing normally consists of several necessary steps and accuracy assessment is the core (Figure 8.1). An obvious advantage of remote sensing is that it can be used in areas where human access is difficult or impossible. This is particularly helpful for detecting habitat in protected areas because most protected areas are located in remote places and are hard to access. This

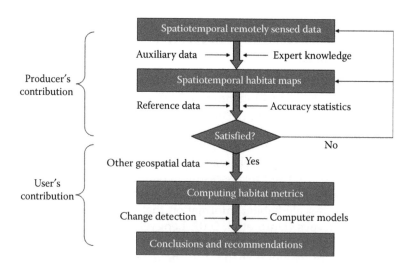

FIGURE 8.1
Flowchart of sound procedures used in detecting habitat with remotely sensed data.

is also a disadvantage because ground data collection is problematic, making accuracy assessment rather difficult if not impossible. Therefore, accuracy assessment perhaps can be the most expensive procedure in remote sensing applications in habitat detection. However, this does not mean that accuracy assessment should be ignored. Information extraction from remotely sensed imagery is always incomplete without accuracy assessment. Once accuracy assessment is made, an analyst can make accuracy-explicit decisions about how to improve technical procedures. From this point of view, the remotely sensed mapping process contains feedback and improvement loops rather than simple one-way steps. In case accuracy assessment is not performed by the analyst/producer, it is up to the end user to assess the data product and decide how to use it, or give up using the data product in the first place. The producers and users sometimes are often the same people in habitat detections because remote sensing data are readily available and image processing software systems are easy to use. This chapter introduces a few technical measures any user can take to enhance habitat detection with remote sensing products, including analyzing imagery and manipulating maps.

8.2 A Multiangle Point of View

Assuming there are n pixels in a project area covered with remotely sensed imagery, and there are c land-use/cover types, a final map product will be one of $N = c^n$ total possible maps. Even if only 1% of the total pixels are difficult to classify between only two land use/cover types, there can be $N = 2^{(c/100)}$ different maps. For example, if $c = 1000$, $N = 2^{10} = 1024$. We know that N could easily exceed 1000, the number of uncertain pixels is at least 15% if overall accuracy is 85%, and there can be more than two land use/cover types that are difficult to separate in real-world applications of remote sensing technology. In addition, remote sensing technology itself normally has a low repeatability. Therefore, it is almost impossible for any two maps derived independently by two analysts to be identical. In contrast, the low repeatability often leads to differences between two maps derived from the same remotely sensed data. A typical example is a low 74% agreement between the two global land-cover datasets of 1-km resolution derived from the same Advanced Very High Resolution Radiometer (AVHRR) imagery by the International Geosphere–Biosphere Programme (IGBP) and the University of Maryland (Hansen and Reed 2000). Such a phenomenon can be explained with a synthetic case as in Figure 8.2. A low agreement between two maps implies that at least one of the datasets has major errors but it is impossible to determine which one is better without explicit error information. Therefore, it is critical to understand error mechanisms associated with land-use and land-cover maps.

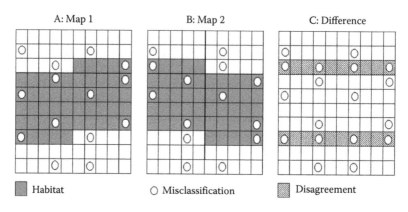

| A: Map 1 | B: Map 2 | C: Difference |

■ Habitat   ○ Misclassification   ▨ Disagreement

FIGURE 8.2
An illustration of the relationships between disagreement and misclassification for two synthetic habitat maps. The overall accuracy of each map is 86%. The overall agreement between the two maps is 80%. Misclassifications are found in both agreement and disagreement areas but 8 out of 20 (40%) disagreement pixels are misclassified while 10 out of 80 (12.5%) agreement pixels are misclassified.

Remote sensing data analysis consists of a variety of technical procedures and quantitative measures on the producer's side but there are no combinations that are universally the best. One fundamental consideration is which spectral bands should be used in a specific project. Wu and Shao (2002) suggested that different band combinations can lead to significantly different land-use and land-cover maps and the best band combination should be determined and used. However, it is common practice to use arbitrary band selections. Here are two considerations. First, more bands are not necessarily better. When statistically sophisticated parametric classification algorithms are used, more bands mean an exponential increase in the number of parameters, and, therefore, require larger and statistically sound training datasets (Landgrebe 2003). Shao and Duncan (2007) introduced a nonparametric index to rank bands possibly available and select a subset of the best bands. Such a technique is effective to simplify band selection processes. Second, a general theory sometimes cannot be applied in specific applications. Shao and Duncan (2007) reported that vegetation burning scars were visually visible from the Landsat Thematic Mapper (TM) imagery but found only in the fourth principal component (PC), though the first three PC bands contained nearly 97% of the total data variance. If one solely relies on eigenvalues to select PC bands, important spectral information could be overlooked and ignored.

Errors and uncertainties exist in every step of remote sensing data analysis. The purpose of remote sensing data analysis is to extract information from the data. As original remote sensing data can be evaluated with signal-to-noise ratio, so can the final mapping products. There is always a balance between classification detail and accuracy level. It is ideal to use maps with high classification detail and accuracy. Between the two considerations, coarse-level classification may be more preferable to low accuracy so that unexpected negative consequences of remote sensing applications are avoided or minimized. The problem is that classification detail is obvious but classification error is largely hidden. Shao et al. (2001) found that the average of multiple maps derived from the same datasets is an accurate alternative of the best map. Such a technique has been rarely used because it is too expensive to produce multiple maps, but can be a practical approach for relatively small geographic areas (Tang et al. 2010). If there are sufficient reference data, an objective and optimized classification can be worked out with the data-assisted labeling approach (Lang et al. 2008).

Habitat detection is not always as straightforward as land-use and land-cover mapping with remote sensing technology. Very often ground information and multiple remote sensing data sources have to be used to accurately monitor, quantify, or evaluable habitat (Schulman et al. 2007; Lengyel et al. 2008; Walters et al. 2008). Tang et al. (2010) demonstrated that the inclusion of ground monitoring and high-resolution data can make a big difference in forest habitat evaluation based on Landsat TM data. This is because many habitat degradations are not detectable with Landsat imagery alone. Certain animal habitat is sensitive to vegetation three-dimensional structure that can be measured with light detection and ranging (or LiDAR) and radar remote sensing (Vierling et al. 2008; Bergen et al. 2009; Frolking et al. 2009).

Although there is always room to improve image data classification and remote sensing technology continues to advance, it is unrealistic to let users wait for perfect habitat maps. Instead, users have to use whatever exists. Even a small error in a habitat map can trigger big uncertainties in decision-making with the map, which is a typical phenomenon of nonlinear error propagations in remote sensing applications (Shao and Wu 2008). Depending on how errors are spatially distributed and what habitat measures are used, the error-propagation phenomenon can be faded or canceled. This can be explained with a hypothetical example. Assuming that the two maps in Figure 8.2 are developed in different times, their differences will reflect changes in habitat over time. All the classification statistics, including overall accuracy as well as user's and producer's accuracies, are the same between the two maps; errors occur in different locations; habitat areas in the two maps are the same. If spatial overlay is performed between the two maps, the changed habitat accounts for 50% of the unchanged habitat in area; the difference in area between the two habitats is zero, suggesting that no change happened to the habitat. Thus, the habitat change can be explained explicitly with the two maps and numbers, that is, a big percentage of habitat area is gained or lost but the gain and loss of habitat are canceled out, resulting in no change in habitat area over time. However, such a conclusion can be wrong because classification errors are not considered in the analysis. If the errors in the disagreement areas are avoided, a different conclusion could be made, that is, habitat has experienced no change spatially over time.

8.3 Enhancing Habitat Maps through Calibration

8.3.1 Signals in Errors

Figure 8.1 is a flowchart that represents somewhat ideal procedures for remote sensing applications in habitat detection, monitoring, or assessment but not every procedure can be completed in practice. Three different outcomes are possible: (1) habitat maps with satisfactory accuracy; (2) habitat maps with unsatisfactory accuracy; and (3) habitat maps with no accuracy assessment. The last two outcomes are not acceptable but not uncommon. It is up to the user if and how the maps are used. Even habitat maps with satisfactory accuracy should be examined and used carefully (Shao and Wu 2008). There are a number of accuracy measures or statistics and each of them has unique implications (Congalton and Green 1999). A satisfactory habitat map by one accuracy measure may not be satisfactory with another accuracy measure. A habitat map can be used for different purposes but we cannot expect a satisfactory habitat map to be acceptable for all purposes. For instance, the detection of habitat fragmentations would require different map qualities from the quantification of habitat extents.

A habitat map often needs to be enhanced before its applications through map calibration. This can apply to all the three outcomes mentioned above in the following cases. (1) A habitat map with satisfactory accuracy at a coarse scale may not be satisfactory at a fine scale. (2) A habitat map with unsatisfactory accuracy may become satisfactory after calibration. (3) A habitat map with no accuracy information needs to be assessed and can be enhanced with necessary reference data if accuracy is relatively low. The premise of map calibration is the understanding of what and where errors exist. The error information contained in the error matrix is more than a limited number of error statistics can describe. For example, a simple index can be derived from producer' and user's accuracy measures so that land area for individual land-use and land-cover types can be adjusted (Shao et al. 2003). Such a technique cannot be used to calibrate maps. The maps in Figure 8.2 demonstrate that classification errors occur more likely along edges of land-use and land-cover patches. This is because edge pixels tend to contain mixed information between adjacent land use and land-cover types and omission or commission errors are inevitable to the mixed pixels if hard classification schemes are used.

8.3.2 Error Matrix Algorithm Design

An error matrix algorithm (EMA) for calibrating various remote sensing classified maps has been developed (Wu 2009). Let X, Y, and $\hat{Y}$ denote an existing habitat map, error-free habitat map, and the estimation of the error-free habitat map, respectively. In other words, X is an actual habitat map with classification errors; Y is an ideal map but cannot be obtained in practice; $\hat{Y}$ is a practically achievable map after calibrating X. Let E denote the error matrix associated with X, EMA can be described with a composite function as follows:

$$\hat{Y} = g(X, f(E)) \tag{8.1}$$

Equation 8.1 explicates that the calibrated habitat map is a function of the existing habitat map and errors in it. It also explains that the calibrated map varies depending on the characteristics of the existing map and its errors.

EMA generally includes the following four steps:

1. Calculate relative omission error (ROE) for each land-use and landcover type based on the error matrix (Shao et al. 2003) and pick-up type (m) with the largest ROE.
2. Scan every pixel in X. If T neighbors within the $W \times W$ window are type m, change the non-m pixel(s) to type m. Pixels on map boundaries are kept unchanged because of the complex neighborhood positions. After scanning the last pixel, a new habitat map is derived.
3. Take the new habitat map as input (X) and repeat step 2 until the overall accuracy starts to decline.
4. Assign the output of the revised habitat map with the highest classification as $\hat{Y}$.

In the EMA procedures above, $W \times W$ is commonly known as window size and the form of $T/(W \times W - 1)$ is the threshold excluding the target pixel in the window center. These concepts are often used in the filters of geospatial data processing. Therefore, EMA can also be called errororiented filter.

8.3.3 A Demonstration with Synthetic Data

The majority filter is a general neighborhood function that can be used for cleaning up salt-and-pepper effects in land use/cover maps by replacing the center pixel of a window (e.g., 3×3, 5×5, or 7×7) with the most common value within the window. The standard threshold of "the most common" is 5/8 for the 3×3 window. As illustrated with Figure 8.3, the majority-filtered map looks cleaner than the original map but its accuracy is not improved. In contrast, the EMA-filtered map not only reduces "salt-and-pepper" noise but has higher accuracy than the original map. Such a test with synthetic data suggests that EMA filter can be effectively used for calibrating habitat or land-use/cover maps.

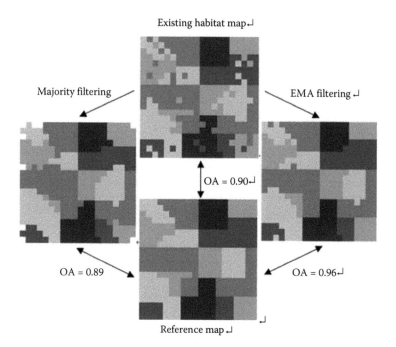

Existing habitat map

Majority filtering   EMA filtering

OA = 0.90

OA = 0.89   OA = 0.96

Reference map

FIGURE 8.3
A comparison between the majority filter and EMA filter with synthetic data.

8.3.4 Calibrating Real-World Maps with EMA

Wardlow and Egbert (2003) compared the 2001 National Land Cover Data (NLCD) with the Kansas Gap Analysis Project (GAP) land-cover data for the entire state of Kansas. A set of reference data consisting of over 3000 sample points was obtained from the Kansas GAP program and used to test EMA. The assessment and comparisons of both land-use and land-cover databases were performed at a classification level of nine land-use and land-cover types including deciduous forest, evergreen forest, shrubland, grassland, herbaceous wetland, barren/rock, cropland, urban, and water. The overall agreement between the two datasets is 75.1%. The overall accuracy is 72.9% for the NLCD and 87.7% for the GAP data. GAP has not only a higher overall accuracy but also a lower variation between producer's and user's accuracies, indicating a better quality (Shao et al. 2003).

Due to limitations of computing power, EMA could not be tested with a single run at the state level. The entire state was divided into 21 blocks and EMA was run for each block at a time. The accuracy range was 40.7%–85.8% and 66.4%–98.8% for the NLCD and GAP datasets, respectively. This indicates that land-use and land-cover data need to be re-assessed when the data are used at a smaller scale than that for which the data were developed. After EMA filtering, 14 out of 21 blocks showed accuracy increase by 0.79%–11.43%

for the NLCD dataset and 6 out of 21 blocks experienced accuracy increase by 0.58%–1.87% for the GAP dataset. By mosaicing the 21 blocks, the EMA-filtered land-use/cover map was assessed at the state level: Overall accuracy was increased (by 2.61%) to 75.5% for the NLCD dataset and (by 0.67%) to 88.5% for the GAP dataset. The overall agreement between the two EMA-filtered datasets became 78.3%, 3.2% higher than the original datasets.

The effectiveness of EMA filtering is not so consistent partially because this filter only considers the mixed-pixel problems along class edges. It may not be effective if classification accuracy is relatively high (no calibration is needed in this case anyway), reference data are insufficient to develop an unbiased error matrix, the window size and threshold parameter may not be the optimum, and/or there are labeling errors in land-use and land-cover maps—a kind of subjective misclassification (Lang et al. 2008). Nevertheless, the enhancement of habitat maps is possible through calibration and such a technique has potential as our understanding about classification errors is improved.

8.4 Concluding Remarks

Remote sensing applications still face two extreme attitudes: (1) because remote sensing techniques cannot accurately obtain as detailed information as one would expect to obtain, all their applications are unreliable; and (2) because remote sensing technology is popular and has been used in all kinds of fields, its application must be reliable. These attitudes represent two logical fallacies: the *perfectionist fallacy* for the former and *appeal to common practice* for the latter. Because remote sensing data contain errors, another fallacy exists and so-called *correlation implies causation* associated with the second fallacy above in any studies that involve remote sensing products. In this case, two completely unrelated variables are linked causally. For example, because habitat became more fragmented after implementing a new conservation policy, the conservation policy caused habitat fragmentation. The relationship sounds interesting but the reality could be that fragmentation was incorrectly detected by misusing remote sensing techniques. Those fallacies must be avoided in habitat detection. It can make a difference by calibrating remotely sensed habitat maps.

8.5 Summary

This chapter provides a concise overview about errors involved in remotely sensed map products and possible measures users can take to reduce the errors. The point is that errors do exist in all habitat maps but often are overlooked by users despite which misleading conclusions have been drawn from low-quality maps. Information from existing literature was summarized to demonstrate users' attitude toward the errors in remote sensing; hypothetical examples were taken to explain inconsistencies in habitat assessment due to classification errors found in habitat maps. This chapter introduces an EMA that can be used for calibrating various remote sensing classified maps and illustrates how it works with figures and real-world data. The use of EMA can contribute to up to 11% of accuracy increase in land-use and land-cover maps, and is an effective measure for improving habitat maps. The increase in habitat map accuracy can help reduce chances for users to commit a fallacy called "correlation implies causation" in habitat studies.

Acknowledgments

This research was supported by the ESE (Lynn) Graduate Fellowships of Purdue University. Land-use and land-cover reference data were provided by the Kansas Gap Analysis Project (GAP).

References

Bergen, K. M., S. J. Goetz, R. O. Dubayah, G. M. Henebry, C. T. Hunsaker, M. L. Imhoff, R. F. Nelson, G. G. Parker, and V. C. Radeloff. 2009. Remote sensing of vegetation 3-D structure for biodiversity and habitat: Review and implications for lidar and radar spaceborne missions. *Journal of Geophysical Research* 114: G00E06.
Congalton, R. G. and K. Green. 1999. *Assessing the Accuracy of Remotely Sensed Data: Principles and Practices*. New York: Lewis Publishers.
Duro, D. C., N. C. Coops, M. A. Wulder, and T. Han. 2007. Development of a large area biodiversity monitoring system driven by remote sensing. *Progress in Physical Geography* 31(3): 235–260.
Fischer, J. and D. B. Lindenmayer. 2007. Landscape modification and habitat fragmentation: A synthesis. *Global Ecology and Biogeography* 16(3): 265–280.
Frolking, S., M. W. Palace, D. B. Clark, J. Q. Chambers, H. H. Shugart, and G. C. Hurtt. 2009. Forest disturbance and recovery: A general review in the context of spaceborne remote sensing of impacts on aboveground biomass and canopy structure. *Journal of Geophysical Research* 114: G00E02.
Hansen, M. C., and B. Reed. 2000. A comparison of the IGBP DISCover and University of Maryland 1 km global land cover products. *International Journal of Remote Sensing* 21: 1365–1373.
Landgrebe, D. A. 2003. *Signal Theory Methods in Multispectral Remote Sensing*. New Jersey: John Wiley & Sons, Inc.

Lang, R., G. F. Shao, B. C. Pijanowski, and R. L. Farnsworth. 2008. Optimizing unsupervised classifications of remotely sensed imagery with a data-assisted labeling approach. *Computers & Geosciences* 34(12): 1877–1885.

Lengyel, S., A. Kobler, L. Kutnar, E. Framstad, P. Y. Henry, V. Babij, B. Gruber, D. Schmeller, and K. Henle. 2008. A review and a framework for the integration of biodiversity monitoring at the habitat level. *Biodiversity and Conservation* 17(14): 3341–3356.

Nagendra, H. 2008. Do parks work? Impact of protected areas on land cover clearing. *AMBIO* 37(5): 330–337.

Newton, A. C., R. A. Hill, C. Echeverría, D. Golicher, J. M. B. Benayas, L. Cayuela, and S. A. Hinsley. 2009. Remote sensing and the future of landscape ecology. *Progress in Physical Geography* 33(4): 528–546.

Schulman, L., K. Ruokolainen, L. Junikka, I. E. Saaksjarvi, M. Salo, S. K. Juvonen, J. Salo, and M. Higgins. 2007. Amazonian biodiversity and protected areas: Do they meet? *Biodiversity and Conservation* 16(11): 3011–3051.

Shao, G., and B. W. Duncan. 2007. Effects of band combinations and GIS masking on fire-scar mapping at local scales in East-Central Florida, USA. *Canadian Journal of Remote Sensing* 33(4): 250–259.

Shao, G., D. Liu, and G. Zhao. 2001. Relationships of image classification accuracy and variation of landscape statistics. *Canadian Journal of Remote Sensing* 27(1): 33–43.

Shao, G. F., W. C. We, G. Wu, X. H. Zhou, and J. G. Wu. 2003. An explicit index for assessing the accuracy of cover class areas. *Photogrammetric Engineering & Remote Sensing* 69(8): 907–913.

Shao, G. F. and J. G. Wu. 2008. On the accuracy of landscape pattern analysis using remote sensing data. *Landscape Ecology* 23(5): 505–511.

Tang, L. N., G. F. Shao, Z. J. Piao, L. M. Dai, M. A. Jenkins, S. X. Wang, G. Wu, J. G. Wu, and J. Z. Zhao. 2010. Forest degradation deepens around and within protected areas in East Asia. *Biological Conservation* 143(5): 1295–1298.

Vierling, K. T., L. A. Vierling, W. A. Gould, S. Martinuzzi, and R. M. Clawges. 2008. Lidar: Shedding new light on habitat characterization and modeling. *Frontiers in Ecology and the Environment* 6(2): 90–98.

Walters, B. B., P. Ronnback, J. M. Kovacs, B. Crona, S. A. Hussain, R. Badola, J. H. Primavera, E. Barbier, and F. Dahdouh-Guebas. 2008. Ethnobiology, socio-economics and management of mangrove forests: A review. *Aquatic Botany* 89(2): 220–236.

Wardlow, B. and S. Egbert. 2003. A state-level comparative analysis of the GAP and NLCD land-cover data sets. *Photogrammetric Engineering and Remote Sensing* 69(12): 1387–1397.

Wu, W. C., and G. F. Shao. 2002. Optimal combinations of data, classifiers, and sampling methods for accurate characterizations of deforestation. *Canadian Journal of Remote Sensing* 28(4): 601–609.

Wu, Y. 2009. Development of an error matrix algorithm (EMA) for calibrating land cover data. Master Degree Thesis No. 55089, Purdue University.

Section II

Remote Sensing for Inventory, Mapping, and Conservation of Protected Lands and Waters

9

Utilization of Remote Sensing Technologies for Matschie's Tree Kangaroo Conservation and Planning in Papua New Guinea[*]

Jared A. Stabach, Lisa Dabek, Rigel Jensen, Gabriel Porolak, and Yeqiao Wang

CONTENTS

9.1 Introduction

Over the past 15 years, the Tree Kangaroo Conservation Program (TKCP) has been working with local landowners at a remote research location in Papua New Guinea (PNG) to gain a better understanding of the habitat requirements and movement patterns of Matschie's tree kangaroo (*Dendrolagus matschiei*). To date, over 70,000 ha (728 km²) of montane rainforest habitat

---

[*] Adapted from Stabach, J. A. et al. 2009. *International Journal of Remote Sensing* 30:405–422.

have been set aside for conservation. Known as the Yupno, Urawa, and Som (YUS) Conservation Area (named after the local drainages in the area), it is the first-ever conservation area established within the nation's borders under the PNG Conservation Areas Act of 1978.

Endemic to the upper montane forests of the Huon Peninsula, *D. matschiei* are arboreal marsupials (Figure 9.1a). However, due to a combination of factors such as increased hunting pressure, loss of habitat from agricultural expansion (Flannery 1998), and a low reproductive rate (Dabek 1994), they

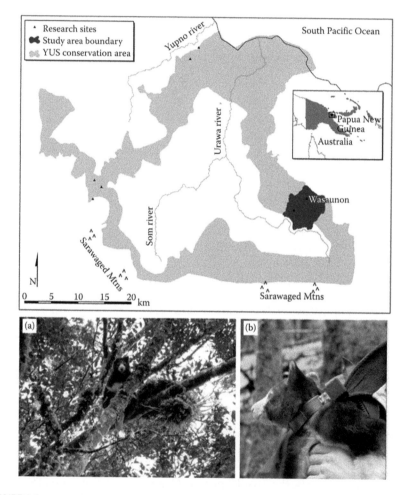

FIGURE 9.1
(**See color insert.**) Map of the Tree Kangaroo Conservation Program research area, Morobe Province, Papua New Guinea. (Adapted from Stabach, J. A. et al. 2009. *International Journal of Remote Sensing* 30:405–422.) Endemic to the upper montane forests of the Huon Peninsula in Papua New Guinea, Matschie's tree kangaroo (*Dendrolagus matschiei*) are arboreal marsupials (a). A Matschie's tree kangaroo was fitted with a Televilt GPS-PosRec™ GPS collar (b). (Photos provided by the Tree Kangaroo Conservation Program of the Woodland Park Zoo. With permission.)

are currently listed by the International Union for the Conservation of Nature (IUCN 2008) as endangered. As the largest mammal on the Huon Peninsula (Flannery 1995), they are an important part of the diet and traditional culture of the people and serve as a focal species for the conservation of large tracts of forest habitat (Pugh 2003).

Tropical rainforests are recognized as some of the richest ecosystems in terms of biodiversity worldwide (Laurance and Laurance 1999; Stibig et al. 2003). Occupying less than 7% of the terrestrial surface on Earth, they contain more than half of all plant and animal species (Raven 1988; Myers 1992). Tropical rainforests, however, are also some of the most endangered ecosystems left on Earth. The Food and Agriculture Organization (FAO) of the United Nations estimates that while deforestation has slowed over the past decade (2000–2010), tropical forests are still being lost at an alarming rate (13 million hectares annually) (FAO 2010).

In PNG, while deforestation has historically been low, recent research indicates that rates of deforestation have dramatically increased over the past two decades (varying between 0.8 and 1.8 ha per year) (Shearman et al. 2009). However, roughly 28 million hectares of tropical forest still remain, making PNG one of the largest reserves of tropical rainforest left in the world (Shearman et al. 2009). Because of (i) the vast amount of pristine forest that remain, (ii) the rich biodiversity and species endemism that occur, (iii) the increasing rate of deforestation, and (iv) the fact that nearly 95% of all land remains in traditional clan-based land ownership (Flannery 1995, 1998; Osborne 1995; Diamond 1997; Shearman et al. 2009), PNG is a high-priority target for conservation (Beehler 1993; Dinerstein and Wikramanayake 1993; Brooks et al. 2006; Shearman et al. 2009).

Conducting field research in PNG, however, is fraught with technical and logistic constraints. Forests are lush with large climax species (such as *Nothofagus starkenborghii* and *Dacrydium nidulum*), but are often located far away from population centers (thus, one of the main reasons why timber extraction has occurred at such low levels). Travel is generally restricted to footpaths and light aircraft, with researchers often spending extended periods of time out at remote research stations. Martin et al. (1998) suggest that the development of species classification methods using remotely sensed data might be the most effective approach for obtaining assessments of forest cover over large areas and at fine spatial scales.

In order to tailor protected areas toward individual wildlife species, specific information about the habitat preferences of these species is also required. While very high frequency (VHF) radiotransmitters have been used for more than 30 years and are the most commonly used devices for tracking the daily movements of animals (Moen et al. 1996; Bowman et al. 2000), tracking animals on a daily basis is a major time/resource investment, especially in remote areas. As an alternative, Global Positioning System (GPS)-based telemetry systems are now available for animals of smaller sizes and have emerged as an increasingly reliable technology to obtain automated animal locations over large geographic areas with great accuracy and precision (Rempel et al. 1995; Bowman et al. 2000).

In this chapter, we provide detailed information about the use of two remote sensing technologies [Landsat-7 Enhanced Thematic Mapper Plus (ETM+) satellite imagery and Televilt GPS-PosRec™ (Post-Recovery) devices] that were used in our study to provide an understanding of the movements of individual animals, along with detailed information about the dominant forest types throughout a subset of the YUS Conservation Area. Combined with extensive field study, these data provide detailed information for the establishment of current protected area boundaries, may aid other research projects with similar objectives, and highlight how remote sensing technologies can be integrated into the creation of protected area and conservation management plans.

9.2 Study Area

Research was conducted in a portion of the YUS Conservation Area, known as the Wasaunon Field Research Area (hereafter referred to as WASU), Morobe Province, Papua New Guinea. The area is located between 6°3′ and 6°10′ south latitude, and 146°51′ and 146°58′ west longitude on the northeastern side of the Huon Peninsula in the Sarawaged mountain range (Figure 9.1). The WASU represents a total of ~4000 ha of land that has been pledged for conservation by the clan landowners of local villages (Yawan, Towet, and Worin). The area is covered extensively (98%) by upper montane tropical rainforest (Pugh 2003) with a moderately dense forest canopy (~84%) (R. Jensen, unpublished field report 2005). The elevation ranges from 2122 to 3067 m with slopes in excess of 60°. June through August are generally the driest months (~150 mm of rainfall per month) with mean annual precipitation estimated to be 2717 mm (Hijmans et al. 2005).

9.3 Datasets and Methods

9.3.1 Land-Cover Classification

9.3.1.1 Satellite Data Sources

A detailed land-cover classification of the dominant forest types throughout the study area was conducted using Landsat ETM+ satellite imagery (path 96, row 64). The scene, acquired on September 8, 2000, was downloaded from the Global Land Cover Facility (GLCF, http://www.landcover.org) and received orthorectified and registered to Universal Transverse Mercator

(UTM) zone 55 South, WGS84 spheroid with a root mean square error of less than 50-m (Tucker et al. 2004). The image consists of six multispectral bands at 28.5-m spatial resolution and one panchromatic band at 14.25-m spatial resolution. The thermal band, however, was not included in this study due to its coarse spatial resolution (60-m). A resolution merge of the six multispectral bands and the single panchromatic band was performed using cubic convolution resampling. The final dataset contains six multispectral bands at 14.25-m spatial resolution.

A digital elevation model (DEM) was used to improve class assignments. Derived from the Shuttle Radar Topography Mission (SRTM), data for Papua New Guinea are distributed in 3 arc-second or 90-m spatial resolution (SRTM 2004). However, numerous data voids exist in the raw SRTM terrain model. These voids were filled with an ancillary data source, the PNG 90-m DEM which was created from the digitized 1:250,000 Joint Operations Graphic charts of PNG and the 1:100,000 Royal Australian Survey Corps series. The PNG 90-m DEM as well as the final SRTM DEM were created using ER Mapper software at the University of Papua New Guinea (P. Shearman, unpublished data 2004).

9.3.1.2 Field Data Collection

One of the most important aspects of any remote sensing image classification is the *in situ* collection of ground reference points (GRPs). From July 1, 2004 through August 12, 2004, a series of GRPs were collected which included (1) nearest-neighbor vegetation plots and (2) field survey points. In total, 168 GRPs were collected.

Seventeen vegetation plots of the 33 nearest neighbors were conducted within the WASU. Plots were placed in what was best delineated on the ground as distinct and homogeneous, since no aerial photographs were available to aid the process. Modified from Webb et al. (1976), the plots were based on a set number of stems from the canopy species of the vegetation type being described rather than a set area (Goosem 1992; Stanton and Fell 2005). In each of the plots (ranging from 0.06 to 0.30 ha), a center tree was identified at random, and the plot separated into four quadrants that coincided with the four cardinal directions (north, east, south, west). The eight nearest canopy species to the center tree location were marked and a voucher specimen was taken. Each of the specimens was later identified to species level with assistance from the staff at the National Herbarium in Lae, PNG and the Queensland Herbarium in Brisbane, Australia. The authority used for plant names and families was the International Plant Name Index (IPNI 2008). Henderson (1997), Mabberley (1997), and Hoft (1992) were also used as references in the herbaria to standardize names. Forest-type descriptions were based on those by Webb et al. (1976) and Tracey (1987). A more detailed description of the plot design can be found in the study by Stabach et al. (2009).

Tree heights, H, were calculated using a simple trigonometric formula:

$$H = \tan\theta x$$

where x is the distance away from the tree measured using a laser range-finder (Bushnell Yardage Pro, Bushnell Corporation, Overland Park, Kansas, USA) and θ is the angle to the top (and bottom, if not level with the tree base) using a Suunto clinometer (PM-5/360PC, Vantaa, Finland). A distance of at least 10-m was maintained for accuracy in angle collection.

Canopy cover was measured with a concave spherical densiometer (model C, Robert E. Lemmon, Forest Densiometers, Bartlesville, Oklahoma, USA). A series of five measurements were collected, one from each of the four cardinal points and one from the center of the plot. Measurements were averaged together to get a more reliable measure of overall canopy cover throughout the plot. The plot data were tabulated in a matrix of species versus plot, analyzed by hierarchical cluster analysis, and grouped into discreet forest types.

Field survey points consisted of visually discernable patches of homogeneous land cover and the precise tree species of radio-tracked *D. matschiei* locations. Televilt GPS-PosRec™ GPS collars, described below, provided the precise location of three individual animals but also emitted a VHF/UHF (ultrahigh frequency) signal that allowed researchers to track animals on a daily basis using traditional radio-telemetry devices (three-element Yagi and/or RX-98H antenna). VHF telemetry was conducted concurrently with the data collected automatically by the GPS collars so that conclusions between the two data sources could be compared and an assessment made (Stabach 2005) and so that the habitat use and movements patterns of the animal could be examined (Stabach 2005; Porolak 2008; Porolak et al. in review). Here, the manual radio-telemetry information provided detailed information about 147 dominant canopy species throughout the study area. A Trimble Geo3c receiver with an external antenna (Trimble Navigation Ltd, Sunnyvale, California, USA) was used to mark the position of each animal. Vegetation information collected included the identification of trees from tracked locations to species level, canopy closure, and the height and diameter-at-breast height of each tree (Stabach 2005).

A series of at least four digital photographs were taken at each field survey point and dynamically linked by their respective time stamps to provide a Virtual Field Reference Database, guiding class assignments. Geographic coordinates were based on a minimum of 10 positions recorded at 1-s intervals and were postprocessed with differential correction from the Cape Ferguson, Australia base station data (1454 km distance). A firmware-imposed positional dilution of precision (PDOP) mask of 8.0 and a signal-to-noise ratio threshold of 4 were used. However, the PDOP was changed to as high as 20.0 on select occasions to accommodate difficulty in signal acquisition. Points were converted from decimal degrees to UTM zone 55 South

(WGS84 spheroid), and entered into a Geographic Information System (GIS) as a point coverage.

9.3.1.3 Image Classification

A multistep, hybrid classification approach was used to separate the upper montane forest into three distinct forest classes identified by the nearest-neighbor vegetation plots. These three distinct forest types are referred to as *D. nidulum* dominant, *Caldcluvia nymanii* dominant, and *N. starkenborghii* dominant, named according to the dominant tree species identified within vegetation plots. An additional "other" forest class was also distinguished based on both visual interpretation and the examination of spectral statistics, but *in situ* data were not collected for verification. Each image was subset to the boundary of the WASU after imposing a 500-m buffer using Arc/Info Workstation 9.0 (Environmental Systems Research Institute, Redlands, California, USA).

Areas covered in clouds were first removed from the image and are generally a necessary preliminary step. Area-of-interest (AOI) polygons were created to minimize misclassification throughout the image and a 50-class unsupervised classification using the Iterative Self-Organizing Data Analysis Technique (ISODATA) was recoded to two classes, effectively separating each of the AOIs into cloud (0) and noncloud (1) areas. The ISODATA technique requires minimal input from the analyst and is the process whereby pixels are grouped together based on their similarity in spectral properties (Jensen 1996). A majority based on a 3×3 pixel window was used to reduce variability and the raster fill tool was used to manually correct remaining misclassified pixels. The resulting classification was used as a mask to produce a cloud-free image and used for all subsequent stages of the vegetation classification.

Grasslands were easily discernable in the image and were separated from the image (since kangaroos are thought to only use grassland habitat as a transition zone between forest patches) using the same procedure as described above for cloud removal. The DEM was used in a rule-based model to separate grassland into two different classes: alpine and anthropogenic. Grasslands above 2750 m in elevation were recoded as alpine (since they would be likely too far away from local villages to serve anthropogenic purposes) while areas below this gradient were coded as anthropogenic (a common feature in PNG).

A 50-class unsupervised classification (ISODATA) was then conducted on the remaining upper montane forest areas and recoded into four separate forest classes: *D. nidulum* dominant, *C. nymanii* dominant, *Nothfagus starkenborghii* dominant, and "other". A supervised classification using these spectral signatures was performed using the maximum likelihood parametric decision rule with equal prior probabilities for each class. Histograms were examined throughout the classification process to assure that classes were normally distributed. Additional spectral signatures where land cover was known from GRPs were added to improve classification results. Both grassland categories were then readded to the final classification results using a simple raster math calculation. Class variability was reduced by a GIS analysis to eliminate "island" polygons below a two-pixel minimum.

The accuracy of the satellite image classification was evaluated using quantitative procedures to compare the "true" land cover with the land cover as determined through the image classification. Four hundred random sampling points were automatically generated throughout the study area, stratified by land-cover class with a minimum of 50 points per class following guidelines suggested by Congalton (1991), Jensen (1996), and Congalton and Green (1999), and added to the GRPs collected during the field study. However, since the "true" land cover of the randomly generated sampling points was not known, the assessment of accuracy using these points was based on our knowledge of the study area and of the satellite imagery. Standard accuracy measures such as commission errors (user's accuracy), omission errors (producer's accuracy), overall accuracy, and κ statistics were generated to create an evaluation of error.

9.3.2 Animal Movements

9.3.2.1 Home Range

In April 2004, three adult female Matschie's tree kangaroos were caught opportunistically by TKCP field staff as part of a pilot study to investigate the home range and movement patterns of animals (Porolak et al. in review). Hereafter, each animal will be referred to as MTK1, MTK2, and MTK3. From these data, we were able to calculate a preliminary fixed-kernel home-range estimate for the species. However, as this estimate was based on only three animals (which has now been increased to a total of 15 animals, both male and female), readers are directed to Porolak (2008) and Porolak et al. (in review) for a more robust and reliable home-range estimate for the species. We provide an estimate here only to demonstrate some of the capabilities of placing GPS collars on animals for protected area establishment. All aspects of animal handling were conducted under the care of a field veterinarian and approved by the International Animal Care and Use Committee at the Roger Williams Park Zoo, Providence, Rhode Island, USA and the University of Rhode Island, Kingston, Rhode Island, USA (Approval No. AN05-03-017).

Each animal was fitted with a Televilt GPS-PosRec™ GPS collar (model C200). Collars weighed 200 g, 2.4–3.0% of total tree kangaroo body weight between 6.7 and 8.2 kg (Figure 9.1b). Devices are "store-on-board" GPS dataloggers that collect and store location information. Each device was programmed to collect two positions daily (6:00 a.m. and 6:00 p.m., local time) for the five-month study period (April 1, 2004–September 10, 2004), release from the animal automatically on a preprogrammed date, and fall to the ground so that the instrument could be retrieved and the data could be downloaded.

Exported Televilt GPS data are accompanied by an accuracy assessment in the form of a position type (3D+ , 3D, 2D, or 1D). 3D+ readings are accurate to within ±15 m from the actual position 95% of the time, 3D readings are typically within ±50 to ±100 m, and 2D or 1D readings have no accuracy parameters; 1D readings, however, should be better than 2D readings as the device needs to acquire information from a previous position in order to pinpoint a location (A. Campos, Telemetry Solutions, pers. comm., 2005).

Kernel methods are not a new statistical technique, but have only recently gained popularity as a home-range estimator (Seaman et al. 1999). The fixed kernel is a nonparametric technique that provides a density estimate, known as a utilization distribution (UD), and is based on the amount of time that an animal spends at one particular location (van Winkle 1975; Worton 1995; Seaman and Powell 1996). Note that the UD describes space use and not necessarily resource use (Anderson 1982). Unlike other common home-range estimators (i.e., minimum convex polygon, Jennrich-Turner), kernel methods are not as sensitive to outliers, require small sample sizes for accurate estimation, and can identify core areas of utilization (Seaman et al. 1999; Kernohan et al. 2001).

Essential to the fixed kernel method is the proper selection of an accurate smoothing parameter or bandwidth (Worton 1995; Seaman and Powell 1996; Seaman et al. 1999). Least-square cross validation (LSCV) is a jackknife procedure that uses an iterative approach to select the amount of smoothing that minimizes the estimated error for a given sample (Silverman 1986) and has been shown to be the most accurate bandwidth to use with the fixed kernel method (Seaman and Powell 1996; Seaman et al. 1999). We used the LSCV smoothing factor without correction, as suggested by Hooge and Eichenlaub (1997), and calculated the 50% (core) and 90% (home range) utilization distributions for each animal. The Animal Movements Extension to ArcView, version 1.1 (Hooge and Eichenlaub 1997), was used for all analyses.

9.3.2.2 Spatial Statistics

Spatial data analyses were conducted to test whether tree kangaroo locations were arranged in complete spatial randomness (CSR) and whether or not there was independence between point processes. These tests were important, as they aided in identifying whether animals exhibit signs of site fidelity and provide information on the level of overlap between each animal. Both first- and second-order properties were explored using MicroSoft Excel (MicroSoft 2000) and the statistical package "R" (R Development Core Team 2010), although only results of the second-order properties will be described here. For a larger description of the analyses conducted, see Stabach (2005). Methods followed those described by Diggle (2003), Diggle and Chetwynd (1991), and Andersen (1992). Statistics were calculated after the removal of suspect observations, although conclusions were the same both before and after removal. In all instances, edge effects were ignored and isotropy was assumed.

Second-order properties refer to the spatial dependence between individual point locations (referred to as events). To test for CSR, we calculated the empirical cumulative distribution of the event–event first nearest-neighbor distances (W) and compared the graphical results with the theoretical cumulative distribution, created by generating 99 simulations of random points and calculating W for each simulation. To provide a quantitative assessment to the visual perception of the graphs, the Kolmogorov–Smirnov test was used to evaluate the difference between the two distributions. A statistical significance level of $\alpha = 0.05$ was used.

Nearest-neighbor distance methods, however, are based on only the closest events. Therefore, only the finest scales of pattern are considered in the analysis (Diggle 2003). To investigate the spatial dependence of the spatial point processes over a wide range of scales, we used the modified L function:

$$L(h) = \sqrt{K(h)/\pi} - h$$

which is derived from Ripley's K:

$$L(h) = \pi h^2$$

The L(h) function uses all event-to-event distances to calculate a measure of spatial pattern at various distances h. $L(h) = 0$ indicates CSR; $L(h) < 0$ indicates regularity; $L(h) > 0$ indicates clustering in the data. Simulation envelopes were calculated to provide a visual indication of the significance in departures from CSR.

To test for spatial relationships between animals, we used the cross L function:

$$\hat{L}_{12}(h) = \sqrt{K_{12}(h)/\pi} - h$$

Essentially, the equation is the same as that for the L(h) function, except that two point processes are concatenated ($L_1 + L_2 = \hat{L}_{12}$) together at a time and randomized using a toroidal shift to calculate simulation envelopes. Positive and negative peaks in $\hat{L}_{12}$ indicate attraction and repulsion between the two processes, respectively. If the estimated $\hat{L}_{12}$ function remains well within the calculated envelopes, however, independence between events is indicated.

9.4 Results

9.4.1 Land-Cover Classification

The remote sensing image classification separated the study area into four dominant forest types (Figure 9.2), with *D. nidulum* dominant forests being the most widespread (39.3%). These results were consistent with field measurements. *C. nymanii* dominant forests were also common (26.0%). A third "other" forest type was also found broadly throughout the study area (25.0%), but for which no vegetation plots were conducted. Clearly, these data highlight the importance of intensive and extended fieldwork. Areas of no data (i.e., clouds) accounted for 3.9% of the total study area (Table 9.1). Descriptions of each of the land-cover classes are found in Table 9.2.

Not surprisingly, highest κ coefficient values were achieved for the two grassland categories, anthropogenic and alpine grassland (91.6% and 83.9%, respectively). The spectral signatures of these two areas were significantly different from the neighboring forest habitats and were easily discernable, both visually and statistically. However, the dominant forest types were much more difficult to separate. *D. nidulum* dominant forests were classified with 79.0% accuracy, *C. nymanii* dominant forests with 30.9% accuracy, and *N. starkenborghii* dominant forests with 59.7% accuracy. The overall κ coefficient for the entire study area was 59.4%. Omission and commission errors for each of the dominant forest classes are also provided (Table 9.3).

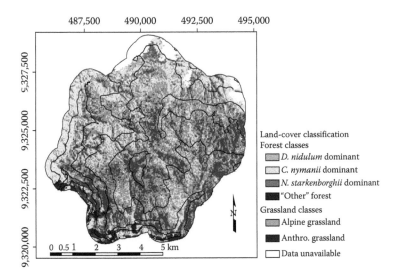

FIGURE 9.2
Land-cover classification results of Landsat-7 Enhanced Thematic Mapper Plus satellite imagery. A 200-m contour interval has been provided for reference. (Adapted from Stabach, J. A. et al. 2009. *International Journal of Remote Sensing* 30:405–422.)

TABLE 9.1

Summary Statistics of Landsat-7 Enhanced Thematic Mapper Plus Land-Cover Classification[a]

Land-Cover Class	Pixels	Hectares	Percentage of Landscape
D. nidulum dominant forest	104,717	2126.4	39.3
C. nymanii dominant forest	69,423	1409.7	26.0
N. starkenborghii dominant forest	7364	149.5	2.8
"Other" forest	66,692	1354.3	25.0
Alpine grassland	1253	25.4	0.5
Anthropogenic grassland	6782	137.7	2.5
No data (clouds)	10,480	212.8	3.9
Total	266,711	5415.9	100.0

[a] Statistics include 500-m buffer.

TABLE 9.2

Descriptions of Land-Cover Classes Derived for Landsat-7 Enhanced Thematic Mapper Plus Satellite Image Classification

Land-Cover Class	Description
D. nidulum dominant forest	Dominant forest type comprising 24 canopy species. Tree species *Quintinia ledermanii, Decaspermum forbesii, D. nidulum,* and *Saurauia capitulata* represent 56.3% of total trees surveyed. Elevation range: 2850–3100 m
C. nymanii dominant forest	Dominant forest type comprising 35 canopy species. Tree species *C. nymanii, Schizomeria serrata, Astronia atroviridis, Dryadodaphne crassa, Macaranga trichanthera,* and *Prunus glomerata* represent 53.3% of total trees surveyed. Elevation range: 2400–2600 m
N. starkenborghii dominant forest	Dominant forest type comprising 6 canopy species. Tree species *N. starkenborghii* represent 87.9% of total trees surveyed. Elevation range: 2800–2850 m
"Other" forest	Dominant forest type unidentified by vegetation plots. Classification based on a difference in spectral reflectance visually identified. Elevation range: unknown
Alpine grassland	Grassland Elevation range: >2750 m
Anthropogenic grassland	Grassland, including fallow and agricultural land. Elevation range: <2749 m

Note: Dominant forest types derived from an analysis of data collected via nearest-neighbor vegetation plots (R. Jensen, unpublished data, 2005). Elevation ranges summarize data from vegetation plots and provide detail to the cut-off ranges used to separate grassland areas. For a complete list of species and a description of the average physiognomic features found throughout each of the forest types, see Stabach et al. (2009).

TABLE 9.3

Confusion Matrix and Summary Statistics for Landsat-7 Enhanced Thematic Mapper Plus Satellite Imagery

| | Classified Results[a] | | | | | | | | | |
	Dacrydium	Caldcluvia	Nothofagus	"Other" Forest	Alpine Grassland	Anthropogenic Grassland	Total	Omission errors (%)	Classification accuracy (%)[b]	κ coefficient (%)[c]
Dacrydium	162	7	1	9		1	180	10.0	90.0	79.0
Caldcluvia	86	57		2			145	60.7	39.3	30.9
Nothofagus	11		35	10			56	37.5	62.5	59.7
"Other" forest	31	1	3	44		2	81	45.7	54.3	48.4
Alpine grassland	8				46		54	14.8	85.2	83.9
Anthropogenic grassland		4				48	52	7.7	92.3	91.6
Total	298	69	39	65	46	51	568	31.0		
Commission errors (%)	45.6	17.4	10.3	32.3	0.0	5.9				

a Percent unclassified pixels = 0.0%.

b Mean classification accuracy = 70.6%, where the mean classification accuracy refers to the average of all class accuracies. Overall classification accuracy = 69.0%, where overall classification accuracy refers to the total correctly classified pixels.

c Overall κ coefficient = 59.4%; variance of κ = 0.001047155.

9.4.2 Animal Movements

Satellite GPS collars remained on MTK1 and MTK3 throughout the study period, fell off during the programmed dates, and properly emitted VHF signals for data collection. MTK2's collar, however, was manually removed on August 20, 2004 after the animal was discovered dead from an apparent New Guinea Harpy eagle (*Harpyopsis novaeguinee*) attack (J. Glick and G. Porolak, pers. comm. 2004) The mortality indicator functioned properly, emitting a change in pulse to aid in collection.

Collars collected 185 fixes out of 923 possible attempts (20.0%) and overall acquisition success was better in the morning (55.7%) than in the evening (44.3%). 2D fixes were collected 45.4% of the time, 3D/3D+ fixes were collected in similar proportions (25.4% and 26.0%, respectively), and 1D fixes were collected 3.2% of the time. MTK3 had the highest proportion of fixes collected (24.5%), followed by MTK2 (18.3%) and MTK1 (17.1%) (Table 9.4). A morning and evening fix (6:00 a.m. and 6:00 p.m., local time) was collected on only 21 occasions (6.6%) within the same day.

The mean fixed-kernel home-range area was 7.3 ± 1.9 ha (50% UD) and 28.3 ± 2.3 ha (90% UD), with MTK3 having the largest home range (30.8 ha) and MTK1 the smallest (26.4 ha) (Table 9.5). Note, again, that these estimates are based on only three adult females over a six-month study period and are reported here to demonstrate the type of information that can be gained for conservation purposes from this remotely sensed technology. For a more robust calculation of home range, readers are directed to Porolak (2008) and Porolak et al. (in review).

While home ranges do overlap on several occasions (Figure 9.3), spatial statistics indicate that point processes are independent (Figure 9.4). This is indicated in Figure 9.4 by the estimated $\hat{L}_{12}$ being well within the calculated envelopes (as is the case between MTK1 and MTK3) and the negative troughs between animals MTK1 and MTK2 and animals MTK2 and MTK3. This repulsion (negative trough) was an expected result as there is no overlap between these animals' home ranges and an approximate distance of 1500 m between MTK1 and MTK2 and 1200 m between MTK2 and MTK3.

Visually, however, each of the point processes seems clustered. Second-order properties confirm this assumption with a high level of significance and over a wide range of spatial scales, illustrated by the fact that the estimated L function is above the simulation envelopes for each animal at all distances (especially for MTK3; Figure 9.5). The first nearest-neighbor distances showed the same result, although these data are not presented here (see Stabach 2005). Animals returned to the same location at multiple times throughout the study period, indicative of a high level of site fidelity. However, the reason behind the level of site fidelity is unknown at this time.

TABLE 9.4

Assessment of Televilt GPS-PosRec™ GPS Collars in the Tree Kangaroo Conservation Program's Wasaunon Field Research Area, Morobe Province, Papua New Guinea, April–September 2004

Animal	Total Fix Attempts	Total Fixes Received	Total Proportion Received (%)	1D Fixes (%)	2D Fixes (%)	3D Fixes (%)	3D+ Fixes (%)
All tree kangaroos	923	185	20.0	3.2	45.4	25.4	25.9
		(103, 82)	(55.7, 44.3)	(1.6, 1.6)	(29.2, 16.2)	(11.4, 14.1)	(13.5, 12.4)
MTK1	315	54	17.1	5.6	53.7	18.5	22.2
		(32, 22)	(59.3, 40.7)	(3.7, 1.9)	(37.0, 16.7)	(9.3, 9.3)	(11.1, 11.1)
MTK2	290[a]	53	18.3	3.8	35.8	30.2	30.2
		(23, 30)	(43.4, 56.6)	(1.9, 1.9)	(17.0, 18.9)	(15.1, 15.1)	(9.4, 20.8)
MTK3	318	78	24.5	1.3	46.2	26.9	25.6
		(47, 31)	(60.3, 39.7)	(0, 1.3)	(32.1, 14.1)	(10.3, 16.7)	(18.0, 7.7)

Note: Results of morning and evening acquisitions (6:00 a.m., 6:00 p.m., local time) are provided in parentheses.
[a] Animal predated two weeks prior to end of study.

TABLE 9.5

Home-Range Size Estimates of Matschie's Tree Kangaroos (*Dendrolagus matschiei*) from Televilt GPS-PosRec™ GPS Collars

		50% of Locations		90% of Locations	
Animal	*n*	$\bar{x}$	SD	$\bar{x}$	SD
All tree kangaroos	3[a]	7.3	1.9	28.3	2.3
MTK1	54[b]	7.6	—	26.4	—
MTK2	53[b]	9.1	—	27.5	—
MTK3	78[b]	5.4	—	30.8	—

Note: Home ranges were calculated using the fixed kernel estimator with the least-square cross-validation (LSCV) smoothing parameter. Both the 50% and 90% utilization distributions have been provided (in hectares).

[a] *n* representative of number of animals.
[b] *n* representative of number of fixes.

9.5 Discussion

The results described here provide detailed information on the use of Landsat-7 ETM+ satellite imagery to discriminate dominant forest types within a heterogeneous tropical rainforest. The low classification accuracies and κ coefficients indicate, however, that additional research is necessary

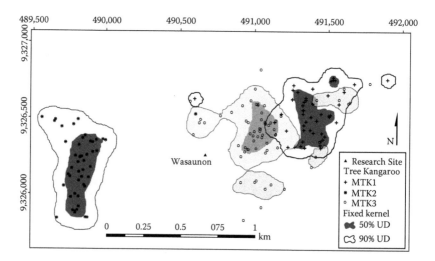

FIGURE 9.3

Fixed kernel home-range estimates. Results compare home ranges calculated from data collected by Televilt GPS-PosRec™ GPS collars in the Tree Kangaroo Conservation Program's Wasaunon Field Research Area, Morobe Province, Papua New Guinea, April–September 2004. Data are projected to the Universal Transverse Mercator zone 55 South coordinate system, WGS84 datum. The 90% and 50% utilization distribution (UD) have been provided.

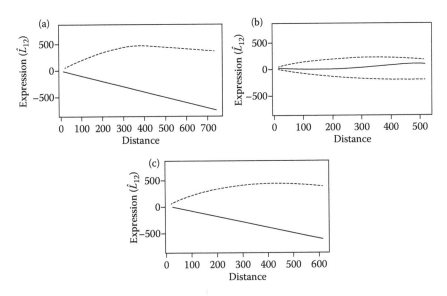

FIGURE 9.4
Tests for independence between spatial point processes. Positive peaks in $\hat{L}_{12}$ indicate attraction, negative troughs repulsion. Independence between point processes is indicated when the estimated $\hat{L}_{12}$ function (solid line) remains well within the calculated envelopes (dashed lines): (a) compares MTK1 and MTK2, (b) compares MTK1 and MTK3, and (c) compares MTK2 and MTK3.

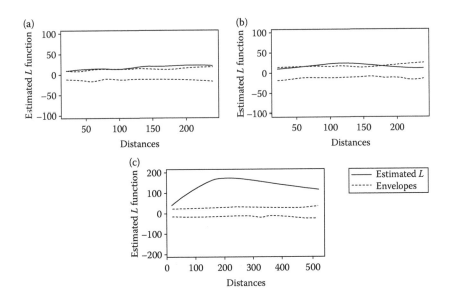

FIGURE 9.5
Investigation of the spatial dependence over a wide range of spatial scales. The estimated L function for each point process is positive and well above the calculated envelopes at all distances, a strong indication of clustering in the data: (a) MTK1, (b) MTK2, and (c) MTK3.

to improve upon the classification. Vegetation plots were fundamental to classification efforts and highlight the need for field sampling to guide class assignments. *D. nidulum* and *C. nymanii* dominant forest types were the most heterogeneous, with many of the tree species identified being common to both forest types. *N. starkenborghii* dominant forests were the most homogeneous and the easiest to visually separate, due in part to the difference in physiognomic features between each of the forest types.

GPS collars provided the position of three animals throughout the study period. However, these devices were designed as "store-on-board" GPS dataloggers. Therefore, manual radio-tracking was still necessary to collect devices and download the information once they had released from the animal. This can be a huge risk for studies with animals that have large home ranges or seasonal migration patterns. However, other types of devices now exist that allow for the remote download of information, even via cellular telephone technology (i.e., Lotek WildCell). Thus, this type of technology allows researchers to receive data in near real time while having only to send field teams out to collect the units when a mortality/drop-off signal has been received.

Data points received were located primarily throughout the *D. nidulum* dominant forest type, thus highlighting the importance in accurately classifying this type of forest. Additional radio-collaring since this time has also shown that the *N. starkenborghii* dominant forest type may also be important, as this forest type was used extensively by a young male that was radio-tracked during the 2006–2007 time period. Having a greater understanding of the habitat types that the animals utilize, along with areas where animals are absent, are essential aspects for conservation and management. In this way, remote sensing can provide detailed information about the areal extent of each of these habitat types and provide important base layers for use with relatively new, presence-only species habitat modeling techniques [i.e., MaxEnt (Phillips et al. 2006), GARP (Stockwell and Peters 1999), among others].

While GPS collars did function properly throughout the study period, the number of points collected by the collars was low. This was undoubtedly due to the diverse topography, dense forest cover, and poor satellite constellation. Even so, data were robust enough to calculate individual home ranges and movement patterns for each of the three animals. Home ranges calculated, however, were considerably smaller than those reported in Porolak (2008) and Porolak et al. (in review). From this more extensive analysis ($n = 15$, three of which were part of this study), Porolak (2008) estimates a home range that is roughly double what is reported here (fixed kernel: 90% UD = 68.7 ± 14.2 ha, 50% UD = 13.8 ± 2.9 ha). These data highlight the need for long-term ecological research, as studies over short periods and with small sample sizes can often have misleading results. However, as the goal of our research was to assess the function of these devices in this remote area, home-range estimates were not expected to be altogether reliable over this short timeframe.

In addition, the home ranges of *D. matschiei* reported by Porolak (2008) are some 40–100 times larger than those of any Australian tree kangaroo or other rainforest macropod reported in the literature. These differences could be due to the impact of hunting reducing the density of the species and thus allowing for greater resource availability per animal, the low productivity of the animals' high-altitude habitat, or a decreased level of fragmentation. All of these factors are important aspects to consider in conservation management plans. Used in combination with remote sensing information, one could calculate the carrying capacity of the species (less interspecific competition) throughout the area.

Movement analyses indicate that animal locations were clustered, implying that animals do not utilize their habitat uniformly. Instead, animals appear to return to specific areas, due either to a food resource that is provided, protection, or some other factor. This has important implications for conservation, as these data indicate that all habitats are not created equal. In addition, point processes (animal movements) were independent of one another. Porolak (2008), however, reports extensive overlap between males and other males/females, as well as between pairs of females. Spatial statistics of these point processes, however, have not been investigated to date, and so results are largely speculative. But, one could imagine that the spatial interactions between animals may vary greatly over time and season. Thus, investigating the interactions of animals (especially for related pairs) could provide interesting information about the structure of these communities and have important conservation implications in the long term.

Understanding how the structure and composition of vegetation influence habitat quality are central aspects to understanding the distribution and abundance of animals (Morrison et al. 1998). Here we describe a pixel-based application, used in combination with GPS radio-collar information, which can be used to aid conservation management plans. Optical remote sensing platforms can provide detailed information about the vegetation throughout tropical areas, but accuracy may be low due to the vast diversity of species that exist in some environments. However, other forms of remote sensing, such as light detection and ranging (or LiDAR), have also been shown to be important and can be used in combination with optical sources for predicting species abundance (e.g., Goetz et al. 2007). Using a combination of different remote sensing tools is likely to improve our knowledge about these threatened habitats and aid in making wise long-term conservation decisions, especially now given that the YUS Conservation Area, a locally owned and managed protected area of over 70,000 ha of rainforest, has been declared for this region.

9.6 Summary

Here we describe the use of two remote sensing data sources (Landsat-7 ETM+ satellite imagery and Televilt GPS-PosRec™ GPS collars) to provide detailed information for Matschie's tree kangaroo conservation in PNG. Used together, these data have improved our knowledge of the vegetation communities throughout the study area and increased our understanding of the movements of these animals. *D. nidulum* dominant forests were the most widespread forest found throughout the study area and were also where tree kangaroos were located. However, additional research has shown that this is are not the only forest type used by the species. GPS collars functioned properly throughout the study period, even though collection su ccess was low. Future GPS collar tracking studies using newer technology may provide more robust results. Animal locations were clustered and independent, indicating that animals do not utilize their habitat uniformly. These data provide vital information toward a better understanding of the habitat, the requirements of the animals, and the long-term conservation of the species.

Acknowledgments

This research was supported by Conservation International, the Roger Williams Park Zoological Society, the Conservation, Food, and Health Foundation, and the Rhode Island Chapter of the Surfrider Foundation. Research was carried out in the Department of Natural Resources Science at the University of Rhode Island. Many thanks go to Phil Shearman, director of the Remote Sensing Centre at the University of Papua New Guinea, Leonardo Salas, former animal population biologist at the Wildlife Conservation Society and now senior scientist at PRBO Conservation Science, and the director and staff of the Lae and Queensland Herbariums. We also thank Joel Glick, former TKCP field coordinator and other Tree Kangaroo Conservation Program staff. Most importantly, we thank the YUS community in Papua New Guinea and the many carriers, assistants, and landowners that made this research possible.

References

Andersen, M. 1992. Spatial analysis of two species interactions. *Oecologia* 91:134–140.
Anderson, D. J. 1982. The home range: A new nonparametric estimation technique. *Ecology* 63:103–112.
Beehler, B. M. 1993. *Papua New Guinea Conservation Needs Assessment*, Vol. II. Waigani, Papua New Guinea: Department of Environment and Conservation.
Bowman, J. L., C. O. Kochanny, S. Demarias, and B. D. Leopold. 2000. Evaluation of a GPS collard for white-tailed deer. *Wildlife Society Bulletin* 28:141–145.
Brooks, T. M., R. A. Mittermeier, G. A. B. Da Fonseca, J. Gerlach, M. Hoffman, J. F. Lamoreux, C. G. Mittermeier, J. D. Pilgrim, and A. S. L. Rodrigues. 2006. Global biodiversity conservation priorities. *Science* 313:58–61.

Congalton, R. G. 1991. A review of assessing the accuracy of classification of remotely sensed data. *Remote Sensing of Environment* 37:35–46.

Congalton, R. G., and K. Green. 1999. *Assessing the Accuracy of Remotely Sensed Data: Principles and Practices*. Florida: Lewis Publishers.

Dabek, L. 1994. Reproductive biology and behavior of captive female Matschie's tree kangaroos, *Dendrolagus matschiei*. PhD dissertation, University of Washington.

Diamond, J. 1997. *Guns, Germs, and Steel: The Fates of Human Societies*. New York: Norton.

Diggle, P. J., and A. G. Chetwynd. 1991. Second-order analysis of spatial clustering for inhomogenous populations. *Biometrics* 1153–1163.

Diggle, P. J. 2003. *Statistical Analysis of Spatial Point Patterns*, 2nd edition. London, UK: Oxford University Press.

Dinerstein, E., and E. D. Wikramanayake. 1993. Beyond "hotspots": How to prioritize investments to conserve biodiversity in the Indo-Pacific region. *Conservation Biology* 7:53–56.

FAO. 2010. *Global Forest Resources Assessment 2010. Main Report*. FAO Forestry Paper No. 140. Rome: FAO, 479pp.

Flannery, T. F. 1995. *Mammals of New Guinea*. New York: Cornell University Press.

Flannery, T. F. 1998. *Throwim Way Leg*. New York: Atlantic Monthly Press.

Global Land Cover Facility (GLCF), U.S. Geological Survey. Available at: www.landcover.org (accessed February 2005).

Goetz, S. J., D. Steinberg, R. Dubayah, and J. B. Blair. 2007. Lidar remote sensing of canopy habitat heterogeneity as a predictor of bird species richness in an eastern temperate forest, USA. *Remote Sensing of Environment* 108(3):254–263.

Goosem, S. 1992. *Cape York Peninsula Rainforest Survey pro forma and Explanatory Notes*. Internal report to the Queensland Department of Environment and Heritage, Cairns.

Henderson, R. F. J. 1997. *Queensland Plants: Names and Distribution*. Brisbane: Queensland Herbarium.

Hijmans, R. J., S. E. Cameron, J. L. Parra, P. G. Jones, and A. Jarvis. 2005. Very high resolution interpolated climate surfaces for global land areas. *International Journal of Climatology* 25:1965–1978. Available at: http://www.wordclim.org.

Hoft, R. 1992. *Plants of New Guinea and the Solomon Islands: Dictionary of the Genera and Families of Flowering Plants and Ferns*. Papua New Guinea: WAU Ecology Institute.

Hooge, P. N., and B. Eichenlaub. 1997. Animal Movement Extension to ArcView, version 1.1. Alaska Science Center—Biological Science Office, United States Geological Survey, Anchorage, AK, USA.

International Plant Names Index (IPNI). 2008. Published on the Internet at: http://www.ipni.org (accessed May 25, 2010).

International Union for the Conservation of Nature (IUCN). 2008. Red list of threatened species. Available at: http://www.redlist.org.

Jensen, J. R. 1996. *Introductory Digital Image Processing: A Remote Sensing Perspective*. Upper Saddle River, NJ: Prentice-Hall.

Laurance, S. G., and W. F. Laurance. 1999. Tropical wildlife corridors: Use of linear rainforest remnants by arboreal mammals. *Biological Conservation* 91:231–239.

Morrison, M. L., B. G. Marcot, and R. W. Mannan. 1998. *Wildlife–Habitat Relationships: Concepts and Applications*, 2nd edition. Madison, WI: University of Wisconsin Press, 435pp.

Kernohan, B. J., R. A. Gitzen, and J. J. Millspaugh. 2001. Analysis of animal space use and movements. In J. J. Millspaugh and J. M. Marzluff (eds), *Radio Tracking and Animal Populations*. San Diego, CA: Academic Press, pp. 125–166.

Mabberley, D. J. 1997. *The Plant-Book*, 2nd edition. Cambridge: Cambridge University Press.

Martin, M. E., S. D. Newman, J. D. Aber, and R. G. Congalton. 1998. Determining forest species composition using high spectral resolution remote sensing data. *Remote Sensing of Environment* 65:249–254.

MicroSoft. 2000. *MicroSoft Excel*. Seattle, WA: MicroSoft Corporation.

Moen, R., J. Pastor, Y. Cohen, and C.C. Schwartz. 1996. Effects of moose movement and habitat use on GPS collar performance. *Journal of Wildlife Management* 60(3):659–668.

Myers, N. 1992. *The Primary Source: Tropical Forests and Our Future*. New York: Norton.

Osborne, P. L. 1995. Biological and cultural diversity in Papua New Guinea: Conservation, conflicts, constraints, and compromise. *Ambio* 24:231–237.

Porolak, G. 2008. Home range of the Huon tree kangaroo (*Dendrolagus matschiei*) in cloud forest on the Huon Peninsula, Papua New Guinea. MS thesis, James Cook University.

Porolak, G., L. Dabek, and A. Krockenberger. In review. Spatial requirements of free-ranging Huon tree kangaroos, *Dendrolagus matschiei* (Macropodidae) in upper montane forest. *Biotropica*.

Phillips, S. J., R. P. Anderson, and R. E. Schapire. 2006. Maximum entropy modeling of species geographic distributions. *Ecological Modelling* 190:231–259.

Pugh, J. A. 2003. Identification of Huon tree kangaroo (*Dendrolagus Matschiei*) habitat in Papua New Guinea through integration of remote sensing and field observations. MS thesis. Kingston, RI: University of Rhode Island.

Raven, P. J. 1988. Our diminishing tropical forests. In E.O. Wilson (ed.), *Biodiversity*. Washington, DC: National Academy Press.

R Development Core Team. 2010. R: A language and environment for statistical computing. Vienna, Austria: R Foundation for Statistical Computing, ISBN 3-9000051-07-0. Available at: http://www.R-project.org.

Rempel, R. S., A. R. Rodgers, and K. Abraham. 1995. Performance of a GPS animal location system under boreal forest canopy. *Journal of Wildlife Management* 59: 543–551.

Seaman, D. E., and R. A. Powell. 1996. An evaluation of the accuracy of kernel density estimators for home range analysis. *Ecology* 77:2075–2085.

Seaman, D. E., J. J. Millspaugh, B. J. Kernohan, G. C. Brundige, K. J. Raedeke, and R. A. Gitzen. 1999. Effects of sample size on kernel home range estimates. *Journal of Wildlife Management* 63:739–747.

Shearman, P. L., J. Ash, B. Mackey, J. E. Bryan, and B. Lokes. 2009. Forest conversion and degradation in Papua New Guinea 1972–2002. *Biotropica* 41:379–390.

Silverman, B. W. 1986. *Density Estimation for Statistics and Data Analysis*. London, UK: Chapman and Hall.

SRTM. 2004. Shuttle Radar Topography Mission summary. Available at: srtm.usgs. gov/mission.html (accessed December 13, 2004).

Stabach, J. A. 2005. Utilizing remote sensing technologies to identify Matschie's tree kangaroo (*Dendrolagus matschiei*) habitat. MS thesis, University of Rhode Island.

Stabach, J. A., L. Dabek, J. Jensen, and Y.Q. Wang. 2009. Discrimination of dominant forest types for Matschie's tree kangaroo conservation in Papua New Guinea using high resolution remote sensing data. *International Journal of Remote Sensing* 30:405–422.

Stanton, P., and D. Fell. 2005. *The Rainforests of Cape York Peninsula*. Cairns: Cooperative Research Centre for Tropical Rainforest Ecology and Management, Rainforest CRC.

Stibig, H.-J., R. Beuchle, and F. Achard. 2003. Mapping of the tropical forest cover of insular Southeast Asia from SPOT-4 vegetation images. *International Journal of Remote Sensing* 24:3651–3662.

Stockwell, D. R. B., and D. P. Peters. 1999. The GARP modelling system: Problems and solutions to automated spatial prediction. *International Journal of Geographical Information Systems* 13:143–158.

Tracey, J. G. 1987. *The Vegetation of the Humid Tropical Region of North Queensland*. Melbourne: CSIRO.

Tucker, C. J., D. M. Grant., and J. D. Dykstra. 2004. NASA's global orthorectified Landsat data sets. *Photogrammetric Engineering & Remote Sensing* 70:313–322.

van Winkle, W. 1975. Comparison of several probabilistic home-range models. *Journal of Wildlife Management* 39:118–123.

Webb, L. J., J. G. Tracey, and W. T. Williams. 1976. The value of structural features in tropical forest ecology. *Australian Journal of Ecology* 1:3–28.

Worton, B. J. 1995. Using Monte Carlo simulation to evaluate kernel-based home range estimators. *Journal of Wildlife Management* 59:794–800.

10

Remote Sensing for Biodiversity Conservation of the Albertine Rift in Eastern Africa

Samuel Ayebare, David Moyer, Andrew J. Plumptre, and Yeqiao Wang

CONTENTS

10.1 Introduction

An important component of biodiversity conservation is understanding how environmental factors influence species abundance and distribution patterns, as well as how the factors change with time (Kerr and Ostrovsky 2003; Turner et al. 2003). The rapidly developing field of remote sensing has been invaluable to biodiversity conservation by providing the means of acquiring data about factors that affect species distribution and on which species distribution depend (Debinski et al. 1999). The field of remote sensing complements traditional field-based methods when collecting data about factors that affect biodiversity at different spatial and temporal frequencies.

In this chapter, we describe how aerial imagery acquired using the EnsoMOSAIC mapping system is being used to support biodiversity conservation in Madagascar, and Eastern and Southern Africa with a focus on the Albertine Rift. The Wildlife Conservation Society (WCS) Flight Program supports conservation projects in Madagascar, and Eastern and Southern Africa through aerial image acquisition and processing. The imagery acquired has been used to map threats to biodiversity, to develop land-use plans for protected area management, to measure vegetation cover and vegetation dynamics. Major threats to biodiversity conservation in the Albertine Rift are increasing human population, civil strife, and industrialization. These have led to large-scale land-use and land-cover change in the region. There has been a tremendous increase in the numbers of people leading to increased pressures on the natural resource base around protected areas. With the conversion of buffer zones to agricultural activities and subsequent competition between people and wild animals for the same resources, there has been an increase in human–wildlife conflict.

10.2 Land-Use and Land-Cover Change and Biodiversity Conservation

The influence of human activities on landscape patterns and processes is the leading cause of biodiversity loss and extinction at local, regional, and global scales (Young et al. 2005; Wenguang et al. 2008; Sawyer et al. 2009; Giam et al. 2010). Zones of conflict for resource use emerge within and adjacent to protected areas due to human population pressure and industrial activities that lead to direct and indirect impacts on biodiversity (Lindermana et al. 2005; Mena et al. 2006; Finer et al. 2008; Wilkie et al. 2000). Zones of conflict in resource use may involve less intense local-scale activities such as fuel wood collection to large-scale industrial activities such as logging and mining. Whereas local-scale activities such as fuel wood collection may have negligible impacts, their gradual accumulation within the landscape will cause changes that influence the way the landscape functions (Theobald et al. 1997; Lindermana et al. 2005). These changes will have a direct effect on biodiversity and on the quality of ecosystem functions provided to humans (Kerr and Ostrovsky 2003; Lindermana et al. 2005). With the rapid development in remote sensing technology, a wide range of information and metrics that contribute to our understanding of species distribution and environmental dynamics are becoming available to support conservation of flora and fauna (Kerr and Ostrovsky 2003; Turner et al. 2003; Leyequien et al. 2007). This has improved our understanding of the drivers of land-cover change within and adjacent to protected areas and has led to a reduction in the global rate of species and habitat loss (Turner et al. 2003; Mena et al. 2006;

Giam et al. 2010). Studies quantifying landscape patterns and dynamics in East and Central Africa have shown that the main drivers of change in land use and land cover are expanding agriculture and industrial activities such as mining and timber harvesting (Laporte et al. 2004; Duveiller et al. 2008; Harter et al. 2010). A major challenge in the application of remote sensing technology in the region is a lack of access to current imagery with high spatial resolution necessary for detecting land-use and land-cover changes at local scales. Aerial image acquisition and processing using the EnsoMOSAIC system provides local-scale high-resolution mapping and auxiliary information that can be used with coarse-resolution satellite imagery for quantifying landscape patterns. The system has been used in support of biodiversity conservation in Madagascar, and Eastern and Southern Africa.

10.3 Remote Sensing of the Albertine Rift

The intense geological process that shaped the East African Rift Valley system is seen from space as two great gashes that cut down through the continent. One starts in Eritrea dividing to run through Ethiopia, Kenya, and Tanzania to Malawi in the Gregory Rift and the other running through Sudan and along the border between the Democratic Republic of Congo (DRC), Uganda, Rwanda, Burundi, and Tanzania down to Zambia and Malawi in the Great Western or Albertine Rift (Figure 10.1). Tectonic movements that have influenced geological processes are also responsible for volcanic activity that has been taking place over the past 20,000 years. As a result, the extreme geological relief of the rift gives rise to spectacular scenery and diversity of habitats, characterized by some of Africa's highest mountains and a mosaic of forests, grasslands, swamps, glaciers, lakes, rivers, hot springs, gorges, waterfalls, and volcanic craters. All of those are sandwiched between valley walls that rise 2–4 km. The elevation range of the Rift Valley is from sea level to 5100 m. Tall mountains, such as the Ruwenzori Mountains that rise to 5100 m and the Virunga volcanoes that rise to 4500 m, together with the valley walls have separated populations of plants and animals and acted as barriers to dispersal, thus, driving vicariant speciation events. Thus, the Rift Valley region is a mega diversity area incorporated within the Eastern Afromontane hotspot, which also includes the Ethiopian Highlands and the Eastern Arc Forests of Tanzania and Kenya (Brooks et al. 2004; Plumptre 2004). The great variety of habitats and major altitudinal relief in the Albertine Rift supports one of the highest diversities of endemic plants and animals in Africa (Brooks et al. 2001; Olson et al. 2001; Kuper et al. 2004; Burgessa et al. 2006; Plumptre et al. 2007). The Albertine Rift, with 1762 species documented and with new species being discovered regularly, is a region with the most biodiversity in Africa for vertebrate

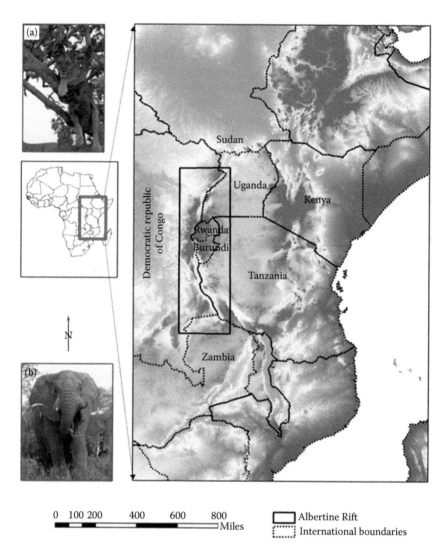

FIGURE 10.1
(**See color insert.**) The Albertine Rift region: (a) the sleeping lion and (b) elephant and acacia.

conservation. It also supports over 5700 plant species (Table 10.1). There are 1181 endemic species and 118 of these are threatened (Table 10.1). However, many more will doubtless be added to the threatened list when more taxa have been evaluated (Plumptre et al. 2007). Global biodiversity assessments have recognized the Albertine Rift as a biodiversity hotspot (Brooks et al. 2004), an endemic bird area (Stattersfield et al. 1998), and an ecoregion (Olson and Dinerstein 1998). It further contains four world heritage sites, two biosphere reserves, and four Ramsar sites (wetlands of international importance).

TABLE 10.1

The Total Number of Species, Number of Endemic Species, and Threatened Species for Seven Taxa in the Albertine Rift

Taxa	Species Richness	Percentage for Mainland Africa	Endemic Species	Threatened Species
Mammals	406	39.3	38	35
Birds	1061	52.3	41	25
Reptiles	175	13.6	16	2
Amphibians	112	19.2	36	16
Butterflies			117	
Fish			366	
Plants	5793	14.5	567	40

Note: Data on endemic fish species for each of the large lakes in the rift and a list of butterfly endemics were also compiled. This provides a minimum estimate of the number of endemic species for these two taxa.

Conservation efforts in the region have been hampered by challenges such as civil strife, a high rate of human population growth, expanding agriculture, logging, and mining. Only 50% of the total area of six landscapes critical for biodiversity conservation is currently protected (Plumptre et al. 2009). The actual percentage set aside for conservation varies greatly between individual landscapes. Some of the most important sites for conservation, such as the Itombwe Massif and Misotshi-Kabogo regions in the DRC, currently lack any protection. Utilizing remote sensing-derived information is of utmost importance in planning and management of the region's biodiversity as well as improving the livelihoods of the surrounding communities.

10.3.1 Data Acquisition and Processing

EnsoMOSAIC is a digital aerial imaging and image processing system developed by Stora Enso Oyj and the Technical Research Centre of Finland (EnsoMOSAIC 2005). It is a complete set of hardware and software from flight planning to producing georeferenced and orthorectified image mosaics. The WCS Flight Program acquired this system in 2006 and it has been used to map protected areas in six countries in Eastern and Southern Africa (Table 10.2). A total of approximately 60,000 km² of aerial image mosaics have been mapped so far (Figure 10.2).

Image acquisition was conducted using a 1974 Cessna 182P aircraft equipped with the EnsoMOSAIC system, a canon EOS-1 DS Mark II camera with a Canon EF 24 mm 1:1.4 L. The first steps of the process of image acquisition involve selecting the desired image resolution and flight planning. Numeric flight plans contain information necessary for imaging and image processing. Normally, several flight plans are calculated in order to evaluate optimal financial, human resource, and time investments. Financial and

TABLE 10.2

Sites Mapped for Each Country

Country	Areas Mapped
Uganda	Queen Elizabeth Conservation Area
	Mount Otzi
	East Madi/Murchison corridor
	Lake Albert shoreline
	Kabwoya-Kaiso Tonya
	Kidepo National Park
	Kidepo Zulia
	Agoro-Agu-lipan
	Kidepo-Nampore Forest Reserve
	Lake George shoreline
	Budongo sugar cane plantations
Democratic Republic of Congo	Virunga National Park
	Virunga Mount Hoyo
	Kabobo
	Itombwe
Madagascar	Masoala National Park
	Makira National Park
	Tsitongambarika
	Ambodiliatry
	Tampolo Reserve
Tanzania	Idodi
	Udzungwa Ruipa South
	Udzungwa MufindiEast
	Udzungwa Kilanzi Kitungulu
	Udzungwa Kigogo
	Udzungwa Ihangana
	Udzungwa Idewa
	Udzungwa Dabaga
	Rubeho Mafwomero
	Mahenge Nawenge
	Mahenge Nambiga
	Handeni Magambazi
	Udzungwa Image
	Rubeho Mangalissa
	Kitulo
	Mbizi Forest
Zambia	Luangwa
	Lukusuzi
	Lundazi
Rwanda	Nyungwe National Park

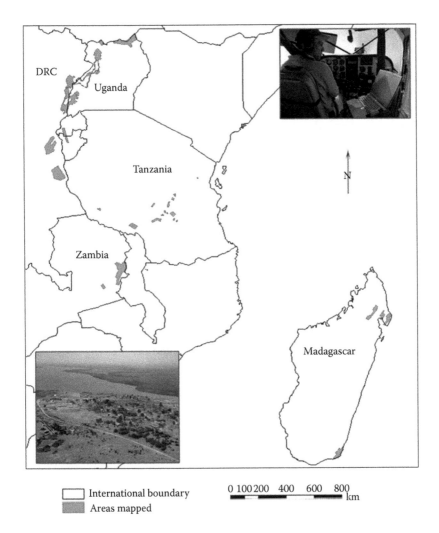

FIGURE 10.2
EnsoMOSAIC-mapped areas in Eastern Africa.

time constraints generally mean that final image resolution for most mapping projects is in the range of 50–100 cm pixels, although it is possible to acquire imagery with a resolution finer than a pixel size of 10 cm.

The EnsoMOSAIC aerial image processing system is designed to create a mosaic of a group of images consisting of many flight lines. Image linking, tie point measurement, and bundle block adjustment are the major steps involved in image rectification. Image linking provides the initial orientation of the images and tie point measurements provide image coordinates for bundle block adjustment which orthorectifies the images. Bundle block adjustment is an iterative mathematical process to solve the orientation and the location of perspective centers simultaneously for the image block. Ground control points can be added during tie point measurement to improve the accuracy of the final mosaic. However, an automatic aerial triangulation algorithm can also be used to rectify automatically thousands of digital images on one run to produce an orthorectified image mosaic or individual orthophotos. After bundle block adjustment is completed, mosaic resampling is carried out from the original images or individual orthorectified images. Spectral corrections can be carried out at various phases of the process and a digital elevation model produced to improve the quality of the final mosaic especially in undulating terrain. The geometrical accuracy ranges between one and two pixels when ground control points have been used in mosaic processing and between three and eight pixels without ground control points depending on terrain and flight conditions. The final output mosaic can be produced as a geocoded TIFF, ERS, and JPG.

10.3.2 Remote Sensing Case Studies

This section summarizes some of the projects in which the aerial image mosaics have been used to support biodiversity conservation in the Albertine Rift. The projects involve mapping threats to biodiversity, land-use planning, and habitat analysis. The approaches of remote sensing used in these case studies have had a great influence in planning research activities, implementation of management programs for protected areas, and aiding the process of creating new protected areas in the region.

10.3.2.1 Mapping Threats to Protected Areas

Remote sensing provides the means of deriving land-use data that describes human interactions and modification of the environment (Kerr and Ostrovsky 2003). The Greater Virunga Landscape (GVL) straddles the borders of Uganda, Rwanda, and the DRC. It is one of the six key landscapes in the Albertine Rift and is among the most species rich of any landscape in the world. It contains eight national parks, four forest reserves, and three wildlife reserves. Since 1996, protected areas in the GVL have suffered from high pressure due to high population density and the effects of civil war in the DRC. Displaced people have moved into Virunga National Park as a result of the insecurity and others have taken advantage of the situation to encroach upon and grab land and resources. In 2006, the EnsoMOSAIC system was used for aerial image acquisition over most of the savanna portions of Virunga National Park and in 2007 images were acquired for the forested northern part of the park. The resulting mosaics were used to digitize areas of human impacts in the landscape. It was possible to identify encroached areas where people were farming, areas where illegal logging had taken place, and individual settlements, including an enumeration of the houses. The results were used to highlight the problems in Virunga National Park and to target conservation interventions. For example, the number of build-ings was used to estimate the number of people that would need to be reset-tled (prior to the survey, the park staff had not been able to visit the settled areas to determine this because of insecurity). This was useful in discussions and planning of interventions with government, the protected area manage-ment authority, and development agencies when highlighting the plight of the park.

10.3.2.2 Mapping Vegetation Cover

Aerial image mosaics obtained in 2006 were used in developing the land-cover map of the GVL. The spatial resolution of the aerial image mosaics used for land-cover classification was 0.5 m. The classification system con-sisted of 29 land-cover classes including alpine, rock, lava, bare earth, heather, herbaceous, euphorbs, bamboo, bamboo-mixed forest, grassland, pastoral-ists grassland, bush/scrub, woodland, wooded grassland, lowland forest, tropical high forest, montane forest, riverine forest, degraded forest, papyrus swamp, swamp forest, other swamp, water/lake, salt water, tea, coffee, tree plantation, other agriculture, and settlements (Table 10.3). The land-cover classes were dependent on the scale of the grid used and were composed of variable proportions. A grid of 250×250 m was overlaid on the GVL mosaic and vegetation types assigned to each cell based on the visual interpretation by one observer. Areas where imagery was not acquired because of cloud cover were mapped from 2006 Advanced Spaceborne Thermal Emission and Reflection Radiometer (ASTER) imagery to fill in the gaps. The resulting veg-etation map (Figure 10.3) was used in the following conservation projects and the results from these studies were incorporated into the regional land-scape management plan and used in individual management plans for each of the protected areas covered.

10.3.2.2.1 Study of Corridors in the Landscape

Historically, the protected in the landscape have been managed indepen-dently increasing the risk of habitat fragmentation. Mapping corridors in the landscape provided the means of better understanding the dispersal and movement of wildlife between different protected areas. Due to civil strife in eastern DRC, the presence of corridors between Virunga National Park and Queen Elizabeth National Park allowed elephants to cross over to Uganda (Plumptre et al. 2007). Use of different vegetation types by elephants and lions could be mapped from Global Positioning System location data for these animals together with the detailed vegetation maps to show important crossing points between protected areas.

10.3.2.2.2 Predicting Ungulate Distribution and Density in the Landscape

Studies of species distribution patterns using remotely sensed data and Geographical Information System involve mapping the habitat and producing

TABLE 10.3

Land-Cover Types and Descriptions for Queen Elizabeth National Park

Land-Cover Type	Description
Papyrus swamp	Dense papyrus—more than 50% cover
Other swamp	Seasonally waterlogged areas with different vegetation—not papyrus
Swamp forest	Forest north of Lake George that is permanently flooded
Grassland	At least 20 m radius of grassland with no trees/shrubs
Wooded grassland	Between 10% and 50% woody cover—grassland under and between trees
Woodland	More than 50% woody cover—grassland between trees
Lowland forest	Trees and shrubs—at least 30% tree cover—trees generally less than 15 m tall
Tropical high forest	Trees only and most canopy trees greater than 15 m
Riverine forest	Narrow strips of trees along streams and rivers
Montane forest	Tropical high forest above 1500 m
Degraded forest	Greater than 50% canopy opening
Euphorbs	Euphorbia candelabra with at least 30% cover
Bush/Scrub	Low-stature bushes with little grass between—at least 50% cover
Tea	Tea forms at least 50% cover
Coffee	Coffee/low-stature tree crops form 50% cover
Tree plantation	Trees planted in rows—eucalyptus or pine
Pastoralists grassland	Grassland used for grazing cattle, goats, and sheep
Other agriculture	Any other short stature crops—cassava, potatoes, etc.
Bare earth	Less than 20% vegetation cover
Settlement	Human habitation and bare earth/roads covers at least 30% of land
Salt water	Crater lakes where water is known to be saline
Water/Lake	Water in major lakes and crater lakes
Lava	Lava erupted from active volcanoes with little vegetation colonizing it
Heather	Areas of giant heather on the mountains above the bamboo zone
Bamboo	Uniform stand of bamboo
Bamboo-mixed forest	Bamboo interspersed with forest
Herbaceous	Dense herbaceous vegetation with no tree cover
Alpine	Small herbs and grasses above the bamboo and heather zones
Rock	Areas of rock with no vegetation cover

predictive habitat distribution models for the species of interest (Nagendra 2001; Wenguang et al. 2008). Understanding the factors that determine the distribution patterns and density of ungulate species helped managers better understand what factors need to be managed to increase the numbers of a particular ungulate species. For example, Uganda Kob (*Kobus kob*) preferred short-grass habitat on sites that had burned frequently.

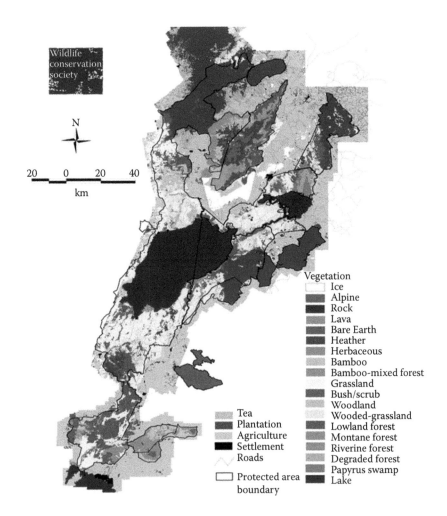

Tea
Plantation
Agriculture
Settlement
Roads

Protected area
boundary

Vegetation
Ice
Alpine
Rock
Lava
Bare Earth
Heather
Herbaceous
Bamboo
Bamboo-mixed forest
Grassland
Bush/scrub
Woodland
Wooded-grassland
Lowland forest
Montane forest
Riverine forest
Degraded forest
Papyrus swamp
Lake

FIGURE 10.3
(See color insert.) Vegetation map of the Greater Virunga Landscape.

10.3.2.2.3 Assessment of Ranging Patterns and Habitat Use of Radio-Collared Elephants and Lions

Species that range widely such as elephants and lions need the whole landscape in order to survive. Understanding the factors that influence their ranging patterns using spatially derived parameters is essential in guiding management decisions of park managers. Assessments were made of habitat associations by elephants and lions in Queen Elizabeth National Park, and in the case of lions associations with prey biomass was included. These led to predictions of potential densities in parts of Virunga National Park where security was precluding access (e.g., Treves et al. 2009).

10.3.2.3 Habitat Change Analysis

Habitat change analysis involves performing repeated inventories of a land-scape, developing models necessary to simulate the processes taking place therein, and evaluating consequences of observed and predicted changes. Understanding the effects of habitat change on species abundance over large geographical areas is essential for biodiversity conservation (Mason et al. 2003). Remote sensing techniques enable habitat characterization and mea-surement of spatial predictors for testing wildlife-habitat models. The EnsoMOSAIC system provides the users with the capability of processing historical aerial images. The imageries can be scanned and processed using EnsoMOSAIC into an orthomosaic. The time series of orthomosaics thus enabled vegetation change analysis. For example, aerial images of Queen Elizabeth National Park obtained in 1955 were scanned and processed using EnsoMOSAIC. The vegetation was mapped in the same way using the same categories as for the 2006 imagery above (Table 10.3). The same grid of vege-tation cells was used in both periods to enable change analyses to be made. The orthomosaic was then compared to the mosaic obtained in 2006 to assess habitat change (Figure 10.4). Data on animal distribution and abundance were integrated into the analysis to assess how habitat change over the past 50 years has affected the species surveyed. During the 1970s, the park lost many of its large mammals to poaching and numbers have only slowly recovered to present-day level. The elephant and the hippopotamus in par-ticular are known to have major impacts on the vegetation and their absence or decline in abundance will lead to rapid changes in cover of woody and herbaceous vegetation. The results show that the woody cover in the park increased since the 1950s because of the decline in the elephant population. However, in some areas there has been a loss in woody cover which is most probably due to changes in the recent past of ranging patterns of large mammal species and their concentration in those areas (possibly to avoid poaching). Similarly, with the decline in the hippopotamus population the aquatic vegetation at the edge of Lake Edward has expanded. Habitat change studies allow continuous monitoring of the factors that affect an areas envi-ronmental evolution and how they change with time.

10.3.2.4 Ground Referencing for Satellite Imagery

Ground referencing of land-cover classes is a fundamental component to any mapping exercise. Land-cover maps obtained from low- and medium-resolution satellite imagery provide general landscape information. However, ground referencing data are necessary for verification of land-cover classes. High-resolution aerial photo mosaics were used to provide auxiliary infor-mation for land-cover maps generated from satellite imagery to compliment traditional labor-intensive and expensive site visits. The Murchison–Semuliki landscape was mapped using ASTER satellite imagery acquired in 2006 to

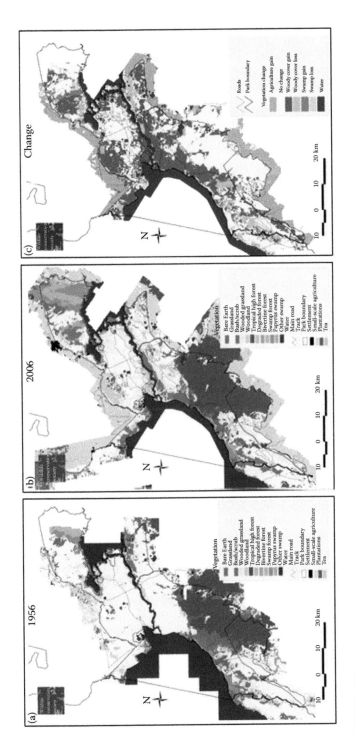

FIGURE 10.4
Vegetation cover and change of Queen Elizabeth National Park between 1956 and 2006.

assess the amount of remaining forest cover. A method that involved physical delineation of the vegetation-cover classes based on an existing vegetation-cover map was used to develop the land-cover map. High-resolution aerial photos obtained in 2007 were used for validation of land-cover types. The land-use and land-cover maps have been used for different applications that include

1. Carbon loss in the region was estimated by measuring the amount of intact forests and the rate of deforestation. Plant data were used to estimate average carbon per hectare.
2. It was also noted from the land-use/land-cover map that there was high forest loss within the landscape due to subsistence farming. However, regrowth was observed in abandoned farm plots.
3. Potential corridors were mapped for animal movements between Murchison and Semuliki landscapes.

EnsoMOSAIC imageries were also used in ground referencing land-cover and land-use maps in northern Uganda. The end of the northern Uganda insurgency that had lasted for 20 years raised several questions about environmental protection as people planned to return to their homes. The results indicated that there was an increase in woody vegetation in northern Uganda and subsequent projects were aimed at developing environmental management plans with district authorities.

10.3.2.5 Assessment and Monitoring of Oil Exploration and Production Activities

Increasing global demand for energy has exacerbated risks to biodiversity conservation from oil and gas exploration, and development and production projects (Copeland et al. 2009; Finer et al. 2008; Sawyer et al. 2009). Recent discoveries of oil and natural gas deposits in the Albertine Rift have raised serious concerns about the impacts of industrial activities and associated land-cover changes on wildlife. Studies quantifying the responses of wildlife behavior to industrial disturbances have shown that the construction of facilities, such as roads, trails, and buildings, and the increased presence of human activities beyond a certain threshold resulted in a direct loss of wildlife, their habitats, and indirect habitat loss following avoidance behavior of affected wildlife species (Theobald et al. 1997; Wilkie et al. 2000; Dyer et al. 2001). In addition to influencing patterns of habitat use, avoidance behavior may result in cumulative social and physiological consequences that may lead to implications on population productivity (Johnson et al. 2005). The impact of oil and gas exploration activities on wildlife may be direct and obvious such as replacement of the native habitats with well pads, access roads, and pipelines but the physical footprints may be insignificant relative to the functional habitat loss due to avoidance (Dyer et al. 2001; Sawyer et al.

2009). The EnsoMOSAIC system is being used for image acquisition in the Albertine Rift to monitor impacts of oil and gas explorations on biodiversity conservation. Kabwoya Wildlife Reserve, located between the rift escarpment and Lake Albert, is an important corridor for wildlife migrations in the region. It is among the protected areas in which oil prospecting and exploratory well drillings are taking place. Aerial imageries were acquired over Kabwoya in 2008 to provide a baseline for future monitoring activities. The planned location of an early production system near Kabwoya with a targeted production capacity of 4000 barrels per day of heavy fuel oil, diesel, kerosene, and naptha was a major cause for concern with regard to direct and indirect impacts on the reserve by this facility. The environmental impacts likely to arise from oil refineries are toxic air and water emissions, accidental releases of chemicals, hazardous waste disposal, thermal pollution, and noise pollution (O'Rourke and Connolly 2003). Maps derived from the EnsoMOSAIC imagery of Kabwoya were used to pinpoint alternative sites for the Early Production System that would minimize the impacts on the area as much as possible.

10.3.2.6 *Participatory Planning of New Protected Areas*

The participatory planning approach for new protected areas is aimed at improving the level to which the surrounding communities participate in, and benefit from wildlife resources. Many protected areas in Africa were created with little or no participation of local communities. As those communities view protected areas as restricting their livelihoods, lack of consultation with the local communities in the establishment of protected areas in many parts of Africa has been among the causes of conflicts between park authorities and adjacent communities. It was recognized that the participation of communities adjacent to protected areas was necessary for the successful creation and management of these areas (Borrini-Feyerabend 1996). Some of the approaches for involving local communities in the management of protected areas in East Africa include collaborative management and protected area outreach. Both approaches attempt to improve long-term conservation goals and livelihoods. Participatory approaches for local communities to be involved in protected area management and creation have been responsible for changes in community attitudes toward wildlife. Maps of vegetation and human settlement derived from EnsoMOSAIC imagery from the Itombwe Massif and Misotshi-Kabogo regions of eastern DRC were used in planning with local communities to agree on new protected area boundaries. The identification of size and distribution of human settlements in the orthomosaic of the area allowed an *a priori* boundary demarcation to be made that minimized the number of settlements included in the new protected area. It also facilitated the work of ground teams sent to local communities to explore the concept of gazettment of a new protected area and to discuss demarcation of different conservation zones for the two new protected areas.

10.4 Concluding Remarks

Based on the remote sensing case studies in the Albertine Rift, EnsoMOSAIC-acquired imageries highlight the importance of examining the spatial–temporal processes for biodiversity conservation. The acquisition of remote sensing data using the EnsoMOSAIC system provides timely spatial and temporal information essential for supporting data corrected by traditional field-based methods for biodiversity conservation. Various factors such as cost, size of the area of interest, spatial resolution, and weather conditions determine the choice between using aerial photography or satellite images. Although the availability and spatial resolution of satellite images have improved over the past decades, the use of the EnsoMOSAIC system provides flexibility during acquisition and processing of the aerial images for biodiversity conservation. The EnsoMOSAIC system addresses some of the challenges of acquiring remote sensing data in the region as follows: by enabling below cloud cover mapping and flexibility in determining the final spatial resolution of the area of interest; aerial reconnaissance surveys allow better planning for ground visits to follow initial aerial exploration of remote sites like in the DRC and we have also found that it can complement satellite image analysis and provide more detailed data at relatively low cost.

Acknowledgments

This study was funded by many different sources including the Wildlife Conservation Society, U.S. Agency for International Development (USAID), the John D. and Catherine T. MacArthur Foundation, Daniel K. Thorne Foundation, Panthera, U.S. Fish and Wildlife Service, Whalesback Foundation, and World Wide Fund for Nature. Many people were involved in the EnsoMOSAIC aerial image acquisition and processing at the WCS GIS and remote sensing hub, Kampala Uganda. We would like to particularly thank Guy Picton Phillipps and Grace Nangendo for their guidance in aerial image processing. We also thank Timothy Akugizibwe, Emmanuel Ourum, Paul Mulondo, and Ivan Buyondo for their work on image processing and vegetation mapping.

References

Brooks, T., Balmford, A., Burgess, N., Fjeldsa, J., Hansen, L.A., Moore, J., Rahbek, C., and Williams, P. 2001. Towards a blueprint for conservation in Africa. *BioScience* 51, 613–624.

Brooks, T., Hoffmann, M., Burgess, N., Plumptre, A., Williams, S., Gereau, R.E., Mittermeier, R.A., and Stuart, S. 2004. Eastern Afromontane. In R.A. Mittermeier, P. Robles-Gil, M. Hoffmann, J.D. Pilgrim, T.M. Brooks, C.G. Mittermeier, J.L. Lamoreux, and G. Fonseca (eds), *Hotspots Revisited: Earth's Biologically Richest and Most Endangered Ecoregions*, 2nd edition. Mexico: Cemex, pp. 241–242.

Borrini-Feyerabend, G. 1996. *Collaborative Management of Protected Areas: Tailoring the Approach to the Context*. Issues in Social Policy. Gland, Switzerland: IUCN.

Burgessa, N.D., Hales. D.J, Ricketts, H.T., and Dinerstein, E. 2006. Factoring species, non-species values and threats into biodiversity prioritisation across the ecoregions of Africa and its islands. *Biological Conservation* 127, 383–401.

Copeland, H.E., Doherty, K.E., Naugle, D.E., Pocewisz, A., and Kiesecker, J.M. 2009. Mapping oil and gas development potential in the US intermountain west and estimating impacts to species. *PloS ONE* 4(10), e7400, doi: 101371/journal0007400.

Debinski, D.M., Kindscher, K., and Jakubauskas, M.E. 1999. A remote sensing and GIS-based model of habitats and biodiversity in the Greater Yellowstone Ecosystem. *International Journal of Remote Sensing* 20(17), 3281–3291.

Duveiller, G., Defourny, P., Desclée, B., and Mayaux, P. 2008. Deforestation in Central Africa: Estimates at regional, national and landscape levels by advanced processing of systematically-distributed Landsat extracts. *Remote Sensing of Environment* 112, 1969–1981.

Dyer, J., S., O'Neill, P.J., Wasel, M.S., and Boutin, S. 2001. Avoidance of industrial development by woodland Caribou. *The Journal of Wildlife Management* 65(3), 531–542.

EnsoMOSAIC. 2005. *EnsoMOSAIC Mosaicking User's Guide*. Finland: Stora Enso Oyj.

Finer, M., Jenkins, C.N., Pimm, S.L., Keane, B., and Ross, C. 2008. Oil and gas projects in the Western Amazon: Threats to wilderness, biodiversity, and indigenous peoples. *PLoS ONE* 3(8), e2932, doi: 10.1371/journal.pone.0002932.

Giam, X., Bradshaw, J.A.C., Tan, T.W., and Sodhi, S. 2010. Future habitat loss and the conservation of plant biodiversity. *Biological Conservation* 143, 1594–1602.

Harter, J., Southworth, J., and Binford, M. 2010. Chapter 12—Parks as a mechanism to maintain and facilitate recovery of forest cover: Examining reforestation, forest maintenance and productivity in Uganda. In H. Nagendra and J. Southworth (eds), *Reforesting Landscapes: Linking Pattern and Process*, Landscape Series, Vol. 10. Dordrecht: Springer Science + Business Media B.V., doi: 10.1007/978-1-4020-9656-3_12.

Johnson, J.C., Boyce, S.M., Case, R.L., Cluff, D.H., Gau, J.R., Gunn, A., and Mulders, R. 2005. Cumulative effects of human developments on Arctic Wildlife. *Wildlife Monographs* (160), 1–36.

Kerr, T.K., and Ostrovsky, M. 2003. From space to species: Ecological applications for remote sensing. *Trends in Ecology and Evolution* 18(6), 299–305.

Kuper, W., Sommer, H.J., Lovett, C.J., Mutke, J., PeterLinder, H., Beentje, J.H., Van Rompaey, R.A.S.R., Chatelain, C., Sosef, M., and Barthlott, W. 2004. Africa's hotspots of biodiversity redefined. *Annals of the Missouri Botanical Garden* 91(4), 525–535.

Laporte, T.N., Lin, S.T., Lemoigne, J., Devers, D., and Honzak, M. 2004. Chapter 6: Towards an operational forest monitoring system for Central Africa. In G. Garik, A.C. Janetos, C.O. Justice, E.F. Moran, J.F. Mustard, R.R Rindfuss, D. Skole, B.L.

Turner, II, and M.A. Cochrane (eds), *Landscape Change Science: Observing, Monitoring and Understanding Trajectories of Change on the Earth's Surface.* Dordrecht, The Netherlands: Kluwer Academic Publishers, pp. 97–110.

Leyequien, E., Verrelst, J., Slot, M., Schaepman-Strub, G., Heitkonig, M.A.I., and Skidmore, A. 2007. Capturing the fugitive: Applying remote sensing to terrestrial animal distribution and diversity. *International Journal of Applied Earth Observation and Geoinformation* 9, 1–20.

Lindermana, A.M., An, L., Bearer, S., He, G., Ouyang, Z., and Liu, J. 2005. Modeling the spatio-temporal dynamics and interactions of households, landscapes, and giant panda habitat. *Ecological Modelling* 183, 47–65.

Mason, D.C., Anderson, G.Q.A., Bradbury, R.B., Cobby, D.M., Wilson, J.D., Davenport, I.J., and Vanepoll, M. 2003. Measurement of habitat predictor variables for organism–habitat models using remote sensing and image segmentation. *International Journal of Remote Sensing* 24(12), 2515–2532.

Mena, F.C., Barbieri, F.A., Walsh, J.S., Erlien, M.C., Holt, L.F., and Bilsborrow, E.R. 2006. Pressure on the Cuyabeno Wildlife Reserve: Development and land use/cover change in the Northern Ecuadorian Amazon. *World Development* 34(10), 1831–1849.

Nagendra, H. 2001. Using remote sensing to assess biodiversity. *International Journal of Remote Sensing* 22(12), 2377–2400.

Olson, D.M., and Dinerstein, E. 1998. The global 200: A representation approach to conserving the earth's most biologically valuable ecoregions. *Conservation Biology* 12, 502–515.

Olson, D.M., Dinerstein, E., Wikramanayake, D.E., Burgess, D.N., Powell, N.V.G., Underwood, C.E., D'amico, A.J. et al. 2001. Terrestrial ecoregions of the world: A new Map life on Earth. *BioScience* 51(11), 933–938.

O'Rourke, D., and Connolly, S. 2003. Just oil? The distribution of environmental and social impacts of oil production and consumption. *Annual Reviews of Environment and Resources* 28, 587–617.

Plumptre, A.J. 2004. Priority sites for conservation in the Albertine Rift and the importance of transboundary collaboration to preserve landscapes. In D. Harmon and G.L. Worboys (eds), *Managing Mountain Protected Areas: Challenges and Responses for the 21st Century.* Italy: Andromeda Editrice, pp. 233–238.

Plumptre, A.J., Kujirakwinja, D., and Nampindo, S. 2009. Conservation of landscapes in the Albertine Rift. In K.H. Redford and C. Grippo (eds), *Protected Areas, Governance and Scale.* Wildlife Conservation Society Working Paper No. 36, pp. 27–34.

Plumptre, A.J., Tim, R.B.D., Mathias, B., Robert, K., Gerald, E., Paul, S., Corneille, E. et al. 2007. The biodiversity of the Albertine Rift. *Biological Conservation* 134, 178–194.

Sawyer, H., Kauffman, J.M., Nielson, M.R. 2009. Influence of well pad activity on winter habitat selection patterns of mule deer. *Journal of Wildlife Management* 73(7), 1052–1061.

Stattersfield, A.J., Crosby, M.J., Long, A.J., and Wege, D.C. 1998. *Endemic Bird Areas of the World: Priorities for Biodiversity Conservation.* Birdlife International Conservation Series No. 7. Cambridge: Birdlife International.

Theobald, M.D., Miller, R.J., and Hobbs, N.T. 1997. Estimating the cumulative effects of development on wildlife habitat. *Landscape and Urban Planning* 39, 25–36.

Treves, A., Plumptre, A. J., Hunter, L.T.B., and Ziwa, J. 2009. Identifying a potential lion *Panthera leo* stronghold in Queen Elizabeth National Park, Uganda, and Parc National des Virunga, Democratic Republic of Congo. *Oryx*, 43(1), 60–66, doi:10.1017/S003060530700124X.

Turner, W., Spector, S., Gardiner, N., Fladeland, M., Sterling, E., and Steininger, M. 2003. Remote sensing for biodiversity science and conservation. *Trends in Ecology and Evolution* 18(6), 306–314.

Wilkie, D., Shaw, E., Rotberg, F., Morelli, G., and Auzel, P. 2000. Roads, development, and conservation in the Congo Basin. *Conservation Biology* 14(6), 1614–1622.

Wenguang, Z., Yuanman, H., Jinchu, H., Jing, J., and Miao, L. 2008. Impacts of land-use change on mammal diversity in the upper reaches of Minjiang River, China: Implications for biodiversity conservation planning. *Landscape and Urban Planning* 85, 195–204.

Young, J., Watt, A., Nowicki, P., Alard, D., Clitherrow, J., Hienle, K., Johnson, R. et al. 2005. Towards sustainable land use: Identifying and managing the conflicts between human activities and biodiversity conservation in Europe. *Biodiversity and Conservation* 14, 1641–1661.

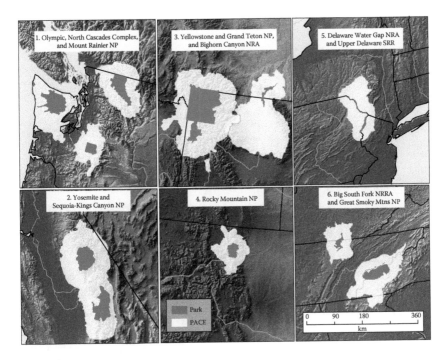

FIGURE 2.2
Maps of protected-area-centered ecosystems (PACEs) delineated in this study for 13 U.S. National Park Service units. PACEs were defined by the criteria in Table 2.4.

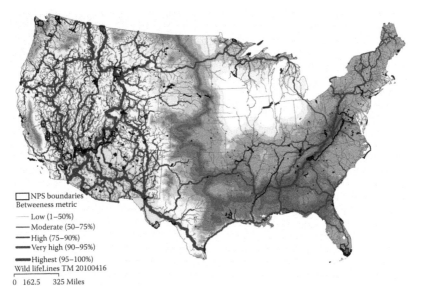

FIGURE 2.4
Map showing connectivity of natural landscapes in the United States. The thickness of red lines indicates magnitude of cumulative movement, assuming that animals avoid human-modified areas. The surface underneath the pathways depicts the averaged cost–distance surfaces, or the overall landscape connectivity surface. Colors range from green through yellow and purple to white, where green is greatest connectivity (lowest travel cost) and white indicated lowest connectivity (highest travel cost). National Park Service units are outlined in black.

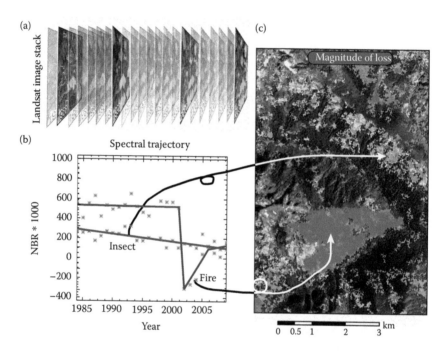

FIGURE 3.3

Temporal segmentation of the spectral trajectory using LandTrendr for data recorded by two pixels in a stack of Landsat images for North Cascades National Park (a). The record of spectral data, here represented by the normalized burn ratio (NBR), bounces up and down from year to year (asterisks), but a set of mathematical algorithms can detect longer-term trends and abrupt changes in those trends (b). The magnitude of change, as captured by the difference of end-points of fitted segments, can be mapped to show patterns of disturbance on a landscape (c).

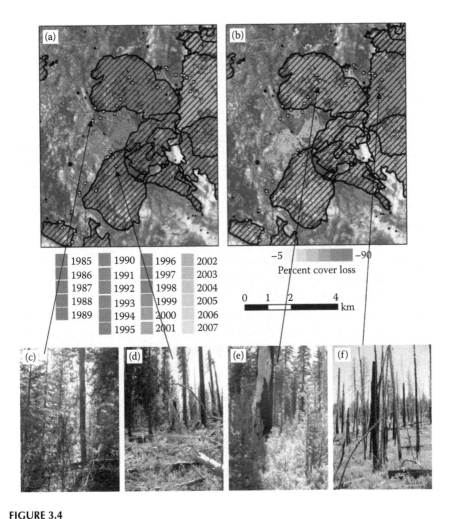

FIGURE 3.4
Fire plays an important role in many national parks of the western United States, including Yosemite National Park shown here. Maps of year of fire disturbance (a) and estimated percent vegetation cover loss (b) show a mosaic of timing and intensity of fires. Field-visited locations show the range of both fire-severity and post-fire recovery, ranging from unburned (c) to recently burned with substantial cover loss (f).

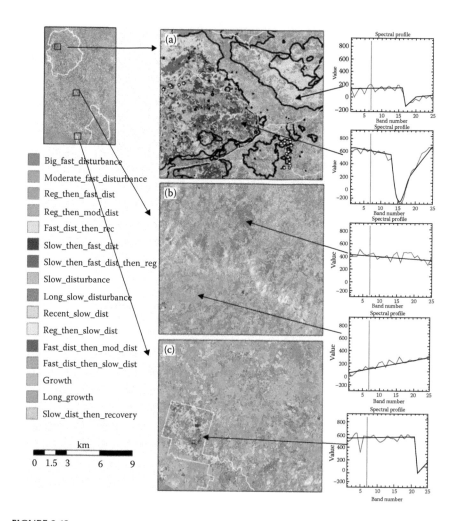

FIGURE 3.12
By grouping pixels with similar sequences of mortality and growth segments, the rich dynamics in and around national parks can be revealed. Here, a suite of change classes show variability in pre- and postfire processes (a), growth of forest after harvest as well as slow decay caused by insects (b), and prescribed burning in a matrix of rapidly growing forest (c).

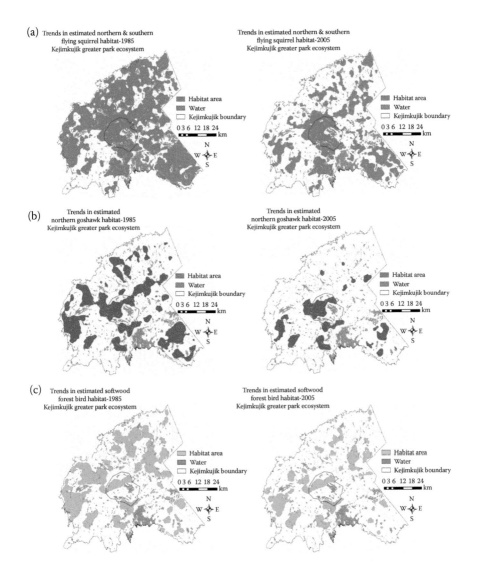

FIGURE 5.2
Estimated habitat amount from 1985 to 2005 for (a) northern and southern flying squirrel, (b) northern goshawk, and (c) softwood forest birds.

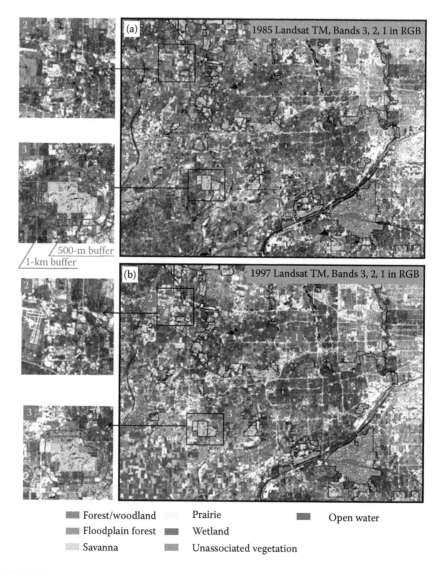

FIGURE 6.3
Comparative display illustrating land-cover change between 1985 and 1997 in the west suburb of Chicago. Site 1 depicts the newly built I-355 tollway which is identifiable in the 1997 image but does not exist in the 1985 image. Site 2 illustrates the area around the West Chicago Natural Preserve and the nearby constructed DuPage County airport. Site 3 indicates the impact of the newly developed residential area on the Springbrook Natural Preserve. Most protected natural areas in this subset are isolated and fragmented by urban land.

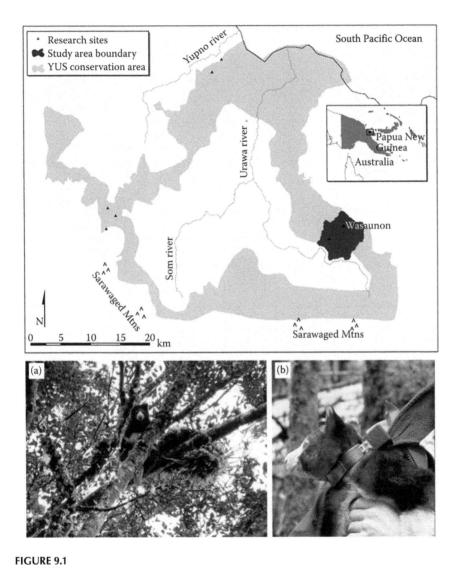

FIGURE 9.1

Map of the Tree Kangaroo Conservation Program research area, Morobe Province, Papua New Guinea. (Adapted from Stabach, J. A. et al. 2009. *International Journal of Remote Sensing* 30:405–422.) Endemic to the upper montane forests of the Huon Peninsula in Papua New Guinea, Matschie's tree kangaroo (*Dendrolagus matschiei*) are arboreal marsupials (a). A Matschie's tree kangaroo was fitted with a Televilt GPS-PosRec™ GPS collar (b). (Photos provided by the Tree Kangaroo Conservation Program of the Woodland Park Zoo. With permission.)

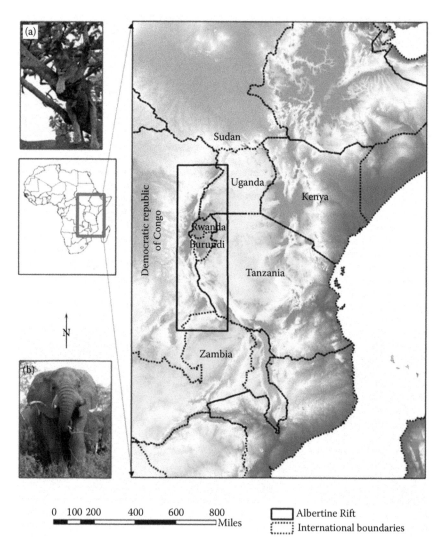

FIGURE 10.1
The Albertine Rift region: (a) the sleeping lion and (b) elephant and acacia.

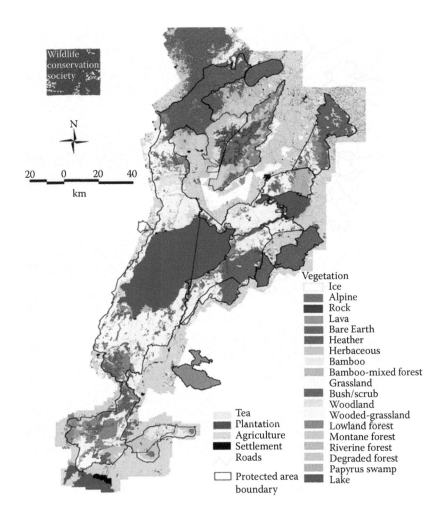

FIGURE 10.3
Vegetation map of the Greater Virunga Landscape.

FIGURE 11.1

The Changbai Mountain range consists of paralleled sections of mountains. It extends toward the west connecting the Qian-Shan Mountains in the Liaodong Peninsula in China and toward the south connecting the Korean Peninsula. In the north is the Northeast China Plain and in the east lie the Sikhote-Alin Mountains in the Russian Far East. The Shuttle Radar Topography Mission illustrates the terrain of the Changbai Mountain range and the volcanic summit in the center (a). The needleleaf and broadleaf mixed forest zone is distributed between 700 and 1100 m (b). The coniferous forest zone is distributed between 1100 and 1700 m, with dominant species of spruce (*Pieca jezoensis, Pieca koreana*) and fir (*Abies nephrolepis*) (c). Between 1700 and 2000 m distributes the zone of subalpine dwarf birch (*Betula ermanii*) forest (d). The alpine tundra zone is distributed above 2000 m, with representative species such as short rhododendron shrubs (*Rhododendron chrysanthum Pall*) and *Vaccinium uliginosum* L. (e).

FIGURE 11.2
Landsat image of the Changbai Mountain region illustrates the mountain range and the highest section where the Changbai Mountain Nature Reserve is situated on the border between China and North Korea. Changbai pines (*Pinus sylvestris var. sylvestriformis*) grow on the northern slope of the volcanic summit of the Changbai Mountain (a). Species inhabiting the forests of the Changbai Mountain region include endangered Amur (Siberian) tiger (*Panthera tigris altaica*) (b, c). (Photos by Yeqiao Wang.)

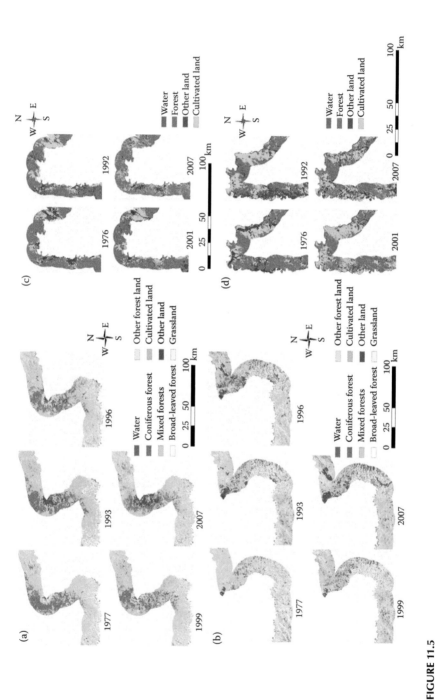

FIGURE 11.5

Land cover maps of 10-km buffer areas along the Yalu and Tumen Rivers for the Changbai Mountain site in China (a) and North Korea (b) and for the Tumen River site in China (c) and North Korea (d) for land-use and land-cover change analysis.

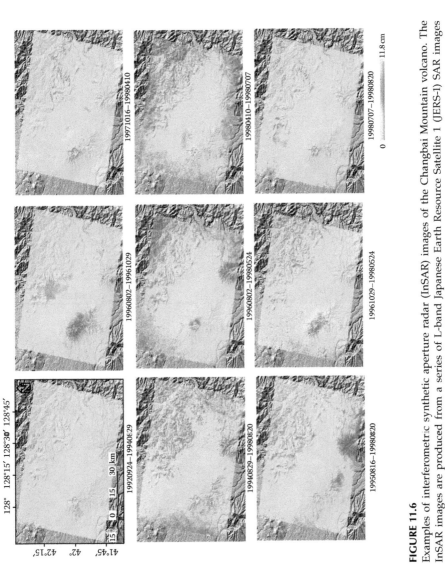

FIGURE 11.6
Examples of interferometric synthetic aperture radar (InSAR) images of the Changbai Mountain volcano. The InSAR images are produced from a series of L-band Japanese Earth Resource Satellite 1 (JERS-1) SAR images acquired between 1992 and 1998.

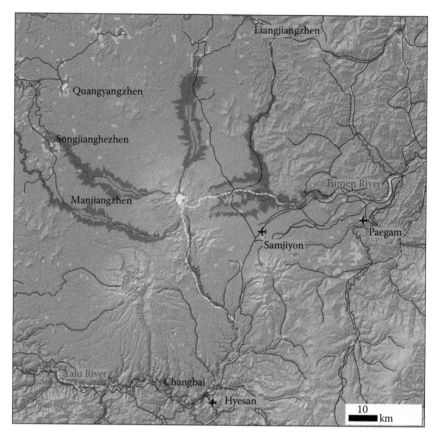

FIGURE 11.8
Distal hazard zones based on three scenarios of lahar volumes of 10 million cubic meters (orange), 100 million cubic meters (yellow), and 1 billion cubic meters (brown) are mapped for the Changbai Mountain volcano with indication of nearby population centers larger than 100 per square kilometers (red) and 1000 per square kilometers (pink).

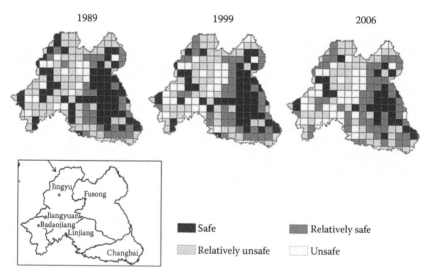

FIGURE 11.10
Spatial variation of ecological security index of Baishan City in grid cells.

FIGURE 12.2
Shuttle Radar Topography Mission and derivative topographic information (elevation, slope, aspects) provide zonal guidance and reference to differentiate vertical vegetation distribution from Landsat and IKONOS data.

Betula ermanii
Tundra/grassy
Tundra/shrub
Tundra/barren
Bared volcanic rock

FIGURE 12.3
Classification of IKONOS data to identify detailed representative species in particular in transition areas between vertical vegetation zones. Field survey confirmed the colonization of pioneer species of subalpine *Betula ermanii* and *Larix olgensis* into the tundra zone at high altitudes of the northern slope.

FIGURE 12.4
Samples of carbonized woods documented the species and number of observations of each species at different altitudes and slopes/aspects of the area surrounding the volcanic summit to reconstruct preeruption vegetation structure and distribution.

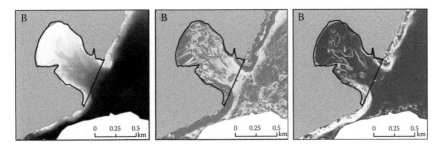

FIGURE 13.4
Bathymetry, slope, and rugosity for Hanauma Bay Marine Life Conservation District on Oahu, Hawaii, derived from light detection and ranging data. (Adapted from Wedding, L. and A. M. Friedlander, *Marine Geodesy* 31, 246–266, 2008.)

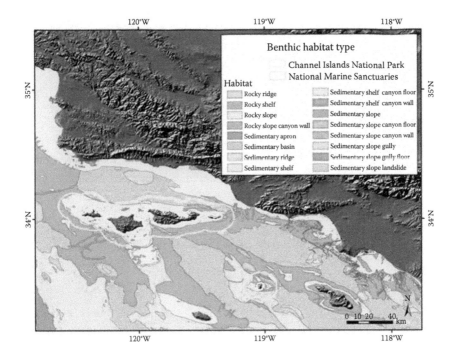

FIGURE 13.9
Benthic habitat map for Channel Islands National Marine Sanctuary, Channel Islands National Park, and surrounding waters.

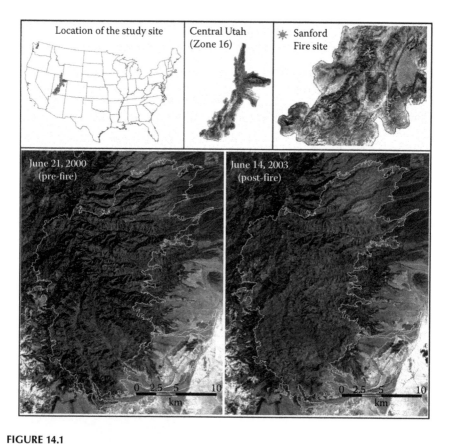

FIGURE 14.1
The Sanford Fire site is located in the southern section of the Central Utah Valley (Zone 16) of the LANDFIRE prototype data products. A comparison of pre- and post-fire Landsat Thematic Mapper reflectance images illustrates the affected areas and fire impacts on vegetation.

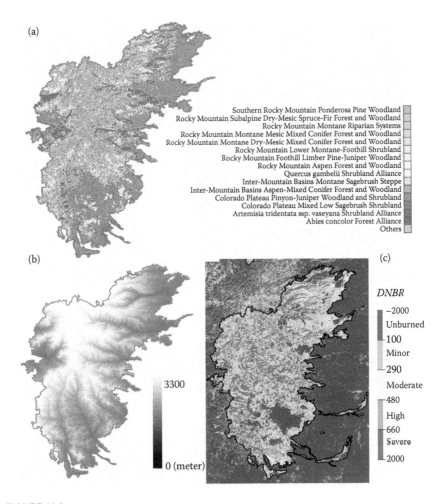

(a)

Southern Rocky Mountain Ponderosa Pine Woodland
Rocky Mountain Subalpine Dry-Mesic Spruce-Fir Forest and Woodland
Rocky Mountain Montane Riparian Systems
Rocky Mountain Montane Mesic Mixed Conifer Forest and Woodland
Rocky Mountain Montane Dry-Mesic Mixed Conifer Forest and Woodland
Rocky Mountain Lower Montane-Foothill Shrubland
Rocky Mountain Foothill Limber Pine-Juniper Woodland
Rocky Mountain Aspen Forest and Woodland
Quercus gambelii Shrubland Alliance
Inter-Mountain Basins Montane Sagebrush Steppe
Inter-Mountain Basins Aspen-Mixed Conifer Forest and Woodland
Colorado Plateau Pinyon-Juniper Woodland and Shrubland
Colorado Plateau Mixed Low Sagebrush Shrubland
Artemisia tridentata ssp. vaseyana Shrubland Alliance
Abies concolor Forest Alliance
Others

(b)

(c)

DNBR

−2000
Unburned
100
Minor
290
Moderate
480
High
660
Severe
2000

3300

0 (meter)

FIGURE 14.2
The existing vegetation type data (a) and digital elevation model data (b) provided pre-fire ecosystem data and environmental setting for the Sanford Fire site. The differenced normalized burn ratio data (c) provided scales of burn severity on each pixel location as estimated measurements of fire impacts.

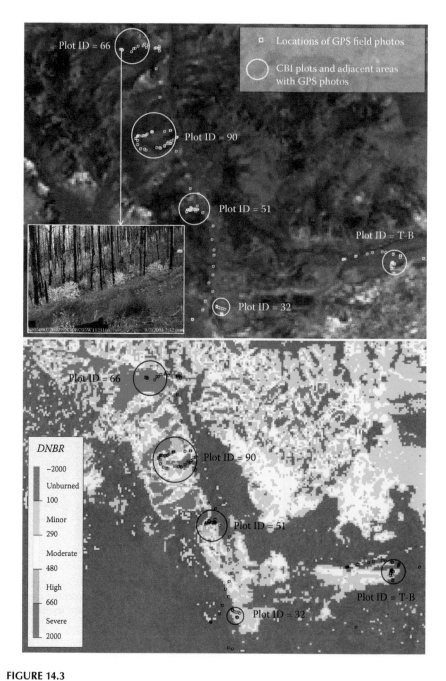

Locations of GPS field photos

CBI plots and adjacent areas
with GPS photos

Plot ID = 66

Plot ID = 90

Plot ID = 51

Plot ID = T-B

Plot ID = 32

DNBR

−2000

Unburned

100

Minor

290

Moderate

480

High

660

Severe

2000

Plot ID = 66

Plot ID = 90

Plot ID = 51

Plot ID = T-B

Plot ID = 32

FIGURE 14.3
An example of plot locations displayed on top of the post-fire Landsat Thematic Mapper image
and differenced normalized burn ratio map.

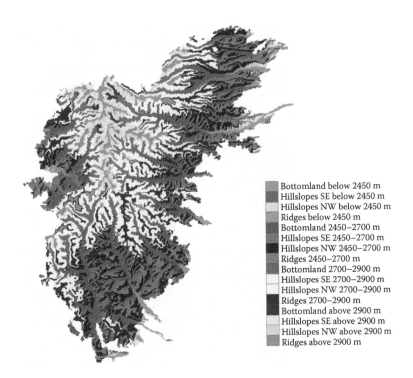

Bottomland below 2450 m
Hillslopes SE below 2450 m
Hillslopes NW below 2450 m
Ridges below 2450 m
Bottomland 2450–2700 m
Hillslopes SE 2450–2700 m
Hillslopes NW 2450–2700 m
Ridges 2450–2700 m
Bottomland 2700–2900 m
Hillslopes SE 2700–2900 m
Hillslopes NW 2700–2900 m
Ridges 2700–2900 m
Bottomland above 2900 m
Hillslopes SE above 2900 m
Hillslopes NW above 2900 m
Ridges above 2900 m

FIGURE 14.4
Land type data developed by integrations of topographic moisture gradients. Slope aspects, and elevations.

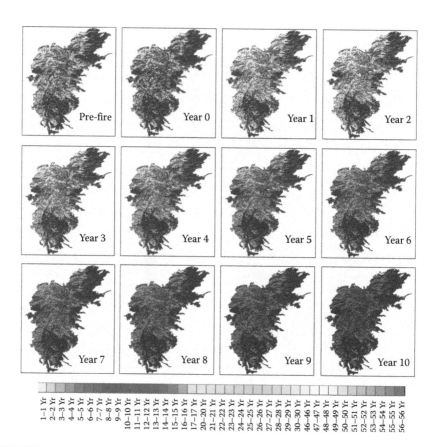

FIGURE 14.6
Examples of simulated post-fire vegetation recovery at species level for quick aspen age cohorts.

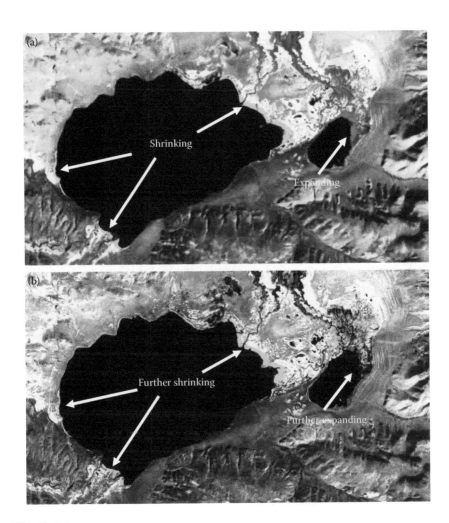

FIGURE 15.8
Lake dynamics at Dawa Co. (a) Lake change between 11/15/1976 MSS and 11/10/1990 TM images; (b) Lake change between 10/10/1990 TM and 10/28/2000 ETM+ images.

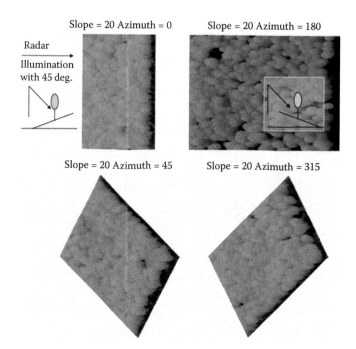

FIGURE 16.5
Simulation of radar images of a forest stand on various slopes.

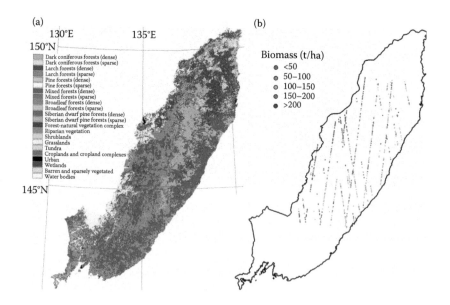

FIGURE 17.4
(a) Land covers of tiger and leopard habitat developed from remotely sensed data, Global Land Cover 2000, and Russian forest maps (Loboda and Csiszar, 2007). (b) Biomass estimates developed from ICESatLiDAR data.

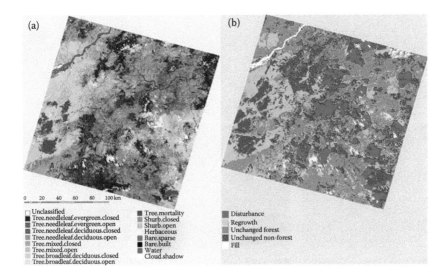

FIGURE 17.5
Land covers (a) and land-cover change, (b) in northern Sikhote-Alin developed from Landsat data.

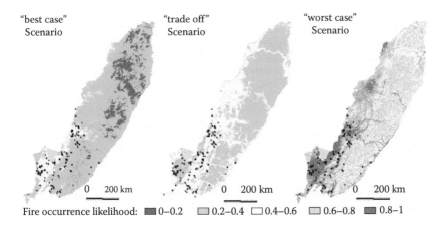

FIGURE 17.12
Likelihood of fire taking place based on the presence of factors that promote fire or inhibit fire occurrence.

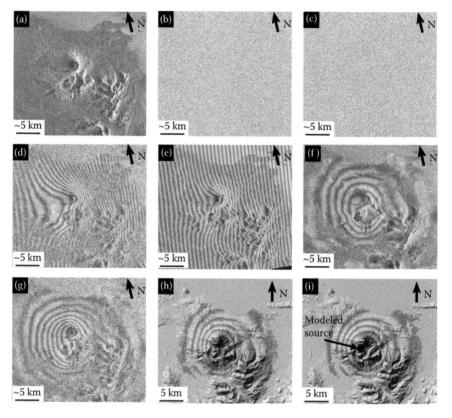

FIGURE 19.1

(a) The amplitude component of an SLC SAR image acquired on October 4, 1995, by ERS-1 satellite over Peulik Volcano, Alaska. (b) The phase component of the SAR image acquired on October 4, 1995, corresponding to the amplitude image in (a). (c) The phase of an SLC SAR image acquired on October 9, 1997, by ERS-2 satellite over Peulik Volcano, Alaska. The amplitude image is similar to that in (a) and therefore is not shown. (d) An original interferogram formed by differencing the phase values of two coregistered SAR images (b and c). The resulting interferogram contains fringes produced by the differing viewing geometries, topography, any atmospheric delays, surface deformation, and noise. (e) An interferogram simulated to represent the topographic contribution in the original interferogram (d). The perpendicular component of the InSAR baseline is 35 m. (f) A topography-removed interferogram produced by subtracting the interferogram in (e) from the original interferogram in (d). The resulting interferogram contains fringes produced by surface deformation, any atmospheric delays, and noise. (g) A flattened interferogram that was produced by removing the effect of an ellipsoidal earth surface from the original interferogram (d). The resulting interferogram contains fringes produced by topography, surface deformation, any atmospheric delays, and noise. (h) A georeferenced topography-removed interferogram (f) overlaid on a shaded relief image produced from a DEM. The concentric pattern indicates ~17 cm of uplift centered on the southwest flank of Peulik Volcano, Alaska, which occurred during an aseismic inflation episode (Lu et al., 2002). (i) A modeled interferogram produced using a best-fit inflationary point source at ~6.5- km depth with a volume change of ~0.043 km^3 on the observed deformation image in (g). Each interferometric fringe (full-color cycle) represents 360° of phase change (or 2.83 cm of range change between the ground and the satellite). Areas of loss of radar coherence are uncolored in (h) and (i).

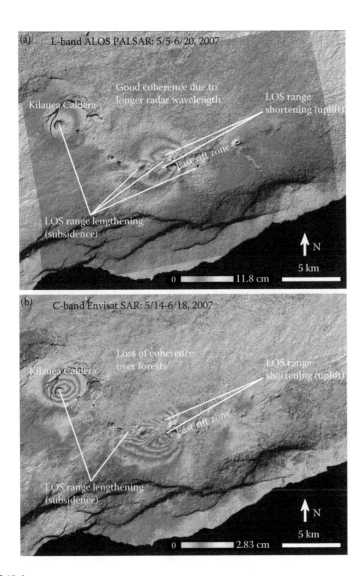

FIGURE 19.4
(a) L-band ALOS and (b) C-band Envisat InSAR images capturing ground-surface deformation associated with the June 2007 eruption at Kilauea volcano. Each fringe (full color cycle) represents a line-of-sight range change of 11.8 and 2.83 cm for ALOS and Envisat interferograms, respectively. InSAR deformation values are draped over the shaded relief map. Areas of loss of coherence are not colored. Note that C-band InSAR image loses coherence over areas of dense vegetation.

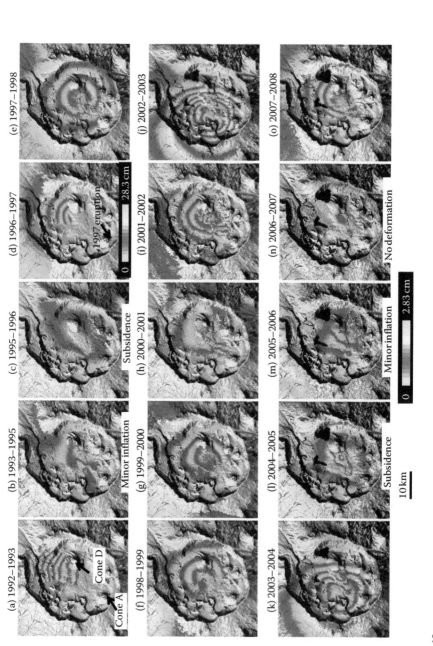

FIGURE 19.10

Deformation interferograms of Okmok Volcano for the periods (a–c) before, (d) during, and (e–o) after the 1997 eruption. Areas of loss of radar coherence are uncolored. Unless otherwise noted, each interferometric fringe (full-color cycle) represents a 2.83-cm range change between the ground and the satellite.

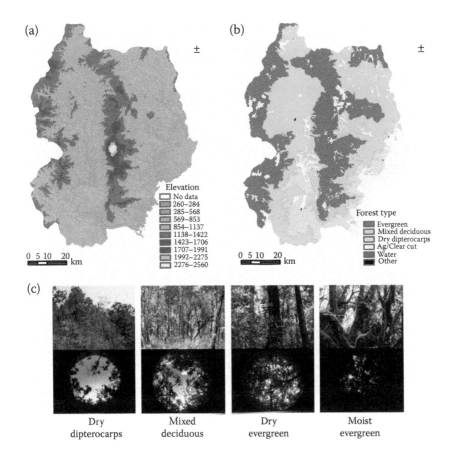

FIGURE 20.3
The DEM map with hillshade effect (a), forest-type map (b), and example field and fisheye photos on the ground (c). The white area in the DEM map is the Doi Inthanon Mount, the highest point in Thailand. Photos are taken during the field trip in January, 2002.

FIGURE 21.7

Visualization of co-registered GIS layers generated using various data integration procedures in EAGLES, a set of decision-support tools integrated into ESRI's ArcGIS environment.
Four geospatial data layers, generated for resource selection analysis (RSPF, Lele and Keim,
2006) of pronghorn summer habitat selection, are (from top to bottom): (a) forage biomass
created by the CASA model assimilating MODIS EVI data (Potter, 2007), (b) coyote utilization
created from kernel density smoothing of relocations of radio-marked adults, (c) elevation
and aspect from a USGS 30 m DEM, and (d) a remotely sensed classification map of forest
(dark green) and sagebrush (light green) using PALSAR and Landsat ETM data. The locations of radio-tracked pronghorn adults (response data used in the RSPF model) are indicated as coregistered purple "spears." The bottom layer is the resource selection probability
surface generated by the EAGLES RSPF model tool where "warmer" temperature colors indicate higher probability of use (ranging from selected to avoided habitat areas) by pronghorn
during the summer.

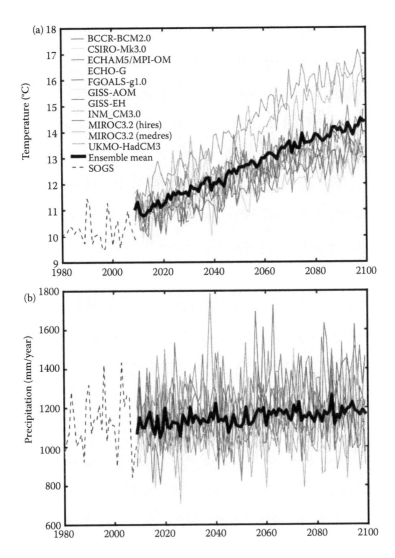

FIGURE 22.6
Projection of mean temperature (a) and precipitation (b) until 2099 downscaled from Coupled Model Intercomparison Project (CMIP3) multimodel data set of SRES A1B scenario. The colored lines correspond to 11 General Circulation Model (GCM) data. The black thick line is the ensemble mean of the 11 GCM data. The dashed line is past time-series data derived from TOPS Surface Observation and Gridding System (SOGS) data.

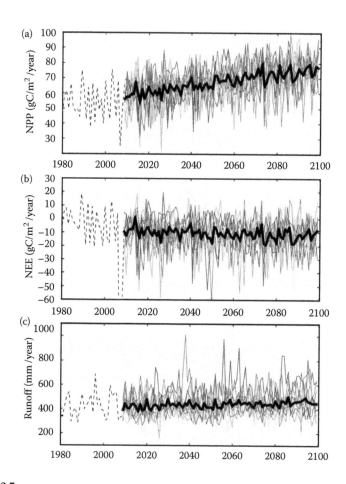

FIGURE 22.7
Same as Figure 21.6 except NPP (a), Net Ecosystem Exchange (NEE) (b), and runoff (c).

11

Remote Sensing Assessment of Natural Resources and Ecological Security of the Changbai Mountain Region in Northeast Asia

Yeqiao Wang, Zhengfang Wu, Hongyan Zhang, Jiquan Zhang, Jiawei Xu, Xing Yuan, Zhong Lu, Yuyu Zhou, and Jiang Feng

CONTENTS

11.1 Introduction

Protected areas have long been the cornerstones of conservation (Wiens et al., 2009). Protected areas provide places in which populations and species can persist and communities and ecosystems can carry out ecological functions. Inventory and monitoring of protected lands and the supporting ecosystems become critically important under the facts of intensified human-induced land-use and land-cover change (LULCC), fragmentation of habitats, and effects of climate changes. This is particularly true for regions that contain protected lands with significant biological diversity and yet fall in the transboundary-sensitive international hotspots of different political regimes and social administration systems. Remote sensing is well documented as an effective tool for mapping and characterizing cultural and natural resources for protected areas (Gross et al., 2009; Kennedy et al., 2009). The multispectral capabilities of remote sensing allow observation and measurement of biophysical characteristics, and the multitemporal and multisensor capabilities allow tracking of changes in those characteristics over time. Remote sensing can play a unique role particularly for monitoring landscape dynamics and LULCC and for evaluating the conditions of natural resources toward a comprehensive understanding of ecological security for maintaining biological diversity and sustainability of natural systems. In this chapter, we introduce a case study in remote sensing assessment of natural resources and ecological security of the Changbai Mountain region in Northeast Asia.

11.1.1 About the Changbai Mountain Region

In a large spatial context, the Changbai Mountain range extends along the border between Northeast China and North Korea in the northeast provinces of Liaoning, Jilin, and Heilongjiang in China and the provinces of Ryanggang and Chagang in North Korea. The range consists of parallel sections of mountains including the Changbai Mountain, Laoye Ling, Wanda Shan, Zhang-Guang-Cai Ling, and Hada Ling. It extends toward the southwest connecting the Qian-Shan Mountains in the Liaodong Peninsula. Toward the north and the northwest is the Northeast China Plain, a major agricultural production base of China. Toward the northeast it connects the Sikhote-Alin Mountains in the Russian Far East. Geologically, the region is on the border of the Pacific competent zone. The Himalayan tectonic movement since the Miocene epoch had resulted in volcanic eruptions and hence the formation of a typical volcanic geomorphology composed of volcanic cones, inclined plateau, and lava table lands (Zhau and Zhau, 1991; Wang et al., 2003). The range has an elevation mostly between 800 and 1500 m and has the typical forest and agricultural ecosystems of Northeast Asia. The highest section and the most representative of this ecoregion is the Changbai Mountain, where the Changbai Mountain Nature Reserve (CMNR) is situated (Figure 11.1).

The CMNR is situated between 41°41′ and 42°26′N latitudes and 127°42′ and 128°17′E longitudes on the China border. The CMNR occupies 196,465 ha of lands with the largest protected temperate forests that support a significant species gene base and biodiversity in Northeast Asia. It is centered by a volcanic summit at 2749 m above sea level. The summit cups a crater lake named Tianchi (Lake of Heaven) with spectacular views and magnificent surrounding landscape. The CMNR was established in 1961 and admitted into the UNESCO's Man and Biosphere Program in 1979.

FIGURE 11.1
(**See color insert.**) The Changbai Mountain range consists of paralleled sections of mountains. It extends toward the west connecting the Qian-Shan Mountains in the Liaodong Peninsula in China and toward the south connecting the Korean Peninsula. In the north is the Northeast China Plain and in the east lie the Sikhote-Alin Mountains in the Russian Far East. The Shuttle Radar Topography Mission illustrates the terrain of the Changbai Mountain range and the volcanic summit in the center (a). The needleleaf and broadleaf mixed forest zone is distributed between 700 and 1100 m (b). The coniferous forest zone is distributed between 1100 and 1700 m, with dominant species of spruce (*Pieca jezoensis, Pieca koreana*) and fir (*Abies nephrolepis*) (c). Between 1700 and 2000 m distributes the zone of subalpine dwarf birch (*Betula ermanii*) forest (d). The alpine tundra zone is distributed above 2000 m, with representative species such as short rhododendron shrubs (*Rhododendron chrysanthum Pall*) and *Vaccinium uliginosum* L. (e).

The most unique feature of the CMNR is its vertical distribution of forest and ecosystem zones due to elevation change. The climate and terrain conditions support four distinctive vertical zones within the core protected areas. According to Huang et al., (1959) the needleleaf and broadleaf mixed forest zone is distributed between 600 and 1600 m. The dominant tree species include Korean pine (*Pinus koraiensis*) and temperate hardwoods such as aspen (*Poplus davidiana*), birch (*Betula platyphylla*), basswood (*Tilia amurensis*), oak (*Quercuc mongolica*), maple (*Acer mono*), and elm (*Ulmus propinqua*), among others. The larch (*Larix olgensis*) and Changbai pine (*Pinus sylvestris var. sylvestriformis*) (Figure 11.2a) form the "bright" coniferous forests on the upper elevation of this vegetation zone. The evergreen "dark" coniferous forest zone is distributed between 1600 and 1800 m with dominant species of spruce (*Pieca jezoensis, Pieca koreana*) and fir (*Abies nephrolepis*). The subalpine dwarf birch (*Betula ermanii*) forest zone is distributed between 1800 and 2000 m with other species such as *L. olgensis*. The alpine tundra zone is distributed between 2000 and 2400 m with representative species such as

FIGURE 11.2
(See color insert.) Landsat image of the Changbai Mountain region illustrates the mountain range and the highest section where the Changbai Mountain Nature Reserve is situated on the border between China and North Korea. Changbai pines (*Pinus sylvestris var. sylvestriformis*) grow on the northern slope of the volcanic summit of the Changbai Mountain (a). Species inhabiting the forests of the Changbai Mountain region include endangered Amur (Siberian) tiger (*Panthera tigris altaica*) (b, c). (Photos by Yeqiao Wang.)

short rhododendron shrubs (*Rhododendron chrysanthum Pall*) and *Vaccinium uliginosum* L. Recent studies separate the "bright" coniferous forest from the mixed forest zone and combined the "bright" and "dark" coniferous forests into one zone of coniferous forest. Current understanding of the distributions of vegetation zones in the CMNR include the Korean pine–broadleaf mixed forest (700–1100 m), the coniferous forest (1100–1700 m), the subalpine forest dominated by *B. ermanii* (1700–2000 m), and the tundra zone (above 2000 m) (e.g., Shao et al., 1996; He et al., 2005). The unique and distinctive vertical zonal pattern of vegetation and the ecosystems showcase a condensed configuration and composition of temperate and boreal forests found across Northeast Asia.

Among the species inhabiting the forests of the Changbai Mountain is the Amur (Siberian) tiger (*Panthera tigris altaica*) (Figure 11.2b, 11.2c), which is extremely endangered. Historically, there may have been more than 4000 Amur tigers in Northeast China. Surveys from the 1970s indicate that the numbers had dropped to ~150 tigers in Northeast China. Recent surveys indicate that fewer than 16 tigers remain in the region (Zhou et al., 2008). However, the vast tracts of forests in the Changbai Mountain and other mountain ranges in Northeast China can provide suitable habitats for Amur tigers. There are ~430–500 Amur tigers inhabiting adjacent similar forests in the Russian Far East (Miquelle et al., 2007). Sherman et al. (2011), in Chapter 17 of this book, introduce a study that uses remote sensing to examine the habitats of Amur tigers and Siberian leopards. Although there is no evidence that a stable reproducing population exists in the Changbai Mountain region and in Northeast China, there are frequent and confirmed reports that Amur tigers regularly cross the border between Russia and China (Yu et al., 2000; Yu et al., 2006; Bing et al., 2008; Zhou et al., 2008). There are increased reports and evidence about casualties of cattle and farmed animals by tiger attacks in particular in the Hunchun-Wangqing areas of the Changbai Mountain range. Amur tigers are habitat generalists. To survive tigers require large areas to ensure population persistence, adequate prey densities, and low mortality rates mostly caused by poaching. Therefore, recolonization of previously occupied Amur tiger habitats in the Changbai Mountain region and in Northeast China is possible if appropriate measures are taken to identify and manage landscapes in an appropriate manner (Li et al., 2010).

11.1.2 Natural Resources and Ecological Security Concerns

Socioeconomic development, aggressive logging, intensified urban and agricultural land use, demographic change, and pollution through air and water systems accelerate the degradation of natural resources of this region. Human-induced land-use and resource change and the uncertainties from potential volcanic eruption and climate change threaten ecosystems of this very unique geographic entity. The CMNR and adjacent lands have been a focus of scientific research in terms of ecosystem structure, function, service, biological diversity (Okitsu et al., 1995; Liu, 1997; Chen and Bradshaw, 1999; Sun et al., 2001; Chen and Li, 2003; Yang and Xu, 2003; Zhang et al., 2008), and, recently, ecological security (Wang, Mitchell et al., 2009; Wang, Wu et al. 2009).

This region has a long and rich history of cultural diversity arising from the coexistence of multiple ethnic groups in China. The Changbai Mountain and the Tianchi (Lake of Heaven) were regarded as the birthplace of the ancestors of the Qing dynasty (1622–1912), and therefore restricted as forbidden areas of a sacred place during the time. Because of such an imperial ban the forests in the Changbai Mountain were protected from being logged and it helped preserve the biodiversity of representative native flora and fauna that inhabited these forests. Therefore, the mountainous ecosystems are mostly intact at the time when the CMNR was established and the regional development in the adjacent areas was relatively late in comparison with the development in other regions in Northeast China.

The Yalu River, the Tuman River, and the Songhua River originate in the Changbai Mountain. In particular, the Yalu and Tumen Rivers serve as the border between China and North Korea. The water resources and water environments are among focal and sensitive international issues of the region. The international collaboration and development of the Tumen River delta region caught the attention of the world. The Tumen River Area Development Project (TRADP) is an ambitious effort by China, North Korea, South Korea, Russia, Mongolia, and Japan to create a free-trade zone in Northeast Asia. The TRADP has been touted as the "future Rotterdam" for Northeast Asia. The involved countries and the United Nation Development Programme (UNDP) envision a 20-year project, costing over $30 billion, which would transform the Tumen River delta area into the transportation and trading hub for Northeast Asia (Figure 11.2). The goal is to make the area a free economic zone for trade to prosper and attract investment into the area. It is estimated that, by 2012, the "Tumen River Plan" will gain breakthroughs in international cooperation throughout the area. By 2020, the Tumen River area will achieve major breakthroughs as a developing and open area. The associated problems, however, would be that the countries involved include long-time adversaries and the agreement could actually lead to further instability in the region if there is significant disagreement on issues. More importantly, much of the area is fragile wetlands and the areas affected by TRADP comprise unique ecosystems and are currently protected such as the CMNR. The hinterland of the TRADP area is also rich in natural resources. It is closely associated with the protection and development of the Changbai Mountain with a major concern about the extraction of these resources.

Social and natural systems have a profound complexity of interconnections. Anthropogenic impacts are always among key issues for regional ecological security concerns. The Changbai Mountain region has experienced dramatic transition and demographic change in the past several decades.

Those changes can be reflected in immigration and human population increase, in socioeconomic development and improved living status, in change of land-use practices, and in changes of government policies regarding socioeconomic development, environmental regulations, and resource management. Effects from human activities are particularly important for such an international conjunction region with cross-border movements of people and commodities, with discrepancies in wealth, and with cultural and linguistic diversity.

Transboundary pollution through air and water systems is one of the many ecological security issues that this region is facing due to rapid economic development, urbanization, and consumption of energy (Schreurs and Pirages, 1998). Transboundary pollutions cause degradation of air, water, and soil, depletion of resources, and loss of biodiversity. Issues related to water resources, water quality, land use, and management within the watersheds that connect to those border rivers need to be closely monitored and studied. The 2010 Korean Peninsula crisis demonstrated the unstable nature of the region, in particular for the ecological security issues in the border region in which the CMNR is situated.

Ecological security is an essential cornerstone for the sustainability of any human and nature systems. It depends on the balance between human demands and actions in consumption and alteration of resource base and the sustainability, vulnerability, and resilience of environmental systems that provide ecosystem services. Since the very beginning human societies have interacted and coevolved with other forms of life (plants, animals, and microorganisms) and learn that their well-being depends on the sustainability of the resource systems. Many ecological security issues, concerns, and actions have been reflected under the framework of "Agenda 21"—the Rio Declaration on Environment and Development, adopted by world leaders in the 1992 U.N. Conference on Environment and Development (World Commission on Environment and Development, 1987; National Research Council, 1999). As pointed out by Pirages (2005), ecological security rests on preserving four interrelated dynamic equilibriums between human and nature: (1) human demands on resources and the ability of nature to provide services; (2) human population and pathogenic microorganisms; (3) human populations and those of other plant and animal species; and (4) the size and growth rates of various human populations. Any type of significant breakdown of the equilibriums will threat ecological security. This is a particular concern for regions with sensitive and fragile ecosystems and ecotones, with transregion and cross-boundary movements such as transportation of pollutants through water and air systems and migration of human populations and other species, and with potential intrastate and interstate conflicts in demographic, environmental, political, and resource issues.

One of the central challenges in understanding the equilibriums is to establish a data framework that can support multidisciplinary studies and quantitative modeling, yet can facilitate inventory and monitoring of the critical components contributing to ecological security issues, such as demographic change, invasive species and biodiversity, deforestation and land degradation, urbanization and intensified land-use change, environmental pollution, food production, freshwater supply, health and disease, rural development, and climate change.

11.2 Remote Sensing Data and Capacities

The improved capacity in Earth observation systems, data products, and availability of the data has profoundly enhanced the knowledge base for understanding and addressing issues of ecological security concerns. Landsat time series data, for example, revealed vegetation patterns and land-cover changes both within the protected CMNR and adjacent areas. As now the historical Landsat data archived by the U.S. Geological Survey have been made available to user communities free of charge, the over 30 years of data facilitate applications in all possible subject areas (Woodcock et al., 2008).

Moderate Resolution Imaging Spectroradiometer (MODIS) data products are among a wide range of Earth observation data for scientific research and management applications. The MODIS derivatives of normalized difference vegetation index (NDVI) (MOD13), land surface temperature (LST) (MOD11), leaf area index (LAI) (MOD15A2), and land-cover (MOD12Q1) data products provide large spatial coverage with high temporal frequencies for inventory and monitoring of the hydrothermal conditions across regional landscapes and for assessing ecosystem conditions and productivity. Advanced Very High Resolution Radiometer (AVHRR) and SPOT-Vegetation data also provide observations from broad spatial scope for inventory and monitoring of the regional landscape dynamic. The upcoming Visible/Infrared Imager Radiometer Suite sensor system, onboard the National Polar-orbiting Operational Environmental Satellite System, could combine the radiometric accuracy of AVHRR with refined spatial resolution and increase the timeliness and accuracy of severe weather event forecasts such as hurricanes and detection of fires, smoke, and atmospheric aerosols.

While high spatial resolution data such as IKONOS and QuickBird are very effective for studying detailed spatial characteristics of the landscape, hyperspectral remote sensing data can provide finer spectral profiles and coverage to reveal ecological conditions and concerns.

Elevation, topography, and soils create habitat diversity and influence distribution and diversity of community types and species. The Shuttle Radar Topography Mission (SRTM) data and digital elevation model data from different sources are available for three-dimensional visualization and modeling. Integration of high spatial resolution airborne and space-born data and the Landsat/MODIS and SRTM data capacity can enhance regional study in multidisciplinary approaches. For example, MODIS data products of LST, NDVI, and land-cover types with SRTM were used to estimate the spatial and temporal changes of soil-moisture status and water conditions in the Greater Changbai Mountain region (Han et al., 2010). The data integration approach reflected the dynamics of landscape variables by MODIS as well as the topographic and terrain effects in this mountainous region.

Ecological applications require data from broad spatial extents that cannot be collected using field-based methods alone. Space-based sensors and computational models are valuable tools for monitoring, reporting, and forecasting ecological conditions at large spatial and temporal scales (Theobald, 2001; Hansen and Rotella, 2002; Jantz et al., 2003; Parmenter et al., 2003; Wessels et al., 2004; DeFries et al., 2005). Remote sensing data and analytical techniques address needs such as identifying and detailing biophysical characteristics of species habitats, predicting the distribution of species and spatial variability in species richness, and detecting natural and human-caused changes (DeFries et al., 1999; Jennings, 2000). Land-cover data have proven especially valuable for predicting the distribution of both individual species (Saveraid et al., 2001) and species assemblages (Kerr and Ostrovsky, 2003) across broad areas too large to directly survey. A remote sensing-based biodiversity index can be designed and applied to detect biodiversity, hotspots with high conservation value (Gould, 2000). Climatic, biophysical, and land-cover data can also be integrated to predict the occurrence of individual species throughout their ranges with the use of genetic algorithms or logistic models (Cumming, 2000).

Mountain ranges often create ecotones and provide habitats and boundaries because of terrain-induced variations of climatic conditions. Ecotonal areas are most sensitive to changes in environmental conditions and are usually where changes first occur (Neilson, 1993; Risser, 1993, 1995; Rusek, 1993; Smith et al., 2001). Integration of Earth observation data with field-based multidisciplinary studies can help overcome a number of challenges and offer a starting point on the path toward ecological security at the ecotone regions. Studies have reported the structure, composition, and succession of vegetation in the CMNR by field investigation and remote sensing observations (Shao et al., 1996; Sun et al., 2001; Liu et al., 2005).

The Tianchi volcanic zone has been active since Cenozonic time (Liu, 1987) and had one of the largest global eruptions in the last 2000 years (Machida et al., 1990; Liu et al., 1998). The volcano in the Changbai Mountain has been quiet since minor eruptions in 1597, 1688, and 1702. It is still a high-risk volcano under intensive monitoring by the Chinese scientists and international volcano observation networks (Stone, 2006). As volcanic eruption is another major regional concern in ecological security, satellite interferometric synthetic aperture radar (InSAR) could be used for monitoring ground surface deformation associated with volcanic and other geological hazards (Lu et al., 2011, Chapter 19 of this book).

11.3 Remote Sensing Assessment of Natural Resources and Ecological Security

11.3.1 LULCC Perspective

Landscape patterns are one of the key measures of ecological security. Landscape patterns that are composed of strategic portions and positions with significance in safeguarding and controlling certain ecological processes are considered ecologically secure (Yu, 1996). Landscape patterns and the dynamics can dramatically influence a host of biological, biophysical, and biochemical resources. The impacts of human-induced LULCC have long been the focus of research for biodiversity and management decisions (Wang and Moskovits, 2001; Hobbs, 2003; Wang, Mitchell et al., 2009). Landscape patterns and change reflect the status of ecosystem connectivity and habitat quality, and are often a central component to assessing changes in other resources such as water quality, flora and fauna, terrestrial vertebrates, and vegetation communities. Understanding the magnitude and pattern of LULCC helps establish a landscape context for the protected areas. This is particularly important for the internationally sensitive border regions.

Remote sensing change detection can discern and simulate areas that have been altered by natural and/or anthropogenic processes. In this case study, we used Landsat remote sensing data to reveal patterns of LULCC of two sites in this region. The core area of the Changbai Mountain that contains the CMNR was taken as one of the study sites, that is, the Changbai Mountain site. The other site was in the middle section of the Tumen River, where human activities strongly interfered with the ecological environment (Figure 11.3).

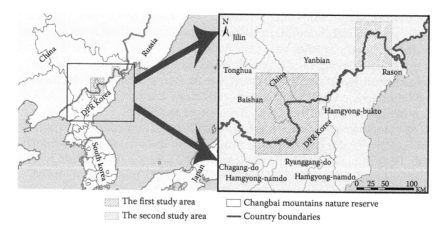

FIGURE 11.3

Location of the Changbai Mountain site (first study area) and the Tumen River site (second study area) as case studies for land-use and land-cover change of the region.

We selected Landsat Thematic Mapper (TM) data acquired between 1976 and 2007 for LULCC mapping and analysis. The classification included eight categories: coniferous forest, mixed broadleaf–conifer forest, broadleaf forest, other wooded land, grassland, cultivated land, water, and other land. We generalized the categories into forest land, cultivated land, water, and other land in the Tumen River study area. The classification was based on an expert knowledge algorithm. A random sampling method was applied to assess the classification accuracy. The classification-derived land-cover maps reveal the landscape features and variations for the two study areas (Figures 11.4). The overall trend and structural change of land cover in both sites are summarized in Tables 11.1 and 11.2. Using a transfer matrix we obtained the general trends of land-cover changes in the study areas.

A particular interest of this study was the LULCC along the Yalu and Tumen Rivers because of the differences in land management practice and socioeconomic development between China and North Korea. We extracted the buffered areas of 10 km from each side along the Yalu and Tumen Rivers to reveal details in land-cover change for the two sites. The results are illustrated in Figure 11.5.

Quantitative analyses reveal the pattern and amount of land-cover change in the two neighboring countries in the past 30 years from 1977 to 2007. For the Changbai Mountain site, the coniferous forest decreased 19%, mixed broadleaf–conifer forest decreased 29%, and cultivated land increased 295%, respectively, on the Chinese side. The coniferous forest decreased 49%, mixed broadleaf–conifer forest decreased 41%, and cultivated land increased 188%, respectively, on the North Korean side. For the Tumen River site, forest and cultivated lands were the dominant land-cover types. In the past three decades, forestland decreased 10% and cultivated land increased 50% on the Chinese side. At the same time, on the North Korean side, forest land decreased 38% and cultivated land increased 191%.

This study adopted the concept of ecological risk index (ERI) (Wei et al., 2008) to evaluate land-cover conditions and changes and the effects on ecological security of the study sites. An ERI defines a relative comprehensive ecological risk in an area and is described as

$$\text{ERI} = \sum_{i=1}^{n} \frac{A_i W_i}{A} \tag{11.1}$$

where i is the number of land-cover types, A_i is the area of land-cover type i, A is the total area of the study site, and W_i indicates the ecological risk intensity parameter of land-use type i (Table 11.3).

The calculated ERIs for the two study sites are summarized in Tables 11.4 through 11.6 (Zhang and Zhang, 2010). The assessment results indicate the status of ecological security concern in the study sites. ERI in both study sites have kept the increasing trend in past decades, which indicate that the

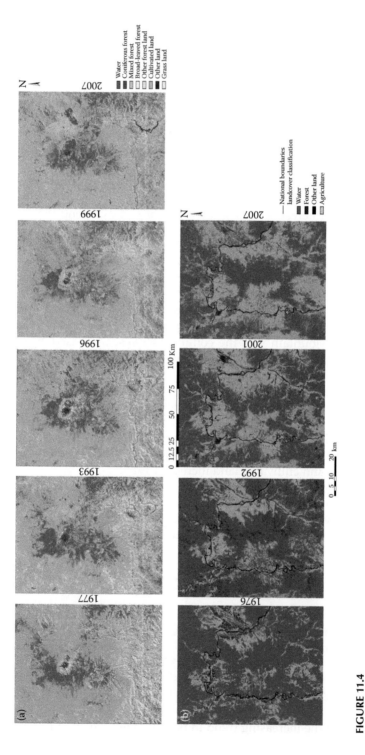

FIGURE 11.4
Land-cover maps derived from classification of Landsat images for the Changbai Mountain site (a) and the Tumen River site (b).

TABLE 11.1
Statistics of Land-Cover Types in the Changbai Mountain Study Site (hm^2)

Land Cover	Year 1977	1993	1996	1999	2007
Coniferous forest	473,524	417,494	396,720	378,352	368,538
Water	46,828	5792	11,738	18,112	12,505
Mixed broadleaf–conifer forest	780,628	669,286	644,035	636,818	616,874
Other wooded land	83,097	111,655	118,375	134,425	144,452
Arable land	108,969	318,897	345,436	369,792	386,970
Broadleaf forest	90,184	81,406	92,268	64,348	66,278
Other land	29,248	36,585	40,646	43,874	50,673
Grassland	108,590	79,954	71,853	75,351	74,779

ecological security of the sites is facing challenges. For the Tumen River site the ERI has already reached the early warning status because of serious human-induced LULCC. The results derived for the buffer areas (Tables 11.7 and 11.8) also present indications of ecological security concerns. The ERI for both countries was similar in the Changbai Mountain site in the 1970s. However, the risk level in North Korea was increasing after the 1990s. For the Tumen River site, the ERI reached the stages of early warning in 1992 and moderate warning in 2007. Further inventory and monitoring are necessary for understanding the threats of LULCC to ecological security of the region and for informed land management decisions.

11.3.2 Remote Sensing of Volcanic Risks

Many volcanic eruptions are preceded by pronounced ground deformation in response to increasing pressure from magma chambers or to the upward intrusion of magma. Therefore, surface deformation patterns can provide important insights into the structure, plumbing, and state of restless volcanoes (Dvorak and Dzurisin, 1997; Dzurisin, 2003, 2007; Lu et al., 2007).

TABLE 11.2
Statistics of Land-Cover Types in the Tumen River Study Site (hm^2)

Land Cover	Year 1976	1992	2001	2007
Water	16,841	17,159	15,853	17,906
Other land	28,736	32,376	54,999	62,373
Forest land	386,458	359,332	323,002	252,968
Cultivated land	60,822	83,990	99,004	159,611

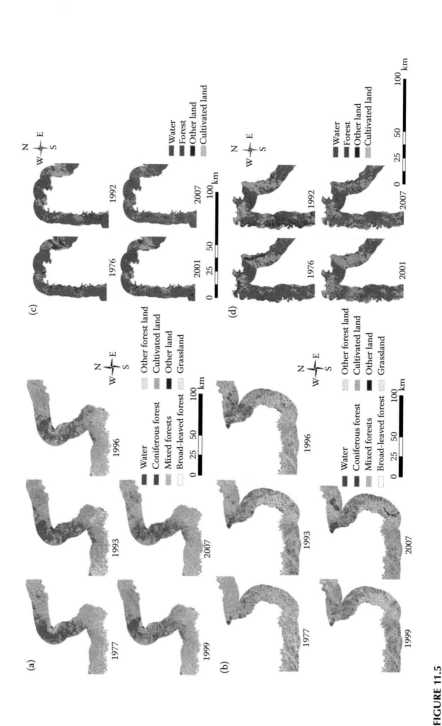

FIGURE 11.5
(See color insert.) Land cover maps of 10-km buffer areas along the Yalu and Tumen Rivers for the Changbai Mountain site in China (a) and North Korea (b) and for the Tumen River site in China (c) and North Korea (d) for land-use and land-cover change analysis.

TABLE 11.3

Ecological Risk Intensity Parameter of Land-Use Type

Land-Use Type	Coniferous	Mixed Broadleaf–Conifer Forest	Broadleaf Forest	Other Woodland	Cultivated Land	Grassland	Other Land	Water
W_i	0.0514	0.0715	0.0768	0.0811	0.2227	0.1118	0.3156	0.0691

TABLE 11.4

Status of the Ecological Securities

Levels	I	II	III	IV	V
ERI threshold	$0.0871 \leq ERI < 0.1094$	$0.1094 \leq ERI < 0.1316$	$0.1316 \leq ERI < 0.1539$	$0.1539 \leq ERI < 0.1761$	$0.1761 \leq ERI < 0.1984$
Status	Safe	Relatively safe	Early warning	Moderate warning	Severe warning

ERI, ecological risk index.

TABLE 11.5

Ecological Risk Index (ERI) of the Changbai Mountain Site

Year	1977	1993	1996	1999	2007
ERI	0.0829	0.1026	0.1056	0.1085	0.1111

TABLE 11.6

Ecological Risk Index (ERI) of the Tumen River Site

Year	1976	1992	2001	2007
ERI	0.1034	0.1119	0.1274	0.1484

TABLE 11.7

Comparison of Ecological Risk Index (ERI) in Buffer Areas of the Changbai Mountain Site

Year	1977		1993		1996		1999		2007	
	China	DPRK	China	DPRK	China	DPRK	China	DPRK	China	DPRK
ERI	0.0808	0.0954	0.0971	0.1139	0.0995	0.1225	0.1026	0.1253	0.1117	0.1341

TABLE 11.8

Comparison of Ecological Risk Index (ERI) in Buffer Areas of the Tumen River Site

Year	1976		1992		2001		2007	
	China	DPRK	China	DPRK	China	DPRK	China	DPRK
ERI	0.1114	0.1093	0.1183	0.1342	0.1225	0.1536	0.1265	0.1598

In short, an "eruption cycle" can be conceptualized as a continuum from deep magma generation through surface eruption, including stages such as partial melting, initial ascent through the upper mantle and lower crust, crustal assimilation, magma mixing, degassing, shallow storage, and final ascent to the surface (Dzurisin, 2003). The timescale for magma generation, ascent, and storage is poorly constrained and variable from one eruption to the next. As a result of such complexities, deformation patterns vary considerably both during the eruption cycle and between different volcanoes. Therefore, understanding the dynamic processes of varied magma systems requires an imaging system capable of characterizing complex and dynamic volcano deformation patterns associated with different volcanic processes. The chance for success in meeting this requirement has increased significantly since the recent advent of satellite InSAR—a remote sensing technique for monitoring ground surface deformation associated with volcanic and other geological hazards (Lu et al., 2011, Chapter 19 of this book).

Figure 11.6 shows examples of InSAR images of the Changbai Mountain volcano. The InSAR images are produced from a series of L-band Japanese Earth Resource Satellite 1 (JERS-1) SAR images acquired between 1992 and 1998. Due to various artifacts in InSAR images (Lu et al., 2011, Chapter 19 of this book), the multi-interferogram InSAR processing is applied to map long-term deformation. Each interferometric fringe (full-color cycle) represents a range change of ~11.8 cm between the ground and the satellite. Most apparent range changes in Figure 11.6 in individual InSAR images are likely due to changes in atmospheric conditions. The InSAR imagery and time-series deformation obtained at four locations indicate that there is no significant ground surface deformation over the Changbai Mountain volcano during the time period (Figure 11.7).

It should be pointed out that some volcanic eruptions are preceded by periods of little or no deformation (e.g., Lu et al., 2010). This suggests that short-term forecasting based on InSAR-mapped deformation alone might be difficult in some cases. InSAR can effectively track surface deformation in the long run, but might not provide a robust indication of an ensuing eruption in the months or weeks beforehand. Experience shows that short-term precursors such as localized deformation, seismicity, and changes in volcanic gas emission commonly are observed when a shallow magma reservoir nears rupture or as magma intrudes surrounding rock (e.g., Sparks, 2003). Such precursors can be detected by *in situ* sensors and they typically provide

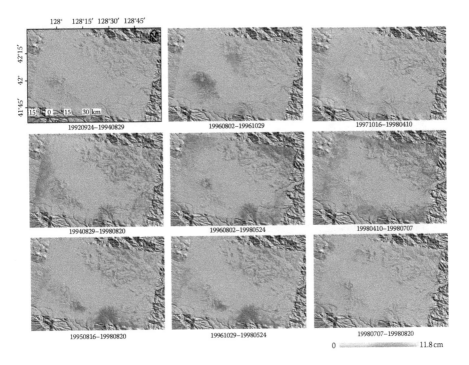

FIGURE 11.6
(**See color insert.**) Examples of interferometric synthetic aperture radar (InSAR) images of the Changbai Mountain volcano. The InSAR images are produced from a series of L-band Japanese Earth Resource Satellite 1 (JERS-1) SAR images acquired between 1992 and 1998.

days to months of warning. Therefore, effective monitoring and hazards mitigation for volcanoes requires the integration and analysis of multiple remote sensing, and geophysical and geochemical datasets in near real time.

Hazardous events at the Changbai Mountain volcano include landslides from steep volcano flanks and floods, which need not be triggered by eruptions, as well as eruption-induced events such as fallout of volcanic ash and lava flows. A proximal hazard zone of ~20 km in diameter centering at the existing Changbai Mountain volcanic cone could be affected within minutes of the onset of an eruption or large landslide. Distal hazard zones could be inundated by lahars (generated either by melting of snow and ice during eruptions or by large landslides) flowing down the slopes of the Changbai Mountain volcano and valleys. Fallout of tephra from eruption clouds can affect areas hundreds of kilometers downwind. The proximal hazard zone extends to a ballistic range of ~10 km in radius. Due to a large crater lake existing on top of the Changbai Mountain, any magma interaction will produce very explosive outputs. Areas prone to inundation by future lahars can be calculated from numerical modeling of an existing digital elevation model (Iverson et al., 1998). Distal hazard zones based on three scenarios of

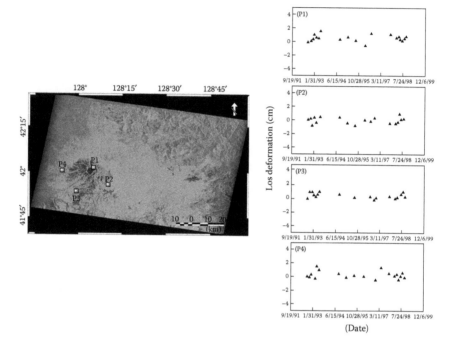

FIGURE 11.7
Interferometric synthetic aperture radar imagery and time-series deformation obtained at four locations indicating no significant ground surface deformation over the Changbai Mountain volcano during the time period.

lahar volumes of 10 million cubic meters (orange), 100 million cubic meters (yellow), and 1 billion cubic meters (brown) are mapped for the Changbai Mountain volcano with indication of nearby population centers (Figure 11.8). Nearby population centers of larger than 100/km² and 1000/km² are shown in red and pink, respectively.

11.3.3 Assessment of Vegetation Succession following Volcanic Eruption Impacts

Spatial pattern of vegetation distribution in the CMNR reflects the trajectories of aftermath regeneration and succession from impacts of significant volcanic eruptions in the most recent several hundred years. Volcanic eruptions not only destroyed the vegetation but also altered the soil contents, landform, and hydrology, which further affected vegetation recovery and succession. Therefore, the timeline of the most recent significant volcanic eruption that devastated the forest ecosystems is the key to understand the history of the vegetation recovery and succession.

Chapter 12 of this book describes a case study that integrated remote sensing-derived patterns of vegetation distribution and *in situ* observations of

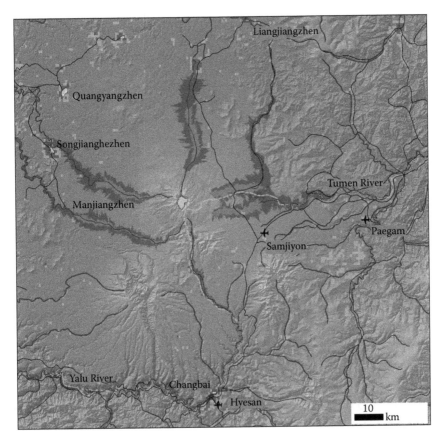

FIGURE 11.8
(See color insert.) Distal hazard zones based on three scenarios of lahar volumes of 10 million cubic meters (orange), 100 million cubic meters (yellow), and 1 billion cubic meters (brown) are mapped for the Changbai Mountain volcano with indication of nearby population centers larger than 100 per square kilometers (red) and 1000 per square kilometers (pink).

species composition by carbonized wood samples that buried in volcanic ash the evidence of preeruption forest structure, species composition, spatial distributions, temporal variations, and the impacted areas. Integrated information from remote sensing and *in situ* survey concluded that the current vegetation distribution and structure in the CMNR reflect the effects of historical volcanic eruptions and succession process, particularly in high-altitude areas. Historical volcanic eruptions not only imposed direct influences on vegetation but also affected indirectly the altered natural conditions such as strata, terrain, and hydrology. Landsat remote sensing data and the derivatives of vegetation maps reveal the pattern of vertical zonal vegetation distribution. The zonal pattern is more complete in the northern slope but not quite obvious in the eastern slope. An abundance of carbonized woods buried in volcanic ash in the eastern slope suggests the areas of direct volcanic impacts.

Remote sensing and *in situ* observations reveal that the distinctive quality of vegetation in the CMNR is heavily affected by volcanic eruptions. The traces of volcanic eruptions exist in the forest ecosystems after 800 years of the last devastated volcanic eruption. The mapping results indicate there was no significant change of tree-line and tundra-line in the CMNR since the early Landsat data acquisition in 1972. Spatial distribution of vertical vegetation zones, forest types, and landscape configuration and composition provide information about the status of vegetation and the impacts from past volcanic eruptions. High spatial resolution vegetation mapping from IKONOS data confirmed in an aerial context that the colonization of subalpine *B. ermanii* into the tundra zone is happening in the high-altitude northern slope of the Changbai Mountain.

11.3.4 Remote Sensing and Modeling for Assessing Ecological Security

Our other study used a five-factor model to assess the ecological security of the selected areas in the Changbai Mountain region. The five factors include *driving forces* (D), *pressures* (P), *states* (S), *impacts* (I), and *responses* (R). The D includes natural and anthropogenic factors to reduce changes in ecosystems and environments. The P includes the needs and functions that could be brought in by driving forces. The S refers to the background value of the ecosystem such as biomass and vegetation coverage. The higher the S value the more secure the ecological condition. The I refers to the change of ecosystem state under the P. The R includes positive actions such as land conservation and protection and ecological restoration. The five-factor model, or *DPSIR* model, has been developed and applied in different cases (Borja et al., 2006; Yi et al., 2010). Figure 11.9 illustrates the conceptual connections among the factors.

The interactive processes of *DPSIR* factors determine the status of regional ecological security. The D factor leads to an increase in disturbances. The P factor generated by the D leads to changes in ecological conditions and in the S factor. Similarly, the changes in S contribute to the I factor. The R factor, or societal response to impacts, depends on how the impacts are perceived and

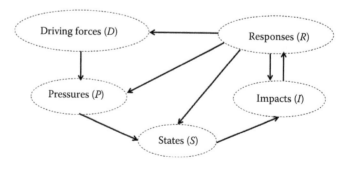

FIGURE 11.9
DPSIR framework for evaluation of ecological security.

evaluated. The five factors can be derived from related indicators as expressed in Equations 11.2a–e.

$$D = \sum_{i=1}^{n} W_{Di}X_{Di} \quad \text{(a)}; \quad P = \sum_{i=1}^{n} W_{Pi}X_{Pi} \quad \text{(b)}; \quad S = \sum_{i=1}^{n} W_{Si}X_{Si} \quad \text{(c)};$$

$$I = \sum_{i=1}^{n} W_{Ii}X_{Ii} \quad \text{(d)}; \quad \text{and} \quad R = \sum_{i=1}^{n} W_{Ri}X_{Ri} \quad \text{(e)}$$

(11.2)

where X_i represents the assessment indicator of i and W_i represents the weight of indicator i. The factors D, P, and I are proportional to the ecological security index (ESI), that is, the higher the values of D, P, and I, the less secure the system. Similarly, ESI varies inversely with S and R factors. ESI can be derived from Equation 11.3.

$$\text{ESI} = \frac{S^{WS} \times R^{WR}}{D^{WD} \times P^{WP} \times I^{WI}}$$

(11.3)

where W represents the weight of factors D, P, S, I, and R.

This study established a quantitative assessment model to evaluate ecological security conditions of Baishan City in the northern slope of the Changbai Mountain. There are six counties and districts including Linjiang, Jingyu, Jiangyuan, Fusong, Changbai, and Badaojiang under the jurisdiction of Baishan City (Figure 11.10). The study selected and defined three categories of indicators that would be useful in assessing ecological

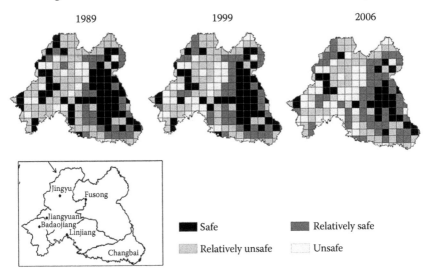

FIGURE 11.10
(See color insert.) Spatial variation of ecological security index of Baishan City in grid cells.

security in natural, social, and economic concerns and that would reflect the comprehensive characteristics of the study area (Table 11.9). The study used supervised classification with visual revision on Landsat TM and Enhanced Thematic Mapper Plus data acquired in 1989, 1999, and 2006. The classification results were then recoded into four general categories of forest, farmland, water, and urban lands. The study area was cut into grid cells of 4 km × 4 km as the assessment unit. The important ecological indicators were extracted and the values of ESI were calculated in an attribute list (Yi et al., 2010). A series of calculated ESI values were classified into four classes of safe ($6 \leq ESI < 10$), relatively safe ($4 \leq ESI < 6$), relatively unsafe ($1 \leq ESI < 3$), and unsafe ($ESI < 1$) to represent discrete regional differences in ESI. Figure 11.10 illustrates a spatial variation of ESIs in the study area

TABLE 11.9

Indices System of Synthetical Ecological Security Assessment of Baishan City

Target	Factors	Indicators	Weight
Ecological security index (ESI)	Driving forces (D)	X_{D1} Annual growth rate of GDP (%)	0.0794
		X_{D2} Annual growth rate of population (%)	0.0505
		X_{D3} Urbanization rate (%)	0.0815
		X_{D4} Losses of natural disasters (million)	0.0842
	Pressures (P)	X_{P1} Population density (p · km^{-2})	0.0434
		X_{P2} Industry pollution index	0.0527
		X_{P3} Road network density (km · km^{-2})	0.0327
		X_{P4} Human activity affection index	0.0956
	States (S)	X_{S1} Forest coverage rate (%)	0.0615
		X_{S2} Drainage density (km · km^{-2})	0.0476
		X_{S3} Annual precipitation anomaly percentage (%)	0.0318
		X_{S4} Annual average temperature anomaly percentage (°C)	0.0334
	Impact (I)	X_{I1} Vegetation cover dynamic degree (%)	0.0676
		X_{I2} Degree of landscape fragmentation	0.031
		X_{I3} Biological richness index	0.0505
		X_{I4} Rate of forest accumulation reduction (m^3 · a^{-1})	0.0605
	Responses (R)	X_{R1} Annual afforestation area (km^2)	0.0405
		X_{R2} Proportion of protected area (%)	0.0343
		X_{R3} Proportion of ecological and environmental protection investment in GDP (%)	0.0213

in 1989, 1999, and 2006 according to land-cover information derived from Landsat data.

The ESI analysis suggested that, overall, Baishan City is in good condition with respect to ecological security concerns. However, there were differences among the counties and district. As for spatial distributions, the ESI levels showed a decreasing trend in the western and southern sections of Baishan City. Time series analysis on ESI values of individual counties and districts indicated degraded status in ecological security for Jingyu and Badaojiang. The main reasons were as follows: The urbanization process put pressures on ecosystems and the environment. Natural disasters caused a series of problems for ecological security concerns. For instance, the disaster caused by the windstorm in 1986 damaged forests and ecological structures were destroyed seriously. Furthermore, the rough and steep terrain and loose volcanic soil that affected vegetation were also concerns. Industrial development activities, such as paper-making, too had an impact on the region as a result of over consumption of natural resources and increased runoff and pollution.

11.3.5 Remote Sensing in Regional Inventory and Monitoring

Coarser resolution remote sensing data and data products can help reveal regional landscape characteristics and hydrothermal conditions for timely observation and analysis. MODIS data products have been used to monitor NDVI variations and soil moisture of the region. For example, MODIS data products of LST (MOD11A2), NDVI (MOD13Q1), and land-cover types (MOD12Q1) were used to estimate the spatial and temporal changes in soil-moisture status and water conditions in the Greater Changbai Mountain region. By integration of SRTM data, the topographic and terrain effects can be integrated for modeling and analysis. A study explored the combination of NDVI and LST approach to develop a regional temperature–vegetation dryness index (TVDI) (Han et al., 2010). TVDI has an advantage over other indexes such as the LST/NDVI curve in that it can be applied to mixed land surfaces and can be used to derive surface-moisture status from satellite information (Nemani et al., 1993; Sandholt et al., 2002). TVDI can be used to analyze a spatial or contextual array of remotely sensed spectral solar-reflective and thermal-infrared measurements to estimate near-surface soil-moisture conditions. The study used the TVDI concept to understand the relationship between NDVI and LST and grouped TVDI into different classes according to spatiotemporal distribution characteristics.

The study concluded that the time series of NDVI, LST, and the derivative of TVDI revealed spatiotemporal characteristics of soil-moisture conditions. The combination of LST and NDVI can enhance the ability of extracting the soil-moisture conditions for a large area and in high time frequency. Using time-series MODIS data products, the relationships between TVDI and land-cover types can be determined. An improved land-cover classification result may help derive moisture information according to the vegetation characteristics of each land-cover type. The results have demonstrated the advantages of extracting soil-moisture condition by remotely sensed data. LST and NDVI are correlated to moisture availability in a large spatial scale. The possibility of inferring surface-moisture conditions from regular remote sensing data products helps significantly in the understanding and monitoring of the ecosystem conditions and the environment (Han et al., 2010).

MODIS NDVI (MOD13Q1) and land-cover types (MOD12Q1) data have been used to monitor regional response from different land-cover types, such as NDVI of deciduous broadleaf forests, mixed forests, and croplands of the Greater Changbai Mountain range covered by the data frame. It could also be used to reveal NDVI change in a particular area of interest such as mixed forests in the CMNR (Figure 11.11). Integration of field observation and regional data cover will help understand landscape variations and the possible causes at a regional scale for informed research and management needs.

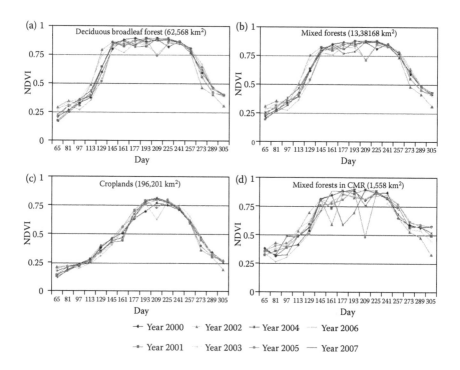

FIGURE 11.11
Normalized difference vegetation index of selected land-cover types of (a) deciduous broadleaf forest, (b) mixed forests, and (c) croplands for the Greater Changbai Mountain region and (d) mixed forests within the Changbai Mountain Nature Reserve derived from MODIS-derived MOD13Q1 and land-cover types (MOD12Q1).

11.4 Concluding Remarks

To study the complexity of geographic systems with human and natural factors, the Northeast Normal University sponsored the "Ecological Security and Data Assemblage of the Changbai Mountains International Georegion" Project under the Science and Technology Innovation Platform Initiative in November 2007. The Northeast Normal University has a long history in the study of geographic systems of the Changbai Mountain region and associated ecosystem constituents, functions, and services. Some of the representative research results are assembled in a book series (Wang et al., 2010; Zhang, 2010). This project strengthened studies about the geographic systems of the Changbai Mountain region, refined and extended scientific understanding of the geographic systems and ecological security issues of this geographic region, and developed and promoted collaborations among research groups. This project was approached by integration of multidisciplinary and interdisciplinary research, combination of field investigations and remote sensing observations, and GIS data assemblage and mapping. To research the geographic systems in micro- and macro-scales, the studies explored the formation of geographic systems, the current landscape patterns and dynamic processes, the disturbance of human activities on natural environments, and their interrelationship at regional scales. In particular, as summarized in this chapter, integration of remote sensing, geospatial modeling, and field-based studies help reveal the structure of the natural resource base, the past and current status, and their change in association with human-induced alterations for informed management of the resource base toward ecological security of this important georegion.

Acknowledgments

This study was supported by the Scientific and Technological Innovation Initiative of the Northeast Normal University under the project "Ecological Security and Data Assemblage of the International Georegion of Changbai Mountains" (Project No. 106111065202) and the National Grand Fundamental Research 973 Program of China (No. 2009CB426305). JERS-1 SAR images are copyrighted Japan Aerospace Exploration Agency and were provided by J.S. Won. We thank J. Griswold for the lahar inundation map. Other individuals contributed significantly to the studies that generated referenced case studies, in particular, Qiang Zhang, Kunbeng Yi, Yu Liang, and the members of the above-mentioned projects.

References

Bing, L., Z. Endi, Z. Zhenhua, and L. Yu. 2008. Preliminary monitoring of Amur tiger population in Jilin Hunchun National Nature Reserve. *Acta Theriologica Sinica* 28:333–341.

Borja, A., G. Ibon, and S. Oihana. 2006. The European Water Framework Directive and the DPSIR: A methodological approach to assess the risk of failing to achieve good ecological status. *Estuarine, Coastal and Shelf Science*, 66 (3): 84–96.

Chen, J., and G.A. Bradshaw. 1999. Forest structure in space: A case study of an old growth spruce-fir forest in Changbaishan Natural Reserve, PR China. *Forest Ecology and Management*, 120:219–233.

Chen, X., and B.L. Li. 2003. Change in soil carbon and nutrient storage after human disturbance of a primary Korean pine forest in Northeast China, *Forest Ecology and Management*, 186:197–206.

Cumming, G.S. 2000. Using habitat models to map diversity: Pan-African species richness of ticks *(Acari: Ixodida)*. *Journal of Biogeography*, 27:425–440.

DeFries, R., C.B. Field, I. Fung, G.J. Collatz, and L. Bounoual. 1999. Combining satellite data and biogeochemical models to estimate global effects of human-induced land cover change on carbon emissions and primary productivity. *Global Biogeochem Cycles*, 13:803–815.

DeFries, R., A. J. Hansen, A.C. Newton, M. Hansen, and J. Townshend. 2005. Isolation of protected areas in tropical forests over the last twenty years. *Ecological Applications*, 15:19–26.

Dvorak, J., and D. Dzurisin. 1997. Volcano geodesy: The search for magma reservoirs and the formation of eruptive vents. *Review of Geophysics*, 35:343–384.

Dzurisin, D. 2003. A comprehensive approach to monitoring volcano deformation as a window on eruption cycle. *Review of Geophysics*, 41(1), 1.1–1.29, doi: 10.1029/2001RG000107.

Dzurisin, D. 2007. *Volcano Deformation: Geodetic Monitoring Techniques*. Chichester, UK: Praxis Publishing Ltd.

Gould, W. 2000. Remote sensing of vegetation, plant species richness, and regional biodiversity hotspots. *Ecological Applications*, 10:1861–1870.

Gross, J.E., S.J. Goetz, and J. Cihlar. 2009. Application of remote sensing to parks and protected area monitoring: Introduction to the special issue. *Remote Sensing of Environment*, 113:1343–1345.

Han, Y., Y. Wang, and Y. Zhao. 2010. Estimating soil moisture conditions of the Greater Changbai Mountains by land surface temperature and NDVI. *IEEE Transactions on Geosceience and Remote Sensing*, 48(6):2509–2515.

Hansen, A.J., and J.J. Rotella. 2002. Biophysical factors, land use, and species viability in and around nature reserves. *Conservation Biology*, 16:1–12.

He, H.S., Z. Hao, D.J. Mladenoff, G. Shao, Y. Hu, and Y. Chang. 2005. Simulating forest ecosystem response to climate warming incorporating spatial effects in northeastern China. *Journal of Biogeography*, 32:2043–2056.

Hobbs, N.T. 2003. Challenges and opportunities for integrating ecological knowledge across scales. *Forest Ecology & Management*, 181:222–238.

Huang, X., D. Liu, and Z. Li. 1959. The natural landscape zones of the northern slope of the Changbai Mountain. *Acta Geographica Sinica*, 25(6):435–446.

Iverson, R., S. Schilling, and J. Vallance. 1998. Objective delineation of lahar-inundation hazard zones. *GSA Bulletin*, 110(8):972–984.

Jantz, C.A., S.J. Goetz, and M.A. Shelley. 2003. Using the SLEUTH urban growth model to simulate the land use impacts of policy scenarios in the Baltimore-Washington metropolitan region. *Environment and Planning*, 31(2): 251–271.

Jennings, M.D. 2000. Gap analysis: Concepts, methods, and recent results. *Landscape Ecology*, 15:5–20.

Kennedy, R.E., P.A. Townsend, J. Gross, W.B. Cohen, P. Bolstad, Y.Q. Wang, and P. Adams. 2009. Remote sensing change detection tools for natural resource managers: Understanding concepts and tradeoffs in the design of landscape monitoring projects. *Remote Sensing of Environment*, 113:1382–1396.

Kerr, J.T., and M. Ostrovsky. 2003. From space to species: Ecological applications for remote sensing. *TRENDS in Ecology and Evolution*, 18(6):299–306.

Li, Z., F. Zimmermann, M. Hebblewhite, A. Purekhovsky, F. Mörschel, C. Zhu, and D. Miquelle. 2010. *Study on the Potential Tiger Habitat in the Changbaishan Area, China*. Beijing: China Forestry Publishing House, ISBN: 978-5038-5544-3.

Liu, J.Q. 1987. Study on geochronology of the Cenozoic volcanic rocks in northeastern China. *Acta Petrologica*, 4:21–31 [in Chinese with English abstract].

Liu, R.X., H.Q. Wei, and J.T. Li. 1998. *Recent Eruptions of the Changbaishan Tianchi Volcano*. Beijing: Scientific Press, pp. 1–159 [in Chinese].

Liu, Q.J. 1997. Structure and dynamics of the subalpine coniferous forest on Changbai Mountain, China. *Plant Ecology*, 132:97–105.

Liu, Q.J., X.R. Li, Z.Q. Ma, and N. Takeuchi. 2005. Monitoring forest dynamics using satellite imagery—A case study in the natural reserve of Changbai Mountain in China. *Forest Ecology and Management*, 210:25–37.

Lu, Z., D. Dzurisin, J. Biggs, C. Wicks, Jr, and S. McNutt. 2010. Ground surface deformation patterns, magma supply, and magma storage at Okmok volcano, Alaska, inferred from InSAR analysis: 1. Inter-eruptive deformation, 1997–2008. *Journal of Geophysical Research*, 115:B00B03, doi: 10.1029/2009JB006969.

Lu, Z., D. Dzurisin, and H.S. Jung. 2011. Chapter 20—Monitoring natural hazards in protected lands using interferometric synthetic aperture radar (InSAR). In *Remote Sensing of Protected Lands*. Boca Raton, FL: CRC Press.

Lu, Z., D. Dzurisin, C. Wicks, J. Power, O. Kwoun, and R. Rykhus. 2007. Diverse deformation patterns of Aleutian volcanoes from satellite interferometric synthetic aperture radar (InSAR). In J. Eichelberger, J. Eichelberger, E. Gordeev, M. Kasahara, P. Izbekov and J. M. Lees (eds), *Volcanism and Tectonics of the Kamchatka Peninsula and Adjacent Arcs*. American Geophysical Union Monograph Series No. 172. Washington, D.C.: American Geophysical Union, pp. 249–261.

Machida, H., H. Morwaki, and D.C. Zhao. 1990. The recent major eruption of Changbai volcano and its environmental effects. *Geographical Reports of Tokyo Metropolitan University*, 25: 1–20.

Miquelle, D.G., Y.M. Pikunov, V.V. Dunishenko, I.G. Aramilev, V.K. Nikolaev, E.N. Abramov, G.P. Smirnov et al. 2005 Amur tiger census. *Cat News*, 46:11–14.

National Research Council. 1999. *Our Common Journey: A Transition Toward Sustainability*. Washington, D.C.: National Academy Press.

Neilson, R.P. 1993. Transient ecotone response to climate change: Some conceptual and modelling approaches. *Ecological Applications*, 3:385–395.

Nemani, R.R., L. Pierce, S.W. Running, and S.N. Goward. 1993. Developing satellite-derived estimates of surface moisture status. *Journal of Applied Meteorology*, 32(3):548–557.

Okitsu, S., K. Ito, and C.H. Li. 1995. Establishment processes and regeneration patterns of montane virgin coniferous forest in northeastern China. *Journal of Vegetation Science*, 6(3):305–308.

Parmenter, A.P., A. Hansen, R. Kennedy, W. Cohen, U. Langner, R. Lawrence, B. Maxwell, A. Gallant, and R. Aspinall. 2003. Land use and land cover change in the Greater Yellowstone ecosystem: 1975–95. *Ecological Applications*, 13:687–703.

Pirages, D. 2005. From limits to growth to ecological security. In D. Pirages and K. Cousins (eds), *From Resource Security to Ecological Security: Exploring New Limits to Growth*. Cambridge, MA: The MIT Press, pp. 1–20.

Risser, P.G. 1993. Ecotones at local to regional scales from around the world. *Ecological Applications*, 3:367–368.

Risser, P.G. 1995. The status of the science examining ecotones. *BioScience*, 45:318–325.

Rusek, J. 1993. Air-pollution-mediated changes in alpine ecosystems and ecotones. *Ecological Applications*, 3:409–416.

Sandholt, I., K. Rasmussena, and J. Andersenb. 2002. A simple interpretation of the surface temperature/vegetation index space for assessment of surface moisture status. *Remote Sensing of Environment*, 79(2/3):213–224.

Saveraid, E.H., D.M. Debinski, K. Kindscher, and M.E. Jakubauskas. 2001. A comparison of satellite data and landscape variables in predicting bird species occurrences in the Greater Yellowstone ecosystem, USA. *Landscape Ecology*, 16:71–83.

Schreurs, M.A., and D. Pirages 1998. Ecological security and the future of inter-state relations in Northeast Asia. In M.A. Schreurs and D. Pirages (eds), *Ecological Security in Northeast Asia*. Seoul: Yonset University Press, pp. 11–22.

Shao, G., Z. Zhao, S. Zhao, H.H. Shugart, X. Wang, and J. Schaller. 1996. Forest cover types derived from Landsat Thematic Mapper imagery for Changbai Mountain area of China. *Canadian Journal of Forest Research*, 26:206–216.

Sherman, N.J., T.V. Loboda, G. Sun, and H.H. Shugart. 2011. Chapter 17—Remote sensing and modeling for assessment of complex Amur (Siberian) tiger and Amur (Far Eastern) leopard habitats in the Russian Far East. In Y. Wang (ed.), *Remote Sensing of Protected Lands*. Boca Raton, FL: CRC Press.

Smith, J.B., H.J. Schellnhuber, and M.M.Q. Mirza (eds). 2001. Chapter 19—Vulnerability to climate change and reasons for concern: A synthesis. Section 3—Impacts on unique and threatened systems; Subsection 3—Biological systems; Unit 3—Ecotones (19.3.3.3). In J. McCarthy, O. Canziana, N. Leary, D. Dokken, and K. White (eds), *Climate Chane 2001: Impacts, Adaptation and Vulnerability*. Contribution of Working Group II to the Third Assessment Report of the Intergovernmental Panel on Climate Change. Cambridge, UK: Cambridge University Press, pp. 932–933.

Sparks, S. 2003. Forecasting volcanic eruptions. *Earth and Planetary Science Letters—Frontiers in Earth Science Series*, 210:1–15.

Stone, R. 2006. A threatened nature reserve breaks down Asian borders. *Science*, 313:1379–1380.

Sun, G., D. William, X. Zhan, Z. Li, J. Masek, K.J. Ranson, and L. Rocchio. 2001. Monitoring forest dynamics using multi-sensor data in Northeastern China, Geoscience and Remote Sensing Symposium, IGARSS apos;01. *IEEE 2001 International*, 2:768–770.

Theobald, D.M. 2001. Land use dynamics beyond the American urban fringe. *Geographical Review*, 91:544–564.

Wang, Y., and D.K. Moskovits. 2001. Tracking fragmentation of natural communities and changes in land cover: Applications of Landsat data for conservation in an urban landscape (Chicago Wilderness). *Conservation Biology*, 15(4):835–843.

Wang, Y., Z. Wu, and J. Feng (eds). 2010. *Research on Geosystems of the Changbai Mountains*, Vol. 2 (1982–1995), Vol. 3 (1996–2006), and Vol. 4 (2007–2010). China: Northeast Normal University Press [in Chinese].

Wang, Y., C. Li, H. Wei, and X. Shan. 2003. Late Pliocene–recent tectonic setting for the Tianchi volcanic zone, Changbai Mountains, northeast China. *Journal of Asian Earth Sciences*, 21:1159–1170.

Wang, Y., B.R. Mitchell, J. Nugranad-Marzilli, G. Bonynge, Y. Zhou, and G. Shriver. 2009. Remote sensing of land-cover change and landscape context of the national parks: A case study of the Northeast Temperate Network. *Remote Sensing of Environment*, 113:1453–1461.

Wang, Y., Z. Wu, J. Feng, X. Yuan, H. Zhang, J. Zhang, and J. Xu. 2009. Earth observation and ecological security: An integrated multidisciplinary approach towards a regional inventory and monitoring. *IEEE Xplore: Geoinformatics*, doi: 10.1109/GEOINFORMATICS.2009.5293501.

Wei, S., C. Wu, Y. Yang, M. Huang, and Z. Yang. 2008. Land use change and ecological security in Yellow River Delta based on RS and GIS technology—A case study of Dongying city. *Journal of Soil and Water Conservation*, 22(1):185–188 [in Chinese with English abstract].

Wessels, K.J., R.S. De Fries, J. Dempewolf, L.O. Anderson, A.J. Hansen, S.L. Powell, and E.F. Moran. 2004. Mapping regional land cover with MODIS data for biological conservation: Examples from the Greater Yellowstone Ecosystem, USA and Pará State, Brazil. *Remote Sensing of Environment*, 92:67–83.

Wiens, J.A., R.D. Sutter, M. Anderson, J. Blanchard, A. Barnett, and N. Aguilar-Amuchastegui. 2009. Selecting and conserving lands for biodiversity: The role of remote sensing. *Remote Sensing of Environment*, 113(7):1370–1381.

Woodcock, C.F., A.A. Allen, M. Anderson, A.S. Belward, R. Bindschadler, W.B. Cohen, T. Gao et al. Landsat Science Team. 2008. Free access to Landsat imagery. *Science*, 320(May):1011.

World Commission on Environment and Development. 1987. *Our Common Future*. New York: Oxford University Press.

Yang, X., and M. Xu. 2003. Biodiversity conservation in Changbai Mountain Biosphere Reserve, northeastern China: Status, problem, and strategy. *Biodiversity and Conservation*, 12(5):883–903.

Yi, K., J. Zhang, Z. Tong, X. Liu, and X. Jiang, 2010. DPSIR-based diagnosis and conceptual assessment model of ecological security in Changbai Mountain area, In Wang, Y., Z. Wu, and J. Feng (eds). Geosystems and Ecological Security of the Changbai Mountains, Vol. 4 (2007–2010), pp. 71–80. Northeast Normal University Press [in Chinese with English abstract].

Yu, K. 1996. Security patterns and surface model in landscape ecological planning. *Landscape Ecology*, 36:1–17.

Yu, X., E. Zhang, and D. Miquelle. 2000. Monitoring the tiger population in Heilongjiang Province, China. *Cat News*, 33:3–4.

Yu, L., Z. Endi, L. Zhihong, and C. Xiaojie. 2006. Amur tiger (*Panthera tigris altaica*) predation on livestock in Hunchun Nature Reserve, Jilin, China. *Acta Theriologica Sinica*, 26:213–220.

Zhang, Z. (ed.). 2010. *Research on Geosystems of the Changbai Mountains*, Vol. 1 (1982–1995). China: Northeast Normal University Press [in Chinese].

Zhang, Q., and H. Zhang. 2010. Land-use and land-cover change along the border of China and North Korea and ecological security. In Y. Wang, et al. (eds), *Geosystems and Ecological Security of the Changbai Mountains*, Vol. 4 (2007–2010). China: Northeast University Press [in Chinese with English abstract].

Zhang, Y., M. Xu, J. Adams, and X. Wang. 2008. Can Landsat imagery detect tree line dynamics? *International Journal of Remote Sensing*, 30(5):1327–1340.

Zhau, S.D., and G. Zhau. 1991. Management of Changbai Mountain Biosphere Reserve: The present conditions, problems, and perspectives. *Mountain Research and Development*, 11:168–169.

Zhou S., H. Sun, M. Zhang, X. Lu, J. Yang, and L. Li. 2008. Regional distribution and population size fluctuation of wild Amur tiger (*Panthera tigris altaica*) in Heilongjiang Province. *Acta Theriologica Sinica*, 28(2):165–173.

12

Integration of Remote Sensing and In Situ Observations for Examining Effects of Past Volcanic Eruptions on Forests of the Changbai Mountain Nature Reserve in Northeast China

Jiawei Xu, Yu Liang, and Yeqiao Wang

CONTENTS

12.1 Introduction

The Changbai Mountain Nature Reserve (CMNR), as introduced in Chapter 11 of this book (Wang et al., 2011), is located on the border between China and North Korea. The CMNR is famous for having a unique set of vertically distributed ecosystems that surround the volcanic summit at 2749 m above sea level. Geologically, the region is situated at the edge of the East Asia continent along the Pacific competent zone. The Tianchi volcanic zone has been active since Cenozoic time (Liu, 1987; Liu et al., 1992a,b) and had one of the largest eruptions in the last 2000 years (Machida et al., 1990; Liu et al., 1998). A study concluded that the volcanic eruption is intracontinental rather than active continental margin or intraplate (Wang et al., 2003; Lei and Zhao, 2005). The volcano in the Changbai Mountain has been quiet since minor eruptions in 1597, 1688, and 1702. It is still a high-risk volcano under intensive monitoring (Stone, 2006) by the Chinese scientists and international volcano observation networks.

CMNR hosts a unique vertically distributed zonal vegetation pattern due to climate and terrain conditions (Wang et al., 2011, Chapter 11 of this book). The vegetation distribution in the CMNR reflects the trajectories of aftermath regeneration and succession following significant volcanic eruptions in the most recent several hundred years. Volcanic eruptions not only destroyed the vegetation but also altered the soil contents, landform, and hydrology, which further affected vegetation recovery and succession. Therefore, the timeline of the most recent significant volcanic eruption that devastated the forest ecosystems is the key to understand the history of the vegetation recovery and succession.

Studies reported that recent intensive eruption at the Changbai Mountain was about 800 years ago (Liu et al., 1999) and between AD 600 and 1000 (Gill et al., 1992). Results from examining the SO_4^{2-} concentration of Greenland ice (GISP2) suggested that the eruption time was about AD 1026 (Zielinski et al., 1994). Results from examining carbonized wood at about 800 m elevation on the northern slope and below 1000 m on the western slope suggested that the eruption time was AD 1050 ± 70, AD 1120 ± 70, and AD 1410 ± 80, respectively (Zhao, 1981). The results from ^{14}C suggested that the eruption time was AD 1215 (Liu et al., 1997). Other studies suggested that time should be about 1200 years ago (Liu, 1999), AD 1199–1200 (Cai et al., 2000) and AD 910–1435 (Guo et al., 2001), respectively. The common understanding is that the eruption that occurred in about AD 1215 resulted in the devastating destruction of the vegetation of the CMNR. Other minor eruptions such as those that occurred in 1625, 1373, 1401, 1573, 1668, 1702, and 1903 affected the eastern slope only but caused no significant disturbance in the vegetation in the current CMNR areas (Kayama, 1943; Zhao, 1980; Liu and Wang, 1992). With the above timeline we analyzed the traces and effects of past volcanic eruptions on forests for an understanding of spatiotemporal patterns of the vegetation in the CMNR. We conducted the study by a combination of intensive field investigation augmented by remote sensing observations.

12.2 Methods

12.2.1 Remote Sensing Data and the Analysis

To reveal spatial distribution and temporal change of forest types, we used time series Landsat data, high-spatial-resolution IKONOS data, and topography information from Shuttle Radar Topography Mission (SRTM) data

TABLE 12.1

Remote Sensing Data Used

Sensor	Time of Acquisition	Note
Landsat MSS	October 31, 1972	Path/row: 125/031
Landsat MSS	September 26, 1977	Path/row: 125/031
Landsat MSS	September 22, 1983	Path/row: 116/031
Landsat TM	May 28, 1993	Path/row: 116/031
Landsat ETM+	September 2, 1999	Path/row: 116/031
Landsat ETM+	June 3, 2001	Path/row: 116/031
Landsat ETM+	October 25, 2001	Path/row: 116/031
Landsat ETM+	October 2, 2007	Path/row: 116/031
IKONOS (9198)	June 4, 2002	Cover the submit
IKONOS (9200)	May 29, 2002	Cover a section of the northern slope
IKONOS (9203)	September 30, 2002	Cover a section of the northern slope
SRTM	A regional coverage with a 3 arc-second resolution	

Note: MSS, Multispectral Scanner; TM, Thematic Mapper; ETM+, Enhanced Thematic Mapper plus; SRTM, Shuttle Radar Topography Mission.

(Table 12.1). Landsat time-series data revealed vegetation patterns and land-cover changes both within the protected CMNR and in adjacent areas (Figure 12.1). The Landsat data from different seasons of the same year, such as June and October 2001, helped identify vertical zonal vegetation distributions. IKONOS data illustrated detailed spatial distributions of forest types. SRTM and the derivatives of elevation and slope/aspect provide guidance and control for information extraction about vertical distribution of forest types and their changes (Figure 12.2).

One of the primary objectives of this study was to differentiate forest categories based on species composition and structural characteristics in different vertical zones using remote sensing data. Integration of Landsat data and SRTM-derived variable has been reported for habitat classification and change detection (Sesnie et al., 2008). We conducted unsupervised and stratified classification on Landsat and IKONOS data, respectively. We first obtained generalized vegetation and land-cover types from classification of Landsat data. The categories include broadleaf forest, needleleaf and broadleaf mixed forest, dark coniferous forest, bright coniferous forest, subalpine birch forest, tundra, barren and volcanic rock, water and volcanic lake, agricultural land, and urban and developed land. We then applied the stratified classification (Wang et al., 2007) using SRTM-derived elevation and slope/aspect as the control to extract forest types within each of the vertical zones. We paid special attention to the dominant and representative species in different vertical zones and on different slopes/aspects. Global Positioning System-guided *in situ* observation and documentation provided guidance for the classification process. We referenced the Geographic Information

FIGURE 12.1
Landsat imagery data reveal vertical zonal vegetation patterns within and adjacent to the protected Changbai Mountain Nature Reserve centered by the volcanic lake.

System boundary of the CMNR to separate land-cover types between protected land and adjacent areas to reveal land-cover changes. For taking advantage of high-spatial-resolution IKONOS data, we conducted resolution merge using 1-m panchromatic and 4-m multispectral data. The resolution-merged dataset helped identify effectively representative forest species, in particular the transition areas between vertical vegetation zones (Figure 12.3). We added additional subcategories to identify detailed vegetation distributions for classification of the IKONOS data.

12.2.2 *In Situ* Survey of Forest Stands and Carbonized Wood

We conducted *in situ* survey by gradient pattern in different vertical zones. We set three 20 × 20-m plots in each of the forest zones, including needleleaf and broadleaf mixed forests from altitudes of 700–1100 m, the coniferous

FIGURE 12.2
(**See color insert.**) Shuttle Radar Topography Mission and derivative topographic information (elevation, slope, aspects) provide zonal guidance and reference to differentiate vertical vegetation distribution from Landsat and IKONOS data.

forest zone from altitudes 1100–1700 m, the subalpine *Betula ermanii* forest zone from altitudes 1700–2000 m; and the Alpine tundra zone from altitudes of 2000–2600 m. We documented the basic status such as altitude, slope degree, species, and density for tree and shrub layers, regeneration layer and herb layer. In addition, we set $8 \times 3\ 20 \times 20$-m plots for eight major vegetation types, including, Korean pine and broadleaf mixed forest, Korean

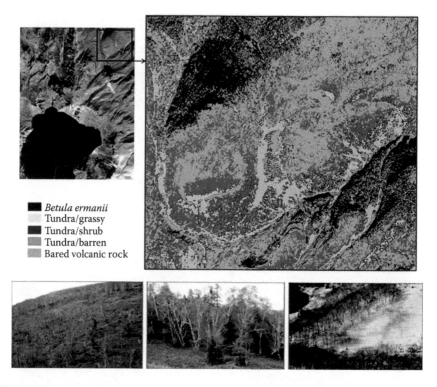

Betula ermanii
Tundra/grassy
Tundra/shrub
Tundra/barren
Bared volcanic rock

FIGURE 12.3
(See color insert.) Classification of IKONOS data to identify detailed representative species in particular in transition areas between vertical vegetation zones. Field survey confirmed the colonization of pioneer species of subalpine *Betula ermanii* and *Larix olgensis* into the tundra zone at high altitudes of the northern slope.

pine and spruce-fir forest, typical spruce-fir forest, *B. ermanii* and spruce-fir forest, oak forest, larch forest, subalpine larch, and low mountain larch for documenting the basic status listed above. We also set 4×3 20×20-m plots in the dark coniferous forest between 1300 and 1400 m in different slopes/aspects for the documentation of species and density for trees, shrubs, regeneration, and herb layers. We then integrated *in situ* and remote sensing observations to reveal the trace of volcanic eruptions and the impacts on vegetation.

Carbonized wood buried in volcanic ash is the evidence of preeruption forest structure, species composition, spatial distribution, temporal variation, and the affected areas (Figure 12.4). To understand the preeruption vegetation structure and distribution, we sampled and documented the species and number of observations for each species found in carbonized wood at different altitudes and slopes/aspects of the volcanic summit of the CMNR. We surveyed the areas at altitudes of 974, 1426, 1731, and 2236 m

FIGURE 12.4
(**See color insert.**) Samples of carbonized woods documented the species and number of observations of each species at different altitudes and slopes/aspects of the area surrounding the volcanic summit to reconstruct preeruption vegetation structure and distribution.

on the northern slope; 951, 1493, 1760, and 2093 m on the western slope; and 936, 1472, 1753, and 2311 m on the southern slope.

12.3 Results

12.3.1 Remote Sensing Vegetation Mapping

The classification of Landsat images and the derivatives of vegetation maps reveal spatial distribution and temporal variation. In particular, when augmented by SRTM-derived elevation model data, the maps illustrate the spatial distribution and pattern of different forest zones in different altitudes and slopes/aspects. Land-cover changes during the periods 1972–1983, 1983–1993, 1993–1999, and 1999–2007 observed from Landsat data mostly resulted from deforestation and agricultural and urban land development. These included significant changes immediately adjacent to the border of the CMNR, which indicate the result of enforced land protection and management of the reserve. There was no change of tree-line and tundra-line in the CMNR since the early Landsat data acquisition in 1972. A recent study reached the same conclusion, reporting an increase in absolute normalized difference vegetation index (NDVI) values (Zhang et al., 2008). The study acknowledged that the increase in absolute NDVI might be attributed to multiple factors, such as climate warming, carbon dioxide fertilization effect, and biases from remote sensing data.

Spatial distribution of vertical vegetation zones, forest types, landscape configuration and composition provide information about the status of vegetation and the impacts from past volcanic eruptions. High-spatial-resolution vegetation mapping from IKONOS data reveal forest structure for specific species. In particular, high-spatial-resolution information confirmed, in an aerial context, the colonization of subalpine *B. ermanii* into the tundra zone in the high altitude of the northern slope (Figure 12.3).

12.3.2 Spatial Variation of Vegetation Distribution

The volcanic eruptions had significant impact on the alpine vegetation compared with those at lower altitudes. Almost all alpine and subalpine vegetation were destroyed by the last devastating eruption that caused a decline in forest timberline between vertical zones. Some species disappeared due to the impacts. For example, although samples of dwarf Siberian pine (*Pinus pumila*) were observed in abundance in carbonized wood, there is still no sign of this species in the high-altitude area of the CMNR. Remote sensing images and *in situ* documentation suggested that although vegetation distribution appears in relatively complete vertical zones, its uniqueness is quite obvious. Landsat images show that the vertical zonal pattern of vegetation is more complete on the northern slope than on the southern and eastern slopes. The *Larix olgensis* and *Pinus koraiensis* forests are widely distributed and often seen in pure stands. The spruce (*Pieca jezoensis*) and fir (*Abies nephrolepis*) forests exist in pure stands on the northern slope. The *B. ermanii* forest zone is extensively established on the northern slope but not on the eastern slope. The difference in spatial distribution results from volcanic impacts.

The study on the carbonized wood samples suggested that the current distribution of forest types has the same pattern as that of forests before eruptions. The *Pinus koraiensis* is the constructive species with mixed broadleaf tree species. The tree-line between needleleaf and broadleaf mixed forests and dark coniferous forests is consistent, which suggested that there was no significant climate change in the past several hundred years. The Landsat data from 1972 to 2007 also illustrate that the tree-line has been consistent in the recent years under other factors that have an impact, such as human disturbances and scenarios of regional warming under global climate change. However, the tree-line and tundra-line were higher than the current ones before volcanic eruptions, especially on the eastern slope. The altitude of current alpine timberline is about 2100 and 2000 m on the northern and eastern slopes, respectively. The current upper tundra-line is 2500–2600 and 2400 m on the northern and eastern slopes, respectively. The highest altitude of residual wood on the eastern slope is 2160 m. There are scattered forest patches above 2100 m. On the northern slope, *B. ermanii* forests can reach altitudes of up to 2200 m. The distributions suggest that the tree-line could be higher than 2200 m on the northern slope, 2100 m on the western slope, and 2300 m on the southern slope. The increasing trend of tree-line altitude is confirmed by the observations that the forest ages are younger at higher altitudes. As simulated by the hydrothermal index (Xu, 1985) the potential timberline is higher than the forest distributions today. The lowered timberline and tundra-line reflect the impacts of volcanic eruptions on vegetation. These also indicate that the vegetation recovery is still in progress.

12.3.3 Extended Subalpine Dwarf-Birch and Larch Forests

The *B. ermanii* and *L. olgensis* are pioneer species and the forests have the characteristics of being in stages of post-eruption succession. Remote sensing images reveal the spatial distribution and locations of the *B. ermanii* forests. The evidence observed in field surveys shows that *B. ermanii* trees are older in age than coniferous species in the mixed forest zone. The remains of *B. ermanii*, however, are rarely found in carbonized wood samples in the volcanic ash. Studies indicate that the pre-eruption spatial distribution of *B. ermanii* was smaller in area than the current extension. *B. ermanii* forests should not be regarded as the product of volcanic eruptions as it exists in the other area of the northeast China and Russia. In comparison with coniferous forests, *B. ermanii* forests can adopt harsh habitats such as poor soil, extreme drought, and cold and windy conditions. The *B. ermanii* forest on the northern slope is largely mixed with *L. olgensis* which has obvious pioneer characteristics. Field survey results indicate that the mixed *L. olgensis* forest is older in age and they are gradually exiting the communities. The *L. olgensis* forests in lower altitudes have been invaded by coniferous forests as the succession progresses. The *Pieca jezoensis* and *B. ermanii* forests will take over after time. Therefore, the extensive distribution of the *B. ermanii* forest zone reflects the impacts of volcanic eruptions and posteruption succession.

The larch species is an important community plant in boreal and high altitude forest. The larch species has a wide extension mainly from altitudes of 500–1950 m (Yu et al., 2005). The areas of larch forest are extensively distributed from needleleaf and broadleaf forest zones to *B. ermanii* forest zones, forming the interzonal vegetation. The wide distribution of *L. olgensis* suggests the uniqueness of vegetation changes affected by volcanic eruptions. *L. olgensis* is a species that prefers light, tends to be resistant to cold and humidity, and coexists with *Pinus koraiensis*, *Pieca jezoensis*, and *A. nephrolepis*, among other species. The number of *Pieca jezoensis* and *A. nephrolepis* trees increase toward high altitudes and the number of *Pinus koraiensis* trees increase toward lower altitudes. *L. olgensis* does not show a clear zonal distribution in the CMNR. The subalpine larch is the secondary forest vegetation formed by the destroyed subalpine coniferous forest, which is mainly distributed at altitudes between 1100 and 1700 m on the eastern and northeastern slopes. The most recent volcanic eruptionsint 1597, 1668, and 1702 had no destructive impacts on the vegetation of the entire area. Those eruptions destroyed part of the forest and produced large bare land on the eastern slope, which is observable from Landsat images. Hence, the *L. olgensis* can result from the recent volcanic eruptions.

12.4 Conclusion and Discussion

The integrated information from remote sensing and *in situ* survey concluded that the current vegetation distribution and structure in the CMNR reflect the effects of historical volcanic eruptions and the process of succession, in particular in high-altitude areas. Historical volcanic eruptions not only directly influenced vegetation but also indirectly affected the altered natural conditions such as strata, terrain, and hydrology. Landsat remote sensing data and the derivatives of vegetation maps reveal the pattern of vertical zonal vegetation distribution. The zonal pattern is more complete on the northern slope but not quite obvious on the eastern slope. However, plenty of carbonized wood samples within volcanic ash are observable in places on the eastern slope, suggesting volcanic impacts.

Spectral features from Landsat data are capable of discriminating forest types and secondary successional stages (Foody and Hill, 1996). However, vegetation classification can be limited when ecologically important differences in forest structure and composition are spectrally similar (Castro et al., 2001; Helmer et al., 2002; Pedroni, 2003). Stratified classification used SRTM-derived topography information to map vegetation and land-cover types that associated with vertical zonal distributions effectively.

With the integrated information we conclude the following.

1. The current vertical zonal vegetation distribution is consistent with that before the volcanic eruptions. The current vegetation base bands follow the same distribution as those about 800 years ago.
2. The rareness of *B. ermanii* in carbonized wood suggests that pre-eruption distribution of *B. ermanii* was smaller in area than the current extension.
3. Volcanic eruptions destroyed high-altitude vegetation, lowered the tree-line, and affected species composition. A species such as dwarf Siberian pine (*Pinus pumila*) was wiped out and vegetation recovery is still in progress.
4. The differences in silvan vegetation composition between the CMNR and similar settings of cold temperate and mid-temperate zones suggest the impacts of past volcanic eruptions on the current vegetation structures. The wide distribution of *L. olgensis* confirms the uniqueness of vegetation change affected by volcanic eruptions.

All the above points demonstrate that the distinctive quality of vegetation in the CMNR is determined by the influence of volcanic eruptions on the ecosystems. Traces of volcanic eruptions exist in the forest ecosystems in the CMNR after 800 years of the last devastated volcanic eruption.

Acknowledgments

This study was funded by the National Accented Foundation Project of China (2002CB111502) and Global Environmental Project CPR/98/ALG/99. The study was supported by the Scientific and Technological Innovation Initiative of the Northeast Normal University under the project "Ecological Security and Data Assemblage of the International Georegion of the Changbai Mountains" (106111065202) and the National Grand Fundamental Research 973 Program of China (No. 2009CB426305). We thank Dr. Guoqing Sun of the University of Maryland for providing the IKONOS data.

References

Cai, Z., D. Jin, and L. Ni. 2000. The historical record discovery of 1199–1200 AD large eruption of Changbaishan Tianchi volcano and its significance. *Acta Petrologica Sinica*, 16(2): 191–193 [in Chinese with English abstract].

Castro, K.L., G.A. Sánchez-Azofeifa, and B. Rivard. 2001. Monitoring secondary tropical forest using space-born data: Implications for Central America. *International Journal of Remote Sensing*, 24: 1853–1894.

Foody, G.M., and R.A. Hill. 1996. Classification of tropical forest classes from Landsat TM data. *International Journal of Remote Sensing*, 17: 2353–2367.

Gill, J., C. Dunlap, and M. McCurry. 1992. Large- volume, mid- latitude, Cl- rich volcanic eruption during 600–1000 AD, Baitoushan, China. American Geophysical Union Chapman Conference (March 23–27), Hito, Hawaii, USA. *Climate, Volcanism and Global Change*, 1–10.

Guo, Z., J. Liu, and S. Sui et al. 2001. Total estimated of effusive volcanic gas of Baitoushan volcano and its significance of 1199–1200 AD. *Science in China Series D: Earth Sciences*, 31(8): 668–675 [in Chinese with English abstract].

Helmer, E.H., O. Ramos, T.D.M. López, M. Quiñones, and W. Diaz. 2002. Mapping the forest type and land cover of Puerto Rico, a component of the Caribbean biodiversity hotspot. *Caribbean Journal of Science*, 38: 165–183.

Kayama, N. 1943. Forest trees in foothills of Baitoushan before volcanic eruptions. *Botanical Magazine (Tokyo)*, 57(679): 258–273.

Lei, J., and D. Zhao. 2005. P-wave tomography and origin of the Changbai intraplate volcano in Northeast Asia. *Tectonophysics*, 397: 281–295.

Liu, J. 1999. *Volcanoes of China*. Beijing: Science Press, pp. 1–219.

Liu, J.Q. 1987. Study on geochronology of the Cenozoic volcanic rocks in northeastern China. *Acta Petrologica*, 4: 21–31 [in Chinese with English abstract].

Liu, Q., and Z. Wang. 1992. Recent volcanic eruption and vegetation history of alpine and subalpine of Changbai Mountain. *Forest Ecosystem Research*, 6: 57–62.

Liu, R., Q. Fan, and H. Wei. 1999. The research of active volcanoes in China. *Geological Review*, 45(Suppl.): 3–15.

Liu, R., S. Qiu, L. Cai, et al. 1997. Age research of the most recent large eruption of Changbaishan Tianchi volcano and its significance. *Science in China Series D: Earth Sciences*, 27(5): 437–441 [in Chinese with English abstract].

Liu, R.X., W.J. Chen, J.Z. Sun, and D.M. Li. 1992a. K–Ar chronology and tectonic settings of Cenozoic volcanic rocks in China. In R.X. Liu (ed.), *Chronology and Geochemistry of Cenozoic Volcanic Rocks in China*. Beijing: Seismological Press, pp. 1–43 [in Chinese].

Liu, R.X., J.T. Li, H.Q. Wei, D.M. Xu, and Q.F. Yang. 1992b. Changbaishan Tianchi Volcano—A recent dormant volcano for potential dangerous eruptions. *Acta Geophysica*, 35: 661–664 [in Chinese with English abstract].

Liu, R.X., H.Q. Wei, and J.T. Li. 1998. *Recent Eruptions of the Changbaishan Tianchi Volcano*. Beijing: Scientific Press, pp. 1–159 [in Chinese].

Machida, H., H. Morwaki, and D.C. Zhao. 1990. The recent major eruption of Changbai volcano and its environmental effects. *Geographical Reports of Tokyo Metropolitan University*, 25: 1–20.

Pedroni, L. 2003. Improved classification of Landsat Thematic Mapper data using modified prior probabilities in large and complex landscapes. *International Journal of Remote Sensing*, 24: 91–113.

Sesnie, S.E., P.E. Gessler, B. Finegan, and S. Thess. 2008. Integrating Landsat TM and SRTM-DEM derived variables with decision trees for habitat classification and change detection in complex neotropical environments. *Remote Sensing of Environment*, 112: 2145–2159.

Stone, R. 2006. A threatened nature reserve breaks down Asian borders. *Science*, 313: 1379–1380.

Wang, Y., C. Li, H. Wei, and X. Shan. 2003. Late Pliocene–recent tectonic setting for the Tianchi volcanic zone, Changbai Mountains, northeast China. *Journal of Asian Earth Sciences*, 21: 1159–1170.

Wang, Y., M. Traber, B. Milestead, and S. Stevens. 2007. Terrestrial and submerged aquatic vegetation mapping in Fire Island National Seashore using high spatial resolution remote sensing data. *Marine Geodesy*, 30(1): 77–95.

Wang, Y., Z. Wu, X. Yuan, H. Zhang, J. Zhang, J. Xu, Z. Lu, Y. Zhou, and J. Feng. 2011. Remote sensing assessment of natural resources and ecological security of the Changbai Mountain Region in Northeast Asia. In Y. Wang (ed.), *Remote Sensing of Protected Lands*. Boca Raton, FL: CRC Press.

Xu, W. 1985. Heat index of Kira and its application of vegetation in China. *Chinese Journal of Ecology*, 3: 35–39.

Yu, D., S. Wang, and L. Tang. 2005. Relationship between tree-ring chronology of *Larix olgensis* in Changbai Mountains and the climate change. *Chinese Journal of Applied Ecology*, 16(1): 14–20.

Zhang, Y., M. Xu, J. Adams, and X. Wang. 2008. Can Landsat imagery detect tree line dynamics? *International Journal of Remote Sensing*, 30(5): 1327–1340.

Zhao, D. 1980. Vertical distribution of the vegetation belts in Changbaishan Mountain. *Forest Ecosystem Research*, 1: 65–70 [in Chinese with English abstract].

Zhao, D. 1981. Preliminary investigation on relation between volcano eruption of Changbai Mountain and the evolution of its vegetation. *Forest Ecosystem Research*, 2: 81–87 [in Chinese with English abstract].

Zielinski, G., P. Mayewski, L. Meeker, et al. 1994. Record of volcanism since 7000 B.C. from the GISP Greenland ice core and implications for the volcano–climate system. *Science*, 264: 948–952.

13

Integration of Remote Sensing and In Situ Ecology for the Design and Evaluation of Marine-Protected Areas: Examples from Tropical and Temperate Ecosystems

Alan M. Friedlander, Lisa M. Wedding, Jennifer E. Caselle, and Bryan M. Costa

CONTENTS

13.1 Marine Ecosystems in Context

13.1.1 Importance and Uniqueness of Marine Ecosystems

Marine ecosystems comprise most of the world's biosphere and provide a wide variety of functions that translate to economic services and value to humans. They provide protein for people throughout the world, especially in developing countries and small island nations (FAO, 2009). The oceans are important in climate regulation and store 20 times more carbon than all of the world's forests and other terrestrial biomass combined (Weber, 1993). Cultural services provided by the oceans include recreation, ecotourism, aesthetic values, as well as spiritual, religious, and cultural heritage values (Beaumont et al., 2007).

The number of animal phyla in the ocean (34) far exceed that on land (11) or in freshwater (14) and of these phyla, 13 are endemic to the ocean realm, whereas only one endemic phylum exists on land and none are found exclusively in freshwater (Norse, 1993). The high biological diversity found in marine ecosystems translates into high chemical diversity that plays an important role in biomedical research and drug development (Hay and Fenical, 1996). Marine organisms are also used directly as fertilizers, biofuels, and building materials (Norse, 1993).

The early life stages of marine organisms are planktonic and travel via ocean currents for days to months before settling onto juvenile and adult habitats. As a result, marine systems are considered more "open" than terrestrial systems and are greatly influenced by both biotic and abiotic factors that vary at a wide range of spatial and temporal scales. Temperate marine ecosystems are characterized by a larger intra-annual variation in water temperature, higher productivity, and lower water clarity than tropical marine ecosystems.

13.1.2 Issues Facing Marine Ecosystems

Despite their ecological significance and economic importance, marine ecosystems are under increasing anthropogenic pressure as a result of climate change, pollution, habitat degradation, invasive species, and overfishing

(Lotze et al., 2006; Jackson, 2008), and there are now no ocean areas that are exempt from these impacts (Halpern et al., 2008). Although fishing is not the only impact on the ocean, it is a substantial one and historical evidence indicates that fishing has led to the wholesale alteration of ocean ecosystems around the globe, with impacts going back to the first human settlement (Lotze et al., 2006). It is now becoming evident that to conserve marine biodiversity, maintain fisheries, and deliver a broad suite of ecosystem services over a long time frame, an ecosystem-based management approach is necessary (Pikitch et al., 2004). To protect marine ecosystem function, biodiversity, as well as the goods and services provided by these systems, marine protected areas (MPAs) have been increasingly implemented as part of an ecosystem-based approach to management in both temperate and tropical systems. By protecting populations, habitats, and ecosystems within their borders, MPAs provide a spatial refuge for the entire ecological system they contain and, if properly designed, implemented, managed, and enforced, provide a powerful buffer against human and natural impacts.

13.1.3 Management Tools to Conserve Marine Ecosystems

The International Union for the Conservation of Nature (IUCN) has defined MPAs as "any area of intertidal or subtidal terrain, together with its overlying water and associated flora, fauna, historical and cultural features, which has been reserved by law or other effective means to protect part or all of the enclosed environment" (Kelleher, 1999, p. XI).The practice of closing a marine area (either seasonally or permanently) is not a modern concept and was common practice in the Pacific Islands for centuries to help sustain healthy populations of marine resources (Johannes, 1978). The poor performance of conventional fisheries management has led to increased interest among marine resource managers in MPAs where human activity is regulated in certain areas of the sea (Lubchenco et al., 2003). MPAs have been shown to conserve fish stocks within their boundaries and provide fisheries benefits outside these protected areas (Lester et al., 2009; Molloy et al., 2009). By increasing sizes and numbers of fishes, reserves increase the reproductive capacity of the species protected within their borders that can enhance adjacent fisheries (Gell and Roberts, 2003; Sladek Nowlis and Friedlander, 2005; Hamilton et al., 2010). MPAs also have many nonfisheries benefits, such as protecting biodiversity and ecosystem structure, serving as biological reference areas, and providing nonconsumptive recreational activities (Bohnsack, 1998; Roberts, 2005).

Individual MPAs need to be networked in order to provide large-scale ecosystem benefits. An MPA network consists of a series of protected areas that are connected by larval dispersal or juvenile and adult movement (Almany et al., 2009; Planes et al., 2009). MPA networks have the greatest chance of protecting all species, life stages, and ecological linkages if they encompass representative portions of all ecologically relevant habitat types in a replicated manner. The concept of essential fish habitat and the principles behind developing MPA networks and ecosystem-based management necessitate examination of greater spatial ranges than those at which typical *in situ* experiments are conducted. MPA networks often span broad geographic regions and the use of remotely sensed imagery has begun to play a critical role in the monitoring and assessment of marine ecosystems within and around these management units. *In situ* data provide important information about ecological communities in these management units; however, this information is greatly enhanced by remotely sensed data which provides broad spatial and temporal coverage of the marine environment.

13.2 Integration of *In Situ* and Remotely Sensed Data

Remotely sensed data that quantifies spatial patterns in habitat type, oceano-graphic conditions and benthic complexity can be integrated with *in situ* ecological data using a spatial framework to design, assess, and monitor MPA networks (Clark et al. 2005, Monaco et al. 2005, Wedding and Friedlander 2008). Remotely sensed data provides the potential to examine MPAs at broader spatial scales that are more appropriate for assessment of large marine management units. Combining remote sensing products (e.g., ben-thic habitat maps) with *in situ* ecological and physical data can support the development of a statistically robust monitoring program of living marine resources within and adjacent to MPAs (Friedlander et al., 2007a). These datasets provide valuable information on seascape metrics (e.g., habitat com-plexity, depth, habitat type) that can inform the design of networks of MPAs (CDFG, 2008).

A suite of remote sensing tools is currently available for the marine envi-ronment (Figure 13.1). The type of data acquired, operational depth, spatial resolution, and cost of acquisition varies greatly depending on the platform and sensor (Table 13.1). Recent advances in remote sensing have provided datasets that span broad geographic extents that support the study of spatial interactions of marine organisms within their environment (Pittman et al., 2007). This broad-scale approach is necessary for conducting monitoring and assessment of MPA networks that span large geographic areas. Several examples of this integrated approach, applied in both temperate and tropical marine environments, are a series of biogeographic assessments conducted by the Biogeography Program of the National Oceanic and Atmospheric Administration (NOAA) in consultation with the U.S. National Marine Sanctuary (NMS) Program to support sanctuary management, planning, and research (Monaco et al., 2005). Typically, a biogeographic assessment consists of three primary activities: (1) compiling individual biogeographic data layers (e.g., Geographic Information System [GIS]/remote sensing data-sets), (2) performing integrated biogeographic analyses, and (3) developing

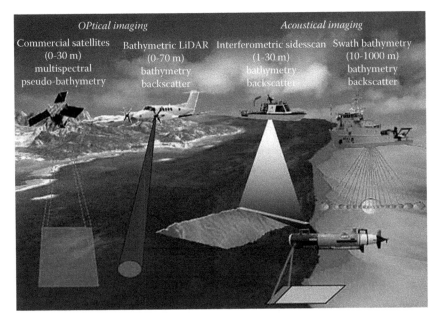

FIGURE 13.1
Suite of remote sensing platforms for marine habitat characterization.

products to aid in management. A key tool used to develop and implement the assessment is the use of GIS technology which aids in data compilation, spatial analyses, and visualization of results to support place-based management needs (Battista and Monaco, 2004). Biogeographic assessments are important because they focus on the large-scale distribution of species and hence provide the basis for predicting biota for a given habitat within a biogeographic province.

13.2.1 Case Study Approach to Sensors and Applications

This chapter highlights several case study examples of marine biogeographic assessments that have been implemented using a combination of remotely sensed and *in situ* datasets. These case studies serve as models to illustrate the integration of remotely sensed data to better understand marine ecosystems at multiple spatial scales and support management actions in both temperate and tropical MPAs. In a general introduction to marine remote sensors, we focus on high-resolution sensors (resolutions of 0.01 to 10s of meters) capable of mapping physical and biological features of benthic habitat, rather than on the coarser resolution sensors of use for broad-scale characterization of pelagic habitats. Ocean surface color, temperature, height, wave, winds, and (soon) salinity can be mapped with other remote sensing platforms, but the resolutions of these data streams are of the order of 0.1–100 km. This is

TABLE 13.1

Data Acquired, Operational Depth, Spatial Resolution, and Cost of Acquisition for
Various Sensors and Platforms Used in Remote Sensing of the Marine Environment

Platform	Sensor	Maximum Operational Depth (m)	Spatial Resolution (m²)	Types of Data Acquired
Space-based	Geoeye-1	≤30	2.7	5 channels
	Quickbird	≤30	6.4	4 channels
	IKONOS	≤30	16	4 channels
	Landsat TM	≤30	900	6 channels
	Hyperion	≤30	900	220 channels
Aerial	Color photography	≤30	1	3 channels
	Hyperspectral	≤30	9	74 channels
	LiDAR	≤30	1–36	Laser bathymetry and intensity
Ship-based	Multibeam (MBES)	5–30	0.5–2	Continuous acoustic bathymetry and intensity
		>30–1000+	2–16+	
	Side-scan (SSS)	5–30	>0.01	Continuous acoustic intensity
		>30–1000+		
	Interferometric (PMBS)	5–30	>0.01	Continuous acoustic bathymetry and intensity
		>30–1000+		
	Single beam (AGDS)	5–1000+	n/a	Discontinuous acoustic bathymetry and intensity

Note: TM, Thematic Mapper; LiDAR, light detection and ranging; MBES, multibeam echo-sounders; SSS, side-scan sonars; PMBS, phase measuring bathymetric sonars; AGDS, acoustic ground discrimination systems.

followed by two case studies (tropical and temperate ecosystems) focusing on the integration of remote sensing and *in situ* observations for the design and evaluation of MPAs. The chapter concludes with constraints and consideration in the use of remote sensing in tropical and temperate marine environments and the use of new technologies, data fusion and decision support tools that serve to synthesize remotely sensed data into a format amenable for marine spatial planning.

13.2.2 Remote Sensing Methods for Marine Applications

Satellites were first used in the 1960s for remote sensing of the earth's surface (Arnold, 1997), but most of these early sensors were not used for marine applications (Green, 2000). Since this time many different types of sensor technologies have come online. For example, marine scientists and managers have had access to a number of satellite data streams providing high spatial resolution panchromatic and multispectral imagery (e.g., IKONOS, Quickbird, etc.) since the late 1990s (Hochberg et al., 2003). By interpreting remotely sensed imagery of benthic habitat, managers can broaden and generalize the information gathered by *in situ* ecological assessments.

13.3 Airborne Remote Sensing

13.3.1 Multispectral Sensors

Some of the first marine remote sensing projects utilized relatively low- to medium-resolution data sources such as early Landsat and SPOT images but these sensors could only classify habitat within a limited number of geomorphologic and benthic cover classes (Hochberg and Atkinson, 2000). IKONOS and Quickbird satellite sensors demonstrate high spatial resolution that is almost equivalent to aerial imagery and is increasingly being used for marine applications (Mumby et al., 2004). Considerable progress has been made in the application of remote sensing technology to support monitoring, assessment, and management of coral reefs (Hamel and Andréfouët, 2010). In the temperate marine environment, both moderate- and high-resolution data from aerial and satellite-based multispectral sensors supported the marine management process for a number of MPAs (e.g., Channel Islands NMS and Stellwagen Bank NMS).

13.3.2 Hyperspectral Sensors

Hyperspectral sensors obtain data in hundreds of different bands of the electromagnetic spectrum (Jenson, 2000) and are increasingly being used for marine benthic habitat mapping (Karpouzli et al., 2004; Kutser et al., 2006). However, one shortcoming is the difficulty in relating spectral reflectance values to biophysical properties on the seafloor in tropical marine environments (Joyce and Phinn, 2003). In temperate ecosystems, hyperspectral sensors have been primarily used to map and characterize coastal habitats and estuaries (Judd et al., 2007) because turbidity limits the utility of these sensors to map nearshore benthic habitats although several successful subtidal mapping projects have been conducted in this biome (Theriault et al., 2006; Vahtmäe et al., 2006).

13.3.3 Light Detection and Ranging Systems

Light Detection and Ranging (LiDAR) systems are active sensors that use laser technology to collect topographic and bathymetric measurements and have become increasingly useful in tropical marine studies (Wedding et al., 2008; Costa et al., 2009; Pittman et al., 2009). LiDAR digital elevation models have been successfully used to map and quantify the morphology of shallow-water coral reefs (Storlazzi et al., 2003; Brock et al., 2006). In addition, LiDAR has been used to measure coral reef complexity and linked to measures of fish assemblage structure (Kuffner et al., 2007; Wedding et al., 2008; Pittman et al., 2009).

LiDAR acquisitions in temperate ecosystems have primarily supported engineering or navigational charting applications (Irish and White, 1998; Young and Ashford, 2006). However, some LiDAR acquisitions have been aimed specifically at supporting the management of MPAs. For example, the Olympic Coast NMS used the SHOALS LiDAR system to acquire depth information as part of a comprehensive effort to map and characterize benthic habitats of the nearshore areas (Intelmann, 2006). LiDAR data were also collected in the Monterey Bay NMS in order to create a seamless bathymetric surface, which was used to help manage erosion in the Elkhorn Slough National Estuarine Research Reserve. In both sanctuaries, LiDAR proved to be a valuable tool for research managers to capture baseline information about, and observed changes to, the seafloor over broad spatial scales within these MPAs.

13.3.4 Radio Detection and Ranging Sensors

Synthetic Aperture Radar (SAR) is a form of Radio Detection and Ranging (Radar) sensor that emits pulses of microwave energy to produce two-dimensional backscatter maps of the roughness of the ocean surface (Jackson and Apel, 2004). Such sensors are capable of detecting small changes in surface roughness (from centimeter to meter) and are independent of sun illumination and cloud cover—two environmental variables that are problematic for optical sensors in temperate ecosystems. An increasingly common approach for measuring ocean surface currents is high-frequency (HF) radar (5–50 MHz). Many years of development have produced a number of HF radars capable of mapping surface current vectors in the coastal zone over spatial scales ranging from meters to hundreds of kilometers, and time intervals ranging from minutes to days. The high spatial and temporal resolution of HF radars enables them to be used for studying coastal circulation processes, which has implications for larval dispersal and retention. Two types of HF radars (direction-finding and beam-forming) are currently used in marine systems and differ in the method used to determine bearing to a sector on the ocean surface. Beam-forming radars, such as the Ocean Surface Current Radar (Hammond et al., 1987) and Wellen Radar (Gurgel et al., 1999), electronically steer a linear array of phased receiver antennas toward a sector of ocean surface. Direction-finding radars, such as the Coastal Ocean Dynamics Application Radar SeaSonde, use directional loop antennas to determine bearing. By observing an ocean area from multiple HF radars distributed over a range of angles along a coastline, a grid of surface current time series may be obtained. Depending on frequency, surface current vectors are interpolated onto grids with spacing ranging from a few hundred meters to several kilometers.

Using HF radar-based surface current mapping in near real time to direct plankton net tows, Nishimoto and Washburn (2002) found much higher abundances of fish larvae inside cyclonic eddies in the Channel Islands NMS. To understand patterns of connectivity between MPAs in central California, Zelenke et al. (2009) estimated back projections of larval trajectories based on observations from a network of HF radars operated by the Southern California Coastal Ocean Observing System and the Central and Northern California Coastal Ocean Observing System. Larval trajectories were determined from velocity time series interpolated on to a 1 km grid and stretching 40 days into the past. The resulting connectivity matrix identified probable larval source and sink locations and can be used to interpret recruitment patterns within MPAs.

13.4 Ship-Based Remote Sensing

13.4.1 Sound Navigation and Ranging Technologies

Sound Navigation and Ranging (Sonar) technologies were first developed in the early 1910s to detect submerged objects and enemies (D'Amico and Pittenger, 2009) but it took another 60 years before sonar systems were able to acquire high-resolution images of the seafloor (Kenny et al., 2003). Sonar gathers information by actively emitting pulses of sound into the water, recording the return, and collecting depth and intensity information about objects in the water column or on the seafloor. Depth is calculated from the time difference from when the sound pulse is emitted and its return, whereas intensity is the quantity of sound that is returned to the sensor from the object or seafloor and is correlated with its density.

Some sonars are down-looking (e.g., single-beam echosounders [SBES] and multibeam echosounders [MBES]) whereas other sonars are side-looking (e.g., side-scan sonars [SSS] and phase measuring bathymetric sonars [PMBS]). SBES measure the depth of the seafloor by emitting a single beam of sound, which travels down to the seafloor and back to the sensor. Some types of SBES, called acoustic ground discrimination systems, may be calibrated to classify seafloor habitats or subsurface sediments (Anderson and Gregory, 2002). Since SBES only collect depth information for a single location, these systems do not provide continuous coverage of the seafloor, leaving spatial gaps in the data and limiting application for large continuous areas.

MBES emit multiple "beams" of sound at the same time, but at different downward-looking angles from the sensor. Unlike SBES, these systems can collect continuous depth and intensity information over large geographic areas. SSS are often towed behind a vessel, just above the seafloor and produce continuous-intensity images of the seafloor by emitting beams of sound sideways (instead of downward) from either side of the sensor. PMBS, also called interferometric sonars, are similar to SSS in that they emit beams of sound sideways from either side of the sensor but also collect bathymetric information by measuring phase elements to calculate the angle from which the acoustic return originated.

Various sonar systems have been used to map and characterize the depth, geology, and associated benthic habitats in many marine environments and, consequently, are important to the ecosystem-based management of MPAs. In the last decade, MBES imagery has been used to map benthic habitats in Stellwagen Bank NMS and Gray's Reef NMS as well as throughout California (USGS, 2000; Kendall et al., 2005), whereas SSS imagery has been used to characterize the seafloor in Channel Islands NMS and Gray's Reef NMS (Greene and Bizzarro, 2003; Kendall et al., 2005). Split-beam echosound is being used to remotely quantify the abundance and distribution of fishes and statistically link these data with habitats to better understand the ecological associations among these marine resources (Koslow, 2009). Together, this suite of acoustic sensors provides resource managers with a valuable toolbox to capture baseline information about the habitats and living marine resources within their MPAs.

Hydroacoustic surveys to assess fish stocks have been conducted for many years but backscatter measurement from the sensors only relay basic information about the reflecting target, largely determined by the swimbladder, and therefore limiting information about individual species recognition (Ona, 1990). Assessments of fisheries using acoustics have been successfully applied primarily to pelagic species in relatively simple systems (Rose, 2003), but recent efforts in Gray's Reef NMS have proven that this technology is an efficient, nonintrusive method for quantifying the abundance and distribution of multiple fish species in the sanctuary (Kracker, 2007). Advances in acoustic technology, especially analysis software, have made these survey methods more powerful in recent years and combining this with information collected at Gray's Reef NMS through diver surveys, video transects, and trapping efforts has enhanced the interpretation of acoustic surveys.

13.5 Tropical Ecosystems

The use of remotely sensed imagery to monitor coral reef health is becoming increasingly important as reefs throughout the world face increasing pressure from coastal populations and natural stresses (Andréfouët et al., 2003; Andréfouët and Reigl, 2004; Mumby et al., 2004). The utility of remote sensing for the study of coral reef ecosystems has been apparent since Landsat was first used in research and exploration of coral reefs on the Great Barrier Reef in 1975 (Mumby et al., 1997). This section will focus on the integration of remote sensing and *in situ* observations for the evaluation of MPAs in the Hawaiian Archipelago.

13.5.1 MPAs in Hawaii

Hawaii established its first MPA over 30 years ago, and since that time numerous protected areas have been established, with varying levels of resource protection, ranging from complete "no-take" areas to those that allow a wide variety of activities to occur. Within the main Hawaiian Islands, there are 34 state-managed areas which limit fishing activities in nearshore marine waters including areas designed to conserve marine life and resolve user conflicts (Friedlander et al., 2008; Figure 13.2), In addition, members of the public have limited or no access to the shoreline and nearshore waters within and around military or security areas on Oahu and Kauai and in Hawaii Volcanoes National Park.

The large number of restricted areas in the main Hawaiian Islands gives the impression of a substantial network of actively managed and protected marine areas, but in reality the majority of these areas are small and most allow some form of fishing within their boundaries. In total, only 0.4% of nearshore main Hawaiian Island waters <60 ft deep (an approximation of the

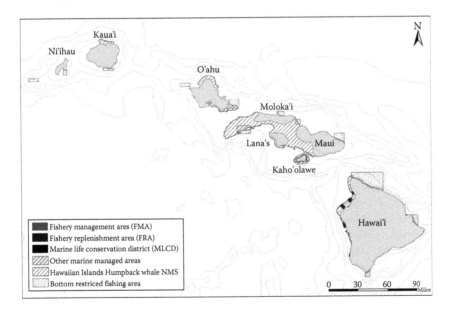

FIGURE 13.2
Marine-managed areas on the main Hawaiian Islands.

inshore habitats which are the primary targets for fishing of reef and reef-associated species) are in no-take MPAs (Friedlander et al., 2008). Owing to the diversity of existing MPAs in Hawaii, it was critical that the efficacies of these areas were evaluated to ensure the effective management of existing areas and inform the design of future MPAs in the state. The first step in this process was to map nearshore benthic habitats (Coyne et al., 2003; Battista et al., 2007) and define species habitat utilization patterns across varying levels of habitat quality and management (Friedlander et al., 2007a,b).

13.5.2 Benthic Map Products of Hawaii

The National Oceanic and Atmospheric Administration (NOAA) acquired and visually interpreted orthorectified aerial photography, IKONOS satellite imagery, and hyperspectral imagery for the nearshore waters (<25 m depth) of the majority of the main Hawaiian Islands (Coyne et al., 2003). The major product of this effort was a series of GIS-based benthic habitat maps that were characterized by a high degree of spatial and thematic accuracy (Figure 13.3) (Coyne et al., 2003; Battista et al., 2007). Habitat features were delineated at a scale of 1:6000, with a minimum mapping unit of 0.004 km². Visual interpretation of the imagery was guided by a hierarchical classification scheme that defined and delineated benthic polygon types based on habitat classifications that included ecologically relevant locational (backreef, forereef, lagoon, etc.) and typological (patch reef, spur and groove, colonized pavement, etc.) strata. A more recent version of these maps that separates biological cover from geomorphic structure presents a significant improvement on the previous maps developed by NOAA for the Caribbean and Hawaii.

13.5.3 Efficacy of MPAS in Hawaii Using Remote Sensing and *In Situ* Ecology

Digital benthic habitat maps were used to evaluate the efficacy of existing MPAs using a spatially explicit stratified random sampling design. These MPAs vary in size, habitat quality, and management regimes, providing an excellent opportunity to test hypotheses concerning MPA design and function using multiple discrete sampling units. The benthic habitat maps were used to guide *in situ* ecological studies of fish assemblages and benthic habitats within a spatial framework, in order to assess the efficacy of these protected areas relative to their adjacent habitats.

Results from this study show that spatial patterns of fish assemblages in Hawaii are largely driven by their habitats and level of protection from fishing. Habitat type and complexity, protected area size, and level of protection from fishing were all important determinates of MPA effectiveness with respect to their associated fish assemblages (Friedlander et al., 2007a,b). This process, using remote sensing technology and sampling across the range of habitats present within the seascape, provided a robust evaluation

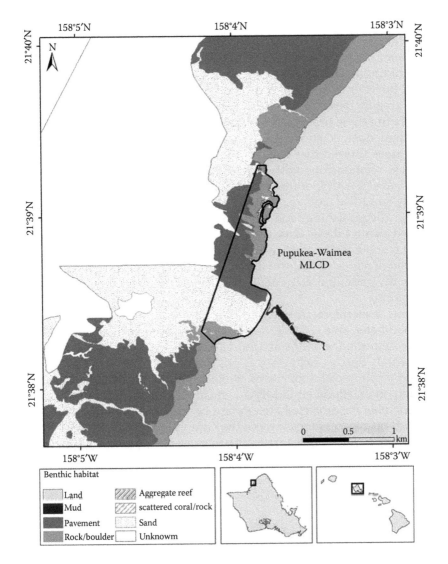

FIGURE 13.3
Benthic habitat map of Pupukea Marine Life Conservation District (MLCD) and adjacent areas on the north shore of Oahu, Hawaii.

of existing MPAs and can help to define ecologically relevant boundaries for future MPA design in a range of locations.

13.5.4 LiDAR and Seascape Metrics

Owing to the important relationships between habitat complexity, ecosystem function, and MPA design in Hawaii (Friedlander et al., 2007a,b),

LiDAR data were used to examine these relationships at larger spatial scales than is possible based on *in situ* measurements alone, and therefore more relevant for management. LiDAR data acquired by the U.S. Army Corps of Engineers Shoals (Scanning Hydrographic Operational Airborne LiDAR Survey) were used to characterize the seascape within each protected area by deriving seascape metrics (e.g., depth, rugosity, and slope) and relating these to fish assemblage characteristics (Figure 13.4; Wedding and Friedlander, 2008; Wedding et al., 2008). An examination of four protected areas showed that variance in depth (within a 75 m radius) had the strongest relationships with most fish assemblage metrics, followed by depth and slope. One of these MPAs on the island of Oahu (Pupukea) was recently modified and expanded owing to community concerns about overuse, illegal fishing, and tourism. A seven-fold increase in area was assessed using LiDAR data, which found a greater range of depths and habitat complexities compared to the original protected area design. The inclusion of a greater range of depth and habitat complexities resulted in a greater diversity and standing stock of reef fishes (Friedlander et al., in preparation).

Various seascape metrics (e.g., habitat diversity, patch fractal dimension, MPA perimeter/area ratio) derived from NOAA's benthic habitat maps were used to characterize MPAs around Hawaii (Friedlander and Wedding, in preparation). Species richness was positivity correlated with habitat diversity and patch fractal dimensions, while larger, more mobile resources species were more abundant in MPAs with small perimeter/area ratios, which are preferable for protected area design because fish are less likely to spill over into open-access areas where they are vulnerable to fishing pressure. These small perimeter/area ratios also increase the core area within the protected area and reduce the amount of boundary perimeter that is available for fishing.

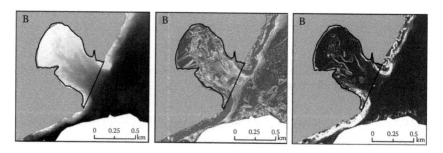

FIGURE 13.4
(See color insert.) Bathymetry, slope, and rugosity for Hanauma Bay Marine Life Conservation District on Oahu, Hawaii, derived from light detection and ranging data. (Adapted from Wedding, L. and A. M. Friedlander, *Marine Geodesy* 31, 246–266, 2008.)

13.5.5 Biogeographic Assessment of the Papahānaumokuākea Marine National Monument

The Papahānaumokuākea Marine National Monument (PMNM) is the largest single protected area in the United States and was recently declared a World Heritage Site. A biogeographic assessment of PMNM was conducted by integrating multiple data types and sources (e.g., oceanography, geology, benthic ecology, fish and fisheries, seabirds, invasive species, and management) into a common spatial and temporal framework using GIS (Friedlander et al., 2009). This integrated assessment identified three unique subregions within the monument based on remotely sensed oceanographic data and derived models, along with information on genetics, distribution patterns, and recruitment dynamics of a diverse group of taxa. This effort contributed to the conservation of the PMNM's marine resources by providing resource managers with a suite of spatially and temporally articulated products (e.g., distribution maps) that identify ecological hotspots and important functional components of the ecosystem.

Analysis was conducted using multibeam data to determine essential fish habitat (EFH) and potential adult habitat on bottomfish in the Northwestern Hawaiian Islands (NWHI) for the PMNM management plan. The U.S. Magnuson–Stevens Fisheries Conservation and Management Act of 1996 refers to EFH as habitat that is recognized as ecologically important to fisheries resources. The bottomfish fishery in PMNM targeted species of deep-slope eteline snappers and one endemic grouper species (WPFMC, 2004) that inhabit depths from 100 to 400 m and are associated with certain features, such as high-relief hard-bottom slopes (Kelley et al., 2006).

The criteria used to delineate potential bottomfish habitat in the NWHI was based on previous analysis done in the main Hawaiian Islands (Parke, 2007). Multibeam data provided the GIS layers for depth, slope, and hardness to identify EFH and potential adult habitat for bottomfish in PMNM. A modeled potential adult habitat included depths from 100 to 400 m, 20% slope, and a hard bottom based on backscatter values from the MBES. An example of "suitable" adult bottomfish habitat based on the above criteria is shown in Figure 13.5 from Kure Atoll, the most northern emergent land mass in the PMNM and the world's highest latitude atoll.

13.6 Temperate Ecosystems

Temperate marine ecosystems pose challenges to the application of remote sensing techniques different from those posed by tropical systems. Among these are the seasonal variability of data streams (due to ice, clouds, storms, etc.) and the limited optical depth of waters. As temperate waters are

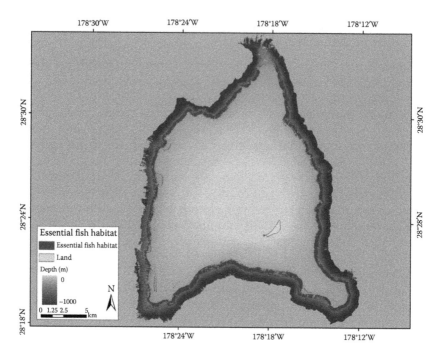

FIGURE 13.5

Potential bottomfish habitat in the Papahānaumokuākea Marine National Monument based on a depth range of 100–400 m, slopes greater than 20%, and hard bottom based on backscatter values. (Adapted from Friedlander, A. M. et al. (eds), *A Marine Biogeographic Assessment of the Northwestern Hawaiian Islands*. NOAA Technical Memorandum NOS NCCOS 84. Prepared by NCCOS's Biogeography Branch in cooperation with the Office of National Marine Sanctuaries Papahānaumokuākea Marine National Monument. Silver Spring, MD: Biogeography Branch, NOAA, 363pp, 2008.)

generally less clear than tropical waters, less spectral signal is available to directly characterize benthic habitats except in emergent areas (e.g., rocky intertidal, beaches, salt marshes). Nonetheless, satellite platforms have been extremely useful in characterizing ocean surface characteristics at coarse scales in these regions, providing insight into regional patterns of temperature, productivity, winds, eddies, and currents at resolutions of 0.25 to 100s of kilometers. Aerial platforms including SAR, infrared cameras, and multispectral imagers and fixed platforms like HF radar have also been extremely useful for particular applications. More recently, advances in sensors and postprocessing technology have led to an increasing role of high-resolution imagery for detailed habitat characterization, including spaceborne multispectral and aerial hyperspectral imagery. Nearshore temperate marine ecosystems also face a diversity of anthropogenic impacts and the use of MPAs as one form of protection is rapidly increasing, resulting in an increasing need for synthetic biogeographic-scale information on temperate coastal marine areas. This section describes several case studies of the use of remote sensing in evaluation, design, and monitoring of nearshore protected areas in California and New England.

13.6.1 MPAs in California

The California Marine Life Protection Act (MLPA), passed by the State of California in 1999, required the design and management of a network of MPAs along California's coast, with multiple goals including protection of diversity, abundance, and ecosystem structure (CDFG, 2008). When designing this network of MPAs, California took a regional approach, dividing the state's coastline into five regions, each with different biogeographic and oceanographic characteristics. In 2003, as part of a separate process, a network of MPAs was implemented in the Channel Islands NMS (Figure 13.6). Both implementation processes were largely stakeholder driven, requiring the use of the best readily available science, and remotely sensed imagery was used extensively throughout the design phase as well as in ongoing monitoring and evaluation of the effectiveness of the protected areas.

The California MLPA utilized a set of science-based guidelines for designing MPA networks. Among the guidelines were that MPAs within each

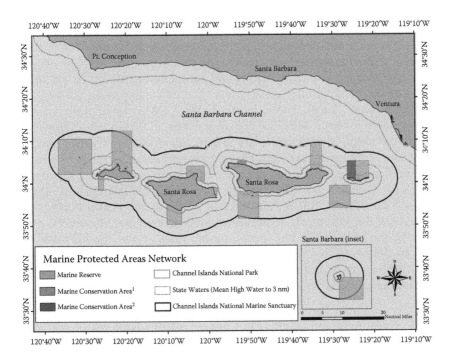

FIGURE 13.6
Marine-protected areas of the California Channel Islands.

particular habitat type must be replicated along the coast (CDFG, 2008). Replication provides stepping stones for dispersal of marine species, insures against localized environmental disaster (e.g., oil spills or other catastrophes), and provides independent experimental replicates for scientific study of MPA effects. Seven habitat types and four depth zones were delineated using benthic habitat maps and bathymetry created using MBES and SSS. Habitat definitions were expanded to include ocean circulation features, recognizing that habitat is not simply defined by the substrate. Seawater characteristics (e.g., temperature, salinity, nutrients, current speed, and direction) are analogous to the climate of habitats on land, and play a critical role in determining biological communities. Delineating these habitat types required the combination of benthic habitat maps and bathymetry (created using MBES and SSS), satellite-derived maps of average sea surface temperature (from the Pathfinder v5 dataset derived from Advanced Very High Resolution Radiometer [AVHRR] instruments on NOAA weather satellites) and phytoplankton concentration (chlorophyll-*a* derived from the Sea-viewing Wide Field-of-view Sensor [SeaWiFS]), and giant kelp cover from aerial (infrared and multispectral imagery) and satellite (multispectral SPOT and Landsat 5 and 7 imagery). All of this information was assembled in an information management system that allowed users to map potential networks of MPAs and assess the adherence of the various proposed maps to scientific guidelines (www.marinemap.com).

Remote sensing data are also being used to monitor these MPAs. For example, over a relatively short geographic scale (~100 km), the Channel Islands NMS encompasses a number of biogeographic regions (Airamé et al., 2003) caused in part by strong environmental gradients in sea surface temperature, productivity, wind stress, and wave exposure (Dever, 2004; Blanchette et al., 2006). The Channel Islands NMS sits at the confluence of two major currents, with the cool, nutrient-rich California current bathing the western islands from the north, whereas the warmer southern California countercurrent bathes the eastern islands from the south. Satellite-derived estimates of sea surface temperature during the past 28 years show a strong gradient throughout the Channel Islands NMS (Figure 13.7) which defines the distribution of species and ecosystem function throughout the sanctuary. To account for environmental variability, differences in the underlying biological communities across the network were used to delineate the appropriate spatial scales for grouping subsets of MPAs in the network (Hamilton et al., 2010), essentially acting as a proxy for environmental variation.

13.6.2 Giant Kelp and Eelgrass in Channel Islands NMS

Giant kelp and other macroalgae form extensive underwater forests and provide food and habitat for many associated fishes and invertebrates in temperate marine ecosystems. The fronds of giant kelp (*Macrocystis pyrifera*)

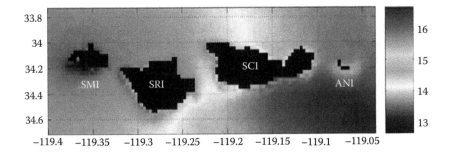

FIGURE 13.7
Sea surface temperature (°C) across the northern Channel Islands averaged over the period 1997–2003 from Advanced Very High Resolution Radiometer satellite. SMI, San Miguel Island; SRI, Santa Rosa Island; SCI, Santa Cruz Island; ANI, Anacapa Island.

can grow >0.5 m per day and float at the surface, making it possible to map surface area of kelp from aerial photographs. In an evaluation of kelp forest cover in and around MPAs within the Channel Islands NMS, scientists used kelp maps and time series derived from aerial surveys, infrared aerial photography, and multispectral imagery, and found that kelp abundance increased substantially throughout the Channel Islands region during the five years since no-take MPAs were established compared to the previous five years (CDFG, 2008). Changes in sea surface temperature throughout the region may account for this increase; however, larger increases in the MPAs may reflect the benefits of protection.

More recently, high spatial resolution and multispectral imagery from the SPOT 5 satellite has been used to quantify both the areal extent and the biomass productivity of giant kelp beds at 10 m spatial resolution (Figure 13.8; Cavanaugh et al., 2010). These observations, together with in-water diver

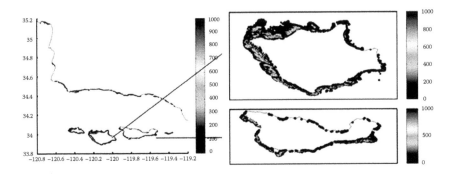

FIGURE 13.8
High spatial resolution, multispectral imagery from the SPOT-5 satellite used to quantify both the areal extent and the biomass productivity of giant kelp beds at 10 m spatial resolution. (Adapted from Cavanaugh, K. C. et al. *Marine Ecology Program Service* 403, 13–27, 2010.)

surveys, have allowed an assessment of temporal changes in kelp forest biomass at multiple spatial scales. The ability to remotely characterize giant kelp cover and biomass on local to regional scales opens many new doors for the study of temperate marine ecosystem dynamics as well as the design and monitoring of MPAs. This technology is an important tool for resource managers of MPAs that encompass temperate coastal and estuarine ecosystems (Kenny et al., 2003) because terrestrial and marine environments are tightly coupled and highly connected.

A biogeographic assessment of Channel Islands NMS utilized multispectral imagery to calculate the combined area of eelgrass (*Zostera marina*) and surfgrass (*Phyllospadix* spp.) inside and outside the sanctuary. Eelgrass and surfgrass beds are important productive habitats because they provide refuge, spawning habitats, and foraging grounds for many marine organisms. Eelgrass and surfgrass maps were used to quantify spatial heterogeneity within different sanctuary boundary concepts, allowing resource managers for the Channel Islands NMS to evaluate the ecological impact of different choices for boundary relocation, contraction, or expansion.

13.6.3 Oceanography and Benthic Habitats for California and New England NMSs

For the assessment of both the Channel Islands and Stellwagen Bank NMSs, radiometric and multispectral data collected by the Pathfinder AVHRR and SeaWiFS instruments, respectively, were used to characterize broad-scale sea surface temperatures, chlorophyll-*a* concentrations, and turbidity. These moderate resolution (1–4 km) oceanographic datasets provided ecological context for both sanctuaries, and helped resource managers better understand how broad-scale oceanographic patterns influenced the fine-scale management units of their MPAs.

A benthic habitat map for portions of the coast of California was created by the National Marine Fisheries Service, in order to identify EFH for Pacific groundfish (NOAA NMFS, 2004). A variety of geological maps from seismic data and acoustic images from sonar sensors were used to partition the seafloor of this temperate area into 33 distinct benthic habitat classes (Greene and Bizzarro, 2003). The different acoustic datasets were collected using several types of MBES and side-scan sonars. The bathymetry and backscatter from these sensors provided a general understanding of the location of seafloor features, such as bedrock types (e.g., sedimentary rocks, crystalline rocks, and carbonate mounds), structures (e.g., faults, folds, and landslides), and bedforms of unconsolidated sediment (e.g., sand waves). This habitat map was later augmented for the biogeographic assessment of the Channel Island with additional benthic data from the Mineral Management Service (now called the Bureau of Ocean Energy Management, Regulation and Enforcement), in order to correct for an underestimation of hard substrate in shallow waters (<30 m) (Figure 13.9; Clark et al., 2005). The augmented map

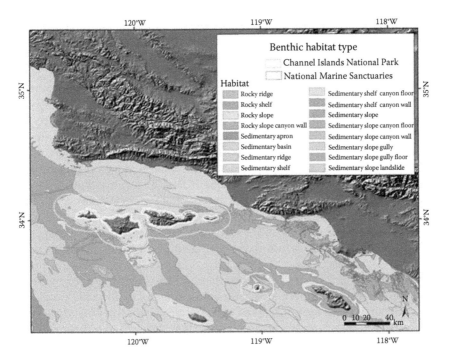

FIGURE 13.9
(**See color insert.**) Benthic habitat map for Channel Islands National Marine Sanctuary, Channel Islands National Park, and surrounding waters.

was then used to quantify (and compare) the spatial heterogeneity of habitats within different sanctuary boundary concepts. The results from this comparison supported the management plan review process for the Channel Islands NMS.

Using the maps mentioned above as well as new surveys, the California Seafloor Mapping Project, a collaborative multi-institutional campaign, is creating the first comprehensive high-resolution base map of California's state waters (shoreline out to 3 nautical miles). The tiered mapping campaign involves the use of MBES surveys, airborne LiDAR and video mapping technologies, computer-aided classification and visualization, expert geological and habitat interpretations, and the creation of an online, publicly accessible, data repository for the dissemination of the project's products (Cochrane and Lafferty, 2002). The creation of a high-resolution 1:24,000 scale geological and habitat base map series, covering all of California's state waters, will support the state's MLPA initiative and its goal to design and manage a statewide network of MPAs.

Similarly, on the other side of the United States, a benthic habitat map for the Gulf of Maine was created in an analogous way by the U.S. Geological Survey and Woods Hole Oceanographic Institution using sediment information (Figure 13.10; Knebel and Circe, 1995). *In situ* benthic sediment point-sample datasets were categorized into nine sediment types (bedrock, gravel, gravel-sand sand, clay-silt/sand, sand-silt/clay, sand-clay/silt, and clay) and integrated to develop a comprehensive map of benthic substrate from the coast of Connecticut to Maine, seaward to the continental shelf-slope interface (USGS, 2000). A higher resolution habitat map was also created for the seafloor inside the Stellwagen Bank NMS using MBES imagery. This finer-scale map revealed highly variable seafloor morphology and sediment textures within the sanctuary, the extent of which were previously unknown. These two habitat maps were later used to develop species–environment models for cetaceans, which quantified the affinity that single cetacean species had for specific ecotones and habitat types (Pittman et al., 2006). The results from these models supported the management plan review process for the Stellwagen Bank NMS and led to several proposed management changes including the alteration of existing shipping lanes to reduce vessel and cetacean interactions.

Radar data, and SAR imagery in particular, have been used for many different purposes in temperate marine environments from describing wind fields and speed and measuring sea surface salinity to ship detection and oil spill recognition and monitoring. It has also been applied to the management of MPAs, notably Channel Islands NMS and Stellwagen Bank NMS. For the Channel Islands NMS, current data were derived from ocean surface altimetry collected by the ERS-2 and TOPEX/POSEIDON radar satellites. These data were used to support the management plan review process for the Channel Islands NMS. SAR imagery was used to characterize the distribution and frequency of internal waves in the ecological assessment of the Stellwagen Bank NMS (Figure 13.10; Battista et al., 2006). Internal waves catalyze the vertical and horizontal mixing, suspension, and transport of sediment, plankton, and larvae in the water column (Haury et al., 1979) and therefore can have a significant influence on coastal ecological patterns (Scotti and Pineda, 2004).

13.7 Moving Forward

13.7.1 Synthesis

This chapter highlighted several MPA case study examples that have successfully integrated remotely sensed and *in situ* datasets for design, monitoring, and assessment of these marine-managed areas. These case studies can serve as models to illustrate the potential for the utilization of remotely sensed data to support marine management actions in both temperate and tropical settings. The case study in the main Hawaiian Islands shows that, by

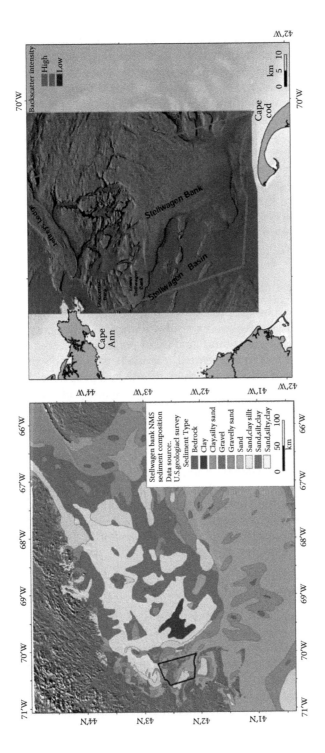

FIGURE 13.10
Benthic habitat map (left) and multibeam imagery (right) for Stellwagen Bank National Marine Sanctuary and surrounding waters.

using remote sensing technology and sampling across the range of habitats present within the seascape, a robust evaluation of existing MPAs can be undertaken. Finally, this approach also served to define ecologically relevant boundaries for future MPA design in a range of locations in Hawaii.

The temperate case study in California detailed the extensive use of remotely sensed imagery throughout the design phase of a recently developed network of MPAs. Remotely sensed imagery was utilized to delineate abiotic and biotic habitat types using both aerial and satellite imagery. Remotely sensed oceanographic datasets provided the ecological context for the MPAs and helped local resource managers better understand how broad-scale oceanographic patterns influenced the finer-scale management units of their MPAs. Remotely sensed imagery is also being utilized for the ongoing monitoring of closed areas in California (e.g., monitoring changes in kelp cover and biomass using imagery from the SPOT 5 and Landsat 5 and 7 satellites).

13.7.2 Data Integration

More holistic approaches to place-based management will require comprehensive ocean zoning if we are to resolve the mismatches between spatial and temporal scales of governance and ecosystems (Crowder et al., 2006). To achieve ecosystem-based management, a spatially explicit GIS approach will be necessary to better understand the patterns and processes that regulate ecosystem function, both to ensure the sustainability of fisheries and to maintain the nonfisheries benefits of the ecosystem to society (Babcock et al., 2005).

An example of the large-scale integrated datasets that can be used to inform MPA design is the Integrated Ocean Observing System, which is the U.S. contribution to the Global Ocean Observing System that is designed to improve weather forecasts and climate predictions. This latter system is the ocean component of the Global Earth Observation System that is working with and building upon national, regional, and international systems to provide comprehensive, coordinated Earth observations from thousands of instruments worldwide.

13.7.3 Decision Support Tools

Decision support systems constitute a class of interactive computer-based information systems that can integrate, share, and contrast many planning options and help managers and stakeholders visualize trade-offs between different management strategies. Airamé et al. (2003) used a computer-based siting tool called SITES to generate potential options for the no-take reserve network in the California Channel Islands NMS and marine reserve designs in the Gulf of Mexico and the Florida Keys demonstrate the effectiveness of combining siting tools and GIS data in designing (Leslie et al., 2003).

Marxan is the most widely used conservation planning software in the world and uses simulated annealing algorithms to minimize the total cost of a reserve network, while achieving a set of conservation goals (Watts et al., 2009). A new extension called Marxan with Zones provides multiple zoning options for each planning unit. Another freely available decision support system, Ecospace, is the spatial component of Ecopath, which is an ecosystem mass-balance modeling approach that has been under refinement for the last quarter of a century (Christensen and Walters, 2004). The most recent version of Ecospace (EwE6) incorporates an optimization module influenced by geospatial information. This is complementary to Marxan as Ecospace provides a robust evaluation of ecological processes, including spatial connectivity, due to its trophic modeling foundation. Proper use of these tools can greatly streamline the Marine Spatial Planning (MSP) process and support its iterative nature.

13.7.4 Data Constraints and Considerations

There are several major issues related to remote sensing of the tropical and temperate marine environments. Broadly speaking, these issues are related to either operational or environmental constraints associated with a particular project or location. Environmental constraints depend on the geographic location and physical conditions of the working environment. Operational constraints relate to the cost and time associated with data acquisition, processing, and analysis, and often depend on the amount of resources allocated for the project. As an example, the way in which operational and environmental conditions affect LiDAR and MBES acquisition are explained in more detail below.

13.7.4.1 Environmental Constraints

When deciding between using LiDAR and MBES, the two most important environmental conditions to consider are depth and water clarity, as these conditions will largely dictate which sensor is best suited for data collection. LiDAR systems are affected by increasing water depths because the intensity of light penetrating the water column decreases exponentially as water depth increases (Mumby and Edwards, 2002). Consequently, the majority of 532-nm blue-green airborne laser systems can only penetrate the full water column where the seafloor is shallower than ~30 m. MBES systems with higher frequencies (100–450 kHz) may be used to map water depths between ~5 and 300 m; systems with medium frequencies (30–100 kHz) may be used to map water depths from 300 to 3000 m; and systems with lower frequencies may be used to map water depths from 3000 to greater than 6000 m (NOAA CSC, 2010).

Although MBES systems may not be constrained by deep water, they can be challenging to operate in depths <30 m, whereas MBES systems in shallow waters may encounter challenging or dangerous navigation conditions. These risks are negligible for LiDAR systems. A survey vessel typically cannot collect data in water shallower than 5 m, leaving a data gap from the 5-m isobath to the shoreline. Topobathy LiDAR systems can fill this data gap and collect seamless coastal datasets. Lastly, the swath width of MBES systems depends on depth—that is, the shallower the water, the less area that is mapped on a single survey line. The swath widths of LiDAR systems, on the other hand, are nearly independent of depth, and therefore make LiDAR systems more efficient at mapping shallow-water (<30 m) marine ecosystems than many MBES systems.

Water clarity also affects the acquisition of remotely sensed information about the seafloor. It affects airborne lasers because they can only penetrate two to three times the secchi disk depth in turbid waters (Piel and Populus, 2007). Consequently LiDAR operations are often conducted when environmental and water clarity conditions are favorable (e.g., neap tides, times of low discharge from coastal rivers, low wave action). Owing to these constraints, LiDAR is most effective in coral reef and other tropical marine ecosystems compared to temperate environments. In temperate ecosystems or in marine environments with poor water clarity, MBES is a more effective tool in advancing seascape ecology.

13.7.4.2 Operational Constraints

When deciding between using LiDAR and MBES, operational constraints to be considered are: (1) size of the area, (2) location and remoteness of the area, (3) temporal resolution of the dataset, (4) data products, (5) spatial (horizontal and vertical) resolution of the dataset, and (6) horizontal and vertical accuracies. Combined, these constraints dictate which sensor is best suited for data collection, as they affect the resources and expertise needed to acquire, process, and analyze the remotely sensed data. In general, it costs more to collect LiDAR or MBES data in large remote locations at high spatial resolutions and accuracies, for both tropical and temperate environments. The first step in a LiDAR or MBES project is to determine sampling frequency as this will determine project cost.

Once the temporal resolution is chosen, the next step is to determine what spatial resolution, accuracy, and products are needed to adequately meet the research objectives. Although LiDAR and MBES systems can both acquire depth information, not all LiDAR systems can acquire and produce meaningful information about return intensity (Costa et al., 2009). Intensity information is particularly important for benthic habitat mapping as it is indicative of sediment properties, including grain size, roughness, and hardness (Hamilton and Bachman, 1982). Future LiDAR systems may be capable of producing intensity surfaces comparable to MBES systems but, for now, MBES sensors may be a more appropriate choice, if physical properties of the sediment are important.

LiDAR and MBES sensors do not necessarily acquire depth information and produce bathymetric products at the same horizontal and vertical resolutions. The sensor's footprint and the along/across track distance between soundings determines the horizontal resolution of the dataset. For full-waveform LiDAR systems, the size of the footprint depends on the scanning geometry of the system and local topography (Sheng, 2008), whereas the distance between soundings depends on the speed of the sortie and sounding rate of the system. The vertical resolution for LiDAR systems depends on the acquisition settings (e.g., pulse length) and waveform postprocessing algorithms. LiDAR surveys conducted at fast speeds with lower sounding rates, longer pulse lengths, and larger scan angles in topographically complex terrain will acquire datasets at coarser horizontal and vertical resolutions.

For MBES systems, the size of the footprint principally depends on the depth of the seafloor and width of the beam, while the distance between soundings depends primarily on the speed of the vessel, acquisition settings (e.g., ping rate), and type (e.g., number and width of beams) of MBES (Kenny et al., 2003). The vertical resolution of MBES bathymetry is affected by the acquisition settings (e.g., pulse length) and type (e.g., frequency) of MBES (NOAA CSC, 2010). MBES surveys conducted in deep water and at fast speeds with acoustic systems that have low frequencies, long pulse lengths, low ping rates, and fewer/wider beams will acquire datasets at coarser horizontal and vertical resolutions. The integration of MBES systems with autonomous underwater vehicles has made the collection of datasets with high horizontal and vertical resolutions in deep-water areas possible, given that these underwater vehicles fly at fixed heights above the seafloor.

Horizontal and vertical resolutions are important parameters to consider since marine communities demonstrate habitat heterogeneity across spatial scales from centimeters to kilometers (Hochberg et al., 2003). The mapping of marine habitats should take place at the same scale as that of the MPA so resulting maps are relevant to management. Stevens (2002) determined that 90% of the world's MPAs are drawn at the local or site scale, and so maps must have sufficient detail to be relevant to these local scales but must also have the ability to be scaled up and applied to a regional setting (Valensi et al., 2003). Currently, most marine habitat mappings occur at the local scale mainly due to cost (Stevens and Connolly, 2004).

13.7.5 New Technologies and Data Fusion

The use of new technologies, data fusion, and decision support tools that serve to synthesize remotely sensed data into a format amenable for marine spatial planning applications will be an important next step in the application of remote sensing to support marine management actions. Scientific research in imagery fusion and integration has grown in the last decade as the number and sophistication of operational sensors increases, providing access to multiresolution, multitemporal, and multifrequency datasets.

Different remotely sensed bands and images may be fused to extract more information than is otherwise derived from a single image alone. Imagery fusion and integration have many purposes (Pohl and Van Genderen, 1998), including: (1) sharpening images, (2) improving geometric corrections, (3) enabling stereoscopy, (4) enhancing features, (5) improving classifications, (6) detecting changes, and/or (7) filling in data gaps or errors. In the marine environment, the fusion or integration of bands or images enhances the ability of scientists to characterize and detect biophysical phenomena on the sea surface (Zielinski et al., 2001), in the water column (Askari, 2001), and on the seafloor (Battista et al., 2007). This information is important to ecosystem-based management and the delineation of MPAs, since these processes affect the spatial distribution of living marine resources in a region (Pittman et al., 2007; Wedding et al., 2008).

Given the importance of benthic habitat mapping and the growing potential of sensor fusion, a collaborative project is underway between the U.S. federal government, private, and academic partners to map the shallow-water (<30 m) seafloor habitats using LiDAR and hyperspectral imagery in Buck Island Reef National Monument in St. Croix, U.S. Virgin Islands. The primary purpose of this project is to determine the combined utility of the LADS LiDAR system and Hyspex VNIR-1600 hyperspectral sensor for identifying and characterizing benthic habitats under variable operational conditions. The spectral signatures of different habitat types will also be measured to assess whether fused LiDAR–hyperspectral imagery can reliably distinguish different genus and/or species of sessile benthic flora and fauna. There is also the potential to compare the map produced from the fused LiDAR–hyperspectral imagery to similar maps produced using sonar and multispectral imagery (http://ccma.nos.noaa.gov/ecosystems/coralreef/stcroix.html).

Advances in remote sensing have enabled us to gather and share information about the marine environment at unprecedented rates and scales. Integrating these data in a geospatial framework along with *in situ* ecological information can help evaluate, monitor, and design MPAs and help to achieve ecosystem-based management. Marine spatial planning is concerned with conflicting usages in marine space and the incorporation of high-quality remotely sensed and ecological data at multiple spatial scales will greatly help in informing this process.

References

Airamé, S., J. E. Dugan, K. D. Lafferty, H. Leslie, D. A. McArdle, and R. R. Warner. 2003. Applying ecological criteria to marine reserve design: a case study from the California Channel Islands. *Ecological Applications* 13: S170–S184.

Almany, G. R., S. R. Connolly, D. D. Heath, J. D. Hogan, G. P. Jones, L. J. McCook, M. Mills, R. L. Pressey, and D. H. Williamson. 2009. Connectivity, biodiversity conservation and the design of marine reserve networks for coral reefs. *Coral Reefs* 28: 339–351.

Andréfouët, S., P. Kramer, T.-P. Damaris, K. E. Joyce, E. J. Hochberg, R. Garza-Pérez, P. J. Mumby et al. 2003. Multi-site evaluation of IKONOS data for classification of tropical coral reef environments. *Remote Sensing of Environment* 88: 128–143.

Andréfouët, S., and B. Reigl. 2004. Remote sensing: A key tool for interdisciplinary assessment of coral reef processes. *Coral Reefs* 23: 1–4.

Anderson, J. T., and R. S. Gregory. 2002. Acoustic classification of marine habitats in coastal Newfoundland. *ICES Journal of Marine Science* 59: 156–167.

Arnold, R. H. 1997. *Interpretation of Air photos and Remotely Sensed Imagery.* Upper Saddle River, NJ: Prentice-Hall.

Askari, F. 2001. Multi-sensor remote sensing of eddy-induced upwelling in the southern coastal region of Sicily. *International Journal of Remote Sensing* 22(15): 2899–2910.

Babcock, E. A., E. K. Pikitch, M. K. McAllister, P. Apostolaki, and C. Santora. 2005. A perspective on the use of spatialized indicators for ecosystem-based fishery management through spatial zoning. *ICES Journal of Marine Science* 62: 469–476.

Battista, T. A., and M. E. Monaco. 2004. Geographic information systems applications in coastal marine fisheries. In W. L. Fisher and F. J. Rahel (eds), *Geographic Information Systems in Fisheries.* Bethesda, MD: American Fisheries Society, pp. 189–208.

Battista, T. A., R. Clark, and S. J. Pittman (eds). 2006. *An Ecological Characterization of the Stellwagen Bank National Marine Sanctuary Region: Oceanographic, Biogeographic, and Contaminants Assessment.* Prepared by NCCOS's Biogeography Team in cooperation with the National Marine Sanctuary Program. NOAA Technical Memorandum NOS NCCOS 45. Silver Spring, MD: Biogeography Branch, NOAA, 356pp.

Battista, T. A., B. M. Costa, and S. M. Anderson. 2007. *Shallow-Water Benthic Habitats of the Main Eight Hawaiian Islands.* NOAA Technical Memorandum NOS NCCOS 61. Silver Spring, MD: Biogeography Branch, NOAA.

Beaumont, N. J., M. C. Austen, J. P. Atkins, D. Burdon, S. Degraer, T. P. Dentinho, S. Derous et al. 2007. Identification, definition and quantification of goods and services provided by marine biodiversity: Implications for the ecosystem approach. *Marine Pollution Bulletin* 54: 253–265.

Blanchette, C. A., B. R. Broitman, and S. D. Gaines. 2006. Intertidal community structure and oceanographic patterns around Santa Cruz Island, CA, USA. *Marine Biology* 149: 689–701.

Bohnsack, J. A. 1998. Application of marine reserves to reef fisheries management. *Australian Journal of Ecology* 23: 298–304.

Brock, J. C., C. W. Wright, I. B. Kuffner, R. Hernandez, and P. Thompson. 2006. Airborne LiDAR sensing of massive stony coral colonies on patch reefs in the northern Florida reef tract. *Remote Sensing of Environment* 104: 31 – 42.

Cavanaugh, K. C., D. A. Siegel, B. P. Kinlan, and D. C. Reed. 2010. Scaling giant kelp field measurements to regional scales using satellite observations. *Marine Ecology Progress Series* 403: 13–27.

California Department of Fish and Game (CDFG). 2008. California Marine Life Protection Act: Master plan for marine protected areas, approved February 2008. 70 pp + appendices.

Christensen, V., and C. J. Walters. 2004. Ecopath with Ecosim: Methods, capabilities and limitations. *Ecological Modelling* 172: 109–139.

Clark, R., J. Christensen, C. Caldow, J. Allen, M. Murray, and S. MacWilliams (eds). 2005. *A Biogeographic Assessment of the Channel Islands National Marine Sanctuary: A Review of Boundary Expansion Concepts for NOAA's National Marine Sanctuary Program*. NOAA Technical Memorandum NOS NCCOS 21. Prepared by NCCOS's Biogeography Team in cooperation with the National Marine Sanctuary Program. Silver Spring, MD: Biogeography Branch, NOAA, 215pp.

Cochrane, G. R., and K. D. Lafferty 2002. Use of acoustic classification of sidescan sonar data for mapping benthic habitat in the Northern Channel Islands, California. *Continental Shelf Research* 22: 683–690.

Costa, B. M., T. A. Battista, and S. J. Pittman. 2009. Comparative evaluation of airborne Lidar and ship-based multibeam sonar bathymetry and intensity for mapping coral reef ecosystems. *Remote Sensing of Environment* 113: 1082–1100.

Coyne, M. S., T. A. Battista, M. Anderson, J. Waddell, W. R. Smith, P. L. Jokiel, M. S. Kendell, and M. E. Monaco. 2003. *Benthic Habitats of the Main Hawaiian Islands*. NOAA Technical Memorandum NOS/NCCOS/CCMA 152. Silver Spring, MD: NOAA, NOS.

Crowder, L. B., G. Osherenko, O. R. Young, S. Airamé, E. A. Norse, N. Baron, J. C. Day et al. 2006. Resolving mismatches in U.S. ocean governance. *Science* 313: 617–618.

D'Amico, A., and R. Pittenger. 2009. A brief history of active sonar. *Aquatic Mammals* 35(4): 426–434.

Dever, E. P. 2004. Objective maps of near-surface flow states near Point Conception, California. *Journal of Physical Oceanography* 34: 441–461.

FAO. 2009. *The State of World Fisheries and Aquaculture*. Rome: Food and Agriculture Organization of the United Nations, FAO Fisheries and Aquaculture Department, 178pp.

Friedlander, A.M., and L. Wedding. In prep. Evaluating the representation of seascape habitat diversity, heterogeneity, and structural complexity in marine protected areas in Hawaii. Environmental Conservation.

Friedlander, A. M., E. K. Brown, and M. E. Monaco. 2007a. Coupling ecology and GIS to evaluate efficacy of marine protected areas in Hawaii. *Ecological Applications* 17: 715–730.

Friedlander, A. M., E. K. Brown, and M. E. Monaco. 2007b. Defining reef fish habitat utilization patterns in Hawaii: Comparisons between MPAs and areas open to fishing. *Marine Ecology Progress Series* 351: 221–233.

Friedlander, A. M., G. Aeby, R. Brainard et al. 2008. The state of coral reef ecosystems of the main Hawaiian Islands. In J. E. Waddell and A. M. Clarke (eds), *The State of Coral Reef Ecosystems of the United States and Pacific Freely Associated States: 2008*. NOAA Technical Memorandum NOS NCCOS 73. Prepared by the NOAA CCMA Biogeography Team. Silver Spring, MD: Biogeography Branch, NOAA, pp. 158–199.

Friedlander, A. M., K. Keller, L. Wedding, A. Clarke, and M. Monaco (eds). 2009. *A Marine Biogeographic Assessment of the Northwestern Hawaiian Islands*. NOAA Technical Memorandum NOS NCCOS 84. Prepared by NCCOS's Biogeography Branch in cooperation with the Office of National Marine Sanctuaries Papahānaumokuākea Marine National Monument. Silver Spring, MD: Biogeography Branch, NOAA, 363pp.

Friedlander, A.M., L. Wedding, M.E. Monaco, E.K. Brown, A. Clark, D. Antoline. In prep. Evaluating the expansion of a marine protected area using a seascape ecology approach: Pupukea Marine Life Conservation District, Hawaii. Conservation Biology.

Gell, F. R., and C. M. Roberts. 2003. Benefits beyond boundaries: The fishery effects of marine reserves. *Trends in Ecology & Evolution* 18: 448–455.

Green, E. 2000. Satellite and airborne sensors useful in coastal applications. In A. J. Edwards (ed.), *Remote Sensing Handbook for Tropical Coastal Management*. Paris: UNESCO, pp. 41–56.

Greene, H. G., and J. J. Bizzarro. 2003. *Final Report: Essential Fish Habitat Characterization and Mapping of the California Continental Margin*. Created for Pacific States Marine Fisheries Commission, 23pp. Available at: http://www.pcouncil.org/resources/archives/briefing-books/april-2004-briefing-book/.

Gurgel, K. W., G. Antonischski, H.-H. Essen, and T. Schlick. 1999. Wellen radar (WERA): A new ground wave radar for remote sensing. *Coastal Engineering* 37, 219–234.

Halpern, B. S., S. Walbridge, K. A. Selkoe, C. V. Kappel, F. Micheli, C. D'Agrosa, J. F. Bruno et al. 2008. A global map of human impact on marine ecosystems. *Science* 319: 948–952.

Hamel, M. A., and S. Andréfouët. 2010. Using very high resolution remote sensing for the management of coral reef fisheries: Review and perspectives. *Marine Pollution Bulletin* 60: 1397–1405.

Hamilton, E. L., and R. T. Bachman. 1982. Sound velocity and related properties of marine sediments. *The Journal of the Acoustical Society of America* 72(6): 1891–1904.

Hamilton, S. L., J. E. Caselle, D. Malone, and M. H. Carr. 2010. Incorporating biogeography into evaluations of the Channel Islands marine reserve network. *Proceedings of the National Academy of Sciences* 107: 18272–18277.

Hammond, T. M., C. Pattiaratchi, D. Eccles, M. Osborne, L. Nash, and M. Collins. 1987. Ocean surface current radar vector measurements on the inner continental shelf. *Continental Shelf Research* 7: 411–431.

Haury, L. R., M. G. Briscoae, and M. H. Orr. 1979. Tidally generated internal wave packets in Massachusetts Bay. *Nature* 278: 312–317.

Hay, M. E., and W. Fenical. 1996. Chemical ecology and marine biodiversity: insights and products from the sea. *Oceanography* 9: 10–20.

Hochberg, E. J., and M. J. Atkinson. 2000. Spectral discrimination of coral reef benthic communities. *Coral Reefs* 19: 164–171.

Hochberg, E., M. Atkinson, and S. Andréfouët. 2003. Spectral reflectance of coral reef bottom-types worldwide and implications for coral reef remote sensing. *Remote Sensing of Environment* 85: 159–173.

Intelmann, S. S. 2006. *Habitat Mapping Effort at the Olympic Coast National Marine Sanctuary—Current Status and Future Needs*. Marine Sanctuaries Conservation Series NMSP-06-11. U.S. Department of Commerce, National Oceanic and Atmospheric Administration, National Marine Sanctuary Program, Silver Spring, MD: Biogeography Branch, NOAA, 29pp.

Irish, J. L., and T. E. White. 1998. Coastal engineering applications of high-resolution LiDAR bathymetry. *Coastal Engineering* 35(1–2): 47–71.

Jackson, C. R., and J. R. Apel. 2004. *Synthetic Aperture Radar Marine User's Manual*. Washington, D.C.: NOAA NESDIS, 464pp.

Jackson, J. B. C. 2008. Ecological extinction and evolution in the brave new ocean. *Proceedings of the National Academy of Sciences* 105(Suppl. 1): 11458–11465.

Jenson, J. R. 2000. *Remote Sensing of the Environment: An Earth Resource Perspective.* Upper Saddle River, NJ: Prentice Hall.

Johannes, R. E. 1978. Traditional marine conservation methods in Oceania and their demise. *Annual Reviews in Ecological Systematic* 9: 349–364.

Joyce, K. E., and S. R. Phinn. 2003. Hyperspectral analysis of chlorophyll content and photosynthetic capacity of coral reef substrates. *Limnology and Oceanography* 48: 489–496.

Judd, C., S. Steinberg, F. Shaughnessy, and G. Crawford. 2007. Mapping salt marsh vegetation using aerial hyperspectral imagery and linear unmixing in Humboldt Bay, California. *Wetlands* 27: 1144–1152.

Karpouzli, E., T. J. Malthus, and C. J. Place. 2004. Hyperspectral discrimination of coral reef benthic communities in the western Caribbean. *Coral Reefs* 23: 141–151.

Kelleher, G. 1999. *Guidelines for Marine Protected Areas.* Gland, Switzerland and Cambridge, UK: ICUN.

Kelley, C., R. Moffitt, and J. R. Smith. 2006. Description of bottomfish essential fish habitat on four banks in the Northwestern Hawaiian Islands. *Atoll Research Bulletin* 543: 319–332.

Kenny, A. J., I. Cato, M. Desprez, G. Fader, R. T. E. Schuttendhelm, and J. Side. 2003. An overview of seabed-mapping technologies in the context of marine habitat mapping. *ICES Journal of Marine Science* 60: 411–418.

Kendall, M. S., O. P. Jensen, D. Field, G. McFall, R. Bohne, and M. E. Monaco. 2005. Benthic mapping using sonar, video transects, and an innovative approach to accuracy assessment: A characterization of bottom features in the Georgia Bight. *Journal of Coastal Research* 21(6): 1154–1165.

Knebel, H. J. and R. C. Circe. 1995. Sea floor environments within the Boston Harbor Massachusetts Bay sedimentary system: A regional synthesis. *Journal of Coastal Research* 11: 230–251.

Koslow, J. A. 2009. The role of acoustics in ecosystem-based fishery management. *ICES Journal of Marine Science* 66: 1–8.

Kracker, L. M. 2007. *Hydroacoustic Surveys: A Non-destructive Approach to Monitoring Fish Distributions at National Marine Sanctuaries.* NOAA Technical Memorandum NOS NCCOS 66. Silver Spring, MD: Biogeography Branch, NOAA, 24pp.

Kuffner, I. B., J. C. Brock, R. Grober-Dunsmore, V. E. Bonito, T. D. Hickey, and C. W. Wright. 2007. Relationship between fish communities and remotely sensed measurements in Biscayne National Park, Florida, USA. *Environmental Biology of Fishes* 78: 71–82.

Kutser, T., E. Vahtmäe, and G. Martin. 2006. Assessing suitability of multispectral satellites for mapping benthic macroalgal cover in turbid coastal waters by means of model simulations. *Estuarine Coastal and Shelf Science* 67(3): 521–529.

Leslie, H., M. Ruckelshaus, I. R. Ball, S. Andelman., and H. P. Possingham. 2003. Using siting algorithms in the design of marine reserve networks. *Ecological Applications* 13: 185–198.

Lester, S. E., B. S. Halpern, K. Grorud-Colvert, J. Lubchenco, B. I. Ruttenberg, S. Gaines, S. Airamé, and R. R. Warner. 2009. Biological effects within no-take marine reserves: A global synthesis. *Marine Ecology Progress Series* 384: 33–46.

Lotze, H. E., H. S. Lenihan, B. J. Bourque, R. H. Bradbury, R. G. Cooke, M. C. Kay, S. M. Kidewell, M. X. Kirby, C. H. Peterson, and J. B. C. Jackosn. 2006. Depletion, degradation, and recovery potential of estuaries and coastal seas. *Science* 312: 1806–1809.

Lubchenco, J., S. Palumbi, S. D. Gaines., and S. Andelman. 2003. Plugging a hole in the ocean: the emerging science of marine reserves. *Ecological Applications* 13: S3–S7.

Molloy, P. P., I. B. McLean., and I. M. Cote. 2009. Effects of marine reserve age on fish populations: A global meta-analysis. *Journal of Applied Ecology* 46: 743–751.

Monaco, M. E., M. S. Kendall, J. L. Higgins, C. E. Alexander, and M. S. Tartt. 2005. Biogeographic assessments of NOAA National Marine Sanctuaries: The integration of ecology and GIS to aid in marine management boundary delineation and assessment. In D. J. Wright and A. J. Scholz (eds), *Place Matters—Geospatial Tools for Marine Science, Conservation, and Management in the Pacific Northwest.* Corvallis, OR, OSU Press, 305pp.

Mumby, P. J., and A. J. Edwards. 2002. Mapping marine environments with IKONOS imagery: enhanced spatial resolution can deliver greater thematic accuracy. *Remote Sensing of Environment* 82: 248–257.

Mumby, P., E. Green, A. Edwards, and C. Clark. 1997. Coral reef habitat mapping: How much detail can remote sensing provide? *Marine Biology* 130: 193–202.

Mumby, P. J., W. Skirving, A. E. Strong, J. T. Hardy, E. F. LeDrew, E. J. Hochberg, R. P. Stumpf, and L. T. David. 2004. Remote sensing of coral reefs and their physical environment. *Marine Pollution Bulletin* 48: 219–228.

Nishimoto, M. M., and L. Washburn. 2002. Patterns of coastal eddy circulation and abundance of pelagic juvenile fish in the Santa Barbara Channel, California, USA. *Marine Ecology Progress Series* 241: 183–199.

NOAA CSC (National Oceanic & Atmospheric Administration Coastal Services Center). 2010. Remote sensing for coastal management. Available at: http://www.csc.noaa.gov/crs/rs_apps/sensors/ (accessed September 9, 2010).

NOAA NMFS (National Oceanic and Atmospheric Administration National Marine Fisheries Service). 2004. *Risk Assessment for the Pacific Groundfish FMP.* Prepared for the Pacific States Marine Fisheries Commission by MRAG Americas, Inc. Tampa, Florida. Available at: http://www.nwr.noaa.gov/Groundfish-Halibut/Groundfish-Fishery-Management/NEPA-Documents/EFH-Final-EIS.cfm

Norse, E. A. 1993. *Global Marine Biological Diversity: A Strategy for Building Conservation Into Decision Making.* Washington, DC: Island Press, 383pp.

Ona, E. 1990. Physiological factors causing natural variations in acoustic target strength of fish. *Journal of the Marine Biological Association of the United Kingdom* 70: 107–127.

Parke, M. 2007. *Linking Hawaii Fisherman Reported Commercial Bottomfish Catch Data to Potential Bottomfish Habitat and Proposed Restricted Fishing Areas Using GIS and Spatial Analysis.* NOAA Technical Memorandum NMFS-PIFSC-11. Silver Spring, MD: Biogeography Branch, NOAA.

Piel, S., and J. Populus. 2007. Chapter 4—Lidar. In R. Coggan, J. Populus, J. White, K. Sheehan, F. Fitzpatrick, and S. Piel (eds), *Review of Standards and Protocols for Seabed Habitat Mapping. Mapping European Seabed Habitats (MESH).* Peterborough, UK: Joint Nature Conservation Committee, pp. 21–42.

Pikitch, E. K., C. Santora, E. A. Babcock, A. Bakun, R. Bonfil, D. O. Conover, P. Dayton et al. 2004. Ecosystem-based fishery management. *Science* 305: 346–347.

Pittman, S., J. Christenson, C. Caldow, C. Menza, and M. E. Monaco. 2007. Predictive mapping of fish species richness across shallow-water seascapes in the Caribbean. *Ecological Modeling* 204: 9–21.

Pittman, S., B. Costa, C. Knot, D. Wiley, and R. D. Kenney. 2006. Chapter 5—Cetacean distribution and diversity. In T. A. Battista, R. Clark, and S. J. Pittman (eds), *An*

Ecological Characterization of the Stellwagen Bank National Marine Sanctuary Region: Oceanographic, Biogeographic, and Contaminants Assessment. NOAA Technical Memorandum NOS NCCOS 45. Silver Spring, MD: Biogeography Branch, NOAA, pp. 265–326.

Pittman, S. J., B. M. Costa, and T. A. Battista. 2009. Using LiDAR bathymetry and boosted regression trees to predict the diversity and abundance of fish and corals. *Journal of Coastal Research* (Special Issue No. 53): 27–38.

Planes, S., G. P. Jones, and S. R. Thorrold. 2009. Larval dispersal connects fish populations in a network of marine protected areas. *Proceedings of the National Academy of Sciences* 106: 5693–5697.

Pohl, C., and J. L. Van Genderen. 1998. Multisensor image fusion in remote sensing: Concepts, methods and applications. *International Journal of Remote Sensing* 19: 823–854.

Roberts, C. M. 2005. Marine protected areas and biodiversity conservation. In E. Norse and L. Crowder (eds), *Marine Conservation Biology: The Science of Maintaining the Sea's Biodiversity.* Washington, D.C.: Island Press, pp. 265–279.

Rose, G. A. 2003. Monitoring coastal northern cod: towards an optimal survey of Smith Sound, Newfoundland. *ICES Journal of Marine Science* 60: 453–462.

Scotti, A., and J. Pineda. 2004. Observation of very large and steep internal waves of elevation near the Massachusetts coast. *Geophysical Research Letters* 31: 1–5.

Sheng, Y. 2008. Quantifying the size of a Lidar footprint: A set of generalized equations. *IEEE Geoscience and Remote Sensing Letters* 5(3): 419–422.

Sladek Nowles, J., and A. M. Friedlander. 2005. Marine reserve design and function for fisheries management. In E. A. Norse and L. B. Crowder (eds), *Marine Conservation Biology: The Science of Maintaining the Sea's Biodiversity.* Washington, D.C.: Island Press, pp. 280–301.

Stevens, T. 2002. Rigor and representativeness in marine protected area design. *Coastal Management* 30: 237–248.

Stevens, T., and R. M. Connolly. 2004. Testing the utility of abiotic surrogates for marine habitat mapping at scales relevant to management. *Biological Conservation* 119: 351–362.

Storlazzi, C. D., J. B. Logan, and M. E. Field. 2003. Quantitative morphology of a fringing reef tract from high-resolution laser bathymetry: Southern Molokai, Hawaii. *Geological Society of America Bulletin* 115: 1344 – 1355.

Theriault, C., R. Scheibling, B. Hatcher, and W. Jones. 2006. Mapping the distribution of an invasive marine alga (*Codium fragile*) in optically shallow coastal waters using the Compact Airborne Spectrographic Imager (CASI). *Canadian Journal of Remote Sensing* 32: 315–329.

United States Geological Survey (USGS). 2000. USGS East-coast Sediment Analysis: Procedures, database, and georef-erenced displays. U.S. Geological Survey Open-File Report 00-358. URL: http://pubs.usgs.gov/of/2000/of00-358/.

Vahtmäe, E., T. Kutser, G. Martin, and J. Kotta. 2006. Feasibility of hyperspectral remote sensing for mapping benthic macroalgal cover in turbid coastal waters— A Baltic Sea case study. *Remote Sensing of Environment* 101: 342–351.

Valensi, F. J., K. R. Clarke, I. Eliot, and I. C. Potter. 2003. A user-friendly quantitative approach to classifying nearshore marine habitats along a heterogeneous coast. *Estuarine, Coastal and Shelf Science* 57: 163–177.

Watts, M. E., I. R. Ball, R. S. Stewart, C. J. Klein, K. Wilson, C. Steinback, R. Lourival, L. Kircher, and H. P. Possingham. 2009. Marxan with Zones: Software for optimal conservation based land- and sea-use zoning. *Environmental Modelling & Software* 24: 1513–1521.

Weber, P. 1993. *Abandoned Seas: Reversing the Decline of the Oceans*, World Watch Paper No. 116. Washington, D.C.: World Watch Institute, 66pp.

Western Pacific Regional Fisheries Management Council (WPFMC). 2004. Bottomfish and seamount grounffish fisheries of the Western Pacific region: 2002 Annual Report. Western Pacific Regional Fisheries Management Council, Honolulu, Hawaii. 17 pp + appendices.

Wedding, L., and A. M. Friedlander. 2008. Determining the influence of seascape structure on coral reef fishes in Hawaii using a geospatial approach. *Marine Geodesy* 31: 246–266.

Wedding, L., A. M. Friedlander, M. McGranaghan, R. Yost, and M. E. Monaco. 2008. Using bathymetric LiDAR to define nearshore benthic habitat complexity: Implications for management of reef fish assemblages in Hawaii. *Remote Sensing of Environment* 112: 4159 – 4165.

Young, A. P., and S. A. Ashford. 2006. Application of airborne LIDAR for seacliff volumetric change and beach-sediment budget contributions. *Journal of Coastal Research* 22(2): 307–318.

Zelenke, B., M. A. Moline, B. H. Jones, S. R. Ramp, G. B. Crawford, J. L. Largier, E. J. Terrill, N. Garfield, III, J. D. Paduan, and L. Washburn. 2009. Evaluating connectivity between marine protected areas using CODAR high-frequency radar. *Oceans 2009, Marine Technology for Our Future: Global and Local Challenges* 1–3(2009): 2261–2270.

Zielinski, O., T. Hengstermann, D. Mach, and P. Wagner. 2001. Multispectral information in operational marine pollution monitoring: A data fusion approach. In *Proceedings of the 5th International Airborne Remote Sensing Conference*, September 17–20, 2001, San Francisco, CA, p. 8.

14

Remote Sensing Assessment of Wildfire Impact and Simulation Modeling of Short-Term Post-Fire Vegetation Recovery within the Dixie National Forest

Yeqiao Wang, Yuyu Zhou, Jian Yang, and Hong S. He

CONTENTS

14.1 Introduction

Wildfires are a growing natural hazard in the United States and around the world. Although as a natural process wildfires can be beneficial to ecosystem functions, direct fire impacts and secondary effects such as erosion, landslides, introduction of invasive species, and changes in water quality are often disastrous. Pattern, severity, and timing of wildfires affect significantly the successional changes of vegetation. Understanding both the effects of fire impacts and the pathway of vegetation recovery are critical for community actions on resource management planning, land-use decision, treatment procedure, habitat restoration, and on studies of ecological and economic complexities in association with wildfires.

Remote sensing data and geospatial modeling have been used to monitor wildfires, assess active fire characteristics, analyze post-fire effects (Fraser and Li, 2002; Lentile et al., 2005), measure forest structure and fuel loads (Skowronski et al., 2007), and estimate fuel moisture for prediction of fire behavior (Dasgupta et al., 2007; Hao and Qu, 2007). Multitemporal remote sensing data have been used to assess fire severity (Brewer et al., 2005; Miller and Thode, 2007; Wimberly and Reilly, 2007) and simulate post-fire spectral response to burn severity (De Santis et al., 2009).

Increasing frequency and extent of wildfires demands significant amount of resources for wildfire management. A variety of programs have been established for wildfire research, management, and education. LANDFIRE, for example, is a multiagency effort to produce consistent and comprehensive maps and data describing vegetation composition and structure, surface and canopy fuel characteristics, historical fire regimes, and ecosystem status across the United States (Rollins and Frame, 2006). In the LANDFIRE project, vegetation is mapped using predictive landscape models based on extensive field reference data, satellite imagery, biophysical gradient layers, and classification and regression trees. The LANDFIRE system and the data products can help identify the extent, severity, and location of wildfire threats to the nation's communities and ecosystems (GAO, 2006). A challenge that any data production in such a capacity needs to consider is updating issues to reflect the effects of landscape-altering events such as wildfires that occurred after the acquisition of remote sensing data and from which the vegetation maps were developed. For example, the Sanford Fire occurred in June 2002 and burned a large area of forest lands. The prototype LANDFIRE data products were developed in 2005 based on Landsat Thematic Mapper (TM) images acquired in 2000. As the Landsat data used did not reflect the major landscape-altering wildfire that occurred between data acquisition and vegetation mapping efforts, updating vegetation data for wildfire impacted areas becomes necessary. Instead of repeating vegetation mapping procedures using a new set of Landsat data, short-term post-fire vegetation recovery simulation that focuses on fire impacted areas only is one of the possible and preferred approaches to address updating requests.

14.2 Forest Simulation Modeling

Different models, such as Vegetation Dynamic Development Tool (VDDT), Tool for Exploratory Landscape Scenario Analyses (TELSA), Forest Vegetation

Simulator (FVS), and LANDscape SUccession Model (LANDSUM), have been developed and used for simulation of long-term forest succession and landscape dynamics (Klenner et al., 2000; Hann and Bunnell, 2001; Hemstrom et al., 2001; Beukema et al., 2003; Merzenich et al., 2003; Swanson et al., 2003; Keane et al., 2004). VDDT provides a nonspatial modeling framework for examining the effects of various disturbance agents and management actions on vegetation change over a long time period. VDDT allows successional pathways to be defined by users and assumes that the landscape is stratified into units with similar successional pathways. Stratification is based on potential vegetation types, which identifies a distinct biophysical setting that supports a unique and stable climax plant community under a constant climate regime. VDDT can team up with TELSA, which is a spatially explicit model, for simulation of vegetative succession, natural disturbances, and management activities. FVS is an individual-tree, distance-independent forest stand projection model. The model simulates tree growth and mortality of a chosen stand, using characteristics such as species, diameter, height, crown length, and relative size. Key outputs are produced at the individual tree and stand level. FVS is able to simulate complex stands composed of many species and many ages. However, FVS may not be suitable to simulate post-fire recovery which involves spatially contiguous processes such as seed dispersal.

LANDIS is a spatially explicit landscape model that is designed to simulate forest change over large spatial and temporal (10^1–10^3 years) scales with flexible spatial resolutions (He and Mladenoff, 1999; Mladenoff and He, 1999). LANDIS can simulate natural and anthropogenic disturbances and their interactions with adequate mechanistic realism and can simulate species-level forest succession in combination with disturbances and management practices. LANDIS modeling assumes that detailed, individual tree information and within-stand processes can be simplified, allowing large-scale questions about spatial pattern, species distribution, and disturbances to be adequately addressed. LANDIS can take remote sensing-derived thematic data as input and the output is compatible with most Geographic Information System (GIS) software for spatial analysis (He et al., 2004; Sturtevant et al., 2004). An improvement in LANDISv4.0a (Syphard et al., 2007) shortened the time interval of simulation from the original design of 10 years to 1 year. These features make LANDIS a candidate model for simulation of short-term post-fire vegetation recovery.

14.3 Remote Sensing Data and the Derivative Variables

A critical challenge toward spatially explicit simulations of short-term post-fire vegetation recovery is data preparation for establishment of initial conditions. These include pre-fire vegetation status, impacted areas and burn severity, topographic locations, and biophysical conditions in an individual spatial unit. One of the critical LANDFIRE data products is the existing vegetation type (EVT). The EVT data represent distribution of terrestrial ecological system classification developed by the NatureServe (Comer et al., 2003). A terrestrial ecological system is defined as a group of plant community types (associations) that tend to co-occur within landscapes with similar ecological processes, substrates, and/or environmental gradients. The LANDFIRE project mapped EVTs using decision tree models, field reference data, Landsat imagery data, digital elevation model data, and biophysical gradient data. The EVT products included existing vegetation type, vegetation canopy cover, and vegetation height. LANDFIRE biogradients are the data layers that describe biophysical gradients that affect the distribution of ecosystem components across landscapes. The indirect gradients include slope, aspect, and elevation and the direct gradients include temperature and humidity. Functional gradients describe the response of vegetation to direct and indirect gradients and include productivity, respiration, and transpiration. Therefore, the biogradients data can be used to define environmental parameters for post-fire vegetation recovery simulation.

The burn severity map is a key factor to quantify fire impacts on vegetation and soil (White et al., 1996; van Wagtendonk et al., 2004; De Santis and Chuvieco, 2007) and to provide baseline information for monitoring restoration and recovery (Brewer et al., 2005). Differenced normalized burn ratio (DNBR) is a continuous index developed from pre- and post-fire Landsat imageries to measure burn severities. The normalized burn ratio (NBR) is defined by Equation 14.1 and the DNBR is defined by Equation 14.2.

$$NBR = \frac{(TM\ Band\ 4 - TM\ Band\ 7)}{(TM\ Band\ 4 + TM\ Band\ 7)} \tag{14.1}$$

$$DNBR = NBR_{pre\text{-}fire} - NBR_{post\text{-}fire} \tag{14.2}$$

The DNBR dataset yields a burn severity with possible values ranging between –2000 and +2000. DNBR can team up with an onsite estimation of the composite burn index (CBI) for impact measurements.

We envisioned that the combination of LANDFIRE data products and the DNBR data would establish initial states and create parameters for spatially explicit and ecological process-based simulation of short-term post-fire vegetation recovery. The specific modeling interests of this study were that (1) spatial resolution of the simulation modeling should be at a 30-m cell size in order to match LANDFIRE vegetation data products; (2) the preferred time interval was 1 year in order to reflect dynamics of vegetation recovery; and (3) the simulation duration should be about 10 years with an assumption that a remapping effort could take place in a 10-year cycle. We focused the simulation on fire-impacted areas only.

14.4 Methods

14.4.1 Data Preparation

14.4.1.1 Study Area

The Sanford Fire site is within the Dixie National Forest close to Cedar City in Utah (Figure 14.1). The Sanford Fire resulted from escapes of two prescribed burns—Adams Head Burn in the south and the Sanford Burn in the north. The prescribed burns were ignited in April and May 2002, respectively, for the purpose of reducing accumulated fuels, keeping pinyon/juniper from expanding further into sagebrush/grasslands, maintaining vegetation at different ages, returning fire to its natural role in the ecosystem and stimulating aspen suckering. Strong winds, low humidity, and increased

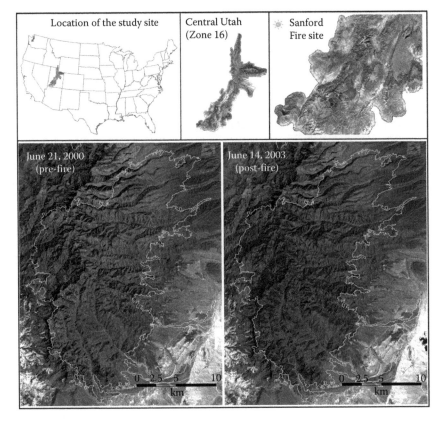

FIGURE 14.1
(See color insert.) The Sanford Fire site is located in the southern section of the Central Utah Valley (Zone 16) of the LANDFIRE prototype data products. A comparison of pre- and post-fire Landsat Thematic Mapper reflectance images illustrates the affected areas and fire impacts on vegetation.

temperatures at the time moved the prescribed burns outside their contain-
ment areas. On June 8, 2002, the two prescribed burns joined, fueled by
strong winds, and the fire was then referred to as the Sanford Fire. The total
amount of land within the perimeter of the Sanford Fire and the two prescribed burns was 31,579 hectares. The fire impact assessment indicated that
the vegetation types affected included 41% sagebrush, 32% mixed conifer,
8% pinyon-juniper, 7% spruce/fir, 4% aspen, and 4% mountain mahogany.
Within the site, about 10% (2995 ha) burned at high severity, 35% (10,996 ha
burned with moderate severity, 5% (1584 ha) burned at low severity, and 50%
(15,720 ha) unburned (Upper Sevier River Community Watershed Project,
2002). A comparison of pre- and post-fire Landsat images illustrates the
affected areas and fire impacts on vegetation (Figure 14.1).

We selected the Sanford Fire site for the following reasons. (1) It is situated
within the Central Utah Valley region where LANDFIRE prototype data
products (zone 16) were completed. (2) The LANDFIRE prototype vegetation
data were developed based on Landsat TM data acquired in 2000 and the
fire occurred in 2002. Therefore, the LANDFIRE data products did not reflect
the damaged vegetation by the Sanford Fire. (3) Burned areas were large
enough with diversified vegetation communities and different severity of
fire impacts.

14.4.1.2 Pre-Fire EVT and DNBR Data

We referred to the LANDFIRE prototype data products as the pre-fire data
as these were developed based on Landsat data two years prior to the Sanford
Fire. We extracted spatial distribution of pre-fire EVT (Figure 14.2a), vegetation height, and percent canopy cover for fire impacted areas and identified
the ecosystem types (Table 14.1). We extracted the biogradients data (e.g.,
Figure 14.2b) from the LANDFIRE prototype data to develop the land type
parameters that are required for LANDIS simulation modeling. We derived
the DNBR data (Figure 14.2c) by the reflectance from pre-fire (June 21, 2000)
and post-fire (June 14, 2003) Landsat TM data. The DNBR suggested that,
in general, a threshold existed between about −100 and +150 DNBR units
that marked an approximate breakpoint between burned and unburned
areas. We considered the areas with DNBR values below this threshold as
unburned. Within the burned area, increased DNBR values would correspond to increased burn severity on the ground.

14.4.1.3 Ground Verification

LANDFIRE project scientists visited the Sanford Fire site in 2003, one year
after the fire, and conducted onsite estimations of the CBI for selected plots.
CBI represented the magnitude of fire effects. Through the numeric scale
between 0.0 and 3.0 CBI described how much a fire had altered the biophysical conditions of a site. The CBI plot data provided reliable references to

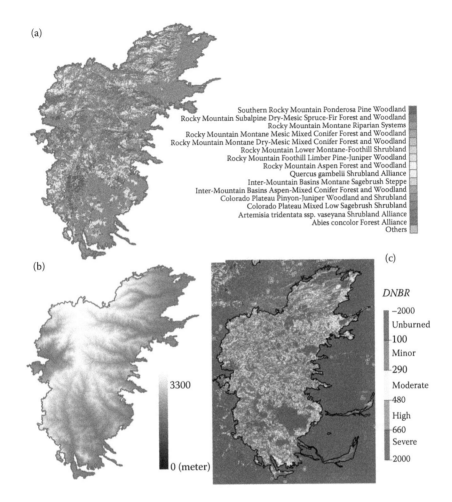

FIGURE 14.2
(See color insert.) The existing vegetation type data (a) and digital elevation model data (b) provided pre-fire ecosystem data and environmental setting for the Sanford Fire site. The differenced normalized burn ratio data (c) provided scales of burn severity on each pixel location as estimated measurements of fire impacts.

connect field measurements with remote sensing-derived DNBR data. We conducted field verification at the Sanford Fire site in September 2005 to observe the status of vegetation recovery three years after the fire and to examine the relationship between burn severity, fire age, and post-fire vegetation recovery.

We employed a Global Positioning System (GPS)-guided navigation to the selected CBI plot sites that were previously examined in 2003. We recorded locations of field transects and points of interests on both Landsat and DNBR data so that the burn severity and recovery status could be evaluated (Figure

TABLE 14.1

Pre-Fire Existing Vegetation Type within the Sanford Fire Site

Code	Existing Vegetation Type from LANDFIRE Prototype Data	Abbreviation
2011	Rocky Mountain Aspen Forest and Woodland	RMA
2016	Colorado Plateau Pinyon-Juniper Woodland and Shrubland	CPP
2049	Rocky Mountain Foothill Limber Pine-Juniper Woodland	RMF
2051	Rocky Mountain Montane Dry-Mesic Mixed Conifer Forest and Woodland	RMD
2052	Rocky Mountain Montane Mesic Mixed Conifer Forest and Woodland	RMM
2054	Southern Rocky Mountain Ponderosa Pine Woodland	SRM
2055	Rocky Mountain Subalpine Dry-Mesic Spruce-Fir Forest and Woodland	RMS
2061	Inter-Mountain Basins Aspen-Mixed Conifer Forest and Woodland	IMA
2159	Rocky Mountain Montane Riparian Systems	RMR
2208	*Abies concolor* Forest Alliance	ACF
2064	Colorado Plateau Mixed Low Sagebrush Shrubland	CPM
2086	Rocky Mountain Lower Montane-Foothill Shrubland	RML
2125	Inter-Mountain Basins Montane Sagebrush Steppe	IMM
2217	*Quercus gambelii* Shrubland Alliance	QGS
2220	*Artemisia tridentata ssp. vaseyana* Shrubland Alliance	ATS

14.3). We paid special attention to locations with apparent signs of past burns for references. We documented the status of vegetation recovery in comparison with estimations from previous field observations (Table 14.2) by georeferenced field photographs.

14.4.2 Parameterization for LANDIS Simulation

In LANDIS simulation, a landscape is divided into equal-sized individual cells. Each cell resides on a certain land type and a disturbance regime type. Each cell is considered to include a unique list of species and their associated age cohorts. The species/age cohort information varies with establishment, succession, and seed dispersal, and interacts with disturbances. Heterogeneity of vegetation, disturbance, and management activities can be modeled at multiple hierarchical levels from the landscape to the cell. For vegetation heterogeneity, LANDIS stratifies the heterogeneous landscape into land types, which are generated from GIS layers of climate, soil, or terrain attributes (slope, aspect, and landscape position). Land types capture the highest level (coarse grain) of spatial heterogeneity caused by various environmental controls. Within a land type, a suite of ecological conditions that results in similar species establishment patterns is assumed, but the stochastic processes such as seed dispersal can result in intermediate level (fine grain,

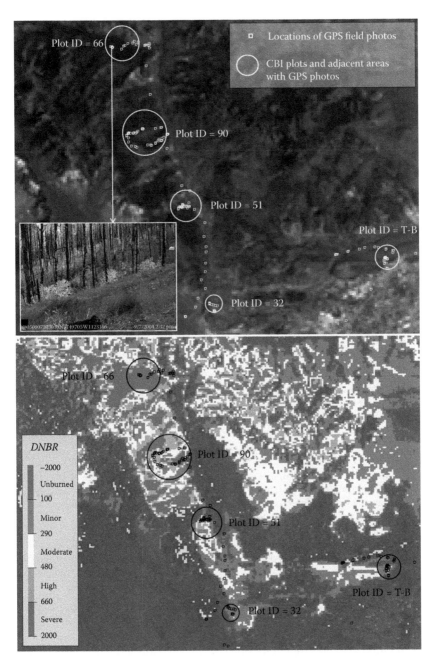

FIGURE 14.3
(**See color insert.**) An example of plot locations displayed on top of the post-fire Landsat Thematic Mapper image and differenced normalized burn ratio map.

TABLE 14.2

Examples of Pre-Fire Existing Vegetation Type (EVT), Differenced Normalized Burn Ratio (DNBR), Composite Burn Index (CBI), and Field Observations for Selected Referencing Plot Sites

Plot ID	Characteristics of Pre-Fire EVT	DNBR	CBI	Observed Recovery after 3 Years
T-B	Colorado Plateau Pinyon-Juniper Woodland and Shrubland: forest canopy >20% and <30%; canopy height (10, 25 m)	369	2.61	65% vegetation cover, predominantly herbaceous species.
51	Colorado Plateau Pinyon-Juniper Woodland and Shrubland: forest canopy >0% and <10%; canopy height (0, 5 m)	− 18	0.74	50%–60% vegetation cover, predominantly herbaceous species. Pinyon-Juniper survived; resprout of quick recovery aspens.
90	Colorado Plateau Pinyon-Juniper Woodland and Shrubland: forest canopy >40% and < 50%; canopy height (10, 25 m)	497	2.75	Less than 50% vegetation cover, 10% grasses, and 30% mixed shrubs, predominantly herbaceous species; no evidence of recovery of Mountain Mahogany that dominated plot area before fire.
66	Rocky Mountain Mountain Mesic Mixed Conifer Forest and Woodland: forest canopy >50% and <60%; canopy height (10, 25 m)	996	2.79	50% vegetation cover, predominantly herbaceous species; dense resprout of aspen; no evidence of fir or coniferous recovery.
32	Colorado Plateau Mixed Low Sagebrush Shrubland: shrub canopy >30% and <40%; shrub height (0, 0.5 m)	193	2.72	60%–70% vegetation cover, predominantly herbaceous species.
86	Rocky Mountain Subalpine Dry-Mesic Spruce-Fir Forest and Woodland: forest canopy >30% and ≤40%; forest height (10, 25 m)	1099	2.92	40%–50% vegetation cover, predominantly herbaceous species; dense resprout of aspen.
86B	Rocky Mountain Subalpine Dry-Mesic Spruce-Fir Forest and Woodland: forest canopy >40% and ≤50%; forest height (10, 25 m)	401	1.42	40% vegetation cover, predominantly herbaceous species; sign of conifer seeding; resprout of aspen.

within land type) heterogeneity of a species distribution. Succession, competition, and probabilistic establishment may result in heterogeneity of species presence and age cohorts even among cells that were initially identical. Disturbance heterogeneity refers to various regimes a disturbance may have on the simulated landscape. LANDIS stratifies the heterogeneous disturbance regimes using disturbance regime maps. For example, fire regimes are characterized by ignition frequency and fire cycle (mean fire return interval) in a fire regime map (Yang et al., 2004). Within-regime heterogeneity is further simulated by a stochastic process of each disturbance regime, and cell-level heterogeneity is simulated through an interaction of disturbance and vegetation in cells.

LANDIS assumes that detailed, individual tree information and within-stand processes can be simplified, allowing large-scale questions such as spatial pattern, species distribution, and disturbances to be adequately addressed. Vegetation succession at each cell is a competitive process driven by species life history attributes, such as longevity, age of sexual maturity, shade tolerance class, fire tolerance class, maximum age of vegetative reproduction and sprouting, sprouting probability, and effective and maximum seeding distance. In contrast to tracking individual trees (Botkin et al., 1972; Botkin, 1993; Urban et al., 1993) LANDIS tracks the presence and absence of species age cohorts. Therefore, succession dynamics is simplified and simulated as birth, growth, and death processes acting on species age cohorts. During a single LANDIS iteration, birth, death, and growth routines are performed on species age cohorts and random background mortality is simulated.

Parameterization for LANDISv4.0a includes three major steps: (1) development of a land type map and the species establishment coefficients for each land type; (2) development of a species vital attribute table; and (3) establishment of initial states for a simulation.

14.4.2.1 Land Type Attribute and Establishment Coefficients

Land type attributes encapsulate environmental variations and can be created from abiotic data sources such as climate, soil, geology, and topography. We used LANDFIRE biogradients data and adopted the method by Wimberly (2004) to express the topographic moisture gradient. We classified the landscape into three categories of *bottomlands*, *hill slopes*, and *ridges*. Bottomlands are distributed on flat terrain adjacent to major streams that include hydric and mesic sites in the study area. Hill slopes are areas with intermediate moisture on hill sides. Ridges are the driest areas and are found on the gently sloping uplands. We considered two slope aspects of the northwest (NW) and southeast (SE) in defining the land type combined with the topographic moisture gradient. Elevation is another factor that defines a land type as spatial distributions of tree species are associated with variations of elevation ranges. We used the pre-fire EVT and the digital elevation model data from the LANDFIRE prototype data products to obtain distributions of vegetation species in different elevations. We calculated the area in number of pixels represented by the EVT data within fire impacted areas. By referencing the calculated number of pixels from the EVT data we divided the Sanford Fire site into four most influential elevation categories on vegetation distributions, that is, <2450, 2450–2700, 2700–2900, and >2900 m. We then combined these features to create the land-type attribute file. As the slope

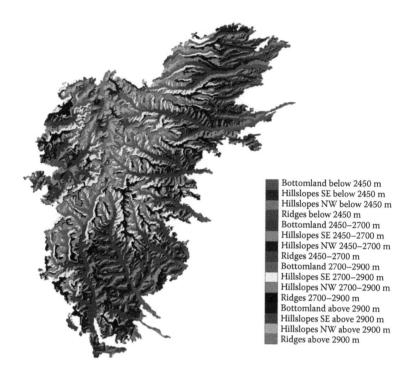

FIGURE 14.4
(**See color insert.**) Land type data developed by integrations of topographic moisture gradients. Slope aspects, and elevations.

and aspects affect mostly the areas on hill slopes rather than the bottomlands and ridges, we applied the slope and aspects to the hill slopes only. The final 16 land types are illustrated in Figure 14.4.

A species establishment coefficient is a number ranging from 0 to 1 that expresses the species' relative ability to grow on different land types. Coefficients are differentiated based on relative responses of species to soil moisture, climate, and nutrients. Reported studies estimated species establishment coefficients from published summaries of species characteristics (Burns and Honkala, 1990; Sutherland et al., 2000) and from studies of community composition in present-day and presettlement forests (Cowell, 1995; Wimberly, 2004). As information about pre-fire vegetation composition within different land types could be extracted from EVT data, it would be efficient and effective to estimate establishment coefficients in each land type based on the existence of species from the pre-fire EVT. We obtained the percentage of vegetation species within defined land types based on EVT and calculated the occurrence frequencies by number of pixels for the main vegetation species within each of the 16 land types. We then obtained the ratio between the pixels of each vegetation species within each land type and the total pixels of that vegetation species, and the ratio between the pixels of each vegetation species within each land type and the total pixels of that land type. Multiplying the two ratios, we derived the establishment coefficients for the 18 main vegetation species for each of the 16 land types.

14.4.2.2 Species Vital Attributes

Species vital attributes include longevity, mature age, shade tolerance, fire tolerance, effective seeding distance, maximum seeding distance, vegetation propagation probability, and maximum sprouting age. We derived the attributes for the simulated species from the Silvics of North America (Burns and Honkala, 1990) and the Plant Database (NRCS, 2009). We defined the effective seed dispersal ranges as between 50 m for gravity-dispersed species, 100 m for large wind-dispersed winged seeds, 150 m for small wind-dispersed winged seeds, and 200 m for small plumed seeds (Sutherland et al., 2000; Wimberly, 2004). For wind-dispersed seeds, we assumed the maximum dispersal distance to be double the effective dispersal distance. For seeds with animal or bird dispersal vectors, we assumed the maximum dispersal distance to be 3000 m. We added the annual grass category in a final species attribute table (Table 14.3) to reflect the fact of significant recovery of annual grass species immediately after a wildfire.

14.4.2.3 Species Composition Map

The species composition map consists of species and their age classes. There are different ways to create species composition map (He et al., 1998). In this study, we referenced the canopy height from the LANDFIRE lifeform data and the descriptions of the vegetation characteristics from the NRCS Plant Database to derive age information and generated the species composition map.

14.4.2.4 Fire Severity Classes and the Fire Regime

In LANDISv4.0a simulation, the fire effect module simulates which species age cohorts are killed on each burned pixel. We referenced the DNBR data to establish the initial state for the simulation. We considered six classes of fire severity as follows:

1. *No fire* (DNBR value <100)
2. *Severity class 1* (DNBR value 100–250)
3. *Severity class 2* (DNBR value 250–400)
4. *Severity class 3* (DNBR value 400–550)
5. *Severity class 4* (DNBR value 550–700)
6. *Severity class 5* (DNBR value >700).

TABLE 14.3

Species Vital Attribute Table for the Main Vegetation Species in the Study Area

Species Name	LONG	MATURE	SHADE	FIRE	EFFD	MAXD	VEG_P	SP_AG	RCLS_COEFF
Pinus edulis	600	25	1	2	50	3000	0	0	0.5
Juniperus osteosperma	600	30	1	2	50	3000	0	0	0.5
Pinus contorta	200	8	1	2	100	3000	0.5	8	0.3
Picea engelmannii	600	40	4	2	100	200	0	0	0.5
Populus tremuloides	85	3	1	5	200	3000	1	1	0.1
Pinus ponderosa	450	15	1	5	50	3000	0	0	0.3
Pseudotsuga menziesii	500	14	2	3	100	3000	0	0	0.3
Abies lasiocarpa	250	20	4	2	100	3000	0	0	0.2
Abies concolor	350	40	4	5	100	200	0	0	0.3
Juniperus scopulorum	300	20	1	2	50	3000	0	0	0.2
Quercus gambelii	120	6	1	5	50	3000	1	1	0.1
Pinus flexilis	600	30	1	2	30	3000	0	0	0.5
Cercocarpus montanus	54	10	3	5	100	200	1	1	0.1
Artemisia tridentata	45	2	1	5	30	60	0	0	0.1
Chrysothamnus nauseosus	35	4	1	2	150	300	1	1	0.1
Artemisia nova	45	2	1	2	50	100	0.2	1	0.1
Amelanchier utahensis	20	3	3	5	50	3000	1	1	0.1
Annual grass	20	1	3	1	200	3000	1	2	0.00005

Note: LONG, maximum longevity (years); MATURE, age of reproductive maturity (years); SHADE, shade tolerance (1: least shade tolerant, 5: most shade tolerant); FIRE, fire tolerance (1: least fire tolerant, 5: most fire tolerant); EFFD, effective seeding distance (m); MAXD, maximum seedling distance (m); VEG_P, vegetation propagation coefficient; SP_AG, maximum age of vegetative propagation; RCLS_COEFF, reclassification coefficient (0–1).

Given that the DNBR data defined burn severity classes at a pixel level, we treated each pixel as a fire regime and set the mean fire size as the pixel size and the standard deviation of fire size as 0. In doing so, we were able to simulate the fire perimeter and severity as those defined by the DNBR data.

14.4.3 Simulation Modeling

To simplify the simulation we combined two ecosystems of the *Artemisia tridentata ssp. vaseyana* Shrubland Alliance (Table 14.1, Code 2220) and the Inter-Mountain Basins Montane Sagebrush Steppe (Table 14.1, Code 2125) into the Colorado Plateau Mixed Low Sagebrush Shrubland (Table 14.1, Code 2064) for the reason that these ecosystems share the similar main sagebrush species. With an added herbaceous category (HBR) to reflect the post-fire growth of annual grass species, the final simulation included 14 ecosystems as showing in the results (see Section 14.5).

The simulation starts at year 0, that is, the time right after the fire occurred. Then the simulation proceeds at a 1-year interval, as *year 'i'* represents the simulated vegetation recovery at the *i*th year after the fire ($i = 1, 2, \ldots, 10$). We assumed that no new fire occurred in the 10-year simulation time period within any burned cell.

14.5 Results

A post-fire summary by the Dixie National Forest suggested that sagebrush would likely become re-established through seeds that are already present in the soil. Aspen was present throughout the burned area as "pure stands" or interspersed with conifers. Although the fire killed many aspen trees, it also enhanced aspen reproduction as the fire stimulated the growth of suckers from the aspen's extensive root system. In many instances the fire left behind bare mineral soil and removed taller plants, which created a suitable condition for aspen seedlings to take root. Areas of mixed conifers, spruce fir, and ponderosa pine will take much longer to become after fires.

The simulation results illustrate post-fire change and spatial distribution of vegetation species within the ecosystems in different years (Figure 14.5). Table 14.4 summarizes the simulated vegetation responses to fire in different severity classes and in percentage areas of the pre-fire ecosystems, year 0 as the fire impacts, and simulated ecosystems 10 years after the fire. For example, areas of HBR increased at different fire severity classes 10 years after the fire. The areas with higher levels of fire severity show a more significant increase of herbaceous vegetation than the areas of a "no-fire" category. The ecosystems that include the aspen species, for example, the Rocky Mountain Aspen Forest and Woodland (RMA) and *Abies concolor* Forest Alliance (ACF) showed recovery trends similar to that of the HBR,

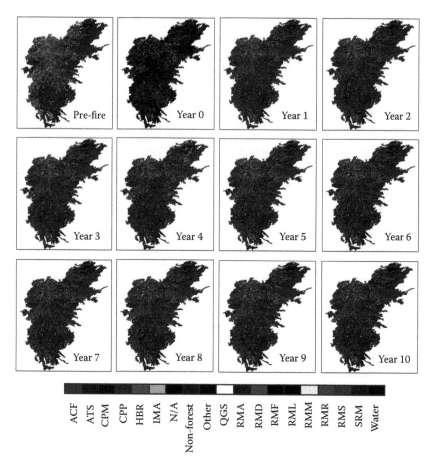

FIGURE 14.5
Examples of simulated post-fire vegetation recovery at ecosystem level.

in particular for the higher burn severity classes. For the ecosystems of the Colorado Plateau Pinyon-Juniper Woodland and Shrubland (CPP) and Rocky Mountain Subalpine Dry-Mesic Spruce-Fir Forest and Woodland (RMS), the recoveries are slower than those ecosystems with aspen species. The higher the fire severity, the slower the recovery would be in percentage of areas. Figure 14.6 illustrates the simulated age cohorts of aspen as an example at the species level. High fire tolerance and resprout capability allow aspens to recover quickly in the succession process. Aspens would have good recovery in land types such as southeast-facing hill slopes.

Species fire tolerance and species establishment are main drivers for post-fire species responses. The simulated results reflect the change of age cohorts in the ecosystems. For example, the simulated RMA shows increases in area after the fire, which reflects the resprout of aspen species in the RMA ecosystem. The

TABLE 14.4

Different Fire Severity Levels for the Pre-Fire, 0-Year, and 10-Year Simulated Ecosystem in Percentage Areas

		No Fire (%)	Severity Class 1 (%)	Severity Class 2 (%)	Severity Class 3 (%)	Severity Class 4 (%)	Severity Class 5 (%)	Total (%)
RMA	Pre-fire	3.34	3.27	3.00	1.94	1.08	0.65	13.28
	0-year	3.34	3.27	6.16	4.95	3.26	0.00	20.97
	10-year	2.70	3.69	5.29	3.85	2.47	3.13	21.15
ACF	Pre-fire	0.45	0.63	0.90	1.10	1.42	2.08	6.57
	0-year	0.45	0.63	1.21	1.46	2.16	0.00	5.91
	10-year	1.16	2.60	3.71	3.48	3.79	3.64	18.38
HBR	Pre-fire	0.00	0.00	0.00	0.00	0.00	0.00	0.00
	0-year	0.00	0.00	0.00	0.00	0.00	0.00	0.00
	10-year	0.03	3.85	3.65	1.85	0.94	1.39	11.71
IMA	Pre-fire	2.32	2.79	3.16	3.02	2.18	1.67	15.13
	0-year	2.32	2.79	0.00	0.00	0.00	0.00	5.10
	10-year	1.79	2.91	1.25	0.98	0.71	0.93	8.58
RMS	Pre-fire	2.53	3.18	3.07	2.41	2.30	4.33	17.81
	0-year	2.53	0.00	0.00	0.00	0.00	0.00	2.53
	10-year	2.53	1.11	0.83	0.57	0.45	0.80	6.29
RMM	Pre-fire	0.19	0.24	0.31	0.36	0.58	1.18	2.87
	0-year	0.19	0.24	0.00	0.00	0.00	0.00	0.44
	10-year	0.76	1.19	0.39	0.32	0.27	0.59	3.52
CPP	Pre-fire	9.08	6.74	6.28	2.79	0.83	0.36	26.08
	0-year	9.08	0.00	0.00	0.00	0.00	0.00	9.08
	10-year	9.09	1.30	1.53	0.68	0.17	0.05	12.81
RMR	Pre-fire	0.83	0.74	0.69	0.42	0.16	0.07	2.91
	0-year	0.83	0.74	0.69	0.42	0.00	0.00	2.68
	10-year	0.86	1.08	0.93	0.52	0.06	0.04	3.48
RMD	Pre-fire	0.30	0.30	0.47	0.38	0.17	0.06	1.68
	0-year	0.30	0.30	0.47	0.38	0.00	0.00	1.45
	10-year	0.56	0.56	0.72	0.56	0.01	0.01	2.42
SRM	Pre-fire	1.37	0.79	0.60	0.35	0.17	0.14	3.42
	0-year	1.37	0.79	0.60	0.35	0.00	0.00	3.11
	10-year	1.11	0.56	0.39	0.20	0.01	0.00	2.26
CPM	Pre-fire	4.20	1.96	0.25	0.02	0.00	0.00	6.44
	0-year	4.20	1.96	0.25	0.02	0.16	0.00	6.60
	10-year	4.21	2.02	0.29	0.04	0.15	0.00	6.71
QGS	Pre-fire	0.66	0.52	0.08	0.01	0.00	0.00	1.27
	0-year	0.66	0.52	0.08	0.01	0.00	0.00	1.27
	10-year	0.66	0.53	0.09	0.01	0.00	0.00	1.29
RML	Pre-fire	0.68	0.34	0.09	0.01	0.00	0.00	1.12
	0-year	0.68	0.34	0.09	0.01	0.00	0.00	1.12
	10-year	0.59	0.37	0.12	0.03	0.01	0.02	1.12

continued

TABLE 14.4 (continued)

Different Fire Severity Levels for the Pre-Fire, 0-Year, and 10-Year Simulated Ecosystem in Percentage Areas

		No Fire (%)	Severity Class 1 (%)	Severity Class 2 (%)	Severity Class 3 (%)	Severity Class 4 (%)	Severity Class 5 (%)	Total (%)
RMF	Pre-fire	0.16	0.18	0.24	0.24	0.14	0.06	1.01
	0-year	0.16	0.00	0.00	0.00	0.00	0.00	0.16
	10-year	0.26	0.00	0.00	0.00	0.00	0.00	0.26
Other	Pre-fire	0.20	0.13	0.05	0.02	0.01	0.00	0.43
	0-year	0.20	10.23	9.64	5.46	3.45	10.61	39.59
	10-year	0.00	0.02	0.00	0.00	0.00	0.00	0.02
Total (%)		26.30	21.79	19.19	13.08	9.03	10.61	100

Note: Refer to Table 14.1 for full forms of all abbreviations.

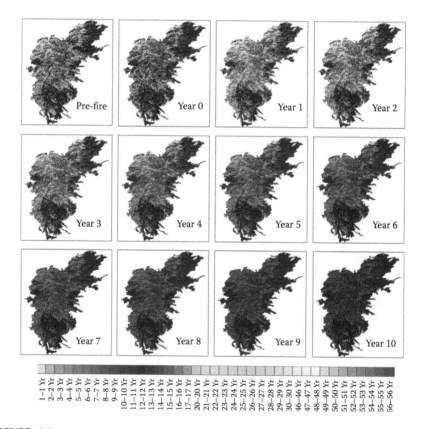

FIGURE 14.6
(See color insert.) Examples of simulated post-fire vegetation recovery at species level for quick aspen age cohorts.

higher the level of fire severity classes, the more significant increase in RMA with time. The increase in areas is gradually reduced through time with the age cohorts changing within the RMA areas. For the other ecosystems such as CPP and RMS, the changes are slow for the 10-year simulation. Changes of ecosystems are negligible for the no-fire areas in 10 years.

14.6 Discussion and Conclusion

The combination of LANDFIRE data products, the DNBR data, and LANDISv4.0a modeling demonstrated a promising approach in simulation of short-term post-fire vegetation recovery. LANDFIRE data products, such as EVT, life-form, and biogradients data, played a unique role in defining the land type, species establishment coefficients, and species attribute table. Pre-fire EVT provided an efficient and effective data source to estimate establishment coefficients for each land type based on the existence of the species. DNBR data made the measurements of fire severity classes possible. DNBR data, teamed up with LANDFIRE data products, defined fire regimes from identified fire impacts. The integration provided critical data to define and establish initial states for the simulation.

In model selection, LANDISv4.0a was capable of simulating the dynamics of forest succession, seed dispersal, and fire disturbance at species level and with a one-year time interval. As a spatially explicit landscape model LANDISv4.0a was able to incorporate finer cell data, such as the 30-m spatial resolution LANDFIRE and DNBR data.

In simulation each species responds differently to the levels of fire severity according to the fire tolerance of species and the level of fire severity. For the cells with a moderate level of fire severity, LANDIS simulation was able to estimate changes in vegetation structure. Fire tolerance of tree species and fire severity determined the post-fire canopy gaps in different ecosystems. For example, as *Pinus ponderosa* could survive at a low level of fire severity, the fire severity would contribute in the simulation to create canopy gaps in ecosystems with *Pinus edulis*. The simulated spatial patterns for ecosystems and species demonstrated the trends of post-fire variation of vegetation, in particular for quick recovery species such as aspens. However, it is difficult to observe changes on slow recovery species and ecosystems dominated by those species since this simulation was limited to 10 years only.

Climatic variations are among the influencing factors for post-fire establishment of vegetation. Climatic conditions in early years after a wildfire should be critical for short-term post-fire vegetation recovery. Precipitations in summer months should increase the survival rate of newly grown vegetation. Varying weather cycle may affect the establishment coefficient, given that seed germination depends on the patterns of precipitations.

The simulated vegetation recovery would be more appropriate with climate-modified establishment coefficients, in particular for short-term simulations.

Acknowledgments

This study was conducted as an Intergovernmental Personal Act Assignment by the USGS/EROS through which the principal author conducted his sabbatical research. Several people assisted in completing this study. In particular, we wish to express sincere thanks to Zhiliang Zhu, Don Ohlen, Xuexia (Sherry) Chen, James Vogelmann, Jay Kost, and Steve M. Howard, who engaged in constructive discussions for model selections, considerations, and provided data.

References

Beukema, S.J., Kurz, W.A., Klenner, W., Merzenich, J., and Arbaugh, M. 2003. Applying TELSA to assess alternative management scenarios. In G. Arthaud and T. Barrett (eds), *Systems Analysis in Forest Resources*. Amsterdam: Kluwer, pp. 145–154.

Botkin, D.B. 1993. *Forest Dynamics: An Ecological Model*. Oxford: Oxford University Press, pp. 101–138.

Botkin, D.B., Janak, J.G., and Wallis, J.R. 1972. Some ecological consequences of a computer model of forest growth. *Journal of Ecology*, **60**, 849–872.

Brewer, C.K., Winne, J.C., Redmond, R.L., Opitz, D.W., and Mangrich, M.V. 2005. Classifying and mapping wildfire severity: A comparison of methods. *Photogrammetric Engineering and Remote Sensing*, **71**, 1311–1320.

Burns, R.M., and Honkala, B.H. 1990. *Silvics of North America. Volume 1: Conifers; Volume 2: Hardwoods. Agriculture Handbook 654*. Washington, DC: USDA Forest Service. Available at: http://na.fs.fed.us/spfo/pubs/silvics_manual/table_of_contents.htm (accessed May 26, 2009).

Comer, P., Faber-Langendoen, D., Evans, R., Gawler, S., Josse, C., Kittel, G., Menard, S. et al. 2003. *Ecological Systems of the United States: A Working Classification of U.S. Terrestrial Systems*. Available at: http://na.fs.fed.us/spfo/pubs/silvics_manual/table_of_contents.htm (accessed May 26, 2009).

Cowell, C.M. 1995. Presettlement Piedmont forests—Patterns of composition and disturbance in central Georgia. *Annals of the Association of American Geographers*, **85**, 65–83.

Dasgupta, S., Qu, J.J., Hao, X., and Bhoi, S. 2007. Evaluating remotely sensed live fuel moisture estimations for fire behavior predictions in Georgia, USA. *Remote Sensing of Environment*, **108**, 138–150.

De Santis, A., and Chuvieco, E. 2007. Burn severity estimation from remotely sensed data: Performance of simulation versus empirical models. *Remote Sensing of Environment*, **108**, 422 – 435.

De Santis, A., Chuvieco, E., and Vaughan, P.J. 2009. Short-term assessment of burn severity using the inversion of PROSPECT and GeoSail models. *Remote Sensing of Environment*, **113**, 126–136.

Fraser, R.H., and Li, Z. 2002. Estimating fire-related parameters in boreal forest using SPOT VEGETATION. *Remote Sensing of Environment*, **82**, 95–110.

GAO (Government Accountability Office). 2006. *Wildland Fire Management: Update on Federal Agency Efforts to Develop a Cohesive Strategy to Address Wildland Fire Threats.* GAO-06-617R. Washington, DC: Government Accountability Office, 19pp.

Hann, W.J., and Bunnell, D.L. 2001. Fire and land management planning and implementation across multiple scales. *International Journal of Wildland Fire*, **10**, 389–403.

Hao, X., and Qu, J.J. 2007. Retrieval of real time live fuel moisture content using MODIS measurements. *Remote Sensing of Environment*, **108**, 130 – 137.

He, H.S., and Mladenoff, D.J. 1999. Spatially explicit and stochastic simulation of forest-landscape fire disturbance and succession. *Ecology*, **80**, 81–99.

He, H.S., Mladenoff, D.J., Radeloff, V.C., and Crow, T.R. 1998. Integration of GIS data and classified satellite imagery for regional forest assessment. *Ecological Applications*, **8**, 1072–1083.

He, H., Shang, B.Z., Crow, T.R., Gustafson, E.J., and Shifley, S.R. 2004. Simulating forest fuel and fire risk dynamics across landscapes—LANDIS fuel module design. *Ecological Modelling*, **180**, 135–151.

Hemstrom, M.A., Korol, J.J., and Hann, W.J. 2001. Trends in terrestrial plant communities and landscape health indicate the effects of alternative management strategies in the interior Columbia River basin. *Forest Ecology and Management*, **153**, 1–3.

Keane, R.E., Cary, G.J., Davies, I.D., Flannigan, M.D., Gardner, R.H., Lavorel, S., Lenihan, J.M., Li, C., and Rupp, T.S. 2004. A classification of landscape fire succession models: Spatial simulations of fire and vegetation dynamics. *Ecological Modelling*, **179**, 3–27.

Klenner, W., Kurz, W.A., and Beukema, S.J. 2000. Habitat patterns in forested landscapes: management practices and the uncertainty associated with natural disturbances. *Computers and Electronics in Agriculture*, **27**, 243–262.

Lentile, L.B., Holden, Z.A., Smith, A.M.S., Falkowski, M.J., Hudak, A.T., Morgan, P., Lewis, S.A., Gessler P.E., and Benson, N.C. 2005. Remote sensing techniques to assess active fire characteristics and post-fire effects. *International Journal of Wildland Fire*, **15**, 319–345.

Merzenich, J., Kurz, W.A., Beukema, S., Arbaugh, M., and Schilling, S. 2003. Determining forest fuel treatments for the Bitterroot front using VDDT. In G. Arthaud and T. Barrett (eds), *Systems Analysis in Forest Resources*. Amsterdam: Kluwer, pp. 47–59.

Miller, J.D., and Thode, A.E. 2007. Quantifying burn severity in a heterogeneous landscape with a relative version of the delta Normalized Burn Ratio (dNBR). *Remote Sensing of Environment*, **109**, 66–80.

Mladenoff, D.J., and He, H.S. 1999. Design and behavior of LANDIS, an object-oriented model of forest landscape disturbance and succession. In D. Mladenoff and W. Baker (eds), *Advances in Spatial Modeling of Forest Landscape Change: Approaches and Applications*. Cambridge: Cambridge University Press, pp. 163–185.

NRCS. 2009. Plant Database. Available at: http://plants.usda.gov (accessed May 26, 2009).

Rollins, M.G., and Frame, C.K. 2006. *The LANDFIRE Prototype Project: Nationally Consistent and Locally Relevant Geospatial Data for Wildland Fire Management.* General Technical Report RMRS-GTR-175. Fort Collins: USDA Forest Service, Rocky Mountain Research Station, 416pp. Available at: http://www.treesearch. fs.fed.us/pubs/24484 (accessed May 26, 2009).

Skowronski, N., Clark, K., Nelson, R., Hom, J., and Patterson, M. 2007. Remotely sensed measurements of forest structure and fuel loads in the Pinelands of New Jersey. *Remote Sensing of Environment*, **108**, 123–129.

Sturtevant, B.R., Gustafson, E.J., and He, H.S. 2004. Modelling disturbance and succession in forest landscapes using LANDIS. *Ecological Modelling (Special Issue)*, **180**, 1–232.

Sutherland, E.K., Hale, B.J., and Hix, D.M. 2000. Defining species guilds in the central hardwood forest, USA. *Plant Ecology*, **147**, 1–19.

Swanson, F.J., Cissel, J.H., and Reger, A. 2003. Chapter 9—Landscape management: Diversity of approaches and points of comparison. In R. Monserud, R. Haynes, and A. Johnson (eds), *Compatible Forest Management*. Amsterdam: Kluwer, pp. 237–266.

Syphard, A.D., Yang, J., Franklin, J., He, H.S., and Keeley, J.E. 2007. Calibrating forest landscape model to simulate high fire frequency in Mediterranean-type shrublands. *Environmental Modelling & Software*, **12**, 1641–1653.

Upper Sevier River Community Watershed Project. 2002 Sanford Fire 2002. Available at: http://www.uppersevier.net/resource/sanfire/sanfire.html (accessed May 26, 2009).

Urban, D.L., Harmon, M.E., and Halpern, C.B. 1993. Potential response of Pacific Northwestern forests to climatic change, effects of stand age and initial composition. *Climatic Change*, **23**, 247–266.

van Wagtendonk, J.W., Root, R.R., and Key, C.H. 2004. Comparison of AVIRIS and Landsat ETM+ detection capabilities for burn severity. *Remote Sensing of Environment*, **92**, 397 – 408.

White, J.D., Ryan, K.C., Key, C.C., and Running, S.W. 1996. Remote sensing of forest fire severity and vegetation recovery. *International Journal of Wildland Fire*, **6**, 125 – 136.

Wimberly, M.C. 2004. Fire and forest landscapes in the Georgia Piedmont: An assessment of spatial modeling assumptions. *Ecological Modelling*, **180**, 41–56.

Wimberly, M.C., and Reilly, M.J. 2007. Assessment of fire severity and species diversity in the southern Appalachians using Landsat TM and ETM+ imagery. *Remote Sensing of Environment*, **108**, 189–197.

Yang, J., He, H.S., and Gustafson, J.E. 2004. A hierarchical statistical approach to simulate the temporal patterns of forest fire disturbance in LANDIS model. *Ecological Modelling*, **180**, 119–133.

Section III

Remote Sensing for Inventory and Monitoring of Frontier Lands

15

Satellite-Observed Endorheic Lake Dynamics across the Tibetan Plateau between Circa 1976 and 2000

Yongwei Sheng and Junli Li

CONTENTS

15.1 Introduction

At an average altitude of ~4500 m, the Tibetan Plateau, sometimes called the "Roof of the World," is considered as one of the last frontier lands of the world. The plateau exerts profound thermal and dynamical influences on the global atmospheric circulation and has great impacts on regional and global climate (Manabe and Broccoli, 1990; Thompson et al., 1997). The plateau is one of the most sensitive areas responding to climate change. Ice core records from the Dasuopu Glacier indicate that the past 50 years have been the warmest in 1000 years (Thompson et al., 2000). Meteorological records from 1955 to 1996 show that the mean annual temperature of the plateau has increased 0.16°C per decade and the mean winter temperature has increased 0.32°C per decade (Liu and Chen, 2000). The accelerated warming is expected to drive an array of significant physical and ecological changes in the region, particularly to the terrestrial water cycle, which plays an integral role in nearly every aspect of the plateau, due to its extremely high altitude, the vulnerable environment, the near-freezing temperatures in this region, and the extensive presence of permafrost and glaciers (Sheng and Yao, 2009).

The plateau is home to the largest group of high-altitude lakes of the world. Most Tibetan lakes are situated in endorheic basins in the heart of the plateau at a density of ~3.5%. These lakes have shrunk significantly since the late Pleistocene (Sheng, 2009), and are currently continuing to experience changes in their distribution and inundation area. Closed to major rivers and oceans, they serve as a sensitive indicator of climate and water cycle variability. These lake changes represent a sensitive and integrative indicator of regional climate and water cycle change, and greatly impact the local and regional ecological environment. However, lake dynamics in the context of climate change are not well understood due to the inaccessibility and harsh environment of this remote plateau. Therefore, identifying and understanding the whole picture of recent lake change across the plateau is vital to our understanding how this crucial region responds to climate change. Though satellite imagery has previously been used to aid Tibetan lake surveys, it mainly has been applied to study small groups of lakes rather than systematic inventory for lake dynamics (Bianduo et al., 2009; Liu et al., 2009). The U.S. Geological Survey (USGS) has released Landsat products, the longest satellite image archive, to the public since 2008 (Woodcock et al., 2008), providing a great possibility to decadal lake survey on the plateau. To this end, this research inventories endorheic lakes across the Tibetan Plateau using multi-temporal Landsat imagery acquired by the early Multi-Spectral Scanner (MSS) and recent Enhanced Thematic Mapper Plus (ETM+) sensors in ca. 1976 and ca. 2000, respectively. Lake dynamics is monitored by comparing these lake surveys approximately 25 years apart. In addition, the spatial patterns of lake dynamics are examined using spatial information technologies in relation to climate records and glacier distributions.

15.2 Study Area and Methods

15.2.1 Study Area

As illustrated in Figure 15.1, the study area of this research covers the entire endorheic lake basins in the core of the plateau with an area of 767,789 km². Averaged at ~5000 m altitude, the study area is underpopulated with extremely low accessibility and difficulty for fieldwork due to the harsh environment (Zheng et al., 1985). Lakes are well distributed in this area and most of them are saline or salt lakes. The study area is mainly in the cold semiarid climate zone and perches mostly on discontinuous and sporadic permafrost. The average temperature of the warmest month (i.e., July) is below 10°C, and annual precipitation is below 300 mm. Lakes in the south are relatively large and formed in geological depressions. The northern region is even colder and dryer, and lakes are greater in quantity but smaller in size. Most lakes in the study area freeze during deep winter but over a short period owing to

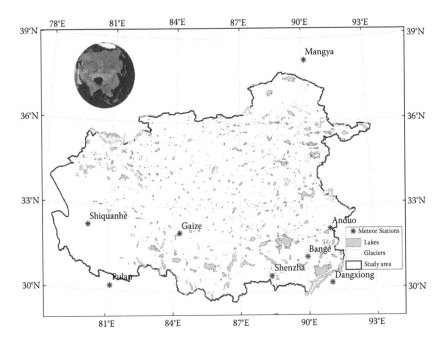

FIGURE 15.1
The study area covers the endorheic basins across the plateau. Available meteorological stations are marked using red asteroids, and lakes and glaciers are displayed in blue and yellow, respectively.

their high salinity. Lake water levels usually start to rise in April and May and reach their peak in July and August.

The Landsat program started to acquire satellite imagery in July 1972, first with the MSS sensor, and later with the TM sensor since 1985 and the ETM+ sensor since 1999. Nearly a hundred scenes of Landsat MSS and ETM+ imagery acquired around 1976 and 2000 were carefully selected to match the ideal monitoring season (i.e., mainly between October and December) when Tibetan lakes are relatively stable to reduce seasonal and inter-annual variability. As such the long-term lake changes can be differentiated from seasonal variations. Table 15.1 lists the acquisition time of 48 MSS images acquired in ca. 1976 and 42 ETM+ images acquired in ca. 2000 used in lake dynamics mapping. In addition, multiple scenes of satellite images are used in areas where a single cloud-free image is not available.

15.2.2 Lake Mapping

Since the goal is to examine the long-term lake dynamics, we need to use early MSS imagery and more recent TM and ETM+ imagery. For regional-scale lake dynamics studies using various Landsat sensors, it is essential to

TABLE 15.1

Landsat Images Used in Lake Dynamics Mapping

ETM+ Images			MSS Images		
Path	Row	Acquisition Date	Path	Row	Acquisition Date
138	35	2001-10-03	148	35	1976-11-11
138	36	2001-10-03	148	36	1976-11-11
138	37	2000-10-16	148	37	1976-11-11
138	38	2000-11-17	148	38	1976-11-11
138	39	2000-11-01	148	39	1976-12-17
139	35	2000-10-07	149	35	1976-11-12
139	36	2000-10-07	149	36	1976-11-12
139	37	2000-10-07	149	37	1976-11-12
139	38	2000-11-08	149	38	1976-11-12
139	39	2000-11-08	149	39	1976-12-18
140	34	2000-10-30	150	34	1976-12-19
140	35	2000-10-30	150	35	1976-12-19
140	36	2000-10-30	150	36	1976-12-19
140	37	2000-10-30	150	37	1976-11-13
140	38	2000-10-30	150	38	1976-11-13
140	39	2000-10-30	150	39	1976-11-13
141	34	2000-11-22	151	34	1976-12-02
141	35	2000-09-19	151	35	1976-12-02
141	36	2000-09-19	151	36	1976-12-02
141	37	2000-10-05	151	37	1976-12-02
141	38	2000-10-05	151	38	1976-12-02
141	39	2000-10-05	151	39	1976-12-02
142	35	2000-10-28	152	34–35	1976-12-21
142	36	2000-10-28	152	36	1976-11-15
142	37	2000-10-28	152	37	1976-11-15
142	38	2000-10-28	152	38	1976-11-15
142	39	2001-10-31	152	39	1976-12-03
143	35	2000-11-04	153	35	1976-11-16
143	36	2000-11-04	153	36	1976-11-16
143	37	2001-10-22	153	37	1976-11-16
143	38	2000-10-03	153	38	1976-11-16
143	39	2000-10-03	153	39	1976-11-16
144	35	2000-08-07	154	35	1976-11-17
144	36	2001-11-14	154	36	1976-11-17
144	37	2000-10-10	154	37	1976-11-17
144	38	2000-10-10	154	38	1976-11-17
144	39	2000-10-10	154	39	1976-11-17
145	36	2001-10-20	155	36	1977-02-16
145	37	2000-11-02	155	37	1976-10-13

TABLE 15.1 (continued)

Landsat Images Used in Lake Dynamics Mapping

ETM+ Images			MSS Images		
Path	Row	Acquisition Date	Path	Row	Acquisition Date
145	38	2000-11-02	155	38	1976-10-13
146	36	2000-10-08	155	39	1976-12-06
146	37	2000-10-08	156	36	1977-02-17
			156	37	1976-11-19
			156	38	1976-11-19
			156	39	1976-11-19
			157	36	1976-11-20
			157	37	1976-11-20

precisely map lakes from different sensors using similar wavelength bands (i.e., MSS bands 4, 5, and 7, TM and ETM+ bands 2, 3, and 4). Theoretically, lakes and other surface water bodies are readily identified in each scene of the satellite data owing to their very low reflectance in near-infrared (NIR) channels of Landsat sensors (0.8–1.1 μm for MSS band 7, 0.76–0.90 μm for TM and ETM+ band 4). However, most lakes in the endorheic basins on the plateau are saline or salt lakes. The diversified mineral contents complicate the lake water spectral responses and impose challenges to remote sensing lake mapping.

The normalized difference water index (NDWI) was developed using green and near-infrared bands to enhance water features and depress disturbing factors (McFeeters, 1996), and has been widely used in water body delineation from satellite images (Lira, 2006; Ouma and Tateishi, 2007; Xu, 2006). NDWI derived as (band4 – band7)/(band4 + band7) for MSS and (band2 – band4)/(band2 + band4) for TM and ETM+ is less sensitive to mineral materials in lake water and demonstrates better capability in delineating water bodies from land than individual bands; therefore, we use it in image segmentation for lake mapping. Top-of-atmosphere (TOA) reflectance of these bands calculated from Landsat DN (Digital number) values (Chander et al., 2009) are used in NDWI computation. Lake maps can be produced by segmenting a NDWI image. However, lakes cannot be properly delineated in the image using a single threshold due to the various conditions of the lakes.

It is desirable to automatically map lakes in regional-scale studies. We developed an automated hierarchical lake mapping approach implemented through image segmentation at both global and local levels, which simulates a human operator's procedures. In the general procedures, lakes in the image are first identified by segmenting the NDWI image using a conservative threshold, which is considered as the global-level segmentation. Then each identified lake is treated as an object and is further refined only in the area surrounding the lake (i.e., a buffer zone) for a fine adjustment at the local level. The local-level segmentation is then implemented iteratively to precisely identify the lake until the iteration converges when the segmentation result becomes stable. This hierarchical method can effectively extract from Landsat imagery lakes under various complex water conditions, and greatly reduces the subsequent lake editing efforts. Following the automated extraction is intensive quality control and assurance to confirm the extracted lake extents are correct and the identified lake changes are real. The output of these procedures is the lake map layer in vector GIS (geographic information systems) format extracted from a Landsat image.

All the lakes extracted from individual Landsat images in the same monitoring episode (i.e., ca. 1976 or ca. 2000) are combined together to produce the lake map for the episode. With an automated tool developed in GIS, validated lake layers extracted from individual satellite images are mosaiced into one lake-map layer for this episode covering the entire study area. In the overlapping areas between adjacent images, the tool can automatically select the lakes from the image acquired at the preferable time.

The above procedures have been applied to map lakes across the study area for the two monitoring episodes. The ca. 2000 lake map is shown in Figure 15.2. About 1200 lakes were identified and mapped from the 1976 MSS images at ~79 m resolution. The 2000 ETM+ images at a higher spatial resolution (i.e., 30 m) produced 2036 lakes, catching a large number of small ponds in size of several pixels. Only 822 and 1109 lakes larger than 0.5 km^2 are found in the 1976 and 2000 lake layers, respectively.

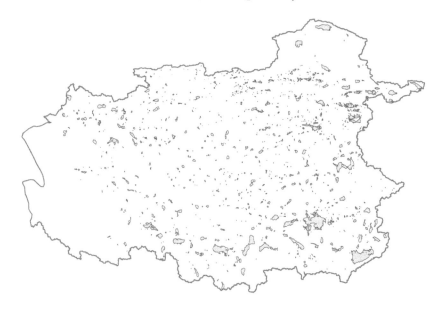

FIGURE 15.2
Ca. 2000 lake map generated from 42 scenes of ETM+ imagery.

15.3 Lake Dynamics

Lake changes between 1976 and 2000 for lakes larger than 16 ETM+ pixels are analyzed using these satellite-derived lake layers. Table 15.2 compares the lake size class distribution between the two monitoring episodes. Almost every single size class in 2000 contains more lakes and in greater area than in 1976. The significantly higher number of lakes in small size class in 2000 is partially because the ETM+ sensor tends to catch smaller lakes at higher resolution than the MSS sensor. We can see that the small size class only contributes marginally to the total lake area statistics. Therefore, we only compare sizable lakes (i.e., larger than 0.5 km²) between the two episodes in the following subsequent analysis to reduce the sensor resolution effect.

A detailed change analysis for individual Tibetan lakes has been conducted in a GIS. Through a lake monitoring tool developed in GIS, a lake on an episode lake map is automatically spatially linked to the same lake on the other episode map. This spatial association establishes a lake-change record for each lake, containing parameters such as lake size at each episode. The percentage areal change of lakes larger than 0.5 km² between 1976 and 2000 is shown in Figure 15.3, where reddish colors present expanding lakes, greenish colors for shrinking lakes, and yellowish colors for stable lakes. Normally, there is a one-to-one correspondence between lakes in two episodes. In dynamic areas, however, it is possible for more than one lake in one episode to correspond to a group of lakes in the other episode, which is a many-to-many correspondence. In case of a disappeared lake or an emergent lake, it becomes a one-to-none or none-to-one correspondence.

A scatter plot of lake area comparison is informative to lake changes at individual level, in which the X and Y axes represent the lake area in the

TABLE 15.2

Lake Size Class Distributions in 1976 and 2000

	1976		2000	
Size Class	Lake Abundance	Total Area (km²)	Lake Abundance	Total Area (km²)
>1000 km²	3	4573	2	3856
Between 100 and 1000 km²	47	11,815	53	13,547
Between 10 and 100 km²	192	7156	230	8020
Between 1 and 10 km²	368	1160	473	1553
Between 0.5 and 1 km²	212	148	351	249
Between 0.1 and 0.5 km²	233	66	352	95
<0.1 km²	145	7	575	23

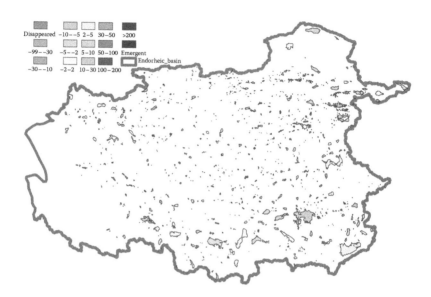

FIGURE 15.3
Lake change map of sizable lakes (0.5 km² or larger) between 1976 and 2000.

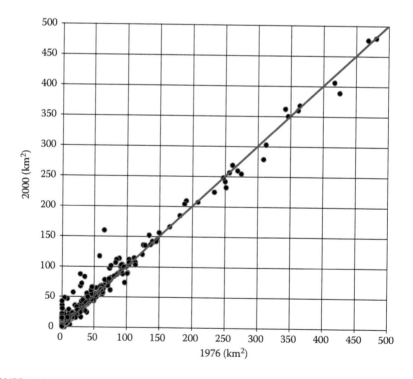

FIGURE 15.4
Lake area comparison between 1976 and 2000 for correspondent lakes larger than 0.5 km².

early and the latter episodes, respectively. Figure 15.4 shows such a scatter plot of Tibetan lakes between 1976 and 2000 for 1125 lake pairs. The red line is the 1:1 reference line of no area change. The dots around the reference line represent stable lakes (i.e., 99 lakes within 2% area change). Expanded lakes are above the line, while shrunk lakes are under the line. Completely disappeared lakes (three small lakes) lie on the X-axis, as the extreme case of shrunk lakes, whereas 160 emergent lakes on the Y-axis. We can see that there are more expanding lakes (i.e., 680) than shrinking lakes (i.e., 183), and the expanding lakes are well scattered all over the study area while the shrinking lakes are mainly located in the northeast and southwest. This analysis reveals that the total lake area has increased by ~9.5% from 24,893 km^2 in 1976 to 27,254 km^2 in 2000. Yibug Caka, the most remarkable outlier in the scatter plot, expanded significantly from 65.5 km^2 in 1976 to 160.1 km^2 in 2000, a 144% increase. The southern half of the lake appeared as dry playa in 1976, but became part of the open-water lake in 2000.

The lake change map exhibits a strong spatial pattern. The lake expansion is found extensively across the plateau, especially in the northwest and southeast. Lakes in the central southern areas have shrunk slightly, less than 5% in general. The northeastern region has experienced lake shrinkage.

Although the observed lake dynamics exhibits a strong spatial pattern in general; however, the general pattern is accompanied by local variations. Figure 15.5 shows a typical mosaic pattern of lake change in the Hoh Xil area in the northeast. This area is considered to be a major lake-shrinking region in general as a result of reduced precipitation in 2000. Four lakes shown in the figure experienced heterogeneous changes. Kekao Lake shrank remarkably at 12% between 1976 and 2000 and Hoh Xil Lake showed a minor shrinkage of 3%. However, Taiyang Lake and Yinma Lake expanded slightly at 1% and 2%, respectively. Since these lakes are close to each other, the climate was largely the same. Additional factors need to be considered in exploring the mechanism of the discrepant changes within the vicinity.

15.4 Mechanism Analysis of Observed Lake Changes

The observed distribution patterns in lake change are analyzed together with the climate records collected from meteorological stations. Available climatology records reveal a drying trend for 40 years before 1998, with five years between 1990 and 1997 experiencing summer/fall drought. Precipitation has increased since 1998, especially in the southern and eastern plateau. Four well-distributed stations among the available stations are selected to represent the climate in the study area: Mangya in the north, Bange in the southeast, Gaize in the central south, and Shiquanhe in the southwest. Figure 15.6

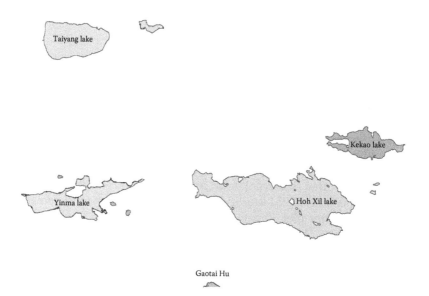

FIGURE 15.5
Lakes Kekao, Hoh Xil, Taiyang, and Yinma showing different changing directions. The color keys are in the legend of Figure 15.3.

shows the annual precipitation records at these four stations. In general, both 1976 and 2000 were normal years at most stations. The precipitation in the southeast (i.e., Bange) was slightly higher in 2000 than in 1976. Western areas were wetter in 2000 than 1976 with a ~30% increase at Shiquanhe, whereas a ~15% decrease of annual precipitation is evident at Mangya, the only available station in the north. There was a warming trend (about 0.5°C) in the

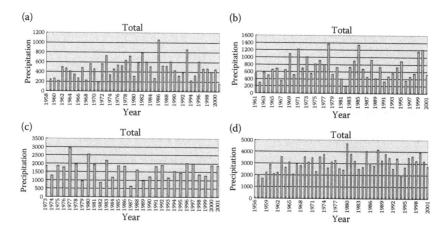

FIGURE 15.6
Annual precipitation records at key meteorological stations. (a) Mangya; (b) Shiquanhe; (c) Gaize; (d) Bange.

southeastern and the western areas of the study site. The southern central area does not exhibit much temperature change. A 3°C temperature increase is found in the north at Mangya. There is a relationship between the climate records and the observed lake change from satellite images. The lake expansion is obvious in the southeastern and western areas, where more precipitation was present in 2000. The northeastern areas experienced significant lake shrinkage at drier climate in 2000. Lakes in the central southern areas where precipitation and temperature stayed stable between 1976 and 2000 shrank only slightly.

Basin-based analysis can explain the mosaiced lake change pattern at local scales. Status of a lake in terms of size and water level is the result of water budget balancing in its basin among precipitation, evaporation/evapotranspiration, and other water sources/sinks. In the context of the climate warming over the plateau, most glaciers are found retreating, releasing excessive melt water to the downstream lakes and providing a significant water source to the lakes in nearby basins. Using glacier distributions derived from the 2000 ETM+ images, we conducted a basin-based water balance analysis with a special attention on the water contribution from glaciers. The basin-based analysis facilitates our understanding on glacial impacts on lake dynamics and explains some local variations.

Figure 15.7 shows the basins of Lakes Yinma, Taiyang, Hoh Xil, and Kekao with Malan glaciers in the vicinity. Overlaying glaciers, stream networks and drainage basins on top of these lakes, we can identify the destination and pathways of glacier melt water. Lake Kekao is not a glacier-fed lake, and shrank greatly under the warmer and dryer climate in 2000. Taiyang Lake

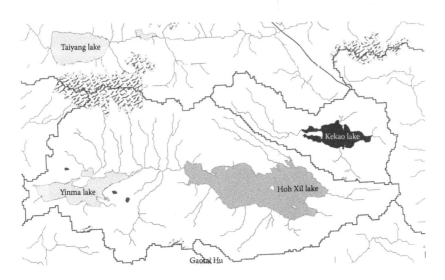

FIGURE 15.7
Lakes Kekao, Hoh Xil, Taiyang, and Yinma in lake basins.

received water supply from Malan glaciers in the south and expanded slightly even in the dryer 2000. Both Hoh Xil and Yinma lakes are in the same endorheic basin fed by the Malan glaciers, and receive excessive melt water from the glaciers under the warmer climate. The stream networks indicate that most glacial melt water damps into the Yinma lake, leading to the 2% expansion. The Hoh Xil lake shrank much less than the Kekao lake owing to the minor glacial melt water supply.

Factors other than glaciers also contribute to the local variations in the general lake dynamics patterns. We here use Dawa Co as an example to illustrate the effect of drainage network change on lake dynamics. As many other Tibetan lakes, Dawa Co is a saline lake located in the heart of the Tibetan Plateau around (31.25 °N, 84.58 °E) with an area of ~110 km². Qingmuke Co, a small satellite lake (~8 km²), is situated on the east. The color composite image (Figure 15.8a), made by assigning the NIR images of the 1976 MSS (November 15) and 1990 TM (November 10) images to the red and

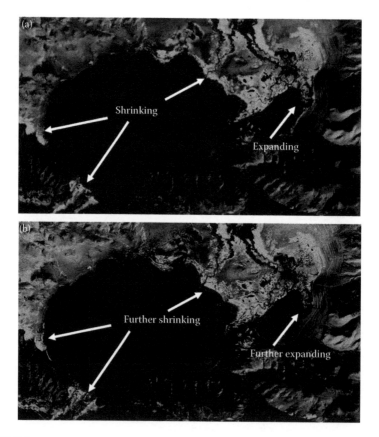

FIGURE 15.8
(**See color insert.**) Lake dynamics at Dawa Co. (a) Lake change between 11/15/1976 MSS and 11/10/1990 TM images; (b) Lake change between 10/10/1990 TM and 10/28/2000 ETM+ images.

green channels, reveals that Dawa Co shrank at its northeastern border by about 500 m (i.e., the light green-colored areas). A similar image (Figure 15.8b) composited using the 1990 TM and 2000 ETM+ (October 28) images shows that the lake continued to shrink during the 1990s by roughly the same amount. On the contrary, Qingmuke Co has expanded (the red areas) constantly from 1976, 1990 to 2000.

The local-scale lake changes found in this area are due to the drainage network change in the upper reaches. Figure 15.9 explains such a mechanism at Dawa Co. Both lakes have been successfully mapped and their extent and size have been extracted for a quantitative dynamics analysis. Dawa Co is found shrinking from 111.4 km^2 in 1976 to 107.0 km^2 in 1990 and further down to 103.7 km^2 in 2000, whereas Qingmuke Co expanded from 6.2 km^2 in 1976 to 7.9 km^2 in 1990 and further to 8.7 km^2 in 2000. The Dawa Co shrank by 7% between 1976 and 2000, whereas the Qingmuke Co expanded about 40% in contrast (Figure 15.9a). The drainage networks suggest that the major glacier melt water flows into the Dawa Co, which cannot explain the shrinkage using the glacier-driving mechanism. Figures 15.9b and c display the near-infrared (NIR) images of the 1976 MSS (November 15) and the 2000 ETM+ (October 28) images, respectively. There are two main streams (marked by **A** and **B**) visible in both satellite images. The stream **A** damps directly

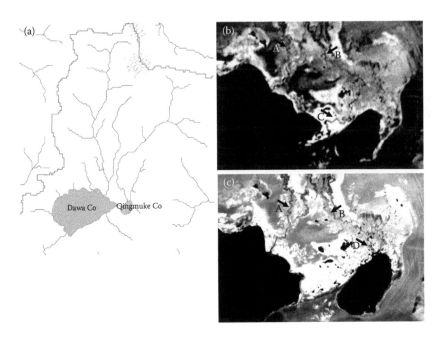

FIGURE 15.9
Lake change mechanism at Dawa Co and Qingmuke Co. (a) Detected lake changes between 1976 and 2000; (b) Drainage networks in 11/15/1976 MSS; and (c) Drainage networks in 10/28/2000 ETM+ image.

into Dawa Co, and the stream **B** also damped water into Dawa Co through the stream segment **C** in the 1976. The stream **A** is distinct on the 2000 image, indicating abundant water supply from the upstream. However, the stream **B** switched its course to damp water into Qingmuke Co through the new segment **D**. Such a stream switch explains the shrinkage of Dawa Co and the expansion of Qingmuke Co. Here the drainage network change before 1990 caused the opposite changes in Dawa Co and Qingmuke Co.

15.5 Conclusions

Tibetan lakes in the endorheic basins are dynamic and sensitive to climate change. This study monitors tibetan lake changes using nearly 100 landsat images acquired around 1976 and 2000. Satellite remote sensing provides a feasible tool to regional-scale multitemporal lake dynamics mapping across the plateau. More than 1100 lakes larger than 0.5 km² located in endorheic basins of the plateau have been monitored ~25 years apart. The lake change analysis reveals that tibetan lakes in general had expanded remarkably (i.e., 9.5%) between 1976 and 2000. Their dynamics varies from region to region and exhibits a strong regional-scale spatial pattern mosaiced together with local variations. both temperature and precipitation play a crucial role in regulating lake dynamics on the plateau. The overall regional-scale pattern is mainly controlled by annual precipitation while the local pattern is extensively caused by the glacier melting accelerated by the tibetan warming as well as other mechanisms such as upstream drainage network change.

Acknowledgments

This research was conducted with the support of NASA NIP project (NNX06AE58G).

References

Bianduo, B., L. Li, W. Wang, and Zhaxiyangzong. 2009. The response of lake change to climate fluctuation in north Qinghai-Tibet Plateau in last 30 years. *Journal of Geographical Sciences* 19(2): 131–142.

Chander, G., B. L. Markham, and D. L. Helder. 2009. Summary of current radiometric calibration coefficients for Landsat MSS, TM, ETM +, and EO-1 ALI sensors. *Remote Sensing of Environment* 113(5): 893–903.

Lira, J. 2006. Segmentation and morphology of open water bodies from multispectral images. *International Journal of Remote Sensing* 27(18): 4015–4038.

Liu, J., S. Wang, S. Yu, D. Yang, and L. Zhang 2009. Climate warming and growth of high-elevation inland lakes on the Tibetan Plateau. *Global and Planetary Change* 67(3–4): 209–217.

Liu, X. D. and B. D. Chen. 2000. Climatic warming in the Tibetan Plateau during recent decades. *International Journal of Climatology* 20(14): 1729–1742.

Manabe, S. and A. J. Broccoli. 1990. Mountains and arid climates of middle latitudes. *Science* 2474939: 192–194.

McFeeters, S. K. 1996. The use of the normalized difference water index (NDWI) in the delineation of open water features. *International Journal of Remote Sensing* 17(7): 1425–1432.

Ouma, Y. O. and R. Tateishi. 2007. Lake water body mapping with multiresolution based image analysis from medium-resolution satellite imagery. *International Journal of Environmental Studies* 64(3): 357–379.

Sheng, Y. W. 2009. PaleoLakeR: A semiautomated tool for regional-scale paleolake recovery using geospatial information technologies. *IEEE Geoscience and Remote Sensing Letters* 6(4): 797–801.

Sheng, Y. W. and T. D. Yao. 2009. Integrated assessments of environmental change on the Tibetan Plateau. *Environmental Research Letters* 4(4), doi: 10.1088/1748-9326/4/4/045201.

Thompson, L. G., T. Yao, M. E. Davis, K. A. Henderson, E. Mosley-Thompson, P.-N. Lin, J. Beer, H.-A. Synal, J. Cole-Dai, and J. F. Bolzan. 1997. Tropical climate instability: The last glacial cycle from a Qinghai-Tibetan ice core. *Science* 2765320: 1821–1825.

Thompson, L. G., T. Yao, E. Mosley-Thompson, M. E. Davis, K. A Henderson, and P. N. Lin. 2000. A high-resolution millennial record of the South Asian Monsoon from Himalayan ice cores. *Science* 289(5486): 1916–1919.

Woodcock, C. E., R. Allen, M. Anderson, A. Belward, R. Bindschadler, W. Cohen, F. Gao et al. 2008. Free access to Landsat imagery. *Science* 320(5879): 1011–1011.

Xu, H. Q. 2006. Modification of normalised difference water index (NDWI) to enhance open water features in remotely sensed imagery. *International Journal of Remote Sensing* 27(14): 3025–3033.

Zheng, D., Q. Yang, and Y. Liu. 1985. *The Tibetan Plateau of China*. Beijing: The Press of Science, 123pp. (in Chinese).

16

Multisensor Remote Sensing of Forest Dynamics in Central Siberia

Kenneth J. Ranson, Guoqing Sun, Viatcheslav I. Kharuk, and Joanne Howl

CONTENTS

16.1 Introduction

In Central Siberia (Figure 16.1), a vast area between the Yenisey and the Lena rivers, the character of taiga changes dramatically. The Central Siberian plateau and Sakha-Yakutia are distinguished by their extreme continental and arid climates. The mean January temperature decreases from –17°C in Krasnoyarsk (56°10′N; 93°00′E) to –43°C in Yakutsk (62°10′N; 129°50′E) while the total annual precipitation decreases from 410 to 200 mm. The Siberian anticyclone dominates the area throughout winter and, with little winter precipitation (34 and 21 mm, respectively), the depth of snow cover is small. A severe climate and continuous permafrost predispose the development of forests composed of cold-resistant species (mainly *Larix gmelinii*) and poor floristic diversity (www.rusnature.info).

It is known that about 70% of the permafrost areas in Siberia are occupied by larch-dominated forests, with the remaining 30% composed of tundra. The maximal depth of permafrost is about 30–100 m in north-western Siberia and 500–1500 m in the northern parts of central and eastern Siberia. Depth of thaw in the summer is typically 5 cm to >1.0 m.

Larch-dominated forests are an important component of the global circumpolar boreal forest. In Russia, larch is the most widespread species and is found from the tundra zone in the north to the steppes in the south. The zone of larch dominance extends from the Yenisey ridge in the west to the Pacific Ocean in the east and from Lake Baikal in the south to the 73rd parallel in the north, where it forms the world's most northern forested stand, and is called Ary-Mas. In Central Siberia, the southern and western margins of the larch forest come in contact with evergreen conifers [Siberian pine

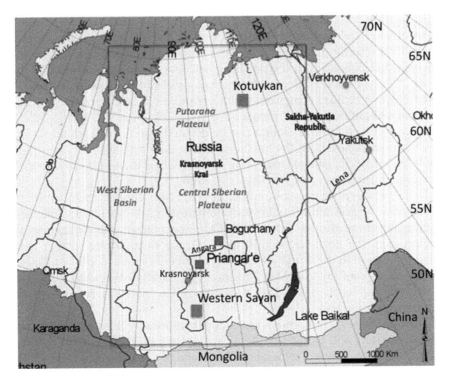

FIGURE 16.1
Study sites in Central Siberia, Russia from south to north: Western Sayan; Boguchany and Priangar'e; and Kotuykan.

(*Pinus sibirica*), pine (*Pinus sylvestris*), spruce (*Picea obovata*), fir (*Abies sibirica*)], hardwoods [birch (*Betula pendula, Betula pubescens*)], and aspen (*Populus tremula*). Larch forms high closure stands as well as open forests, and can be found primarily growing over permafrost in locations where other tree species barely survive. Wildfires are typical for this territory, and they most commonly occur as ground fires due to low crown closure. Due to the surface root system (caused by permafrost) and dense lichen–moss cover, ground fires are primarily stand-replacing fires. The vast area of larch forests, including the forest–tundra ecotone, is generally considered a "carbon sink." However positive long-term temperature trends at higher latitudes result in an increase in fire frequency and an increase of greenhouse gas emissions, and may convert this area to a source for greenhouse gases.

Litter decomposition in the larch communities is reduced by low summer temperatures, resulting in increased litter thickness. The thick litter layer, together with moss and lichens, becomes a thermal insulator that promotes permafrost formation at an increasing depth in the soil. In addition, during low-precipitation years, the ground cover layer dries and becomes a fire fuel source. This facilitates the spread of fires over tens to hundreds of kilometers.

The burning fires emit greenhouse gases, which in turn can increase temperature and decrease the permafrost depth. The fire return intervals (FRI) within the interior of larch forests were found to be about 82 ± 7 years, and increased to 300 years in the northern extreme of the larch forest (Kharuk et al., 2008). There is evidence of decreasing FRI (fire frequency increase) in the twentieth century compared with the nineteenth century caused by both natural and anthropogenic impacts.

Taiga forest in Central Siberia can be divided into northern, middle, and southern subzones. Both the northern and middle subzones are dominated by larch forests (*L. gmelinii* and *Larix sibirica*), while the southern subzone is dominated by Scotch pine (*P. sylvestris*) stands.

In the northern subzone of Central Siberia, forest–tundra and open forests reach their largest latitudinal extent. Relatively high (about 13°C) summer temperatures on the lee side of the Putorana plateau, which shelters the region from the northerly winds, allow forests to penetrate further north than anywhere in the world. On the Taymyr Peninsula, woodlands (known as "forest islands" or "Ary Mas") formed by *L. gmelinii* extend to the world's northernmost location at 72°30′N in the valley of the river Novaya, the Khatanga's tributary.

The tundra–taiga transition area is dynamic because it is very sensitive to human activity and climate change. During the past 6000 years in northern Eurasia, there has been a general cooling trend of about 2–4°C, and larch and birch stands have retreated between 400 and 500 km southward during this period (Callaghan et al., 2002). Temperatures have warmed by as much as 2°C in the past three decades in parts of the Northern Hemisphere (Hansen et al., 1999). Reports on modern changes of the tundra–taiga boundary associated with climate warming are rare (e.g., Kharuk et al., 2004a, 2007–2010, Sturm et al., 2001), but observations within the northernmost forest stand showed regeneration advance into tundra and stand densification (Kharuk et al., 2004b, 2006).

The northward movement of the tundra–taiga boundary may be the eventual outcome if climatic warming persists over centuries or millennia (Skre et al., 2002). The situation, however, is complicated by human activities that have led to ecosystem degradation in this area. In some Russian case studies, southward displacement of the taiga–tundra boundary was reported due to human disturbance and increasing waterlogging, which led to paludification and the death of tree line trees (Skre et al., 2002; Vlassova, 2002). Local variations in climate and human activities require continued monitoring and research.

In the middle taiga, forests composed of *L. sibirica* prevail in the relatively warmer and moister western areas, with some dark taiga species (spruce, Siberian pine, and fir). East of the Yenisey, much of the taiga is represented by monospecies larch stands with the presence of Siberian pine and spruce along the rivers. Because of the severity of the environment and the remoteness of the area, these forests remain virtually untouched by humans. In the drier eastern part with low seasonal permafrost thawing, *L. gmelinii* dominates with the sporadic appearance of *P. sylvestris*.

For the middle taiga, it was shown that "dark needle" coniferous forest (DNC), made up of Siberian pine, spruce, and fir, is expanding into the habitat of larch. The age structure of the regeneration (with mortality control) showed that it was 20–30 years old. The results obtained indicate climate-driven migration of Siberian pine, spruce, and fir into traditional larch habitat. On the western and southern margins of the larch-dominated forest, DNC regeneration formed a second layer in the forest canopies, which could eventually replace the larch in the overstory. With stand densification, Siberian pine received an additional advantage since larch is a shadow-intolerant species (Kharuk et al., 2007).

In the southern taiga, vegetation is more varied because of the higher diversity of climatic and soil conditions. The *P. sylvestris* forests are dominating in the west, *L. sibirica* and *L. gmelinii* forests are dominating in the east, and the dark taiga (i.e., fir, spruce, and Siberian pine) are dominating along the high watersheds that have cool summers and ample precipitation. The Sayan Mountains are a system of deeply eroded ridges. These mountains have an average elevation of 1000–2000 m, but the highest summits, Munku-Sardyk and Mongun-Taiga, reached 3492 and 3976 m, respectively. Permafrost occupies about 50% of the total area of the Western Sayan and almost the whole of the eastern Sayan except for its westernmost part. Vertical zonality is well expressed in the distribution of soils and vegetation in the Sayans. Although topographic and climatic variabilities create differing vertical sequences, all are dominated by taiga vegetation. In the Western Sayan Mountains, there is a considerable difference between the altitudinal sequences of the northern and southern macroslopes. The most prominent feature of the northern Western Sayan sequence is the large extent of the dark taiga belt. This is composed mainly of Siberian pine (*P. sibirica)* and fir (*A. sibirica*) with an admixture of *L. sibirica* in the upper regions. Spruce (*P. obovata*) is another important dark taiga tree species, especially in river valleys. The lower part of the forest belt has been substantially modified by human activity. In disturbed areas, taiga has been replaced by secondary birch–aspen forests and patches of pristine taiga survive only locally. The high mountainous zone is represented by floristically rich subalpine and alpine meadows and mountainous tundra communities. The southern macroslope of the Western Sayan receives more insolation and is much drier. This sequence begins with the steppe (which is intensively cultivated at present) developing on chernozem and in drier regions on southern chernozem and chestnut soils. The steppe zone is succeeded by a narrow belt of birch–aspen forest–steppe. The forest belt is composed mainly of *L. sibirica* forests with well-developed undergrowth and the herbaceous cover enriched by steppe species. The dark taiga is confined to higher elevations of this belt and reaches highest elevations on north-facing slopes. Taiga, the largest biome in northern Eurasia, accounts for a quarter of the world's pristine forests (Dirk et al., 1997). It has been affected by development, in particular by the production of timber and oil. Although the annual industrial production of timber has declined since the 1980s, many areas experience problems with respect to illegal cutting of forests, fragmentation of mature stands, and unacceptable forest-harvesting practices. Global climate and land use changes have multiple effects on forests worldwide. However, the multiscaled interactions among climate change, disturbance regimes, and land use change make it difficult to predict key ecosystem characteristics except by coarse, generalized estimates. Many Siberian forests are facing the twin pressures of rapidly changing climate and increasing timber harvest activity. Mean temperatures have risen significantly over the past 40 years, and this trend is expected to continue. The frontier of timber harvest is pushing into previously uncut areas.

Mean temperatures have risen significantly over the past 40 years, and this trend is expected to continue. There are reports showing that Siberian pine and larch growing in the alpine forest–tundra ecotone are strongly responding to warming by an increase of growth increments, stand densification and regeneration density, upward tree line shift, and transformation of krummholz to arboreal forms. Climate-induced waves of upslope and downslope tree migration were reported for the alpine forest–tundra ecotone in the southern Siberian Mountains. Observations show that tree mortality was observed during the Little Ice Age, but lagged behind the initial cooling. Living tree natality dates showed that tree line advance began at the end of the nineteenth century, but lagged behind the warming temperatures. Larch and Siberian pine regeneration now survive at elevations up to 160 m higher in comparison with the maximum observed tree line recession during the Little Ice Age and surpasses the historical maximum during the last millennium by up to 90 m. A 1°C change promoted an upward shift in the tree line of about 80 m. The tree line advance rate was estimated at 0.90 ± 0.22 m/year (Kharuk et al., 2010a). At present, at high elevation, seedlings are still in the vulnerable stage and could be killed by cold winters consisting of low temperatures and strong winds, and this could result in the recession of the tree line.

Studies within Altai-Sayan Mountain upper forest belt sites showed an increase of the dense forest stands of about 1.5 times during the past four decades. An increase of tree growth increment starting in the mid-1980s was observed, which was strongly correlated with mean summer temperatures. Stand densification was also observed along rivers and streams due to earlier snowmelt that increases the growing period. Substantial densification in tree line populations seems to be a common phenomenon in northern and high-elevation environments and occurs more frequently than actual elevational tree line advance (Kharuk et al., 2008). Forest response to climate variables at high elevations is nonuniform because tree establishment and survival depends on the availability of sheltered (wind protected) areas. The forest spatial distribution is dependent on azimuth, elevation, and slope steepness and this pattern changed over the past decades. A typical upper boundary is a mosaic because tree and regeneration survival depends on the availability of sheltered relief that is provided by rocks or local depressions (Kharuk et al., 2010c).

Milder climate also promotes changes in tree morphology, that is, transformation of mat and prostrate krummholz into vertical form (Kharuk et al., 2006, 2010a). The past decades of warming caused a widespread transformation of larch and Siberian pine mat and krummholz to a vertical form beginning in the 1980s. This date approximately coincides with the period when winter temperatures surpassed the mean value during the twentieth century. Larch is much less likely to be found in krummholz forms than Siberian pine. This species surpasses Siberian pine in frost and wind resistance and was observed in arboreal forms where Siberian pine was still prostrate.

In a warming climate, Siberian pine should enjoy a competitive advantage due to its higher-temperature response. Stand densification is also beneficial for Siberian pine since larch is a shade-intolerant species. Thus, current climate change should lead to an increase in the proportion of Siberian pine in the upper canopy. Substitution of "light-needle" deciduous larch by evergreen conifers decreases albedo and provides positive feedback for even greater warming. The other expected consequence is an increase of biodiversity since Siberian pine-dominated communities provide a better food base for animals and birds. Larch will continue to maintain its advantage in drier areas and in zones of temperature extremes.

The Siberian forests are the habitat of many insect species. Periodic outbreaks of certain insect pests cause a decrease in growth increment, forest decline or mortality over vast areas. The Siberian silkmoth (*Dendrolimus superans sibiricus*, Tschetw) feeds heavily on needles of certain tree species, defoliating and killing large stands rapidly. This is one of the primary factors of taiga succession. The preferred pest host species are fir and Siberian pine but spruce and larch are also sometimes affected. Outbreaks are encouraged by favorable weather conditions: low summer precipitation, relatively mild winters with stable, dense snow cover, and lack of late spring and early autumn frosts. In contrast cold, rainy summers and severe low-snow winters are not favorable for the Siberian silkmoth. Outbreaks have a periodicity, occurring about every 15–25 years. Between 1878 and 2004, 10 Siberian silkmoth outbreaks were observed in the Yenisey River watershed area. The largest outbreak (1954–1957) resulted in tree damage of over about 4 million ha and tree mortality of about 1.5 million ha (Kharuk et al., 2003, 2004a,b). The last catastrophic outbreak occurred in the Priangar'e area and caused damage of about 0.7 million ha stands and killed 300 thousands ha stands between 1993 and 1995. Outbreaks of insect pests promote wildfires, because pest-killed stands accumulate combustible material in the form of dead wood, grass, and bush communities.

Global change is likely to significantly change forest composition of south-central Siberian landscapes, with some changes taking ecosystems outside the historic range of variability. Direct climate effects generally increased tree productivity and modified probability of establishment, but indirect effects on the fire regime generally counteracted the direct effects of climate on forest composition. Harvest and insects significantly changed forest composition, reduced living above-ground biomass, and increased forest fragmentation. Gustafson et al. (2010) studied the relative effects of climate change, timber harvesting, and insect outbreaks on forest composition, biomass (carbon), and landscape pattern in south-central Siberia and found that a global change is likely to significantly change forest composition of south-central Siberian landscapes, with some changes taking ecosystems outside the historic range of variability. Remote sensing provides a useful tool for monitoring forest disturbances and estimation of forest parameters in Central Siberia because the area is broad and hard to reach. Twenty years ago, two authors of this chapter, Dr. Ranson and Dr. Kharuk, committed to collaborate to study the remote forests of Siberia. Since then, with Sukachev Institute of Forest and NASA's support, we have been conducting research on forest mapping, disturbance characterization, and parameter retrieval using multisensor data in Central Siberia. For example, in the Western Sayan Mountains, radar data were used to map forest above-ground biomass (Sun et al., 2002). In the middle subzone of Central Siberia, around Boguchany and Prianger'e, multisensor data, including Landsat ETM+ data, radar data from the Japanese Earth Resources Satellite (JERS-1), European Remote Sensing satellite (ERS-1), and Canada's Radarsat-1, were used to detect fire scars, logging, and insect damage in the boreal forest (Ranson et al., 2003). From July 10 to July 25, 2008, a team of American and Russian scientists conducted an expedition to study an extremely remote and harsh section of northernmost Central Siberia. The expedition started from a flat spot near the headwaters of the Kotuykan River, above the Arctic Circle, and the team, using three inflatable rubber boats packed with survival gear and scientific instruments, traveled the river, stopped frequently to make observations and collect data for several ongoing studies. Data from the field, collected by Geoscience Laser Altimeter System (GLAS), Phased Array type L-band Synthetic Aperture Radar (PALSAR), and Landsat MSS, TM, and ETM+, were used for monitoring changes of forest cover, and estimation of above-ground biomass. The following sections will describe these studies in detail.

16.2 Forest Biomass Estimation from SAR Data in the Western Sayan Mountains

16.2.1 Introduction

Methods and algorithms have been developed for mapping above-ground biomass in the boreal forest (Dobson et al., 1992; Le Toan et al., 1992; Beaudoin et al., 1994; Rignot et al., 1994; Dobson et al., 1995; Ranson et al., 1995; Saatchi et al., 1995; Ranson and Sun, 1997a; Bergen et al., 1998; Kurvonen et al., 1999; Paloscia et al., 1999; Saatchi and Moghaddam, 2000). These studies concentrated on relatively flat areas, where terrain effects were not significant. Estimation of forest biomass using synthetic aperture radar (SAR) data can be complicated by topography that influences radar backscatter (Rauste, 1990; Bayer et al., 1991; van Zyl, 1993; Luckman, 1998), particularly through local incidence angle and shadowing. Changes in radar incidence angle caused by terrain slope can have several effects on radar image data. For example, radar backscattering varies with incidence angle, which varies with terrain slope and aspect. Foreshortening is also a terrain-induced effect where a smaller incidence angle results in more ground surface area being illuminated. Another effect of terrain on the backscatter is the apparent change of the forest spatial structure in the radar field of view. For example, when trees of a relatively uniform stand grow on a slope, a portion of the sides of these trees will be directly exposed to the radar beam.

Terrain correction techniques are designed to reduce effects of incidence angle and illuminated target area. For correction of the illuminated pixel area, simple algorithms can be used if a suitable digital elevation model (DEM) exists (Kellndorfer et al., 1998). Correction of the backscattering dependence on incidence angle requires knowledge of the land cover type within a pixel. A few attempts have been made to correct terrain effects by using simple radar backscattering models and a DEM. For example, Goering et al. (1995) used a DEM and empirical radar backscatter models to reduce terrain effects from ERS-1 SAR images. However, Goyal et al. (1998) found that the small-scale topographic features resolved by SAR could not be resolved by a DEM in rugged terrain. Periodic artifacts due to the terrain model generation methodology were observed in the derived variables (e.g., slopes). Other methods, such as image ratios, were used to reduce the effects of radar incidence angle caused by topography (Ranson et al., 1995; Shi and Dozier, 1997; Wever and Bodechtel, 1998; Ranson et al., 2001). Wever and Bodechtel (1998) proposed the use of L-band hv (Lhv) and X-band vv (Xvv) ratio or difference images for radiometric rectification.

A method was developed to correct for the backscatter dependence on terrain using an algorithm derived from simulated radar backscattering of a forest stand on various slopes. The derived dependence of the L-band hh (Lhh) and Lhv backscattering on radar local incidence angle was used to remove the terrain effect from the Lhv data. Finally, a biomass map was produced from the corrected Lhv data.

16.2.2 Study Area and Data

The study area, in the Western Sayan Mountains, covers a 50×25 km area with center coordinates of 53°4.2′N latitude and 93°14.3′E longitude (Figure 16.1). The area is the site of the Ermakovsky Permanent Study Area established in 1959 and is used for research by the Sukachev Institute of the Siberian Branch of the Russian Academy of Sciences.

The Western Sayan Mountains are a complex of ridges dissected by a widespread drainage network. Topographically, the territory is very hetero-geneous and therefore climatic conditions and ecosystems are very diverse. The elevation varies from 2400 m above sea level to 1300–1500 m in the basins. The slopes are covered by dark needle coniferous forest with a pre-dominance of Siberian pine (*P. sibirica*) and fir (*A. sibirica*). Spruce (*P. obovata*) is an admixture within the drainage network. In a southward direction, with the decrease in precipitation, larch (*L. sibirica*) appears within canopy (mostly on the south-facing slopes). Within the study site, forest types are arranged in elevational belts, including pine–birch forest–steppe (up to 250–300 m), a narrow belt of light coniferous and mixed stands (up to 400–450 m), and dark needle coniferous stands (up to 1600–1700 m) that gradu-ally transform into a subalpine belt of meadows and sparse fir–Siberian pine stands (1600–1800 m). At higher elevations, there are very sparse stands mixed with mountain tundra, bushes, alpine meadows, and stony areas. The mean annual, summer, and winter temperatures are –3°C, +12°C, and –16°C, respectively. Precipitation totaled about 1200 mm/year, which mainly occurs in May–September.

A Russian forest inventory map (1:50,000) compiled from aerial photo-graphs and site visits between 1993 and 1995 was used as ground truth infor-mation (Figure 16.2). The map is typical of forest inventory maps with forest units related to the economic value of the stands. A total of 56 biomass plots ranging from 1.16 to 24.0 kg/m² were prepared from field measurements and forests inventory data. The field plots were sorted according to biomass val-ues. Even-numbered plots were used for developing biomass estimation model and the odd-numbered plots were used for validation.

Shuttle Imaging Radar-C (SIR-C) data were used in this study. The SIR-C/X-SAR missions were flown during April 9–19, 1994 and September 30–October 10, 1994 (Stofan et al., 1995). The instrument had quad-polarized (hh, hv, vv, vh) L-band (wavelength = 23 cm) and C-band (5.6 cm), and vv polar-ized X-band (3 cm) radar channels. The mission was a cooperative experi-ment between NASA's Jet Propulsion Laboratory (JPL), the German Space Agency, and the Italian Space Agency. The SIR-C image data used in this study were acquired on April 16, 1994 with an image center incidence angle of 46.4°. The original image is single-look complex (SLC) data with line spac-ing (azimuth) of 5.8 m and pixel spacing (slant range) of 13.3 m. The images were processed with six looks in azimuth and two looks in range direction resulting in images with a pixel size of ~35 m.

Since previous work (Dobson et al., 1995; Ranson et al., 1995; Ranson and Sun, 1997b) had shown that Lhv data were especially sensitive to forest above-ground biomass, only L-band data were used in this study. Figure 16.3a is the Lhv image of the study area. In the image, the Sayan Mountains can be seen on the left side of the image. Forested mountains appear bright

FIGURE 16.2
(a) Western Sayan Mountains and (b) authors (from left: Guoqing Sun, Kenneth Jon Ranson, and Viatcheslav I. Kharuk) in the field. (Photograph by Viatcheslav I. Kharuk.)

or dark depending on the slope and aspect with respect to the radar illuminating direction (from right of this image). Some deforested areas can be seen in the center of the image. A broad-level plain to the right has large wetland areas such as the dark object in the lower right corner and bare agricultural fields. The village of Ermakovsky and the Sukachev Institute of Forest field camp are located at the upper right in this image.

SIR-C SAR image   Corrected using DEM of coarser resolution

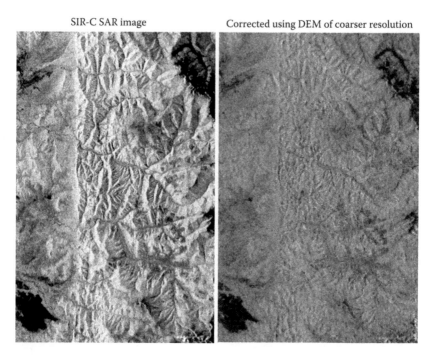

FIGURE 16.3
(a) Original L-band hv radar image. (b) Corrected using DEM of coarser resolution.

16.2.3 Methods

16.2.3.1 Terrain Effects Correction Using a Digital Elevation Model

Slope and aspect were generated from elevations and used to calculate the local incidence angle for a pixel of the radar image:

$$\cos(\vartheta) = \sin(s)\cos(\alpha)\sin(s+\varphi) + \cos(\vartheta_0)\sin(\alpha) \tag{16.1}$$

where θ is the local incidence angle, s is local slope, θ_0 is radar incidence angle at the center of the image, φ is aspect of the slope, and α is azimuth angle of the radar look direction.

Radiometric distortion due to the illumination areas was corrected using the local incidence angle with an equation of the form used by Kellndorfer et al. (1998).

$$\sigma^0_{corr} = \sigma^0 \frac{\sin(\vartheta)}{\sin(\vartheta_0)} \tag{16.2}$$

where σ^0_{corr} is radar backscatter coefficient after correction and σ^0 is original backscatter coefficient.

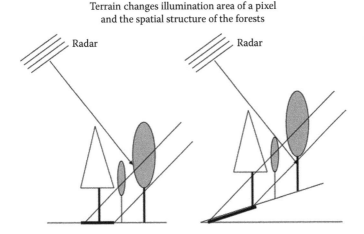

FIGURE 16.4
Terrain changes illumination area of a pixel and the spatial structure of the canopy.

Kellndorfer et al. (1998) found that this correction was adequate for land cover classification purpose. However, for the estimation of biomass from radar backscattering, the effect of terrain on scattering mechanisms needs to be considered. The radar backscattering is a function of incidence angle even in a flat area. With a fixed radar-looking direction, terrain changes the local illumination direction (both zenith and azimuth, see Figure 16.4) of radar beam interactions with the forest canopy. Radar backscattering models may be used to model the dependence of backscattering on local radar incidence angle if detailed knowledge of the forest types and their structure information are known.

16.2.3.2 Modeling Radar Backscatter of Forest Stands on Slopes

The three-dimensional radar backscatter model (Sun and Ranson, 1995) was modified for this study to include the effect of slopes on backscatter. The modified model accepts a stem map (with location, diameter at breast height (dbh), height, species, and crown shape for each tree), an elevation map (height for each surface pixel), and a soil surface roughness and dielectric constant map as inputs to simulate high-resolution polarimetric radar images of the forest stand. The scattering components are also available from the modeling outputs. Detailed measurements of the forest structure were not available for the Western Sayan Mountain study area. A 100×100 m stem map of conifer stand was made during the Boreal Ecosystem-Atmosphere (BOREAS) Study (Sellers et al., 1997) in 1994. The measurements included the stem locations and dbh for every tree in the stand. This stand is a typical boreal conifer forest with above-ground dry biomass of about 10 kg/m². In addition to the stem map, total height, crown length, and crown width were measured. The relationships between these parameters and dbh were developed from the field measurements. These relationships were then used to infer tree crown characteristics for each tree from its dbh. When a slope was introduced, the horizontal position of a tree was not changed, but the tree was moved vertically depending on its position within the stand. Tree crowns were modeled as cones. The backscattering from the ground surface was calculated using the IEM (integral equation model) model (Fung, 1994). The parameters for the simulation are listed in Table 16.1.

High-resolution radar images of the stand on various slopes and azimuth directions were simulated. The radar backscattering coefficients were obtained by averaging the simulated images. Three slopes (10°, 20°, and 30°) and eight azimuth directions (45° increments from 0 to 360°) for each slope were simulated. Local incidence angle was calculated for each case.

Since Lhh backscatter is more sensitive to slope (because of greater canopy penetration and seeing more ground), it was used to estimate the local incidence angle (θ) from the Lhh image. Then, it was used to correct the Lhv image. To do so, a simple analytical relationship between backscattering coefficients and incidence angle needed to be established for both hh and hv polarizations.

The simple backscattering models for vegetation-like media take the form of (Ulaby et al., 1982):

$$\sigma^0 (\theta) = \sigma_0 \cos^p \theta \qquad (16.3)$$

TABLE 16.1

Parameters Used in Radar Backscatter Simulations

	Radius	Length	Probability (%)
Geometry			
Needle	0.3 mm	2.7 cm	100
Branch	0.2 cm	15 cm	90
	0.8 cm	50 cm	8
	1.6 cm	150 cm	2
Orientation			
Needle	Vertical preferred ($p(\theta) = 4 \sin^2\theta / \pi$, $\theta \in (0°, 90°)$ is the zenith angle of the long axis)		

Branch zenith angle	10°		20°	30°	40°		50°	60°	70°	80°	90°
Probability (%)	1.5		2	1.5	3		14	25	22	14	17

Density

Needle: 23000/m³ Branch: 120/m³

Dielectric Constants (real, imaginary)

Needle	(18.03, 6.10)
Branch	(15.38, 5.29)
Trunk	(6.68, 2.07)
Soil surface	(10.0, 2.0) Roughness: standard deviation of surface height $\sigma = 2.5$ cm, surface correlation length $l = 18$ cm

where θ is the local incidence angle. Both σ_0 and p are polarization dependent. When $p = 1$, the model means that the scattering coefficient (scattering per unit surface area) is dependent on cos θ, which is the ratio of projected area (normal to the incoming rays) to the surface area. When $p = 2$, the model is based on the Lambert's law for optics. Ulaby et al. (1982) pointed out that although either $p = 1$ or 2 seldom closely approximate the real scattering, sometimes $p = 1$ or 2, or a value between 1 and 2, may be used to represent scattering from vegetation. The model simulation results were fit to this simple model to estimate σ_0 and p for both L-band hh and hv polarizations.

16.2.3.3 Biomass Estimation

The biomass parameters of 56 stands were defined from forest inventory tables based on age and site index. These tables and methods are in operational use for Russian forest inventory and management. Positions of these stands were located on corrected Lhv radar images and the backscattering signatures were extracted. These stands were sorted according to biomass and selected alternately for either model development or testing. Regression relationships were developed between the cube root of total biomass and the averaged radar signature similar to the method described by Ranson and Sun (1997a). The derived equation was used to convert the Lhv images to biomass maps.

16.2.4 Results

16.2.4.1 Terrain Correction with a DEM

In this study, we first corrected the dependence of illuminated pixel area within the SIR-C image on incidence angle using the DEM available from NIMA. We found that the spatial resolution and accuracy of this DEM were not suitable for terrain effect correction of SIR-C imagery. Figure 16.3b is the Lhv image that was corrected using the local incidence angle derived from the DEM. While the correction for large slopes appears to be appropriate, the smaller slopes have not been corrected due to the lower resolution of the DEM data than that of the SIR-C image. Consequently, the method of using backscatter modeling to account for terrain effects on backscatter was used.

16.2.4.2 Modeling of the Terrain Effect

Figure 16.5 shows simulated radar images of the stand on a 20° slope with four different azimuth directions. The one on the upper left represents a stand on a 20° slope facing the radar. The radar looks from left to right. The image on the upper right represents the same scene, but the surface is facing to the right and away from radar. Trees are still growing vertically, and so the images of the tree crowns did not change, but crowns project longer shadows upon the ground surface. The backscattering from the ground (a

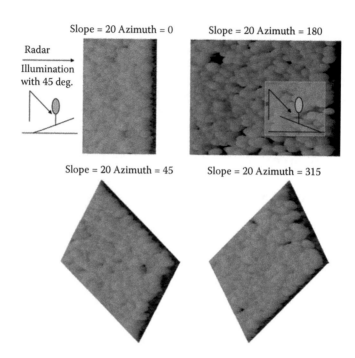

FIGURE 16.5
(**See color insert.**) Simulation of radar images of a forest stand on various slopes.

rough surface) decreases when the slope faces away from the radar (resulting in a larger local incidence angle effect).

The pixel size of these images in Figure 16.5 is 0.5 m × 0.5 m. The radar incidence angle used was 46.4° (illumination from left), the same as the SIR-C image used for the modeling. The observed change of the backscattering coefficient (the brightness of the images) with changing slope was caused by three major factors. The first is the change of the illuminated area per pixel. This is easily seen in Figure 16.5 as the increase in the number of range pixels (resulting in less area illuminated by a pixel) of the scenes. The second major factor is the change in tree shadowing. There are more shadows cast by trees visible in the images for slopes facing away from the radar. The third major factor is change in contribution of surface backscattering because of local incidence angle.

The simulated dependence of radar backscatter on local incidence angle shown in Figure 16.6 was used to correct terrain effects. The best fits of the simulated data yield the following two equations of the form suggested by Ulaby et al. (1982):

$$\sigma^0_{hh}(\theta) = 0.361 * \cos^{1.78}\theta \quad r^2 = 0.93 \tag{16.4}$$

$$\sigma^0_{hv}(\theta) = 0.203 * \cos^{1.50}\theta \quad r^2 = 0.95 \tag{16.5}$$

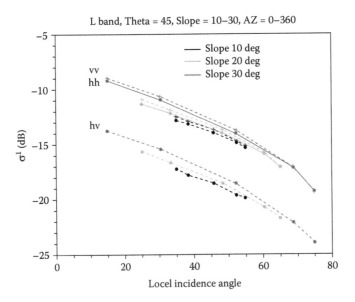

FIGURE 16.6
Relation between backscattering coefficient and local incidence angle for L-band.

Ideally, equations of this form should be developed for different land cover types, which will be part of our future modeling efforts. For this study, we used the pair of equations to make the terrain correction of the L-band hv (Lhv) image. Here, we assume that Equation 16.4 is applicable to the SIR-C Lhh image.

16.2.4.3 Terrain Correction from Modeling

For each pixel, Equation 16.4 was used to estimate cos0 from the Lhh image data:

$$\cos\theta = \left(\frac{\sigma^0_{hh}(\theta)}{0.361} \right)^{1/1.78}$$

(16.6)

If the actual SAR data are different than the simulated data and give a different value for σ^0_i other than the 0.361, the resulting cosθ will be

$$\cos\theta = \left(\frac{\sigma^0_{hh}(\theta)}{\sigma^0_i} \right)^{1/1.78} = \left(\frac{\sigma^0_{hh}(\theta)}{0.361} \right)^{1/1.78} a$$

(16.7)

where $a = (0.361/\sigma^0_i)^{1/1.78}$ and accounts for the difference between σ_0s from the simulation (0.361) and radar image (σ^0_i).

The purpose of the terrain correction is to bring the Lhv backscattering coefficients at incidence angle θ to a reference incidence angle θ_0. Using

Corrected LHV SIR-C images   Biomass map from corrected LHV images

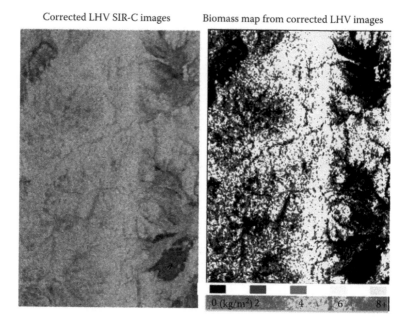

FIGURE 16.7
(a) Corrected L-band hv image. (b) Biomass map from corrected Lhv image.

Equation 16.5 for both θ and θ_0 and taking the ratio of the two result in the following equation:

$$\sigma_{hv}^0(\theta_0) = \sigma_{hv}^0(\theta)(\cos\theta_0/\cos\theta)^{1.50} \qquad (16.8)$$

The reference incidence angle θ_0 can be of any value, but the natural choice will be the SIR-C radar incidence angel at the image center (46.4°). It can be seen that the uncertainty of $\cos\theta$ caused by the factor a (Equation 16.7) only causes a relative scaling to the corrected Lhv image (Equation 16.8).

The corrected Lhv image using this method (Figure 16.7a) shows that the terrain pattern was removed (compared with Figure 16.3a, uncorrected data, and Figure 16.3b corrected with the DEM). The low-biomass areas, such as clear-cuts, top of high mountains, and bare valleys, are still identifiable. A threshold limit for Lhv backscattering was set in the correction, and so the pixels with Lhv backscattering lower than this limit (shadowing, water surface, or other very low backscattering targets) will not be "corrected."

16.2.4.4 Biomass Estimation

The equation developed from corrected Lhv data is

$$B^{1/3} = 8.45 + 0.67\,\sigma°, \quad r^2 = 0.78, \quad N = 28 \qquad (16.9)$$

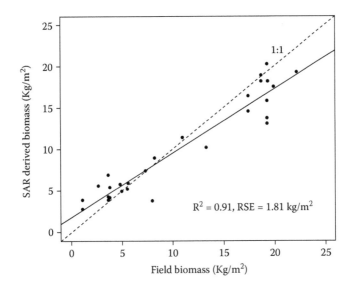

FIGURE 16.8
Comparison of SAR-derived biomass with field biomass: SAR biomass = 1.84 + 0.77 * field biomass, $r^2 = 0.91$, $n = 28$, RSE = 1.81 kg/m^2.

Figure 16.7b is the biomass map developed from corrected data. Biomass differences shown are mostly related to logging, disturbance, or natural vegetation communities, such as wetlands. The biomass map has a continuous level of biomass ranging from 0 to ≥25 kg/m^2. This upper value range was specified since a small number of points in the image exceeded the maximum biomass levels in the training and testing data.

The comparison between field biomass test data and predicted biomass developed from the terrain-corrected Lhv data is shown in Figure 16.8. The accuracy for the independent set of validation points, given as the root mean square error, was acceptable at 1.81 kg/m^2. The r^2 of 0.91 was also very good, but the predicted values did not follow a one-to-one relationship with the field measurements. Statistical tests performed on the regression coefficients showed that the slope (0.77) was significantly different from 1.00 and the intercept (1.84) was significantly different from 0.00. Consequently, the prediction model overestimated areas with low biomass levels and underestimated areas with higher biomass levels.

16.2.5 Conclusions

The effect of terrain on SAR backscatter and subsequent biomass estimation was discussed. We have demonstrated a model-based method for terrain effect correction of SAR images without using a DEM. However, this method requires multiple polarization SAR data. It seems that if general information on forest structure is available, this method could be used in other areas.

The terrain slope changes the local radar incidence angle as well as the forest structure perceived by the radar. The dependence of radar backscattering on the slope and aspect (azimuth of the slope) is very complex. Regardless of the methods to be used for terrain effect correction (using DEM or not), certain assumptions have to be made about the nature of the backscatter. The 3D radar model used in this study provides a tool to simulate complex structure of the forest stand in mountainous areas. If land cover information is available, this method can be applied to reduce the terrain effect for different cover type using different equations. If a very good DEM is available, the 3D model can be used to simulate radar backscattering dependence on terrain, and then it can be used to correct single polarization SAR data.

In this work, we based our correction on the model results from a biomass stand of 10 kg/m^2, which resulted in better estimates of midrange biomass values. Methods to improve the biomass estimates over the full range by simulation of low- and high-biomass cases need to be explored.

16.3 Disturbance Detection Using Radar and Landsat Data

16.3.1 Introduction

Disturbance is an important factor in determining the carbon balance and succession of forests. Fires and resulting scars can be detected using measured changes in temperature during the fire and the vegetation changes immediately after the burn (Kasischke et al., 1993; Martin, 1993). Michalek et al. (2000) also reported on the utility of TM data for assessing stand density and fire severity in Alaska. Woodcock et al. (2001) showed that Landsat 7 could be used over large areas to detect change, especially logging in the western United States. Defoliation of forest stands results in changes in reflectance and can also be used to detect insect damage in forests. Early work by Dottavio and Williams (1983) and Nelson (1983) demonstrated the utility of Landsat data for gypsy moth and spruce budworm damage in U.S. forests. Landsat has also been studied to provide information on other insect outbreaks (Royle and Lathrop, 1997; Radeloff et al., 1999). A number of papers in the Russian literature describe successful use of airborne and satellite systems to monitor insect outbreaks (e.g., Peretyagin et al., 1986; Kharuk, et al., 1989, 2003, 2004a,b, 2009a,b).

A problem with optical systems for northern forest studies was a lack of available data caused by cloud cover and low solar illumination in winter. The launch of the SAR systems, European Resource Satellite (ERS) 1 and 2, Japanese Earth Resources Satellite, and Canada's Radarsat, provided the availability of data in all weather conditions. Kasischke et al. (1992) found that ERS data could be used to detect fire scars in the boreal forest because the fire scars were 3–6 decibels (dB) brighter than the rest of the landscape. This brightness is a result of physical changes that occur due to fire, including increased surface roughness, removal of tree canopies, and alteration of soil moisture patterns (Bourgeau-Chavez et al., 1993). While optical and thermal sensors are sensitive to the initial changes in temperature and vegetative cover, SAR is sensitive to the longer-term roughness and moisture patterns that occur postfire.

Landsat has provided long-term, high-resolution optical data (30 m). The high-resolution data available from orbiting SARs also provide a closer look at disturbance patterns. While large-area frequent coverage may not be practical, the detailed reflectance and backscatter can provide information useful for identifying type and extent of disturbances on a local to a regional scale. This information can then be used with the coarser resolution systems to identify disturbance over large remote areas such as Siberia. This section describes work toward understanding the use of remote sensing to detect important disturbance factors (fire scars and insect damage) in Siberia and explore the use of combined data from Landsat and SAR systems.

16.3.2 Study Sites

The study area is located in Central Siberia within 88–92° East longitude and 50–70° North latitude (Figure 16.1). Within this larger site are Landsat image-sized (~180 × 180 km) intensive study sites identified by their predominant disturbance. The Boguchany wildfire test site was selected because of the presence of large fire scars and logged areas. The site is located at 97°25′E and 59°2′N, 75 km north of the Angara River and 350 km east of the Yenisey River in Eastern Siberia. The Priangar'e insect site is located to the west of the Boguchany site at 94°30′E and 57°30′N, and was plagued by a severe insect outbreak between 1992 and 1995.

The Boguchany test area, named after the nearby town, is located within an important region for timber logging in Siberia (Kharuk and Ranson, 2001). The elevation of the study site ranges from 300 to 500 m. The growing season in the region is short, ranging from late May to early September. In the summer, smoke plumes from burning wildfires obscure the sky; fire is the principal factor that determines ecosystem dynamics in this region and therefore most of the stands are of pyrogenic origin (Kharuk and Ranson, 2000). Scotch pine (*P. sylvestris*) and larch (*Larix sibirica*) cover most of this landscape. However, other conifers, such as Siberian pine (*Pinus sibirica*), spruce (*P. obovata*), and fir (*Abies siberica*), can also be found in patches in the area. Deciduous stands such as birch (*B. pendula*) and aspen (*P. tremula*) cover the areas of lower elevation in this region. Several methods of logging are practiced in the area, including the Finland technique (logging with seedlings preserved) and complete clearing where no vegetation is left on the site. These sites are covered with live grasses in the summer and covered with dry, dead grasses in the fall.

The fires that caused the burn scars in this study were ignited by lightning and extinguished by rainfall. This study will focus on the two largest fire scars in the area (see Figure 16.9). Fire scar 1 is the product of two fires that were detected on July 16 and 19, 1996 and merged into one fire on July 21, 1996. One of the two fires is known to have started on a 1979 clearcut in an area of regenerating pine, birch, and aspen when a large volume of dead wood ignited. The fire was a strong surface and crown fire and by the time it was extinguished on August 8, 1996, 32 thousand ha of forest, old clear–cuts, and dense regenerating stands were burned. The second fire contributing to fire scar 1 started in an approximately 100-year-old pine–larch stand that also included some regenerating pine and larch trees. Fire scar 2 burned in an undisturbed coniferous forest 60 km northwest from fire scar 1 also in 1996. The fire scars were located using satellite imagery and verified by field surveys in the fall of 1999 conducted by scientists

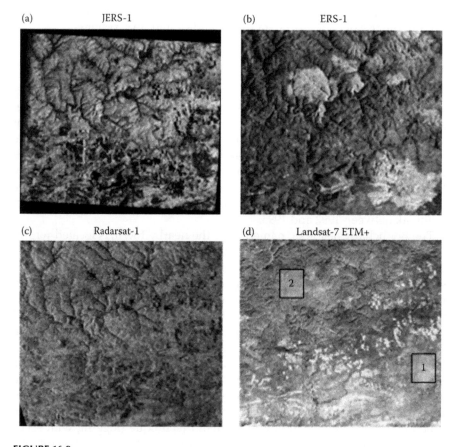

FIGURE 16.9
(a) The JERS (Lhh), (b) ERS (Cvv), (c) Radarsat (Chh), and (d) Landsat 7 images over the Boguchany site.

from the Sukachev Institute of Forest. Ground location was determined and survey plot measurements and digital on-ground photos were taken. IKONOS Carterra imagery was also available from the summer of 2001 for field checking.

The insect damage study site, as shown in Figure 16.10, is within the Niznee Priangar'e region where a severe Siberian silkmoth (*Dendrolimus sibiricus*) outbreak occurred between 1993 and 1995 (Kharuk et al., 2003). The topography of the area consists of a plateau with low hills. Soils are mainly spodosols (podzols). Climate is continental with cold dry winters and warm moist summers. Annual precipitation is 400–450 mm. Mean annual temperature is +2.6°C with an absolute minimum of –54°C recorded during December and a maximum of +36°C recorded in July. Vegetative growth period is about 100 days. Forests cover 95% of the area. The dominant species is fir (*A. sibirica*); other species include Siberian pine (*P. sibirica*), also

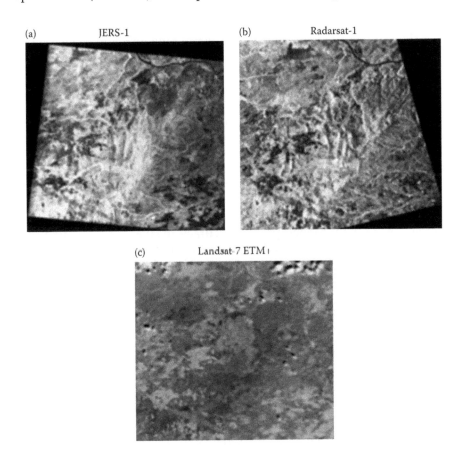

FIGURE 16.10
(a) The JERS (Lhh), (b) Radarsat (Chh), and (c) Landsat 7 images over the Priangare site. ERS data were not available for this site at the time of the analyses.

known locally as Siberian cedar, Siberian spruce (*P. obovata*), Scotch pine (*P. sylvestris*), larch (*L. sibirica*), aspen (*P. tremula*), and birch (*B. pendula*). Stands are of average productivity with a wood stocking density of 200–230 m³/ha and a mean age of 135 years. Typical insect damage is character- ized by complete defoliation and death of conifer stands, or death of only conifer trees within mixed stands.

16.3.3 Data and Preprocessing

Available JERS, ERS-1, Radarsat, and Landsat 7 satellite data were analyzed to determine to what extent these sensors could detect the presence of fire scars, clear-cuts, and insect damage. Table 16.1 summarizes important parameters of the sensors used.

The JERS data were resampled to 25 m pixel size, reoriented, and filtered using a 3 × 3 Frost filter (Frost et al., 1982). The ERS-1 data were received from the Alaska SAR Facility (ASF). These data were then multilooked to 25 m pixel size, reoriented, converted to ground range, wrapped into a longitude/ latitude grid using corner coordinates, and filtered using a 3 × 3 Frost filter. The Radarsat standard beam data received from ASF were previously con- verted to ground range. The data were then ingested, resampled to 25 m pixel size, wrapped into a longitude/latitude grid using corner coordinates, and filtered using a 3 × 3 Frost filter. The Landsat 7 scenes were ordered and received from the EOSDIS EROS Data Center Distributed Active Archive Center (DAAC). There was no radiometric terrain correction applied to the images because neither area had a steep topographic gradient (the elevation difference was less than 250 m). No additional atmospheric corrections were applied to the Landsat 7 data.

To attain greater geometric accuracy and to ensure that the data sets were coregistered with the highest possible accuracy, the JERS, ERS, and Radarsat data were registered to the Landsat 7 scene. Landsat 7 data were selected as geometric ground information for this site because these data have good geometric calibration. Ideally, orthorectification of radar images using a DEM should be performed before registration, but there was no high- resolution DEM available. Instead, we used a large number of control points to register the images. Because of the low terrain relief, the results seem satisfactory.

The SAR images were manually registered to the Landsat 7 scene. To accomplish this, points at the intersection of linear features were selected such as on roads, rivers, and clear-cuts when applicable. In Boguchany, 103 points were used to register the ERS data, 74 to register JERS, and 65 to reg- ister the Radarsat data. In Priangar'e, 70 control points were used to register the JERS data and 90 to register the Radarsat data. The same procedure was used for both sites. After registration, images were subset to the area covered by each of the four sensors. Figures 16.9 and 16.10 show the SAR and Landsat 7 images used for the analysis.

16.3.4 Methods

16.3.4.1 Vegetation Classes

The following land cover classes were identified for the two sites: *coniferous forest (CF), broadleaf deciduous forest (DF), regeneration/sparse forest (RS), bare surfaces (BS), and clear-cut (CC).* For the Boguchany site, the following disturbance classes were added: *burned coniferous forest (BC), burned deciduous forest (BD), and burned logged areas (BL).* Additionally, two classes of insect damage were identified in the Priangar'e area: *severely damaged (SD)* with complete defoliation of a stand and *moderately damaged (MD)* with only conifer trees defoliated. Since the insect outbreak had occurred in 1996 and subsequently subsided, the two classes represent severity of damage rather than stage of insect attack. Table 16.2 provides a list of classes and descriptions for the two study sites.

TABLE 16.2

Radarsat and Landsat Data Used for Boguchany and Priangar'e Sites

	Boguchany Site			Priangar'e Site	
Sensor	JERS	ERS-1	Radarsat ST4	JERS	Radarsat ST4
Frequency (GHz)	L-band (1.275)	C-band (5.3)	C-band (5.3)	L-band (1.275)	C-band (5.3)
Wavelength (cm)	23.5	5.66	5.66	23.5	5.66
Polarization	hh	vv	hh	hh	hh
Incidence angle (°)	38.9	23	34	38.9	34
Image center	58.01°N, 97.43°E	59.49°N, 97.55°N	59.10°N, 97.33°E	57.27°N, 94.16°E	58.01°N, 93.86°E
Orbital direction	Descending	Descending	Ascending	Descending	Ascending
Image swath (km)	75	100	100	75	100
Altitude (km)	580	785	798	580	798
Data take date	March 31, 1997	June 7, 1998	August 21, 1999	May 19, 1997	August 18, 2000
Pixel size (m)	12.5	12.5	12.5	12.5	12.5
Sensor	Landsat 7			Landsat 7	
Data take date	October 3, 1999			July 22, 2000	
Image center	58.71°N, 96.81°E			57.31°N, 94.36°E	
Path and row	P141 R19			P140 R20	
Resolution (m)	30			30	
Sensor	ETM+			ETM+	
Cloud cover (%)	0			9%	
Bands	7 + pan			7 + pan	

16.3.4.2 Training Site Selection

Field campaigns were conducted in the Boguchany area in the fall of 1999 and Priangar'e area in the summer of 2000. During this field campaigns, tree species were identified. GPS measurements were acquired and in Boguchany, plot measurements pertaining to the successional stages of the burned and logged areas were obtained. Figure 16.11a shows the burned

FIGURE 16.11
(a) Regenerations on a burn site at Boguchany. (b) Aftermath of logging in Boguchany area. (c) Trees (fir and pine) killed by a Siberian silkworm (*Dendrolimus sibirica*) outbreak. (d) Insect-damaged forest at Priangar'e seen from air. (Photograph by Viatcheslav I. Kharuk.)

FIGURE 16.11 Continued.

area in Boguchany. A field visit to Priangar'e included aerial overflights to obtain photography of damaged areas (Figure 16.11b). Information gathered during these field campaigns along with the existing local ecological knowledge of the staff at the Sukachev Institute of Forest provided a good basis for determining and locating the different vegetation classes on the Landsat and radar images.

The training sites for the classes mentioned above were determined based on the information gathered in the field, the multiyear and multiseason coverage provided by other Landsat scenes, and the contextual information provided by the individual Landsat scenes. Once the training sites were determined, histograms were examined for each class in each radar band. If the data were normally distributed, the class was left intact. If, however, the histogram showed a multimodal distribution, these training sites were displayed using the radar bands and training sites assigned to a more or less homogeneous subclass. Then the histograms for these subclasses were once again reviewed to make sure that the distribution of the values was normal. This way, the deciduous forest and bare ground classes were split into three subclasses on the Priangar'e site, and the burned logged class was split into two subclasses on the Boguchany site. Approximately one-third of the training sites were set aside for testing the classification and two-thirds were used for training the classifier.

The clear-cuts in both Boguchany and Priangar'e sites appear as rectangles with straight edges cut out of the forest cover in a checkerboard fashion revealing their man-made nature. The older clear-cuts are clearly overgrown with deciduous trees, whereas the most recent ones have exposed bare soil. Because of the time difference between JERS and Landsat 7 data, only those logged sites were included in the clear-cut class that were at least 3 years old and had grasses and seedlings growing on them. Based on *a priori* work, it was determined that the older, now tree-covered clear-cuts could not be separated from the natural deciduous forest cover. The fire scars in the Boguchany area are spatially quite distinct from the clear-cuts. The fire scars have lobe-like edges that are at times discrete and at other times more transitional.

It is worth noting that if an unburned area is spectrally, structurally, and texturally heterogeneous, it is likely that the fire scar visible in the landscape after burning will also be spectrally, structurally, and texturally heterogeneous. This is to say that fire scars are not monolithic features at a 30 m resolution. The patterns observable within a fire scar provide valuable information of the history of the site. When anthropogenic disturbance (such a logging) has occurred in the area prior to the burn, the burned area will be a patchwork of spectral, structural, and textural features shaped by a combination of anthropogenic and natural disturbance factors. This texture information was used to identify these sites but was not explicitly included in the classification.

On the Priangar'e site, the logged areas do not appear in juxtaposition with the insect damage. In this case, the anthropogenic (logging) and natural disturbances (insect infestation) are spatially separate. Insect damage appears on the landscape as patchy "thinned-out" forested areas since insect only damaged the needles of the coniferous trees and left the leaves of other trees intact. The degree of the damage they caused partly depends on the species composition of the stands: if a stand was composed of coniferous species (food species), the stand was severely damaged. If the stand consisted of a mix of food and nonfood species, the damage was more moderate. It is important to mention that severely and moderately damaged classes are not thematically distinct. Instead, they are two, somewhat arbitrarily defined overlapping areas on a thematic continuum between completely healthy and completely damaged forest stands.

Training sites for each class were chosen, keeping in mind that the radar data available were acquired over a period of 3 years. The changes that have occurred within the landscape during this period had to be eliminated or at least minimized within the training sets. For example, some of the Boguchany training sites were eliminated from the training set because on 1991 Landsat 5 images they appeared as coniferous forest, and by the time the 1999 Landsat 7 scene was taken, the site became a clear-cut. Since there was no additional information available on this particular site, it could not be determined at what point between the two dates the site was logged and whether or not the date of its logging fell within the 3-year period the radar data was acquired.

16.3.4.3 Data Analysis

The purpose of this analysis was to determine whether or not and how each sensor was detecting each land cover class and whether or not the radar sensors were capable of separating the classes from one another based on backscatter information alone. Once the training sites were carefully selected and split into subclasses as described above, backscatter values were extracted from each class for each radar sensor, and descriptive statistics were generated. The analysis procedure consisted of two parts: (1) Transformed Divergence (Richards and Jia, 1999) analysis and (2) maximum likelihood classification. Transformed Divergence Measure (TDM) is a measure of separability between classes and may therefore be used to assess the quality of the class spectral mean vectors and covariance matrices. A high TDM (>1.80) indicates good statistical separation of the classes and indicates how well each sensor or sensor combination detected each land cover class. Maximum likelihood classification provides the means to examine the separability of classes in a mapping or thematic sense. After classification, the subclasses were merged into their original parent class.

16.3.5 Results and Discussion

16.3.5.1 Radar Data Analysis

16.3.5.1.1 Burned Site

Figure 16.12a presents the average backscatter and standard deviations for each radar sensor for the seven classes from the Boguchany fire scar study area. For JERS data, the coniferous and deciduous forest classes, as well as the burned deciduous and coniferous forest classes, have very similar back-scattering coefficients. This is probably because at L-band (0.23 m wavelength), larger tree branches and trunks are the primary scatterers. After surface and crown fires, many of the tree trunks still remained standing as seen on the images of the burned forest sites. This might explain why the returns are so bright for both unburned and burned forest types in the L-band. It is also clear that the regeneration/sparse, clear-cut, and burned logged areas classes all have lower backscattering coefficients, which is likely due to the absence of large branches and trunks. Classes with little or no tree cover (RS, CC, and BL) also have similar backscatter and, as a group, have lower backscatter than the classes with standing trees.

In Figure 16.12a, it can also be seen that the unburned classes (CF, DF, RS, and CC) all have lower ERS-1 brightness values than the burned classes (BC, BD, and RS). The postfire regeneration class seems to have intermediate values. C-band radar is scattered by structures about 5 cm in size such as leaves and small twigs on trees or grasses. Field observations revealed that small structures such as leaves and twigs were no longer present on burned trees; however, grasses having leaves of similar sizes are abundant on the

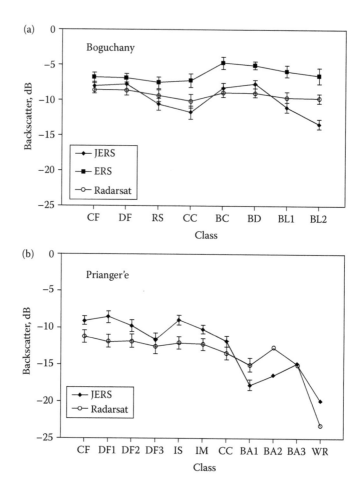

FIGURE 16.12
Mean and standard deviation backscatter coefficient for land cover classes at (a) Boguchany burn scar site and (b) Priangar'e insect damage site.

fire scar during the summer months. Based on this, the burned and unburned vegetation should be difficult to distinguish. There must be some other factor such as soil moisture (Bourgeau-Chavez et al., 1993) influencing the CVV backscatter that causes the burned areas to be brighter than the unburned ones.

The plotted Radarsat backscatter shows very little difference between any of these classes (Figure 16.12a). Only the clear-cut class has backscatter values that are a bit lower than the others. These areas also appear dark on the radar image (Figure 16.12b). There is no obvious explanation as to why burned and unburned classes are so clearly separable using CVV ERS data and why the CHH Radarsat backscatters for these same classes are so similar.

Only 1 year passed between the acquisition of the two data sets; therefore, land cover change is unlikely be the answer. There is an 11° difference in incidence angle between the two sensors (ERS = 23°, Radarsat = 34°), but it is not well understood exactly how incidence angle influences radar backscatter from burned areas. Soil moisture could have changed over the 1-year period and it is also possible that at a larger incidence angle, the differences in soil moisture between burned and unburned areas are less pronounced.

The separability of classes using the radar data was quantitatively examined with the use of TDM. For JERS data, high TDM values exist between logged classes (CC and BL) and unburned and burned conifer (CF and BC, respectively) and unburned and burned deciduous stands (DF and BD, respectively). In this case, unburned forest stands are not separable from burned forest stands. High TDMs also exist between RS and BD classes. This indicates that forested classes and classes lacking tree cover are easily separable from each other using JERS data regardless of their burned state. ERS and Radarsat TDM values were generally lower than those for JERS. The exceptions were for ERS data that had much higher separability values for burned forest (BC and BD) and unburned forest (CF and DF). From these results it is clear that any single radar sensor used alone cannot be used to discriminate between burned and unburned forest classes, between deciduous and coniferous forest classes, and between unburned and burned nonforested classes. However, JERS data can be used to discriminate between forest and nonforest classes regardless of burning and between postlogging regeneration and forest classes also regardless of burning.

ERS data appear most useful for discriminating between burned forest areas and unburned forest, and regeneration and clearings. Other class pairs with relatively high TDMs include RS and BC (1.49) and BD (1.73). This indicates that postcutting regeneration is easily separable from the burned forest classes. However, the separability between the RS and the unburned forest classes (CF and DF) is very poor (≤ 0.20). Low TDMs were found between CF and DF classes indicating that CVV data cannot be used to distinguish between coniferous and deciduous forest classes. TDM values were also minimal between CF and CC.

From these data it is clear that the CVV band alone cannot be used to discriminate between coniferous and deciduous stands, between clear-cuts and forest classes, and between clear-cuts and postfire regeneration classes. However, ERS data can be used to discriminate between the burned and unburned land cover classes, regardless of other characteristics of the site, and between postcutting regeneration classes and burned forest classes. JERS data at the L-band seem to detect larger structural differences between forest types that are caused by logging (i.e., removal of large trunks). At the same time, ERS C-band data seem to detect soil moisture differences and perhaps structural and moisture differences at a leaf level associated with burning. This indicates that the combination of the two sensors should provide improved results in discriminating logged and burned areas. The maximum separability from the Radarsat data is 0.72 and occurs between the CF and the CC classes. This value is quite low and indicates that the Radarsat data alone are not suitable for distinguishing any of these classes from each other.

For the TDM values generated based on the three sensor data combined, the average separability increased to 1.55. Although this is an increase from using each sensor alone (JERS average separability: 1.23, ERS: 0.64, and Radarsat: 0.16), on the whole, combining the three sensors does not provide very good distinction between these eight classes since TDM values under 1.8 are considered poor. Combining the radars provided the greatest increases in useful separability ($\geq$1.80) over individual radars between burned classes (BC, BD, BL) and regenerating forest (RS). Overall, forest (CF, DF) could be separated from disturbance classes (RS, CC, and BL), but not from burned forest (BC, BD). Burned forest could be separated from regeneration and clear-cut. The common theme among class pairs is that classes can be separated successfully that have different structural characteristics determined by the presence or absence of large trunks and branches, such as forest and nonforest classes. This is mostly due to the Lhh band JERS data, since these class pairs had reasonably high TDM values (around 1.7) using JERS data alone. ERS contributes the most in separating burned forest from other classes; however, TDMs never reached 1.80 for any class pair.

Table 16.3 lists the maximum likelihood classification results of the combined radar data for the Boguchany site. Only the burned logged class (BL) was identified with accuracy greater than 80%. Forest classes were confused with each other as were burned forest classes. Regeneration and clear-cut classes were mostly confused with each other. These classification results indicate that using this combination of radars might provide useful classification of forest classes (CF + DF), logging (CC + RS), burned forest (BD + BD), and burned logged areas (BL). The overall classification accuracy for all classes was about 66%.

16.3.5.1.2 Insect Damage Site

Figure 16.12b presents the average backscatter and standard deviations for the insect-damaged site. Neither JERS nor Radarsat backscatter differs much across the forested sites (CF, DF, IS, IM). JERS backscatter decreases slightly for clear-cuts and drops off for the bare surface class and water. Radarsat does not show this decrease in backscatter except for the water class. Apparently CC and BS surfaces are sufficiently rough to the C-band radar beam to maintain backscatter levels similar to forested sites.

Two trends in the radar separability values for the Priangar'e site are obvious that both JERS and Radarsat can distinguish water from the land cover classes very successfully, including the bare surface subclasses. JERS and, for the most part, Radarsat are also successful at distinguishing bare surfaces from the vegetated classes (1.92–2.00). Radarsat has low TDM values between bare surfaces and clear-cuts and fails to separate the BA-2 class from all the

TABLE 16.3

Vegetation Class and Training Set Information

Class	Training Pixel Number	Testing Pixel Number	Class Name	Description
Boguchany Site				
CF	4184	1723	Coniferous forest	Predominantly needle leaf species, including larch
DF	4544	1361	Deciduous forest	Predominantly broadleaf leaf species
RS	3593	1797	Regeneration/ sparse	Site logged over 10 years ago, mixture of pine and deciduous seedlings
CC	3371	1210	Clear-cuts	Recently logged stands with low vegetation cover of grasses and forbs
BC	3679	1810	Burned coniferous	Burned needle leaf species, including larch
BD	3754	1675	Burned deciduous	Burned broadleaf leaf species
BL	7459	1889	Burned logged	Burned logged stands
Priangar'e site				
CF	7934	3836	Coniferous forest	Predominantly needle leaf species, including larch
DF	7119	2663	Deciduous forest	Predominantly broadleaf leaf species
IS	6774	4082	Severe insect damage	Defoliated stands, few live trees
IM	3373	1809	Moderate insect damage	Stand with defoliated and undamaged trees
CC	6191	3864	Clear-cut	Recently logged stands with low vegetation cover of grasses and forbs
BS	3384	1154	Bare surface	Nonvegetated areas may include roads, bare soil, fresh clear-cuts, rock outcropping, bogs
WR	975	467	Water	Taseyeva River, tributary of the Angara River

vegetated classes. For JERS, TDM values are very low between coniferous forest (CF) and insect damage classes (IS, IM) and the deciduous forest subclasses and the moderate insect damage class. This might be because the insects only damage the leaves of the trees and the L-band radar does not detect leaves, only major branches and trunks. Radarsat values are low for

these classes, but higher than JERS for separating conifer forest from disturbance classes (IS, IM, and CC).

Radarsat separability of the forests classes was poor ($\leq$1.31), as was separability of damaged forest classes from each other and with undamaged conifer forest ($\leq$0.97). The TDM values between deciduous subclasses and both
damaged classes were also extremely low ($\leq$0.28). However, the separability
between coniferous forest and clear-cuts was higher (1.77). This may be
because there is volume scattering occurring within the tree canopies while
volume scattering back to the radar is absent from the grassy clear-cuts.

With the combined use of the two radars the distinction between clear-
cuts and coniferous forest (TDM = 1.96) and clear-cuts and severe insect
damage (1.86) increased. There was no large increase in the separabilities
between the other classes. In addition, there was good separability of conifer
forest and the deciduous subclass (DF3). Low TDMs were found for CF and
the other two deciduous subclasses suggesting a possible mixture of conifer
and deciduous trees or forest density differences among these deciduous
classes. Overall, the combination of JERS and Radarsat may be useful for
separating clear-cuts from other forest types, but is not useful for separating
insect-damaged stands from undisturbed forest.

The results of classification of the training sites using the JERS and Radarsat
backscatter show 61% correct classification of conifer forest and 77% correct
classification of the deciduous forest (combined subclasses). Reasonable classification results (>80%) were obtained for clear-cuts and bare areas and
water (Table 16.4). Only 29% of the severely insect damaged and 46% of the
moderately damaged classes were classified correctly. Misclassifications
were primarily with deciduous forest (51% and 41%, respectively), indicating
that the combination of JERS and Radarsat is not useful for recognizing this
disturbance.

TABLE 16.4

Classification Confusion Table for Boguchany Area Classes and Combined JERS,
ERS, and Radarsat Data

	Percent Classified As						
Name	CF	DF	RS	CC	BC	BD	BL
CF	62.21	26.94	1.65	0.00	1.60	7.60	0.00
DF	46.96	45.69	2.29	0.00	0.11	4.78	0.18
RS	3.79	1.36	72.25	10.97	0.39	0.58	10.66
CC	0.09	0.18	19.25	52.65	0.00	0.00	27.82
BC	7.23	0.68	0.52	0.00	52.35	36.99	2.23
BD	3.76	2.53	0.03	0.00	17.05	76.27	0.37
BL	0.20	0.20	4.95	9.40	1.03	0.00	84.22

Note: Average accuracy = 63.7%, overall accuracy = 65.8%.

16.3.5.2 Landsat and Combined SAR

16.3.5.2.1 Burned Site

Mean spectral reflectance digital numbers (DN) from burned area training sites are shown in Figure 16.13a. Only Bands 3 (0.63–0.69 μm), 4 (0.76–0.90 μm), and 5 (1.55–1.75 μm) are shown for illustration. Because of the postsenescence timing of the acquisition, deciduous trees are bare and ground vegetation is dead, reducing near-infrared (NIR) reflectance. Conifer forest has higher NIR response than burned conifer forest. Deciduous forest and burned deciduous forest exhibit a similar trend but with higher responses. Clear-cuts have unique spectral characteristics in this fall image with overall higher responses, especially in the SWIR (band 5).

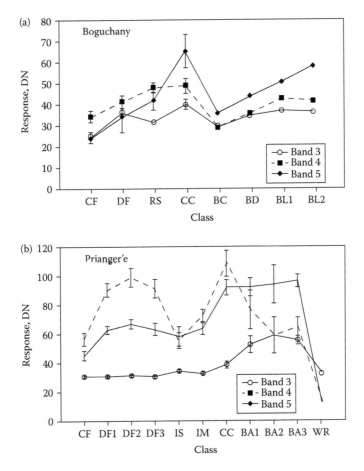

FIGURE 16.13
Means and standard deviations of Landsat 7 spectral digital numbers (DN) for land cover classes at (a) Boguchany burn scar site and (b) Priangar'e insect damage site.

The TDMs from Landsat ETM+ data are greater than 1.80 for all classes except between burned forest classes (BC and BD) and between burned logged (BL-1 and BL-2) and clear-cut (CC) classes. Regeneration (RS) and BL-1 TDM was slightly less than 1.80. Even though this Landsat 7 image was acquired in 1999, 3 years after the burn, many dead, burned trees still remained standing on the burned forested sites casting their shadows on the regenerating vegetation forest floor. This is why there is good distinction between the live and burned forest classes. One exception to this good separation between live and burned vegetation classes are the clear-cut (CC) and the burned logged (BL) classes (TDM ≤ 1.45). The burned logged sites were logged prior to the burn in 1996. When the burn occurred, there were no trees standing on these sites, only grasses and seedlings. Since there were no mature trees on the site, there were no burned trunks left standing either; therefore, no trunks could cast their shadows on the regenerating grasses and seedlings after the burn and lower the site's reflectance in the NIR. This is why 3 years after the fire the burned logged site seems spectrally similar to a clear-cut class and the regenerating/sparse class.

Using the Landsat spectral statistics to classify the Boguchany burned area produced generally good accuracy. As shown in Table 16.5, conifer and deciduous forest classes, and regeneration and the two burned forest classes had classification accuracies greater than 89%. Clear-cut and burned logged areas were confused with each other, resulting in lower classification accuracies of 83% and 84%, respectively. Overall accuracy was 90% and Kappa coefficient was 0.88, indicating that Landsat reflective bands should perform well in discriminating the burned area classes.

Combining the three radars and Landsat data increased the TDM values for those classes that the optical and microwave sensors alone could not distinguish well. The largest increase occurred in the case of the burned logged and burned deciduous class where TDM increased from 1.35 to 1.97

TABLE 16.5

Classification Confusion Table for Priangar'e Area Classes and JERS and Radarsat Combined Data

	Percent Classified As							
Name	**NULL**	**CF**	**DF**	**IS**	**IM**	**CC**	**BA**	**WR**
CF	0.00	61.44	36.50	0.52	1.54	0.00	0.00	0.00
DF	0.00	4.77	77.42	4.09	8.51	5.21	0.00	0.00
IS	0.00	9.71	51.34	29.57	9.24	0.13	0.00	0.00
IM	0.00	0.80	40.76	10.91	46.13	1.39	0.00	0.00
CC	0.00	0.08	14.00	0.19	4.28	81.39	0.05	0.00
BA	0.09	0.00	0.00	0.00	0.00	0.89	99.03	0.00
WR	0.00	0.00	0.00	0.00	0.00	0.00	0.00	100.00

Note: Average accuracy = 62.48%, overall accuracy = 70.71%.

when the spectral and structural information was combined. However, there was only a minor increase in the TDM values between the clear-cut and burned logged subclasses since in this case both classes were both spectrally (regenerating grasses and seedlings) and structurally (lack of trunks) similar.

In summary, L-band radar data provided structural information of the vegetation such as the presence or absence of large trunks and C-band radar data seem to provide information on soil moisture conditions while Landsat data provide spectral information on the vegetation cover such as whether or not the vegetation is reflective in the NIR. This synergistic interaction between the optical and microwave sensor is key to distinguishing disturbed sites from nondisturbed ones since they might look extremely similar using one or the other type of data alone.

Classification with combined data sets of Landsat, JERS, ERS, and Radarsat resulted in classification accuracies above 90% for all classes (tables not shown for brevity). The overall classification accuracy was nearly 94% with a Kappa coefficient of 0.93. The classes with the most improvement were the burned logged class (BL) from 84% to 93% and the CC class from 83% to 90%. The reduction of confusion between these two classes resulted in the higher classification accuracies. The added information on surface roughness condition available with the radar likely contributed here.

16.3.5.2.2 Insect Damage Site

Figure 16.13b shows that NIR reflectance is high for broadleaf trees and ground vegetation for the midsummer acquisition of the insect-damaged area. The shorter wavelength reflectance (bands 1, 2, and 3, *vis* 3) did not vary much across the classes with vegetation in them. NIR and SWIR reflectance (bands 4, 5, and 7, *vis* 4 and 5), however, are quite different and appear suited for discriminating forest classes from disturbed classes. Notice the overlapping spectral responses for the three deciduous subclasses. This is also apparent for the bare surface subclasses except for TM band 4, suggesting a sparse vegetation cover on BA1 (more than on BA1 and BA2, but less than the clear-cut (CC)).

The widely varying spectral reflectance shown indicates that Landsat data can be used very successfully to distinguish among water, bare ground, clear-cuts, and all the vegetation classes. Very high TDM values were observed between all class combinations from Landsat data, indicating good separation of the forest classes and disturbances. Even TDM values between severe and moderate insect-damaged classes were high (1.83). The only low TDM values were found among deciduous subclasses or among bare surface subclasses.

TDM results obtained after combining JERS, Radarsat, and Landsat 7 for the insect disturbance area show that there was only modest improvement in TDM when adding the radar data with the Landsat over the Landsat alone for most of the classes. However, combining the radar data with the Landsat

TABLE 16.6

Classification Confusion Table for Boguchany Area Classes and Landsat 7 Data

	Percent Classified As						
Name	CF	DF	RS	CC	BC	BD	BL
CF	98.35	1.60	0.05	0.00	0.00	0.00	0.00
DF	1.01	94.89	2.68	0.00	0.04	0.07	1.30
RS	0.22	2.59	92.32	3.73	0.00	0.00	1.14
CC	0.00	0.12	2.17	83.06	0.00	0.00	14.65
BC	0.03	0.49	0.16	0.00	95.90	2.58	0.85
BD	0.00	0.32	0.43	0.00	3.30	89.00	6.95
BL	0.00	0.44	0.87	8.48	1.56	4.45	84.21

Note: Average accuracy = 91.10%, overall accuracy = 90.55%.

data increased the separability between the severe and the moderately severe insect-damaged classes from 1.83 to 1.93. Of interest is the increase in separability between DF-1 and DF-3 deciduous forest subclasses. Recall that the subclasses were selected from training sets originally selected from Landsat data, but yielded multimodal histograms with radar backscatter. DF-3 then is spectrally similar to DF-2 in the Landsat bands but apparently structurally dissimilar as inferred from the JERS backscatter. Based on this, DF-3 is likely a deciduous conifer or larch (*L. sibirica*).

The classification results with Landsat 7 data were excellent as shown in Table 16.6. Every class had at least 95% classification accuracy. Overall accuracy was 98.6% with a Kappa coefficient of 0.98. Adding the additional radar channels (results table not shown) offered only slight improvement in class accuracy with an overall 99% accuracy and a Kappa coefficient of 0.99.

16.3.6 Conclusions

This study was designed to examine the utility of using different radar systems and Landsat 7 for identifying forest landscape classes, especially those related to disturbance. We found that the results were limited when using each single-channel radar alone; however, JERS and ERS were found to be useful for identifying certain classes. JERS was most useful for separating forest from disturbed classes with no standing trees. ERS was more useful for separating forest classes from disturbed classes where trees are left standing. Radarsat, on the other hand, was the least effective individual radar for this study. Combining the radars improved the identification of classes over results obtained with any single radar. Generally, if one radar sensor was found to have high separability for a pair of classes, adding additional radars did not greatly increase the separability. If all radars had low separability, combining the radars had very little benefit. In both sites, the low separabilities found between CF and DF and burned forest and insect damaged forest classes indicate that it is not possible to separate classes that have both large trunks and leaves present on them even by using combined radar sensor data.

Regarding the detection of disturbance, the available data were acquired over a 2-year period; therefore, careful comparison of radars for burn scar detection was not possible. Changes in surface soil moisture can greatly change the backscatter from burn scars as shown and verified by other researchers (e.g., Kasischke et al., 2011). Landsat 7 data proved the most useful of any single remote sensing system for recognizing forest type and discriminating between disturbance types. Even with nongrowing season images, as was the case for the fire-damaged site, the results were very promising. Combining the Landsat data with the available radar data improved the separability of classes and the overall classifications. The results also indicate that the combination of radar and Landsat 7 may be especially useful for recognizing other forest types by utilizing the structural information of radar and spectral information of Landsat 7. As radar and Landsat 7 data become more widely available, combining these data sets should improve the accuracy of forest mapping activities. However, there is extra effort and cost involved in registering different image types.

This work underscores the importance of using multichannel SAR data for forest studies. When combined with optical data, the SAR appears to offer potential for improving classification. The future multichannel systems may contribute greatly to improved results in forest analysis and disturbance mapping.

16.4 Characterization of Forest–Tundra in the North of Central Siberia

16.4.1 Introduction

Northern Siberia is a climatic hot spot—an area that is warming faster than the rest of the planet. In the past 30 years, average temperatures across the region have risen by 1–3°C, while the worldwide average increase in that time was about 0.6°C. The region remains fiercely cold. The average winter time low in Khatanga, a small village in Northern Siberia, is −37°C and can drop to −59°C. Yet the warming trend is so rapid here that scientists are curious to watch the effects on the land. Scientists from all over the world are now looking at Siberia.

Starting July 10, 2008, the authors Ranson and Kharuk led a team of American and Russian scientists (Figure 16.14) to study an extremely remote and harsh section of northernmost Central Siberia. A Russian MI 8 helicopter flew the team from Khatanga above the Arctic Circle, and landed at a flat

FIGURE 16.14
At their first campsite, the team assembles for a group photo in front of one of the not-yet-inflated rafts. Back row from left to right: Guoqing Sun, Ross Nelson, Vaitcheslav I. Kharuk, Kenneth Jon Ranson, Mukhtar Naurzbaev, and Sergei Im. Front row from left to right: Pasha Oskorbin and Paul Montesano. (Photograph by Kenneth Jon Ranson.)

spot (70°41′34″N, 105°38′46″E) near the headwaters of the Kotuykan River and the team made a rapid exit from the helicopter (Figure 16.15) into the rain. In the following 2 weeks, the team used three rubber boats to travel the river, stopping frequently to make observations and collected data in support of several ongoing studies.

16.4.2 Study Region and Data

The study area is located around 70°20′ – 71°15′N, 102°40′ – 105°50′E (Figure 16.16). The river flowing from south to north in the western part of the area is the Kotuy, which flows northward into the Kheta River and then north into the Arctic Ocean. Figure 16.17 shows the lower reach of the Kotuy River as seen from the helicopter (the upper-left corner of Figure 16.16). In the middle of Figure 16.16 is the Kotuykan River. The white triangle near the lower right of the image was the landing place of the Russian MI 8 helicopter. Figure 16.18 shows a scene where the Kotuykan River joins with the Kotuy River. The mountains, the Siberian Traps, were formed from basaltic lava flows during massive eruptions about 250 million years ago. The freeze/thaw cycle

FIGURE 16.15
In steady rain, a Russian MI 8 helicopter drops the scientists off on the banks of the Kotuykan River (70°41′34″N, 105°38′46″E) in northern Siberia. In the foreground, scientists cover gear with plastic. This is the first campsite of the expedition, and it will not be a soft one. The beach is covered with marble- to microwave-sized stones. (Photograph by Kenneth Jon Ranson.)

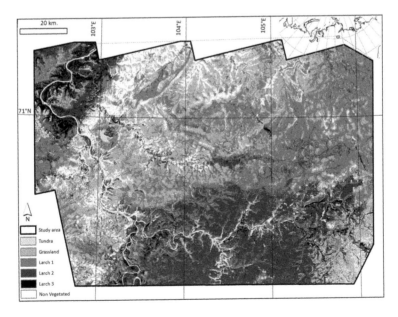

FIGURE 16.16
Black and white rendition of Landsat ETM+image acquired on July 23, 2009.

FIGURE 16.17
The low reach of Kotuy River seen from the helicopter, near Khatanga. (Photograph by Kenneth Jon Ranson.)

FIGURE 16.18
A nice spot for lunch, overlooking the Kotuy River. The mountains, the Siberian Traps, were formed from basaltic lava flows during massive eruptions about 250 million years ago. The freeze/thaw cycle cracks and crumbles the rocks. The weather and the river have eroded the mountains into spectacular formations and sheer drop-offs. (Photograph by Kenneth Jon Ranson.)

cracks and crumbles the rocks. The weather and the river have eroded the mountains into spectacular formations and sheer drop-offs.

On the river bank across from the campsite (Figure 16.19), there are larch trees growing along the river; these are gradually replaced by tundra and bare rocks as the elevation increases. In the background are the flat-topped mountains known as the Siberian Traps. The slope in the foreground is littered with basaltic rocks formed from lava flows about 250 million years ago. Figure 16.20 is a picture taken on the top of a trap from one side of the river looking across to the opposite river bank. At the top of these traps, few trees can survive.

The harsh climate of Siberia is a challenging one for larch trees. The photo (Figure 16.21) shows the fates of several trees. A tree without bark or branches leans across the center of the photo. This tree died centuries ago, but the frigid and arid climate has kept it from decaying. In the foreground, a tree that broke at the trunk and toppled managed to survive when a side branch grew into a vigorous new tree. In front and to the right of the "reborn" tree is a small dead tree that still has branches and bark. It is an ancient tree that died recently. In its last years, it put energy into making seed. Also present are the well preserved remains of fossil trees that died to the extremely frigid climate of the Little Ice Age, between the fourteenth

FIGURE 16.19
A view of the campsite taken from across the Kotuykan River. In the background are the flat-topped mountains known as the Siberian Traps. The slope in the foreground is littered with basaltic rocks formed from lava flows about 250 million years ago. The campsite was originally set up next to the riverbank. It is now on high ground; the river dropped about 2 m overnight. (Photograph by Kenneth Jon Ranson.)

FIGURE 16.20
Tundra on the top of the mountains with few trees. (Photograph by Guoqing Sun.)

FIGURE 16.21
View of a sparse stand of larch trees in extreme northern Siberia. See text for a description of the factors at work in this stand. (Photograph by Kenneth Jon Ranson.)

and eighteenth centuries (Figure 16.22). Although they died hundreds of years ago, the frigid climate has prevented them from decaying. New trees that are colonizing the area as the climate warms. These trees are growing far above the "fossil" tree line, which is evidence that the current warming trend is very strong. These data on the ages of both old and new trees will be used in future analysis to create a timeline of climate change in this part of Siberia.

Landsat data (MSS, TM, and ETM+) from 1973 to 2009 were acquired for this study. These images were used for classification and comparisons of vegetation status. Japan's PALSAR on the Advanced Land Observing Satellite (ALOS) (Shimoda et al., 2009), data acquired in 2007 by JAXA's ALOS mission, was used for showing the vegetation cover and information related to above-ground biomass. The lidar waveform data acquired by Geoscience Laser Altimeter System (GLAS) on board NASA's ICESat satellite were used to predict biomass.

Table 16.7 provides the classification confusion table gor Priangar'e area classes from Landsat-7 data. Table 16.8 lists the PALSAR and Landsat data acquired for this study. Landsat data (MSS, TM, and ETM+) were acquired in 1973, 2002, and 2009 and these images were used for the classification and comparisons of the vegetation status. Dual-pol PALSAR data acquired in 2008 by JAXA's ALOS mission (http://www.eorc.jaxa.jp/ALOS/en/obs/palsar_strat.htm) were acquired to study the vegetation cover and

FIGURE 16.22
Remnants of larch trees that grew in this area prior to the Little Ice Age. The wood is well preserved by the cold dry climatic conditions. Trees in background are young larch recolonizing this area. (Photograph by Kenneth Jon Ranson.)

TABLE 16.7

Classification Confusion Table for Priangar'e Area Classes and Landsat 7 Data

	Percent Classified As						
Name	CF	DF	IS	IM	CC	BA	WR
CF	99.04	0.03	0.53	0.40	0.00	0.00	0.00
DF	0.01	99.61	0.00	0.32	0.06	0.00	0.00
IS	0.52	0.00	96.22	2.23	0.01	1.02	0.00
IM	0.06	1.46	1.87	95.61	0.03	0.98	0.00
CC	0.00	0.02	0.00	0.00	98.34	1.65	0.00
BA	0.00	0.00	0.03	0.00	1.45	98.43	0.00
WR	0.00	0.00	0.00	0.00	0.00	0.00	100.00

Note: Average accuracy = 98.18%, overall accuracy = 98.14%.

TABLE 16.8

PALSAR Data Used for the Study

ALPSRP132821410/1420	7/22/2008	hh, hv
ALPSRP131071410/1420	7/10/2008	hh, hv
ALPSRP129321410/1420	6/28/2008	hh, hv
ALPSRP127571410/1420	6/16/2008	hh, hv
LANDSAT 1 MSS	7/23/1973	Bands 4,5,6,7
LANDSAT 7 ETM+	6/26/2002	Bands 1,2,3,4,5,7
LANDSAT 5 TM	7/23/2009	Bands 1,2,3,4,5,7

information related to above-ground biomass. The lidar waveform data acquired by the Geoscience Laser Altimeter System (GLAS) on board NASA's ICESat satellite were used to predict biomass from waveform data. Although it is not ideal from a vegetation measurement standpoint, GLAS data have been used for forest studies (e.g., Harding and Carabajal, 2005; Lefsky et al., 2005; Ranson et al., 2007; Boudreau et al., 2008; Sun et al., 2008). GLAS systematically samples the forest vertical structure and provides top canopy height in addition to the surface elevation. GLAS illuminates/ measures an area on the ground ~65 m in diameter, although the footprint size and degree of circularity changed significantly over the course of the mission (Abshire et al., 2005). Sequential GLAS footprints are spaced 172 m apart (Schutz et al., 2005). GLAS waveform data (GLA01), land products (GLA14), and associated documentation are available at the National Snow and Ice Data Center (NSIDC) website (http://nsidc.org/data/icesat/data.html).

16.4.3 Data Processing

16.4.3.1 PALSAR Data Processing

The L-band dual-polarization (hh/hv) ALOS PALSAR data in Level 1.1 were converted from digital number to sigma0 using the revised calibration coefficients (Shimada et al., 2009). The Repeat Orbit Interferometry Package (ROI_PAC) version 3.0 released on October 4, 2007 was used to geo-locate the SAR data. After the image data were imported into ENVI, these images were geographically mosaiced. Figure 16.23 shows the mosaic of PALSAR L-band hv data for the region. A ratio of hv to hh was also generated.

16.4.3.2 Landsat Image Processing and Classification

A dark-object subtraction technique was used to correct for varying atmospheric conditions for each subscene; subsequently, Landsat DN was converted to top-of-atmosphere (TOA) radiance in $W/(m^2\ sr\ \mu m)$ and then to surface reflectance. The DN-to-reflectance conversion is important because DN is an inappropriate index of change over time, given differences in sensor calibration, solar zenith angle, and sensor viewing angle, among others (Slater, 1980). The Normalized Difference Vegetation Index (NDVI) has been in use for many years to measure and monitor plant growth (vigor), vegetation cover, and biomass production from multispectral satellite data (Tucker, 1979). NDVI was calculated from Landsat data acquired in 1973, 2002, and 2009.

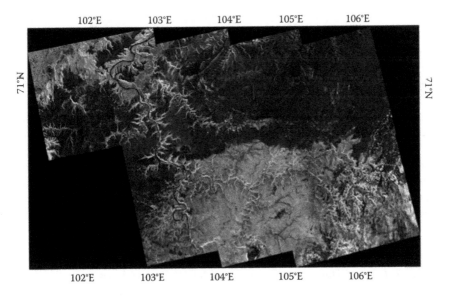

FIGURE 16.23
Mosaic of PLASAR l-band hv images (seven scenes) of the northern Siberia study site.

The Landsat data acquired on July 23, 2009 were classified using the unsupervised method (isodata) of ENVI package. The clusters were then combined into seven classes based on field observations.

16.4.3.3 GLAS Data Processing

The GLAS data used in this study were the level-1 product GLA01 (waveform) and level-2 product GLA14 (Land/Canopy Elevation). Among the elevations reported in the GLA14 products are the heights of up to six Gaussian peaks fit sequentially to a given waveform. The last, that is, lowest fitted Gaussian peak included in GLA14 data is assumed to be the ground peak. The difference between signal beginning and this peak will be the top or maximum canopy height. From the waveform data, many additional variables were generated (Sun et al., 2008) that captured characteristics of the canopy structure. The indices derived from GLAS waveform and used in the study include the total extent of waveform (wlen) and top canopy height (h14) from GLA14 data; the ratio of waveform energies returned from canopy to ground (eratio), the heights of four energy quartiles (rh25, rh50, rh75, rh100), and additional eight heights (rh10, rh20, rh30, rh40, rh60, rh70, rh80, rh90) where 10 – 90% of total waveform energy were cumulated above ground surface, calculated from waveform.

16.4.3.4 Field Timber Volume Data

The equation used to estimate stem volume of larch trees was

$$V = 0.00001 * H * (3.24 * D^2 + 6.601 * D + 3.361) \qquad (16.10)$$

where D is the dbh in cm, H is the height of the tree (m), and V is the volume of the tree (m^3). This equation is based on the data at Lucunskoe of the Ary-Mas Reserve (72°34′) (Bondarev, 1989).

During the 2-week field campaign, 100 GLAS footprints were sampled. The stem volume was calculated for all of these footprints using the above equation. It was found that the rh50 of some footprints was less than zero, which was probably caused by noisy signal. Also, a few footprints had rh50 greater than 15 m, which was problematic because the highest tree we measured in the field was less than 15 m. After excluding these abnormal points, a total of 53 points were left for development of a prediction model using the step-wise regression in S-plus.

16.4.4 Results and Discussions

16.4.4.1 Landsat Data Classification and NDVI Comparisons

Figure 16.24 is the classification from Landsat 5 TM data acquired on July 23, 2009. These classes are water (blue), larch forests with three levels of densi-

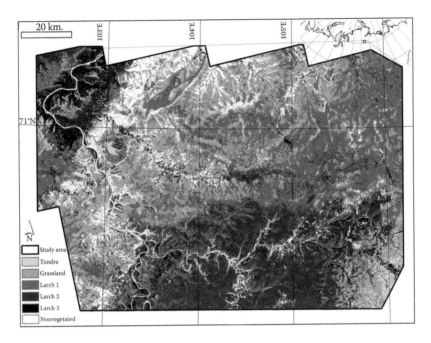

FIGURE 16.24
The vegetation map derived from classification of Landsat images.

ties (bright green, green, and dark green), grass tundra (yellow), wet tundra (maroon), and bare surface (pink). Dense larch forests are growing in the south or along the valley.

Figure 16.25 is a false color composite using NDVI of July 23, 2009, June 26, 2002, and July 23, 1973 as red, green, and blue. It can be seen that in the area of know larch forests, the NDVI was stable across imaging period so the composite color is white or light gray. The red color in the grass-covered areas indicates that NDVI on July 23, 2009 was higher than on other dates.

16.4.4.2 Timber Volume Assessment

The regression model was created by step-wise regression in S-plus. The procedure picked nine indices (h14, rh10, rh20, rh25, rh30, rh40, rh70, rh80, and rh100). The relation between field stem volume and GLAS predicted volume is (Figure 16.26):

$$Y = 0.26 + 0.75 * X \tag{16.11}$$

with an R^2 of 0.75 and residual standard error of 4.89 m^3, and where F-statistic is 159 and p-value is zero.

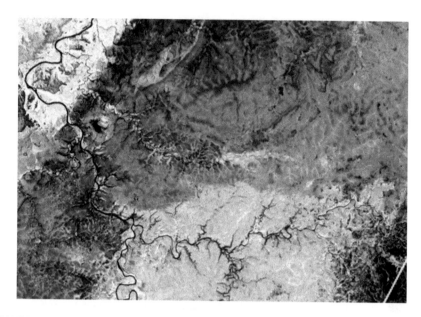

FIGURE 16.25
NDVI image derived from Landsat data. Brighter areas depict larch forest, darker areas have less vegetation, that is, tundra, bare rock, and water.

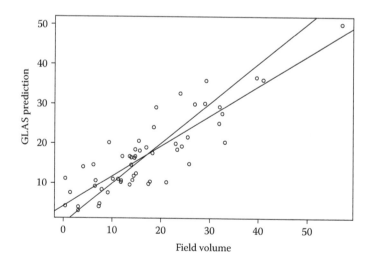

FIGURE 16.26
The comparison of GLAS predicted timber volume with field data: $B_{pred} = 4.26 + 0.75B_{field}$, $R^2 = 0.75$, RSE = 4.89 m^3/ha. F-statistic: 159 on 1 and 52 degrees of freedom, and the p-value is 0.

FIGURE 16.27
GLAS footprints overlaid on a PALSAR image. Bright tones along rivers are from scattering from bank edges and larch trees growing along the river. Medium tones are sparse larch forest. Dark tones depict tundra and grassland areas.

Figure 16.27 shows GLAS orbits for the data acquired in one data-take period. The background image is composed of three channels of PALSAR L-hh (red), L-hv (green), and the ratio of hv to hh (blue). The GLAS data of L3F (May–June 2006) and L3G (October–November 2006) were used in this study. The stem volume prediction model was applied to the footprints with the rh50 between zero and 15 m. Then the footprints falling in the larch forests were compiled, and the mean stem volumes were calculated. The mean stem volume was 17.88 m³/ha (standard deviation 11.10 m³/ha) derived from L3F data ($n = 185$), and 21.61 m³/ha (standard deviation 13.47 m³/ha) from L3G data. Most GLAS footprints sampled in the field were from L3G data.

16.4.5 Conclusion

Landsat multispectral data, PALSAR L-band SAR data, and GLAS lidar waveform data along with the field sampling data were used to characterize the land cover and stem volume in the area along the Kotuykan River in the extreme north of the Central Siberia. The analysis shows some capabilities of long-term observations with Landsat and the capability of observing forest structure with lidar and forest cover type with SAR. The data sets used herein represent the state of the technology available to measure frontier lands. The U.S. National Research Council advised NASA on the important measurements and technologies for answering pressing science questions. Among the several missions recommended is the Deformation, Ecosystem Structure and Dynamics of Ice or DESDynI. This mission will orbit a multi-beam lidar and a multipolarization SAR on separate platforms to provide the best data ever for quantifying forest height and above-ground biomass (Hall et al., 2011).

Acknowledgments

This work was supported in part by NASA's Science Mission Directorate and Russian Academy of Sciences.

References

Abshire, J.B., Sun, X., Riris, H., Sirota, J.M., McGarry, J.F., Palm, S.,Yi, D., and Liiva., P. 2005. Geoscience Laser Altimeter System (GLAS) on the ICESat Mission: On-orbit measurement performance, *Geophysical Research Letters*, 32, L21S02. doi:10.1029/2005GL024028.

Bayer, T., Winter, R., and Schreier, G. 1991. Terrain influences in SAR backscatter and attempts to their correction, *IEEE Transactions on Geoscience and Remote Sensing* 29(3), 451–462.

Beaudoin, A., Le Toan, T., Goze, S. Nezry, E., Lopes, A., Mougin, E., Hsu, C.C., Han, H.C., Kong, J.A., and Shin, R.T. 1994. Retrieval of forest biomass from SAR data, *International Journal of Remote Sensing* 15:(14), 2777–2796.

Bergen, K., Dobson, M.C., Pierce, L.E., and Ulaby, F.T. 1998. Characterizing carbon in a northern forest by using SIR-C/X-SAR imagery, *Remote Sensing of Environment*, 63(1), 24–39.

Bondarev, A.I. 1989. Forest measurement guide for most northern forest in the world. Forest measurement and forest regulation. Krasnoyarsk. Siberian Technology University, pp. 35–39.

Boudreau, J., Nelson, R.F., Margolis, H.A., Beaudoin, A., Guindon, L., and Kimes, D.S. 2008. Regional aboveground forest biomass using airborne and spaceborne LiDAR in Quebec, *Remote Sensing of Environment*, 112(10), 3876–3890, ISSN 0034-4257, doi: 10.1016/j.rse.2008.06.003.

Bourgeau-Chavez, L.L., Kasischke, E.S., and French, N.H.F. 1993. Detection and interpretation of fire disturbed boreal forest ecosystems in Alaska using space born SAR data, *Proceedings of the Topical Symposium on Combined Optical-Microwave Earth and Atmosphere Sensing*, Albuquerque, New Mexico, New York: IEEE.

Callaghan, T.V., Crawford, R.M.M., Eronen, M., Hofgaard, A., Payette, S., Rees, W.G., Skre, O., Sveinbjornsson, B., Vlassova, T.K. and Werkman, B.R. 2002. The dynamics of the tundra-taiga boundary: An overview and suggested coordinated and integrated approach to research, AMBIO, Special Report 12, Tundra-Taiga Treeline Research, pp. 3–5.

Dirk, B., Nielson, D., and Tangley, L. 1997. *The Last Frontier Forests.* World Resources Institute, Washington, USA.

Dobson, M.C., Ulaby, F.T., LeToan, T., Beaudoin, A., Kasischke, E.S., and Christensen, N. 1992. Dependence of radar backscatter on coniferous forest biomass, *IEEE Transactions on Geoscience and Remote Sensing*, 30(2), 412–415.

Dobson, M.C., Ulaby, F.T., Pierce, L.E., Sharik, T.L., Bergen, K.M., Kellndorfer, J., Kendra et al. 1995. Estimation of forest biophysical characteristics in northern Michigan with SIR-C/X-SAR, *IEEE Transactions on Geoscience and Remote Sensing*, 33(4), 877–895.

Dottavio, C.L., and Williams, D.L. 1983. Satellite technology—An improved means for monitoring forest insect defoliation, *Journal of Forestry*, 81(1), 30–34.

Gustafson, E.J., Shvidenko, A.Z., Sturtevant, B.R., and Scheller, R.M. 2010. Predicting global change effects on forest biomass and composition in south-central Siberia, *Ecological Applications*, 20(3), 700–715.

Frost, V.S., Stiles, J.A., Shanmugan, K.S., and Holtzman, J.C. 1982. A model for radar images and its application to adaptive digital filtering of multiplicative noise, *IEEE Transactions on Pattern Analysis and Machine Intelligence*, 4(2), 157–166.

Fung, A.K. 1994. *Microwave Scattering and Emission Models and Their Applications.* Norwood, MA: Artech House. 573pp.

Goering, D.J., Chen, H., Hinzman, L.D., and Kane, D.L. 1995. Removal of terrain effects from SAR satellite imagery of Arctic tundra, *IEEE Transactions on Geoscience and Remote Sensing*, 33(1), 185–194.

Goyal, S.K., Seyfried, M.S., and O'Neill, P.E. 1998. Effect of digital elevation model resolution on topographic correction of airborne SAR, *International Journal of Remote Sensing*, 19(16), 3075–3096.

Hall, F.G., Bergen, K., Blair, J. B., Dubayah, R., Houghton, R., Hurtt, G., Kellndorfer, J. et al. 2011. Characterizing 3D vegetation structure from space: Mission requirements, *Remote Sensing of Environment* (in press).

Hansen, J., Ruedy, R., Glascoe, J., and Sato, M. 1999. GISS analysis of surface temperature change, *Journal of Geophysical Research*, 104, 30997–31022.

Harding, D.J., and Carabajal, C.C. 2005. ICESat waveform measurements of within-footprint topographic relief and vegetation vertical structure, *Geophysical Research Letters*, 32(21): Art. No. L21S10 Oct. 15.

Kasischke, E.S., Tanase, M.A., Bourgeau-Chavez, L.L., and Borr, M. 2011. Soil moisture limitations on monitoring boreal forest regrowth using spaceborne L-band SAR data, *Remote Sensing of Environment*, 115(1), 227–232.

Kasischke, E.S., Bourgeau-Chavez, L.L., French, N.H.F., Harrel, P.A., and Christensen, N.L. Jr. 1992. Initial observations on using SAR to monitor wild fire scars in the boreal forest, *International Journal of Remote Sensing*, 13, 3495–3501.

Kasischke, E.S., French, N.H.F., Harrel, P.A., Christensen, N.L. Jr., Ustin, S.L., and Barry, D. 1993. Monitoring wild fires in boreal forests using large area AVHRR NDVI composite image data, *Remote Sensing of Environment*, 45, 61–71.

Kellndorfer, J., Pierce, L.E., Dobson, M.C. and Ulaby, F.T. 1998. Toward consistent regional-to-global-scale vegetation characterization using orbital SAR systems, *IEEE Transactions on Geoscience and Remote Sensing*, 36(5), 1396–1411.

Kharuk, V., Ranson, K., and Dvinskaya, M. 2007. Evidence of evergreen conifer invasion into larch dominated forests during recent decades in central Siberia, *Eurasian Journal of Forest Research*, 10(2), 163–171.

Kharuk, V.I., Dvinskaya, M.L., Im, S.T., and Ranson, K.J. 2008. Tree vegetation of the forest–tundra ecotone in the western Sayan Mountains and climatic trends, *Russian Journal of Ecology*, 39(1), 8–13.

Kharuk, V.I., Im, S.T., and Dvinskaya, M.L. 2010b. Forest-tundra ecotone response to climate change in the Western Sayan Mountains, Siberia, *Scandinavian Journal of Forest Research*, 25(3), 224–233.

Kharuk, V.I., Im, S.T., Dvinskaya, M.L., and Ranson, K.J. 2010a. Climate-induced mountain treeline evolution in southern Siberia, *Scandinavian Journal of Forest Research*, 25(5), 446–454.

Kharuk, V.I., Im, S.T., Ranson, K.J., and Naurzbaev, M.M. 2004a. Temporal dynamics of larch in the forest-tundra ecotone, *Doklady Earth Sciences*, 398(7): 1020–1023.

Kharuk, V.I., Peretyagin, G.I., and Palnikova, E.N. 1989. Automatic indication of pest outbreaks using false-color images, *Russian Journal of Remote Sensing*, 5, 70–73 (in Russian).

Kharuk, V.I., and Ranson, K.J. 2000. Microwave fire scar detection. In: *Biodiversity and Dynamics of Ecosystems in North Eurasia. V.1. Part 2: Biodiversity and Dynamics of Ecosystems in North Eurasia: Informational Technologies and Modeling.* (Novosibirsk, Russia, August 21–26, 2000). IC&G, Novosibirsk, pp. 174–176.

Kharuk, V.I., Ranson, K.J., and Dvinskaya, M.L. 2008. Wildfires dynamic in the larch dominance zone, *Geophysical Research Letters*, 35, 1–6.

Kharuk, V.I., Ranson, K.J., Im, S.T., and Naurzbaev, M.M. 2006. Forest-tundra larch forests and climatic tends, *Russian Journal of Ecology*, 37(5), 291–298

Kharuk, V.I., Ranson, K.J., Kuz'michev, V.V., and Im, S.T. 2003. Landsat-based analysis of insect outbreaks in southern Siberia, *Canadian Journal of Remote Sensing*, 29(2), 286–297.

Kharuk, V.I., Ranson, K.J., Kozuhovskaya, A.G., Kondakov, Y.P., and Pestunov, I.A. 2004b. NOAA-AVHRR satellite detection of Siberian silkmoth outbreaks in eastern Siberia, *International Journal of Remote Sensing*, 20(24), 5543–5555.

Kharuk, V.I., Ranson, K.J., and Im, S.T. 2009b. Siberian silkmoth outbreak pattern analysis based on SPOT VEGETATION data, *International Journal of Remote Sensing*, 30(9), 2377–2388.

Kharuk, V. I., Ranson, K. J., Im, S.T., and Dvinskaya, M.L. 2009a. Response of *Pinus sibirica* and *Larix sibirica* to climate change in southern Siberian alpine forest-tundra ecotone, *Scandinavian Journal of Forest Research*, 24(2), 130–139.

Kharuk, V.I., Ranson, K.J., Im, S.T., and Vdovin, A.S. 2010c. Spatial distribution and temporal dynamics of high elevation forest stands in southern Siberia, *Global Ecology and Biogeography Journal*, 19, 822–830.

Kurvonen, L., Pulliainen, J., and Hallikainen, M. 1999. Retrieval of biomass in boreal forests from multitemporal ERS-1 and JERS-1 SAR images, *IEEE Transactions on Geoscience and Remote Sensing*, 37(1), 198–205.

Lefsky, M.A., Harding, D.J., Keller, M., Cohen, W.B., Carabajal, C.C., Espirito-Santo, F.D., Hunter, M.O., and de Oliveira, R. 2005. Estimates of forest canopy height and aboveground biomass using ICESat, *Geophysical Research Letters*, 32(22), Art. No. L22S02.

Le Toan, T., Beaudoin, A., Riom, J., and Guyon, D. 1992. Relating forest biomass to SAR data, *IEEE Transactions on Geoscience and Remote Sensing*, 30(2), 403–411.

Luckman, A.J. 1998. The effects of topography on mechanisms of radar backscatter from coniferous forest and upland pasture, *IEEE Transactions on Geoscience and Remote Sensing*, 36(5), 1830–1834.

Martin, P.M. 1993. Global fire mapping and fire danger estimation using AVHRR images, *Photogrammetric Engineering and Remote Sensing*, 60, 563–570.

Michalek, J.L., French, N.H.F, Kasischke, E.S., Johnson, R.D., and Colwell, J.E. 2000. Using Landsat TM data to estimate carbon release from burned biomass in an Alaskan spruce forest complex, *International Journal of Remote Sensing*, 21(2), 323–338.

Nelson, R.N. 1983. Detecting forest canopy change due to insect activity using Landsat MSS, *Photogrammetric Engineering and Remote Sensing*, 49, 1303–1314.

Paloscia, S., Macelloni, G., Pampaloni, P., and Sigismondi, S. 1999. The potential of C- and L-band SAR in estimating vegetation biomass: The ERS-1 and JERS-1 experiments, *IEEE Transactions on Geoscience and Remote Sensing*, 37(4), 2107–2110.

Peretyagin, G.I., Kharuk, V.I., and Mashanov, A.I. 1986. Automatic classification of false-color images of larch stands damaged by pests, *Russian Journal of Remote Sensing*, 2, 110–118 (in Russian).

Radeloff, V.C., Mladenoff, D.J., and Boyce, M.S. 1999. Detecting jack pine budworm defoliation using spectral mixture analysis: Separating effects from determinants, *Remote Sensing of Environment*, 62(2), 156–169.

Ranson, K.J., Kimes, D.S., Sun, G., Kharuk, V., and Nelson, R. 2007. Using MODIS and GLAS data to develop timber volume estimates in central Siberia. *IGARSS 2007 IEEE Int.*, Barcelona, Spain, July 23–28, 2007. pp. 2306–2309, doi:10.1109/IGARSS.2007.4423302.

Ranson, K.J., Kovacs, K., Sun, G., and Kharuk, V.I. 2003. Disturbance recognition in the boreal forest using radar and Landsat-7, *Canadian Journal of Remote Sensing*, 29(2): 271–285.

Ranson, K.J., Saatchi, S., and Sun, G. 1995. Boreal forest ecosystem characterization with SIR-C/X-SAR, *IEEE Transactions on Geoscience and Remote Sensing*, 33(4), 867–876.

Ranson, K.J., and Sun, G. 1997a. Mapping of boreal forest biomass from spaceborne synthetic aperture radar, *Journal of Geophysical Research*, 102(D24), 29,599–29,610.

Ranson, K.J., and Sun, G. 1997b. An evaluation of AIRSAR and SIR-C/X-SAR images for mapping Northern forest attributes in Main, USA, *Remote Sensing of Environment*, 59, 203–222.

Ranson, K.J., Sun, G., Kharuk, V.I., and Kovacs, K. 2001. Characterization of forests in Western Sayani mountains, Siberia from SAR data, *Remote Sensing of Environment* 75(2),188–200, ISSN 0034-4257, doi: 10.1016/S0034-4257(00)00166-8.

Rauste, Y. 1990. Incidence-angle dependence in forested and nonforested areas in Seasat SAR data, *International Journal of Remote Sensing*, 11(7), 1267–1276.

Richards, J.A., and Jia, X. 1999. *Remote Sensing and Digital Image Analysis, An Introduction*. New York: Springer, pp. 241–247.

Rignot, E., Way, J.B., Williams, C., and Viereck, L. 1994. Radar estimates of aboveground biomass in boreal forests of interior Alaska, *IEEE Transactions on Geoscience and Remote Sensing*, 32(5), 1117–1124.

Royle, D.D., and Lathrop, R.G. 1997. Monitoring hemlock forest health in New Jersey using Landsat TM data and change detection techniques, *Forest Science*, 42(3), 327–335.

Saatchi, S.S., and Moghaddam, M. 2000. Estimation of crown and stem water content and biomass of boreal forest using polarimetric SAR imagery, *IEEE Transactions on Geoscience and Remote Sensing*, 38(2), 697–709. ISSN: 0196-2892.

Saatchi, S., van Zyl, J., and Assar, G. 1995. Estimation of canopy water content in Konza Prairie grasslands using synthetic aperture radar measurements during FIFE, *Journal of Geophysical Research*, 100(D12), 25481–25496.

Schutz, B.E., Zwally, H.J., Shuman, C.A., Hancock, D., and DiMarzio, J.P. 2005. Overview of the ICESat mission, *Geophysical Research Letters*, 32, L21S01.

Sellers, P.J., Hall, F.G., Kelly, R.D., Black, A., Baldocchi, D., Berry, J., Ryan, M. et al. 1997. BOREAS in 1997: Experiment overview, scientific results, and future directions, BOREAS special issue, *Journal of Geophysical Research*, 102(D24), 28731–28769.

Shi, J.C., and Dozier, J. 1997. Mapping seasonal snow with SIR-C/X-SAR in mountainous areas, *Remote Sensing of Environment*, 59(2), 294–307. Shimoda, H. 2009). Overview of Japanese Earth Observation programs, Proc. SPIE, 7474, 74740G, doi: 10.1117/12.831136.

Shimada, M., Tadono, T., and Isono, K. 2009. PALSAR radiometric and geometric calibration, *IEEE Transactions on Geoscience and Remote Sensing*, 47(12), 3915–3932.

Skre, O., Baxter, R., Crawford, R.M.M., Callaghan, T.V., and Fedorkov, A. 2002. How will the tundra-taiga interface respond to climate change? AMBIO, Special Report 12, Tundra-Taiga Treeline Research, pp. 37–46.

Slater, P.N. 1980, *Remote Sensing: Optics and Optical Systems*. Reading, MA: Addison-Wesley Publishing (575 pp).

Stofan, E.R., Evans, D.L., Schmullius, C., Holt, B., Plaut, J.J., van Zyl, J., Wall, S.D., and Way, J. 1995. Overview of results of Spaceborne Imaging Radar-C, X-band Synthetic Aperture Radar (SIR-C/X-SAR), *IEEE Transactions on Geoscience and Remote Sensing*, 33(4), 817–828.

Sturm, M., Racine, C., and Tape, K. 2001. Climate change—Increasing shrub abundance in the Arctic, *Nature*, 411, 445–459.

Sun, G., and Ranson, K.J. 1995. A three-dimensional radar backscatter model of forest canopies, *IEEE Transactions on Geoscience and Remote Sensing*, 33(2), 372–382.

Sun, G., Ranson, K.J., and Kharuk, V.I. 2002. Radiometric slope correction for forest biomass estimation from SAR data in Western Sayani mountains, Siberia, *Remote Sensing of Environment*, 79, 279–287.

Sun, G., Ranson, K.J., Kimes, D.S., Blair, J.B., and Kovacs, K. 2008. Forest structural parameters from GLAS: An evaluation using LVIS, SRTM data and field measurements, *Remote Sensing of Environment*, 112(1), 107–117.

Tucker, C.J. 1979. Red and photographic infrared linear combinations for monitoring vegetation, *Remote Sensing of the Environment*, 8, 127–150.

Ulaby, F.T., Moore, R.K., and Fung, A.K. 1982. *Microwave Remote Sensing, Active and Passive*, Volume III. Artech House, Norwood, MA, USA. 1064 pp.

Van Zyl, J.J. 1993. The effect of topography on radar scattering from vegetated areas, *IEEE Transactions on Geoscience and Remote Sensing*, 31(1), 153–160.

Vlassova, T.K. 2002. Human impacts on the tundra-taiga zone dynamics: The case of the Russian lesotundra, AMBIO, Special Report 12, Tundra-Taiga Treeline Research, pp. 30–36.

Wever, T., and Bodechtel, J. 1998. Different processing levels of SIR-C/X-SAR radar data for the correction of relief induced distortions in mountainous areas, *International Journal of Remote Sensing*, 19(2), 349–357.

Woodcock, C.E., Macomber, S.A, Pax-Lenney, M., and Cohen, W.B. 2001. Monitoring large areas for forest change using Landsat: Generalization across space, time and Landsat sensors, *Remote Sensing of Environment*, 78, 194–203.

17

Remote Sensing and Modeling for Assessment of Complex Amur (Siberian) Tiger and Amur (Far Eastern) Leopard Habitats in the Russian Far East

N. J. Sherman, T. V. Loboda, Guoqing Sun, and H. H. Shugart

CONTENTS

17.1 Introduction

Remote sensing is a valuable tool for characterizing the vast habitat of wide-ranging endangered species, such as the Amur tiger (*Panthera tigris altaica*), also called the Siberian or Ussurian tiger (Cushman and Wallin, 2000; Prynn, 2004), and the critically endangered Amur (Far Eastern) leopard (*P. pardus orientalis*). Mapping, monitoring, and modeling tiger and leopard habitat related to vegetation and terrain most often used by tigers and leopards can help target resources toward locations that are most important for ensuring a future in the wild for these species, as well as to identify new areas that would be appropriate to connect good habitat, including that in established reserves. Fewer than 400 adult and subadult Amur tigers (Chundawat et al., 2010; Miquelle et al., 2010b, 2007) and 14–20 adult Amur leopards (Jackson and Nowell, 2008) (Figure 17.1) persist in the wild in a topographically complex and biologically diverse landscape of about 270,000 km^2 in the Russian Far

FIGURE 17.1
(a) Amur tiger. (Reprinted with permission from WCS Russia Program.) (b) Amur leopard. (Reprinted with permission from WCS Russia Program and ISUNR.)

East (RFE) (Sanderson et al., 2006) and in northeastern China near the Russia–China border (Miquelle, 2010b; Smirnov and Miquelle, 1999) (Figure 17.2).

Tiger and leopard range in Russia is bounded by the Sea of Japan to the east and extends from about 42° N latitude in the south to slightly north of the Amur River (about 50°5′N), and westward into the Changbaishan area in

FIGURE 17.2
Present and potential Amur tiger habitat in RFE and northeastern China based on winter track surveys. (Reprinted with permission from WCS Russia Program.)

Jilin Province and Heilongjiang Province in northeastern China (Figure 17.2). Sightings of tigers and leopards in China and contiguous North Korea are very rare. A 2007 report by the Chinese government asserts that 18–22 tigers live within its borders (Miquelle et al., 2010c).

Climate, topography, and natural history combine to support a large number of rare and endemic plant and animal species in this area (Newell, 2004), which represents the intersection of the Siberian boreal forest and the East Manchurian mixed forests of northeastern Asia (Krestov, 2003; Miquelle, 2010b). Tiger and dry soil conditions mostly in the Russian province of Primorskiy Krai (or Primorye) and the southern end of Khabarovskiy Krai. The Sikhote-Alin Mountains extend along the spine of this narrow landscape, from southwest to northeast for about 1300 km. The highest peak in the Sikhote-Alin Mountains is 2078 m, but mountains generally do not exceed 1200 m (Rosenberg et al., 1998). Foothills that reach the coast create significant changes in elevation and vegetation types across short distances. Lower mountain ranges and dry soil conditions west of the Sikhote-Alin range and in the south contribute to a mosaic of vegetation types that provide food and shelter for the primary prey of Amur tigers and Amur leopards: ungulates such as red deer (*Cervus elaphus*), sika deer (*Cervus nippon*), wild pigs (*Sus scrofa*), and roe deer (*Capreolus pygargus*) (Krestov et al., 2006; Miquelle et al., 2010b) (Figure 17.3). However, compared to many tropical biomes where tigers and leopard are also found in the wild, these forests support much lower prey densities and consequently lower tiger and leopard densities (Smirnov and Miquelle, 1999). The average home range of a female Amur tiger, based on studies in a protected area, is 440 km². Male tigers establish territories that overlap with one to three females, and average more than 1000 km² (Miquelle et al., 2010a,b).

Forests of the RFE are among the richest temperate forests in the world and their value in terms of conservation of biodiversity has been long recognized at both national and international levels (Miquelle et al., 2010b; Newell, 2004). A wide range of protected areas have been established in the RFE during the twentieth century (Table 17.1), and the number of proposed and newly considered protected areas continues to grow even after the economy has shifted toward a resource-oriented emphasis. Internationally, central Sikhote-Alin has been recognized by the UNESCO World Heritage Convention as one of the world's most distinctive natural areas (UNESCO, 2011).

Similar to other regions in the world, the status of protected areas in Russia varies depending on specific designation. *Zapovedniks* are strictly protected reserves that are used for conservation, educational and research purposes only. These nature reserves play a particularly critical role in studying protected species as well as in species conservation. Most information available to date about the ecology and biology of tigers and leopards was collected within protected areas (Smirnov and Miquelle, 1999). *Zakazniks* are national parks that are used for recreation as well as conservation and education purposes.

FIGURE 17.3
(a) Mixed broadleaf/pine/conifer forests in central PrimorskiyKrai. (b) Coastal oak forests in Sikhote-Alin Biosphere Reserve. (Photos by N. J. Sherman.)

Nature reserves and protected areas play roles of paramount importance in tiger and leopard conservation. Winter censuses have shown that density of Amur tigers and Amur leopards is higher in protected areas, confirming the generally accepted notion that the parks are acting as a source for meta-populations of these species (Miquelle et al., 2010b). Outside park boundaries,

TABLE 17.1

Russian Federal Protected Areas in Tiger and Leopard Habitat

Protected Area	Type	Size (km²)	Establishment Date
Anyuiskiy	National Park	4293	2007
Bolonskiy	Zapovednik[a]	1036	1997
Bolshekhehtsirskiy	Zapovednik	454	1963
Botchinskiy	Zapovednik	2673	1994
Kedrova Pad	Zapovednik	180	1916
Lazovsky	Zapovednik	1209	1935
Leopardovyi	National Wildlife Refuge	1694	2008
Sikhote-Alin	Zapovednik	4014	1935
Udegeyskaya Legenda	National Park	886	2007
Ussuriskiy	Zapovednik	404	1934
Zov Tigra	National Park	821	2007

Source: Data from WWF—Russia, 2011; Miquelle, D. G., et al. 1999. In J. Seidensticker, S. Christie, and P. Jackson (eds), *Riding the Tiger*, pp. 273–295. Cambridge: Cambridge University Press.

[a] A *Zapovednik* is a strictly protected scientific nature reserve.

tigers and leopards are more vulnerable to poaching, encounters with hunters, collisions with cars, and retaliation for livestock predation (Goodrich et al., 2008). However, for large carnivores with very low population densities on the landscape (roughly 1.5 adult and subadult tigers per 1000 km² based on reported numbers of tiger and habitat available), conservation efforts confined to the boundaries of protected area are not likely to be effective. A broader framework of landscape-focused conservation of tigers has been proposed and accepted for all wild tigers (Dinerstein et al., 2006).

Historical variability of the Amur tiger population throughout the twentieth century showed that unlimited hunting of tigers and ungulates during the first half of the century led to the near decimation of the tiger population (Matyushkin, 2006). Tigers were seen extremely Rarely between 1963 and 1966, even within the Sikhote-Alin Nature Reserve, the largest protected area in the RFE, established nearly 30 years earlier (Table 17.1). With nearly 610,000 hectares under protection within tiger habitat by the mid-1960s, tiger population continued to decline until a change in state policy and the designation of the Amur tiger as a "protected species" helped to boost the number of tigers, first within the nature reserves and then outside them (Matyushkin, 2006).

Considering that Amur tiger and Amur leopard conservation efforts need to be supported at the scale of a large (nearly 300,000 km²) region, the role of remote sensing and modeling efforts in habitat monitoring and assessment becomes apparent. Remote sensing permits the accurate assessment of land-cover types and forest characteristics at a detailed scale (Hyde et al., 2006; Justice et al., 1998, 2002; Vina et al., 2008). Moreover, remote sensing is currently the only viable source of information regarding habitat conditions across the entire range of the Amur tiger and Amur leopard that can be continuously acquired at daily, monthly, and annual scales. In addition, remotely sensed data help to improve predictive capabilities regarding habitat quality and availability. Satellite-based information products are frequently used to validate, drive, and enhance available ecosystem models. In this chapter, we present a range of applications of remote sensing for monitoring and modeling present and future habitat availability for the Amur tiger and the Amur leopard in the RFE.

17.2 Monitoring and Assessment of Habitat

In the relatively undisturbed forests of the Sikhote-Alin Biosphere Reserve, Amur tiger home ranges vary from an average of 440 km^2 for females to more than 1000 km^2 for males (Goodrich et al., 2005; Miquelle et al., 2010a and b). This low density of tigers in the highest quality habitat available is directly related to the relative scarcity of ungulate prey (Miquelle et al., 2010b). Further conversion and degradation of available tiger and leopard habitat necessitate an expansion of home ranges of individual tigers and lead to further decrease in large carnivore density on the landscape. Since suitable habitat is a limited resource for both tigers and leopards, conversion and degradation of tiger and leopard habitat may lead to a decline in tiger population.

Remote sensing offers various applications for characterization and monitoring of vast remote and mountainous landscapes of the RFE. In this work, we employ the opportunities for land-cover mapping and characterization provided by optical instruments at different resolutions, as well as spaceborne radar and Light Detection and Ranging (LiDAR) data.

17.2.1 Mapping and Characterization of Forested Biomes from Multiple Remotely Sensed Data Sources

In the past decade, numerous coarse resolution maps of land cover from various spaceborne instruments have become available to the user community. However, due to differences in map legends, spatial resolution, time-frame of data acquisition, and inherent errors in mapping algorithms, these maps show a considerable amount of disagreement in their land-cover assessment. In order to minimize the mapping errors of individual products, we fused three coarse-resolution data sets, with a specific focus on identifying the areas of agreement in tree-dominated landscapes of the RFE (Loboda, 2009).

In this approach, we combined the map of Russian forests (Bartalev et al., 2004), Global Land Cover 2000 (GLC2000) map (Bartalev et al., 2003), and the Moderate Resolution Imaging Spectroradiometer (MODIS) land-cover product (MOD12Q1 collection4) (Friedl et al., 2002). This approach allowed identification of areas of multisource data agreement, with a focus on tree-dominated landscapes, and population of vaguely defined classes (e.g., recent burns) with meaningful land-cover information (e.g., shrub or grass cover). The output map produced a consensus assessment of forest-dominated landscapes by six dominant forest types in dense (40–100%) and sparse (10–39%) crown cover categories and 10 classes within nontree-dominant landscapes, including tundra, riparian vegetation, shrublands, grasslands, croplands and cropland complexes, forest–natural vegetation mosaic, wetlands, barren and sparsely vegetated, water, and urban areas (Figure 17.4).

The recent release of the full Landsat archive opened new opportunities in characterization of land covers using moderate resolution optical satellite imagery over a nearly 40-year time period. First images of the RFE from the Multi-Spectral Scanner (MSS) date to the early 1970s. The continuation of the Landsat missions with improved Thematic Mapper (TM) on Landsat 4 and 5 and Enhanced Thematic Mapper plus (ETM+) on Landsat 7 have provided a unique opportunity to map land covers in the RFE at 30 m and observe land-cover change over several decades. We used a Landsat ETM+ image acquired in 2002 to develop a detailed map of land cover following a Land

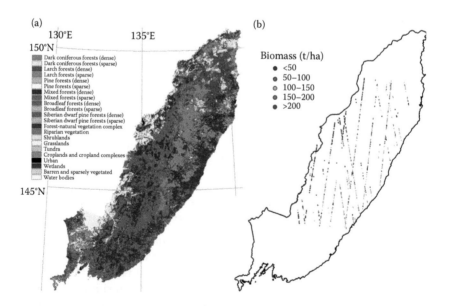

FIGURE 17.4
(**See color insert.**) (a) Land covers of tiger and leopard habitat developed from remotely sensed data, Global Land Cover 2000, and Russian forest maps (Loboda and Csiszar, 2007). (b) Biomass estimates developed from ICESatLiDAR data.

Cover Classification System (LCCS)-compatible, regionally defined classification scheme (Figure 17.4).

Our decision tree-based classification produced a map of overall accuracy ~84% and Kappa 0.83 for the full 15 classes (NELDA, http://www.fsl.orst.edu/nelda/sites/sd_sikh.html) (Figure 17.5). We also used this scene to assess the accuracy of the coarse resolution products used in the development of the coarse resolution land-cover data set described above. It showed that both GLC2000 and the MODIS land-cover product provided reasonably close estimates of the amount of tree cover in the area (72%, 70%, and 79%, respectively for Landsat, GLC2000, and MOD12Q1); however, both products underestimated the amount of shrub cover (23%, 2%, and 9%, respectively for Landsat, GLC2000, and MOD12Q1, which provides critically important habitat for moose and red deer, and thus plays an important role in habitat quality for tiger prey species.

LiDAR represents an important frontier in depicting and understanding the structure of forests and wildlife habitat. LiDAR allows the three-dimensional characterization of tree form at a detailed and accurate scale. The Geoscience Laser Altimeter System (GLAS) a LiDAR instrument on board NASA's Ice, Cloud, and Land Elevation Satellite (ICESat) was designed to monitor ice sheet elevations. Although the space LiDAR's engineering and operational specifications are not ideal from a vegetation measurement standpoint, GLAS data have been used for forest studies (e.g., Boudreau et al., 2008; Harding and Carabajal, 2005; Lefsky et al., 2005; Ranson et al., 2007; Sun et al., 2008). We utilized the GLAS LiDAR data set to gain additional insights

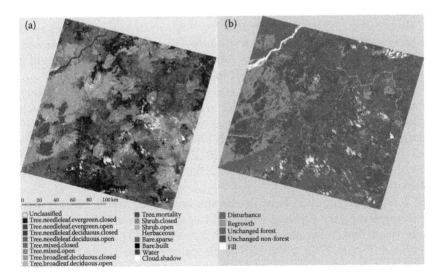

FIGURE 17.5
(See color insert.) Land covers (a) and land-cover change, (b) in northern Sikhote-Alin developed from Landsat data.

into the vertical structure of forest stands and to estimate their biomass as a proxy for identification of old growth forests.

GLAS systematically samples the forest vertical structure and provides top canopy height in addition to surface elevation. GLAS illuminates/measures an area on the ground of about 65 m in diameter. Because LiDAR provides an opportunistic sampling scheme along the orbital track on the ground the results of this analysis were used mostly to inform the FAREAST vegetation model described later in this chapter.

The primary GLAS data used in this study were the level-1 product GLA01 (waveform) and level-2 product GLA14 (Land/Canopy Elevation). Among the elevations reported in the GLA14 products are the heights of up to six Gaussian peaks, fit sequentially to a given waveform. The last, that is, the lowest, fitted Gaussian peak included in GLA14 data is assumed to be the ground peak. The difference between the signal beginning and this peak will be the top or maximum canopy height. From the waveform data, many additional variables were generated using software written by one of the co-authors (Sun et al., 2008), which captured characteristics of the canopy structure. Nonlinear models were developed to infer biomass from the GLAS waveform data.

In order to fully exploit the large information content of the GLAS waveform data, we used neural networks to find the most accurate relationships between GLAS data and biomass. Kimes et al. (1998, 2006) present an overview of the architecture, learning rules, and mathematical analyses of neural networks and their applications. Neural networks attempt to find the best nonlinear function based on the network's complexity without the constraints of linearity or pre-specified nonlinearity. An exhaustive search for the subset of GLAS waveform variables was conducted, which resulted in the highest neural network accuracy (highest R^2 and lowest root mean square error) of biomass. The model was then used to predict canopy height and biomass at all GLAS footprints used in this study.

Finally, we evaluated the advantages offered by active microwave remote sensing in providing three-dimensional characterization of forest structure in the RFE. ALOS-PALSAR data were acquired for the field research site Kabaniy (45.139106N, 135.884459E) in Sikhote-Alin Biosphere Reserve. Dual-pol L-band SAR data acquired on July 18, 2007 and September 2, 2007, form a pair of InSAR data with vertical and horizontal baselines of 457.98 and 274.43 m, respectively.

ROI_PAC (https://openchannelsoftware.com/projects/ROI_PAC), a Repeat Orbit Interferometry Package, was used to process the InSAR data. Two SAR images were first co-registered, and then coherence and phase difference were calculated. Figure 17.6a is the LHH backscattering coefficient and the right image (b) is the coherence image. Coherence parameters were used to characterize vegetation cover and biomass (Strozzi et al., 2000). The combination of incoherence, the averages, and differences of radar backscattering for the two different acquisition dates provided additional differentiation between land

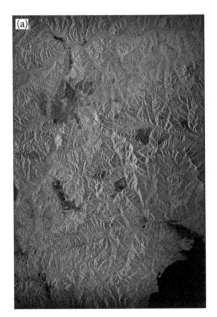
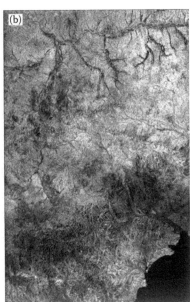

FIGURE 17.6
(a) LHH backscattering coefficient, (b) coherence of LHH InSAR data.

covers and can form a color image showing different land-cover types (Strozzi et al., 2000) (Figure 17.7a). Figure 17.7b is a false color (bands 4, 3, 2) Landsat Thematic Mapper image for the same location. It can be seen that the radar images provide complementary information to Landsat data.

These data were used to develop a processing methodology chain and extract experimental characteristics of land cover at the test site. The complex mountainous terrain in the region created challenges in the use of all remotely sensed data sources, but had a particularly pronounced effect on LiDAR and radar data aimed at evaluating tree height and biomass. Shuttle Radar Topography Mission (SRTM) Digital Elevation Model (DEM) data were used to perform geo-coding of the radar data.

Another polarimetric PALSAR image was acquired on May 18, 2007 centered at 48.12° N and 136.96° E. Figure 17.8a is a false color image from the polarimetric SAR in geo-coded format and Figure 17.8b is the Landsat image of the same area.

17.2.2 Monitoring Habitat Disturbance from Remotely Sensed Data

The landscape of the RFE is shaped by disturbance, such as wildfire, logging and human settlements. Historically, fire was not common in current leopard and tiger range, and most fires are associated with burning of agricultural

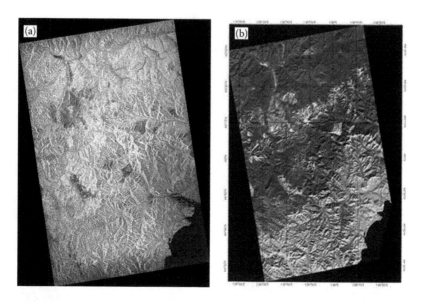

FIGURE 17.7
(a) Interferometry Land Use data (R—coherence, G—average backscattering of the two images, B—difference between the two images), developed from InSAR data of the Dual-pol L-band Synthetic Aperture Radar (SAR). (b) False color (bands 4, 3, 2) Landsat Thematic Mapper image for the same location.

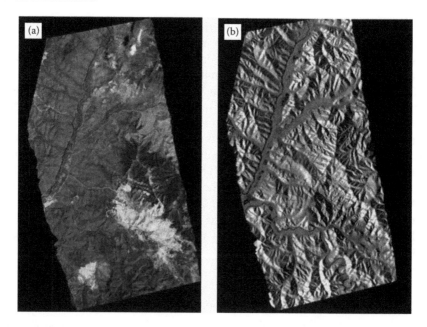

FIGURE 17.8
Polarimetric SAR data and ETM+ data: (a) Geo-coded L-band HH, HV, VV; (b) ETM+ image (bands 4, 3, 2).

areas in the spring and fall. Apart from agricultural burning, most wildfires are anthropogenic in origin, occurring close to human settlements, roads, and railroad tracks (Loboda and Csiszar, 2007). Fire occurrence is largely limited by the arrival of the summer monsoon, which brings ample (500–700 mm per year) precipitation (Stolbovoi and McCallum, 2002) and a distinct rainy season during late June–mid-August, the warmest part of the year. However, a disruption of the monsoonal weather pattern can lead to extremely large and catastrophic fires. For example, nearly 1 million hectares of forested and nonforested landscapes were burned during the 2003 fire season in the RFE (Loboda, 2009). Many large and catastrophic fire events in the RFE lead to the conversion of forest into non-tree-dominated landscape for an extended amount of time. However, even low intensity, nonstand-replacing fires influence forest structure by removing understory in mature forests and impacting nuts, pine cones, and other elements of the seed bank (Miquelle et al., 2004).

Massive logging activities started in the second half of the twentieth century in the highest-quality tiger habitat—Korean pine stands (Matyushkin, 2006). By the mid-1980s, nearly all Korean pine stands of the RFE were affected by industrial logging. Although the Korean pine harvest became illegal by early 1990s, industrial logging did not diminish but rather was redirected to harvesting spruce, fir, and larch stands, further contributing to conversion of tree-dominated habitat into open landscapes.

The results of multitemporal analysis of forest cover change in the RFE, developed using a multidated stack of Landsat TM and ETM+ images over path 111 row 25 (1990–2002–2005), show an alarming rate of disturbance by both natural and anthropogenic drivers over this 15-year time period (Figure 17.9). This forest change detection is based on the Disturbance Index (DI) method developed by Healey et al. (2005). The method tracks the DI change over time on a per-pixel basis. This is a well-established method that allowed us to identify negative and positive change in forest cover with the overall accuracy of 84% (Kappa = 0.8). This analysis indicates that only 55% of forests within the test area remained undisturbed between 1990 and 2005. The results also show that 13% of nonforested area in 1990 returned to a forested state by 2005 and, overall, in this site, restoration of tree-dominated landscapes surpassed conversion into nontree-dominated landscapes. Although visual analysis of forest cover change confirmed that the landscapes have been altered by fire and logging activities, fire appears to be the major contributor to the landscape mosaic of vegetation patterns. Visibly discernable scars at various stages of recovery are frequently found adjacent to each other across a burned site.

The longevity of burn signatures in these landscapes and the availability of long-term Landsat data records allowed the development of a new approach to reconstructing disturbance history in the RFE from the present-day distribution of land covers. This method allows for combining the capabilities offered by the long-term but spatially limited Landsat

FIGURE 17.9
Extensive disturbance from fire, logging and development is characteristic of Amur tiger and Amur leopard habitat, as shown in the map developed from Landsat TM and ETM+ images from 1990 to 2005.

data and the temporally limited (2000–present) but spatially contiguous MODIS data. Landsat data over the 1972–2002 period were used to develop training and validation data sets to drive a MODIS-based decision tree classification (Figure 17.10). The classification reliably differentiated between disturbed and mature forests with an overall accuracy of 88%

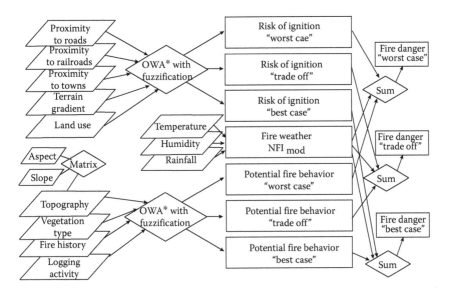

FIGURE 17.10
A decision-tree-based classification system for fire danger in forested areas developed using Landsat data for 1972–2002.

(Kappa 0.73). The type and specific decade of disturbance were estimated with a lower overall accuracy of 70% (Kappa 0.64). As expected, the method was better at identifying and timing burns compared to logged areas. However, this is a promising new method that has the potential to evaluate trajectories of land-cover change in boreal and temperate biomes over a 30-year time period and to provide better understanding of long-term tiger habitat dynamics and habitat vulnerability to climatic and policy-driven change.

Finally, the daily observations of on-going fire activity provided by the MODIS active fire product (Giglio et al., 2003) and the multiyear record on burned area developed from the MODIS surface reflectance product (Loboda and Csiszar 2007; Loboda, 2009) provide the bases for continuous monitoring and mapping of fire-related disturbances in tiger habitat. The analysis of fire occurrence in the RFE between 2001 and 2005 brought to light a number of important findings (Loboda, 2009). First, most burning in these areas occurs during the spring season and, most frequently, it affects broadleaf forests, grasslands, and croplands. While spring fires are the most numerous and frequent, they generally are low intensity, nonstand-replacing fires with typical managed fire characteristics. Late season (autumn) fires have characteristics very similar to spring fires; however, the total amount of fire occurrence in autumn is generally lower. Summer fires are unusual and can be supported only during atypical weather patterns—for this region, absence of monsoon rains in June and July. However, these fires result in large, contiguous, stand-replacing events in coniferous stands, similar to those of the 2003 fire season.

17.3 Modeling Future Habitat Availability

Remote sensing also plays an important role in developing predictive capabilities for modeling tiger and leopard habitat. Modeling provides an environment that supports long-term species conservation planning as well as a tool for evaluating the efficiency of specific management scenarios or policy implications. In this study, we used remotely sensed data products to analyze modeling outputs and enhance predictive capabilities for an ecological model, FAREAST (Yan and Shugart, 2005).

FAREAST is an individual, gap-based model, meaning that it simulates the establishment, growth and death of each individual tree in a small, defined area, similar to the clearing created in a forest when a large tree falls (Yan and Shugart, 2005). Loss of a major tree crown allows sunlight to penetrate to the forest floor, which sets in motion a multitude of processes, including the sprouting of seeds, the sudden growth or "release" of trees suppressed by low light levels, the demise of shade-dependent species, and competition among species to dominate the next generation (Shuman and Shugart, 2009; Shugart, 1984, 1998).

FAREAST simulates changes in forest species and biomass over time in response to parameters such as local climate, topography, elevation, aspect and species characteristics (Yan and Shugart, 2005). The model incorporates detailed biological characteristics for 58 tree species occurring in northeastern China and Russia. In addition to Korean pine (*Pinus koraiensis*) and Mongolian oak (*Quercus mongolica*), which are associated with the occurrence of tiger and leopard prey (Miquelle et al., 2010a), FAREAST includes elm (*Ulmus* spp), maple (*Acer* spp), larch (*Larix* spp), birch (*Betula* spp), spruce (*Picea* spp), and fir (*Abies* spp).

17.3.1 Linking Observed and Modeled Characterization of Vegetation in the RFE

In order to compare remotely sensed with modeled results, 1000 random points across the study area were selected. For each point, land cover and forest type derived from basal area calculated by FAREAST were compared with a coarse-resolution estimate of land cover and forest type across the entire tiger habitat (described in Section 17.2.1). As a first step, we compared only points that were within the boundaries of strictly protected reserves, because these locations are most likely to represent old growth, mature forests, which FAREAST is designed to simulate. After processing 300 years of succession, forest-type results for FAREAST matched MODIS-derived

TABLE 17.2

Comparison of Observed and Modeled Forest Types within
Protected Areas

Reserve Name	No. of Sites	No. of Plots	Match %
Bolshekhehtsirskiy	3	600	67
Lazovsky	4	400	50
Sikhote Alin	15	3000	40
Ussuriskiy	1	200	0

land covers most closely at Bolshekhehtsirskiy Zapovednik (Table 17.2). This reserve is the farthest north of those tested, at 48°12′ N, 134°49′ E, near the city of Khabarovsk. At the Sikhote-Alin Biosphere Zapovednik, the match percentage for the 15 sites tested was 40%. At Lazovsky Zapovednik, the rate was 50%. At Ussuriskiy Zapovednik, the southernmost reserve, the MODIS-derived forest type for the one point within the reserve was larch, whereas FAREAST predicted a broadleaf forest.

In Sikhote-Alin and Bolshekhehtsirskiy, at each site where FAREAST results matched those derived from remote sensing, both the model and remotely sensed data predicted broadleaf forests. At Lazovsky, a broadleaf forest was predicted at one site where results matched, and a mixed forest was expected at the second site. At sites where forest types did not match, remotely sensed results often suggested broadleaf forests, while the model predicted broadleaf/pine/conifer, suggesting site disturbance, such as snow or wind damage, fire, or permitted clearing, which remotely sensed data would detect, but would not be indicated in model results.

A more detailed assessment of FAREAST modeling of forest structure was carried out in comparison with LiDAR estimates of biomass from GLAS (described in Section 17.2.1). Consistency in the relationship between GLAS-derived canopy height and biomass, and model-produced canopy height and biomass, is significant because it affirms the reasoning used to develop biomass metrics from remotely sensed data. The relationship between canopy height and biomass (tC/ha) was compared with the FAREAST-derived tallest tree and biomass (tC/ha) for the two strictly protected reserves that were within the track of ICESat, Sikhote-Alin Biosphere Reserve and Botchinskiy Reserve. At each point, or "site," the FAREAST model ran 200 times, representing 200 random plots. Temperature, precipitation, and seed growth varied randomly at each plot. The model was run for 300 years at each point, representing the approximate age of primary, old-growth forests in the reserves. In both cases, FAREAST results for biomass after 300 years showed a similar relationship to tallest tree height, a proxy for canopy height, as biomass, developed from canopy height using GLAS data (Figure 17.11).

Overall, FAREAST biomass and maximum tree height averages were greater than the canopy heights and biomass developed from GLAS data,

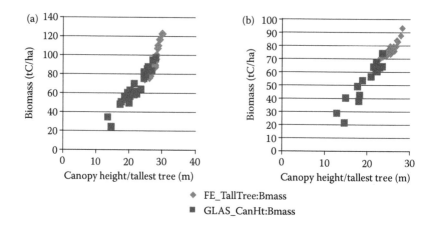

FIGURE 17.11

Biomass estimates relative to canopy height derived from GLAS data were similar to model-derived estimates of plot biomass compared with each plot's tallest tree. (a) GLAS-based vs model-derived canopy height: biomass relationship, Sikhote-Alin Zapovednik. (b) GLAS-based vs model-derived canopy height: biomass relationship Botchinskii Zapovednik.

possibly because FAREAST simulates an old growth primary forest, whereas the GLAS LiDAR instrument data reflect existing vegetation, which may have undergone logging, fire or other disturbance.

17.3.2 Developing Predictive Capabilities for Forecasting Fire Disturbance

The daily coverage of the RFE provided by MODIS data and advanced data products, including active fires (Giglio et al., 2003) and burned area (Loboda and Csiszar, 2007), have provided the first consistent and unbiased view of fire occurrence across the entire tiger habitat for nearly a decade. Previous records of fire occurrence provided by fire management agencies in the Russian Federation lacked both correct estimates on the amount of area burned and specific locations of burning (Conard et al., 2002). Additional availability of high-level satellite data products, including land cover, terrain, and disturbances, and nonsatellite spatially explicit information about transportation and settlements, can support the development of regional models of fire occurrence (Loboda and Csiszar, 2007; Loboda, 2009).

A fuzzy logic-driven model of fire occurrence in the RFE, based on the information provided by satellite products, was developed using the MODIS data over the 2001–2005 time period and validated using fire occurrence during the 2006 fire season (Loboda, 2009). The fire occurrence model (FOM) presents a part of a larger modeling framework of fire threat aimed at understanding the impacts of fire occurrence on Amur tiger habitat (Loboda and Csiszar, 2007). The FOM is evaluated as a sum of equally weighted components evaluating: (1) risk of ignition (ROI), (2) potential fire behavior (PFB), and (3) fire weather (FW).

These components operate at the same spatial scale (1 km) but varying temporal scales (monthly, seasonally, and daily for ROI, PFB, and FW, respectively). The FOM outputs daily spatially explicit projections of the likelihood of fire occurrence as a fuzzy set of three potential scenarios. The "best case" scenario is driven by factors mitigating against fire occurrence, the "worst case" scenario is driven by factors enhancing fire occurrence, and the "trade off" scenario assigns equal weights to mitigating and enhancing factors (Figure 17.12). All scenarios produce quantitative estimates ranging between 0 and 1, where 0 indicates extremely unlikely fire occurrence and 1 indicates a condition of near certain fire occurrence. Model validation using MODIS fire detections from 2006 fire season showed that, for all three scenarios, actual fires were detected within areas of elevated threat specific for each modeling scenario. The "best-case" scenario has the shortest dynamic range of values, gravitating toward lower fire occurrence likelihood estimates, while the "worst case" scenario has the widest dynamic range, gravitating toward moderate and higher values. Validation results led to the conclusion that predictive capabilities of the FOM are well-suited to be used for both operational fire monitoring and scientific research applications. At present, the FOM is adapted to model present and future fire occurrence under the influence of climate change in the Amur tiger and leopard habitat, Mediterranean-type ecosystems of Southern California, and North American tundra.

While fire disturbance is included in the modeling flow of the FAREAST model, in this study it was represented by a static probability developed from the literature sources and held constant across the entire landscape

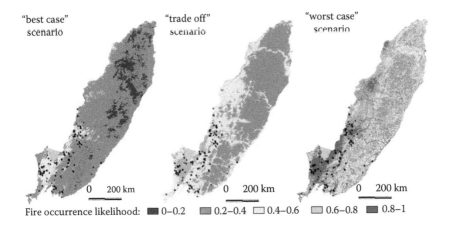

FIGURE 17.12
(See color insert.) Likelihood of fire taking place based on the presence of factors that promote fire or inhibit fire occurrence.

irrespective of forest types, terrain and human presence of the landscape. Coupling the FAREAST model with the FOM is complex because the two models operate on vastly different temporal scales (daily for the fire model and monthly over hundreds of years for FAREAST). Combining them would slow vegetation modeling down to unsustainable processing speed. However, because previous research has demonstrated stark differences in fire occurrence within various land covers of the RFE (Loboda and Csiszar, 2007; Loboda, 2009), improvements to the assumed probabilities of fire occurrence within the FAREAST model were necessary.

Fire frequency estimates were developed from the observed frequency of fire occurrence combined from the active fire observations obtained from the MODIS active fire product (MCD14) (Giglio et al., 2003) and the burned area estimates obtained using the regionally adjusted algorithm for the RFE (Loboda and Csiszar, 2007). These products include complementary information regarding fire occurrence in the RFE. The burned area product provides a more complete estimate of large contiguous burning events, whereas the active fire product provides additional information on spatial distribution of short episodic fire events. These two products were merged into monthly estimates of fire activity in the RFE for the period of 2001–2008, and the areas of overlap between the two were eliminated to avoid double counting. The following specific steps were taken to reduce the commission error from merging the two products:

1. Active fire detections from both Terra and Aqua satellites were buffered to 1-km-diameter circles to simulate the native resolution of the MODIS active fire product resolution. As it is impossible to determine where within the 1 km zone the fire occurred, the entire area was considered affected.

2. Burned areas were time-stamped using the information on the first date of burn area detection and the overlapping active fire detections. Repeatedly burned areas (areas that were detected as recently burned during March–May and then again during October–November) were assigned only one date.

3. Buffered active fire detections were subsequently intersected with burned area maps at a monthly time step. Active fire buffers that overlapped with burned areas were removed from further assessment.

4. The remaining buffered active fires were merged with the burned area maps to produce "fire affected area" coverage monthly.

The resultant binary burned/unburned data set was evaluated using a regression tree methodology against terrain (slope, aspect, elevation) from SRTM DEM, coarse-resolution land-cover map (see Section 17.2.1), previous disturbance to account for fuel modifications, land use, proximity to transportation routes (major roads and railroads), and proximity to human settlements. Decision trees were built at a monthly time step for each year between 2001 and 2008, and later the results were averaged at a monthly time step (as a mean value over 2001–2008) to preserve the seasonal variation in spatial distribution of fire occurrence as a function of landscape parameters (Figure 17.13a,b, and c). A cumulative yearly fire probability of monthly

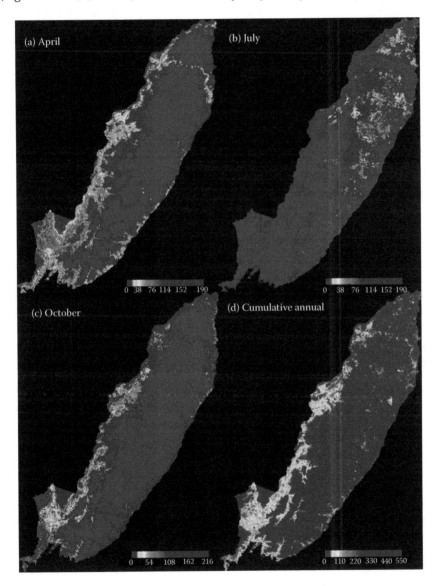

FIGURE 17.13
Fire occurs most frequently in spring and autumn and is associated with human activity. Lower summer fire occurence is associated with dark coniferous forests, previously disturbed sites, and moisture from summer monsoons.

means was developed for the RFE in support of the current modeling framework of the FAREAST model (Figure 17.13d).

The frequency of fire occurrence for forests that are remote and unaffected by human activity is within the range previously reported in the literature and is reflected in previous parameterization of the FAREAST model. However, these fire occurrence probability surfaces highlight spatio-temporal dynamics of landscape susceptibility to fire occurrence in the RFE. The results confirm the linkage between human activity and presence on the landscape, as well as spring and fall fires in the RFE. They indicate that mean annual fire occurrence probability in areas of high population density or agricultural use reaches 20% during spring seasons. Fire occurrence probability is considerably lower during summer months due to the presence of the summer monsoon and is directly linked to the distribution of dark coniferous forests and previously disturbed sites.

17.4 Implications of Study Findings for Long-Term Habitat Protection and the Role of Protected Areas

The RFE currently represents a stronghold for tiger conservation (Dinerstein et al., 2006). The RFE contains two of 20 Global Priority Tiger Conservation Landscapes (TCL), described as regions offering a high probability of long-term persistence of at least 100 individual tigers, with evidence of breeding and minimal–moderate threat levels. The strategic document for tiger conservation defines preservation of whole landscapes, including core areas, buffer zones, and dispersal routes, as the goal of tiger conservation in the wild (Dinerstein et al., 2006). This document cites habitat destruction and degradation as one of the greatest threats to tigers and underlines the importance of preserving every remaining portion of the habitat necessary to facilitate the movement of tigers across landscapes. A habitat protection plan for the Amur tiger was developed to ensure the long-term existence of interlinked core conservation units spread across the RFE with an intent to "guard against catastrophic events and minimize the effects of long-term habitat and genetic erosion" (Miquelle et al., 1999, p. 277).

Our research has demonstrated that "protected area" status does not limit habitat disturbance in core areas of tiger habitat. Considering that the contribution of fire-to-habitat modification in the RFE is greater than human activity and, despite a close link between fire ignitions and human presence in general, large fires that occur during uncharacteristically dry years burn in remote areas with limited human access. These fires can result in extensive (over 500,000 ha in the summer of 2003) conversion of tree-dominated land into open landscapes, thus considerably modifying the available tiger habitat. Over 3% of the area within the habitat protection plan burned during the 2001–2005 period with nearly 2% burned during the single catastrophic season of 2003 (Loboda, 2009).

The repeated burning in broadleaf forests of the southern tip of Amur tiger range (see Section 17.2.1) may be contributing to the decline in tiger numbers in these areas reported by field surveys (Miquelle, 2006). Repeated low-intensity surface fires observed in these forests do not kill the dominant trees, but remove the understory and surface layers, impeding forest regeneration and degrading the browsing base for large prey species such as red deer (~140–250 kg (Heptner et al., 1989)). Although the resultant increase in herbaceous cover within these forests is beneficial for smaller prey species (e.g., sika deer, ~60–131 kg (Heptner et al., 1989)), the increase in prey biomass associated with distribution density for sika deer in the oak forests of the RFE (from ~0.3 to ~1 individuals/km^2 between 1992 and 2002) is not comparable to the decrease in densities of red deer (from ~4 to ~2 individuals/km^2 between 1992 and 2002) (Stephens et al., 2005).

In certain cases habitat disturbance has a positive effect, improving tiger habitat by converting larch and spruce/fir forests, which provide a standard habitat for tigers, into shrub-dominated communities supporting higher prey densities. Such habitat modification frequently occurs during large fire seasons (exceptionally dry conditions) and can lead to extensive habitat conversion with short-term negative but long-term positive influence on tiger habitat quality. However, the impact of these fires on spatial habitat connectivity and, subsequently, tiger dispersal is not well understood and requires further investigation.

Early results of habitat availability modeling under the influence of climate change (Loboda, 2008) indicate that several core areas included in the habitat protection plan are likely to see more frequent and severe wildland fires. In particular, wildland fires are likely to impact on the established protected areas, including Lasovsky and Ussuriy State Reserves, Kedrovaya Pad' Zapovednik, and Barsovy Zakaznik, which may subsequently lose their role as a stronghold and source of tiger population in the tiger conservation landscapes. The existing and proposed protected areas (Upper Ussuriy National Park, Southern Primorye Nature Park, and Borisovskoe Plateau Zakaznik) and proposed ecological corridors (Lazovsky Nature Park, and Southern Sikhote-Alin) in the southern section of the RFE are small (~25,000–180,000 ha) and narrow (15–44 km across). The width of these narrow protected areas is not sufficient to ensure habitat connectivity under the current levels of fire occurrence and is likely to become less sufficient under the 10–20% annual increase in fire occurrence, projected for the end of the twenty-first century. Based on the sizes of observed fire scars in the catastrophic year of 2003, protected areas larger than 50,000 ha and broader (in their narrowest part) than 35 km may be required to maintain habitat connectivity under the projected increase in wildland fire occurrence. While no significant change in fire threat to the tigers is expected over the northern part of the RFE, these areas currently represent low-quality habitat types (larch and dark coniferous forests) and thus are unlikely to provide a reasonable substitute for the lost range in the south.

Since the designation of additional protected areas may not be practical or possible due to competing land uses and economic development, habitat connectivity can be ensured through strategic planning of forest use aimed at maintaining a sufficient portion of connected forested landscapes at various stages of regrowth necessary to support sufficient prey densities and ensure distribution of tigers across landscapes. However, this strategic planning will require close collaboration from various stakeholders and support from remotely sensed monitoring and habitat modeling activities.

17.5 Future Research Needs and Directions

The pressing nature of the decline in the number of tigers and extent of their habitat (Dinerstein et al., 2006) consumes much of current tiger conservation operations and leaves little room for long-term studies aimed at evaluation of climate change impacts on future habitat availability and sustainability of the species. This research presents one of the first contributions to the extension of the strategic framework for tiger conservation and represents a step toward understanding the long-term potential for habitat availability and connectivity.

This research has laid a foundation for optimizing the FAREAST model by comparing modeled outputs and observational data and by enhancing representation of fire disturbance within the FAREAST modeling framework. The next step will be to merge the FOM and the FAREAST predictive capabilities within the fire threat model to assess the impact of climate change on long-term availability, suitability, and connectivity of the Amur tiger and Amur leopard habitat under the influence of climate change. Remote sensing will continue to play a central role in future research activities as the longer satellite record of landscape modification will support advancements in parameterization of both fire and vegetation models.

While remote sensing has been seen as a valuable source of information in many regions of the world, including those containing rich historical information about landscape modification, the importance of remote sensing is far greater in frontier landscapes. In these locations, remote sensing is frequently the only reliable source of information available to support research and monitoring of habitat modification over a large spatial extent, and thus is an indispensable resource in conservation of large carnivores. Future advancements in satellite-borne instruments and data-acquisition strategies will permit better characterization of the landscape and its suitability to support endangered Amur tigers and critically endangered Amur leopards. Our findings highlight the advantages offered by merging optical remotely sensed data with LiDAR, and radar characterization of three-dimensional forest structure. These advantages go beyond supporting habitat characterization and will contribute strongly to further understanding of carbon cycling in terrestrial ecosystems and the role of protected areas in carbon sequestration.

Acknowledgments

This study was supported by the National Aeronautics and Space Administration (NASA) through grants from the Interdisciplinary Science (GG10906128162), Earth System Science Fellowship (NNG04GR15H), and the Northern Eurasia Land Dynamics Analysis activities (PI - Dr. Krankina, Oregon State University) under NASA Land Cover Land Use Change program grant (NNG06GF54G).

References

Abshire, J. B., X. Sun, H. Riris et al. 2005. Geoscience Laser Altimeter System (GLAS) on the ICESat Mission: On-orbit measurement performance. *Geophysical Research Letters* 32, LS1S02.

Bartalev, S., A. Belward, D. Ershov et al. 2003. A New SPOT4-VEGETATION derived land cover map of Northern Eurasia. *International Journal of Remote Sensing* 24: 9, 1977–1982.

Bartalev, S., D. Ershov, A. Isaev et al. 2004. *Russia's Forests*. TerraNorte Information System. RAS Space Research Institute. <terranorte.iki.ru> 08/05/2005.

Botkin, D. B., J. F. Janak, and F. R. Wallis. 1972. Some ecological consequences of a computer model of forest growth. *J. Ecol.* 60: 849–873.

Boudreau, J., R. F. Nelson, H. Margolis et al. 2008. Regional aboveground forest biomass using airborne and spaceborne LiDAR in Quebec. *Remote Sensing of the Environment* 112: 10.

Chapin, F. S. III., M. Sturm, M. C. Serreze et al. 2005. Role of land-surface changes in Arctic summer warming. *Science* 310: 657.

Chundawat, R. S., B. Habib, U. Karanth et al. 2010. *Panthera tigris*. In: IUCN 2010. *IUCN Red List of Threatened Species*. Version 2010.4. www.iucnredlist.org. Downloaded on 03 February 2011.

Conard, S. G., A. Sukhinin, B. Stocks et al. 2002. Determining effects of area burned and fire severity on carbon cycling and emission in Siberia. *Climatic Change* 55: 197–211.

Cushman, S. A. and D. O. Wallin. 2000. Rates and patterns of landscape change in the Central Sikhote-alin Mountains, Russian Far East. *Landscape Ecology* 15: 643–659.

Dinerstein, E., C. Loucks, and A. Heydlauff. 2006. *Setting Priorities for the Conservation and Recovery of Wild Tigers: 2005-2015. A User's Guide*. WWF, WCS, Smithsonian, and NFWF-STF, Washington, DC—New York.

Friedl, M., D. McIver, J. Hodges et al. 2002. Global land cover mapping from MODIS: Algorithms and early results. *Remote Sensing of Environment* 83: 287–302.

Giglio, L., J. Descloitres, C. Justice et al. 2003. An enhanced contextual fire detection algorithm for MODIS. *Remote Sensing of Environment* 87 (2–3): 273–282.

Goodrich, J., L. Kerley, D. Miquelle et al. 2005. Social structure of Amur tigers on Sikhote-Alin Biosphere Zapovednik. In D. Miquelle E. Smirnov, and J. Goodrich (eds), *Tigers of the Sikhote-Alin Zapovednik: Ecology and Conservation*, pp. 50–60. Vladivostok: PSP.

Goodrich, J., L. Kerley, E. Smirnov et al. 2008. Survival rates and causes of mortality of Amur tigers on and near the Sikhote-Alin Biosphere Zapovednik. *Journal of Zoology* 276: 323–329.

Goodrich, J., I. Seryodkin, D. G. Miquelle et al. 2011. Conflicts between Amur (Siberian) tigers and humans in the Russian Far East. *Biological Conservation* 144: 584–592.

Harding, D. J. and C. C. Carabajal. 2005. ICESat waveform measurements of within-footprint topographic relief and vegetation vertical structure. *Geophysical Research Letters* 32(21): Art. No. L21S10.

Healey, S. P., W. Cohen, Y. Zhiqiang et al. 2005. Comparison of Tasseled Cap-based Landsat data structures for use in forest disturbance detection. *Remote Sensing of the Environment* 97: 301–310.

Heptner, V. G., A. Nasimovich, and A. Bannikov. 1989. *Mammals of the Soviet Union: Ungulates*. New Delhi: Amerind Publishing Co. Pvt. Ltd.

Hyde, P., R. Dubayah, W. Walker et al. 2006. Mapping forest structure for wildlife habitat analysis using multi-sensor (LiDAR, SAR/InSAR, ETM+, Quickbird) synergy. *Remote Sensing of Environment* 102(1–2): 63–73.

IPCC. 2007. *Climate Change 2007: The Physical Science Basis. Working Group I Contribution to the Fourth Assessment Report of the Intergovernmental Panel on Climate Change*. Eds. S. Solomon, D. Qin, M. Manning. Cambridge: Cambridge University Press.

Jackson, P. and Nowell, K. 2008. *Panthera pardus* ssp. *Orientalis*. In IUCN 2010. *IUCN Red List of Threatened Species. Version 2010.4* www.iucnredlist.org. Downloaded on 03 February 2011.

Jolly, W. M., J. M. Graham, A. Michaelis et al. 2004. A flexible, integrated system for generating meteorological surfaces derived from point sources across multiple geographic scales. *Environmental Modeling and Software* 20(7): 873–882.

Justice, C. O., E. F. Vermote, and J. R. G. Townshend. 1998. The Moderate Resolution Imaging Spectroradiometer (MODIS): Land semote sensing for global change research. *IEEE Transactions on Geoscience and Remote Sensing* 36(4): 1228–1249.

Justice, C. O., J. R. G. Townshend, E. F. Vermote et al. 2002. An overview of MODIS Land data processing and product status. *Remote Sensing of Environment* 83: 3–15.

Kerr, J. T. and M. Ostrovsky. 2003. From space to species: Ecological applications for remote sensing. *Trends in Ecology and Evolution* 18(6): 299–305.

Keyser, A. R., J. S. Kimball, R. R. Nemani et al. 2000. Simulating the effects of climate change on the carbon balance of North American high-latitude forests. *Global Change Biology* 6 (Suppl. 1): 185–195.

Kimes, D., R. Nelson, M. Manry et al. 1998. Attributes of neural networks for extracting continuous vegetation variables from optical and radar measurements. *Int. J. Rem. Sens.* 19: 2639–2663.

Kimes, D., K. J. Ranson, G. Sun et al. 2006. Predicting LiDAR-measured forest vertical structure from multi-angle spectral data. *Remote Sensing of the Environment* 100(4): 503–511.

Krestov, P. V. 2003. Forest vegetation of Easternmost Russia (Russian Far East). In T. Kolbek, M. Strutek, and E.O. Box (eds), *Forest Vegetation of Northeast Asia*. New York: Kluwer.

Krestov, P. V., J.-S. Song, Y. Nakamura et al. 2006. A phytosociological survey of the deciduous temperate forests of mainland Northeast Asia. *Phytocoenologica* 36(1): 77–150.

Lefsky, M. A., D. J. Harding, M. Keller et al. 2005. Estimates of forest canopy height and aboveground biomass using ICESat. *Geophysical Research Letters* 32(22): Art. No. L22S02.

Lefsky, M. A., M. Keller, Y. Pang et al. 2007. Revised method for forest canopy height estimation from Geoscience Laser Altimeter System waveforms. *Journal of Applied Remote Sensing* 1: 013537.

Loboda, T.V. 2008. *Impact of Climate Change on Wildland Fire Threat to the Amur Tiger and its Habitat*. PhD dissertation, University of Maryland. <http://hdl.handle.net/1903/8253>

Loboda, T. V. 2009. Modeling fire danger in data-poor regions: A case study from the Russian Far East. *International Journal of Wildland Fire* 18(1): 19–35.

Loboda, T. V. and I. A. Csiszar. 2007. Assessing the risk of ignition in the Russian Far East within a modeling framework. *Ecological Applications* 17(3): 791–805.

Matyushkin, E. N. 2006. Carnivora: the Amur tiger. In A. Astafiev and M. Gromyko, (ed), *Flora and Fauna of the Sikhote-Alin Nature Reserve*, pp. 335–341. Vladivostok: Primpoligraphcombinat Ltd. [In Russian].

Miquelle, D. G. 2006. Monitoring Amur Tigers. 2005–2006. Final Report to the National Fish and Wildlife Foundation, Save The Tiger Fund from the Wildlife Conservation Society (Grant#2005-0013-026). <http:www.savethetigerfund.org/AM/Template.cfm?Section=Home&CONTENTID=8564&TeMPLATE=/CM/ContentDisplay.cfm>04/02/2008.

Miquelle, D., Y. Darmin, and I. Seryodkin. 2010c. Panthera tigris ssp. Altaica. In: IUCN 2010. *IUCN Red List of Threatened Species*. Version 2001. (www.iucnredlist.org) Downloded 08 July 2011.

Miquelle, D. G., J. M. Goodrich, L. L. Kerley et al. 2010a. Science-based conservation of Amur tigers in the Russian Far East and Northeast China. In R. Tilson and P. J. Nyhus (eds), *Tigers of the World*, pp. 403–423. London: Elsevier Inc.

Miquelle, D. G., J. M. Goodrich, E. N. Smirnov et al. 2010b. Amur tiger: A case study of living on the edge. In D. W. MacDonald and A. J. Loveridge (eds), *Biology and Conservation of Wild Felids*, pp. 325–339. Oxford: Oxford University Press.

Miquelle, D. G., I. Nikolaev, J. Goodrich et al. 2004. Searching for the co-existence recipe: A case study of conflicts between people and tigers in the Russian Far East. In R. Woodroffe, S. J. Thirgood, A. Rabinowitz (eds), *People and Wildlife: Conflict or Co-existence?* pp. 305–356. Cambridge: Cambridge University Press.

Miquelle, D. G., T. W. Merrill, Y. M. Dunishenko et al. 1999. A habitat protection plan for the Amur tiger: Developing political and ecological criteria for a viable land-use plan. In J. Seidensticker, S. Christie, and P. Jackson (eds), *Riding the Tiger*, pp. 273–295. Cambridge: Cambridge University Press.

Miquelle, D. G., D. G. Pikunov, Y. M. Dunishenko et al. 2007. 2005 Amur tiger census. *Cat News* 46: 11–14.

Newell, J. 2004. *The Russian Far East: A Reference Guide for Conservation and Development*. McKinleyville: Daniel & Daniel.

Prynn, D. 2004. *Amur Tiger*. Edinburgh: Russian Nature Press.

Ranson, K. J., D. S. Kimes, G. Sun et al. 2007. *Using MODIS and GLAS Data to Develop Timber Volume Estimates in Central Siberia*. IGARSS 2007 IEEE Int., Barcelona, Spain, July 23–28, pp. 2306–2309.

Rosenberg, V. A., V. N. Bocharnikov, and S. M. Krasnopeev. 1998. Biological diversity in the Sikhote-Alin forests and measures of its conservation. In M. Hansen and T. Burk (eds), *Integrated Tools for Natural Resources Inventories in the 21st Century*. Gen. Tech. Rep. NC-212. St. Paul, MN: U.S. Dept. of Agriculture, Forest Service, North Central Forest Experiment Station, pp. 326–333.

Sanderson, E., J. Forrest, C. Loucks et al. 2006. *Setting Priorities for the Conservation and Recovery of Wild Tigers: 2000–2015. The Technical Assessment*. WCS, WWF, Smithsonian and NFWF-STF. New York, Washington, DC.

Schutz, B. E., H. J. Zwally, C. A. Shuman et al. 2005. Overview of the ICESat mission. *Geophysical Research Letters* 32: L21S01.

Shugart, H. H. 1984 (reprint 2003). *A Theory of Forest Dynamics: The Ecological Implications of Forest Succession Models*. Caldwell: The Blackburn Press.

Shugart, H. H., S. Saatchi, F. G. Hall. 2010. Importance of structure and its measurement in quantifying function of forest ecosystems. *Journal of Geophysical Research* 115: G00E13.

Shugart, H. H. and D. C. West. 1977. Development of an Appalachian deciduous forest succession model and its application to assessment of the impact of the Chestnut Blight. *Journal of Environmental Management* 5: 161–179.

Shugart, H. H. 1998. *Terrestrial Ecosystems in Changing Environments*. Cambridge: Cambridge University Press.

Shuman, J. K. and H. H. Shugart. 2009. Evaluating the sensitivity of Eurasian forest biomass to climate change using a dynamic vegetation model. *Environmental Research Letters* 4: 045024.

Smirnov, E. V. and D. G. Miquelle. 1999. Population dynamics of the Amur tiger in Sikhote-Alin Zapovednik, Russia. In J. Seidensticker, S. Christie, and P. Jackson (eds), *Riding the Tiger: Tiger Conservation in Human-Dominated Landscapes*, pp. 61–70. Cambridge: Cambridge University Press.

Soja, A. J., N. M. Tchebakova, N. J. F. French, et al. 2007. Climate-induced boreal forest change: Predictions versus current observations. *Global and Planetary Change* 56: 274–296.

Stephens, P. A., O. Zaumyslova, G. Hayward et al. 2005. Ungulate abundance in Sikhote-Alin Zapovednik: Trends and causal factors. In D. Miquelle, E. Smirnov, J. Goodrich (eds), *Tigers of the Sikhote-Alin Nature Reserve: Ecology and Conservation*, pp. 113–124. Vladivostok: Wildlife Conservation Society Russia Program [In Russian].

Stolbovoi, V. and I. McCallum (eds.) 2002. CD-ROM *Land Resources of Russia*. Laxenburg, Austria: International Institute for Applied Systems Analysis ans the Russian Academy of Science.

Strozzi, T., P. B. G. Dammert, U. Wegmuller, et al. 2000. Land use mapping with ERS SAR interferometry. *IEEE Transactions on Geoscience and Remote Sensing* 38 (2): 776–775.

Sun, G., K. J. Ranson, D. S. Kimes, et al. 2008. Forest structural parameters from GLAS: An evaluation using LLVIS, SRTM data and field measurements. *Remote Sensing of the Environment* 112(1): 107–117.

Thornton, P. E., S. W. Running, M. A. White. 1997. Generating surfaces of daily meteo-rological variables over large regions of complex terrain. *Journal of Hydrology* 190 (3–4): 214–251.

United Nations Educational, Scientific and Cultural Organization (UNESCO). 2011. World Heritage Convention. World Heritage List. <http://whc.unesco.org/en/list/766> Accessed 04/02/11.

Vina, A., S. Bearer, H. Zhang et al. 2008. Evaluating MODIS data for mapping wildlife habitat distribution. *Remote Sensing of the Environment* 112: 2160–2169.

Wolfe, R. E., M. Nishihama, A. J. Fleig et al. 2002. Achieving sub-pixel geolocation accuracy in support of MODIS land science. *Remote Sensing of Environment* 83: 31–49.

Wolfe, R. E. and N. Soleous. 2002. MODIS land products and data processing. In J. J. Qu and W. Gao et al. (eds), *Earth Science Satellite Remote Sensing*, pp. 110–122. Beijing: Tsinghua University Press.

Yan, X. and H. H. Shugart. 2005. FAREAST: A forest gap model to simulate dynamics and patterns of eastern Eurasian forests. *Journal of Biogeography* 32: 1641–1658.

Zhang, N., H. H. Shugart, and X. Yan. 2009. Simulating the effects of climate changes on Eastern Eurasian forests. *Climatic Change* 95: 341–361.

18

The Influence of Realistic Vegetation Phenology on Regional Climate Modeling

Lixin Lu

CONTENTS

18.1 Introduction

Land is a key component of the climate system. Most land surfaces are covered by vegetation, and there is general recognition of the importance of vegetation control on the exchange of energy, mass, and momentum between the land surface and the atmosphere. For example, Gash and Nobre (1997) reviewed the climatological measurements from the Anglo-Brazilian Amazonian Climate Observational Study (ABRACOS) and found that the difference in radiation and energy balance between forests and clearings produces higher air temperatures in the clearings, particularly in the dry season. In areas of substantial deforestation, higher sensible heat fluxes from the cleared forests produce deeper convective boundary layers, with differences in cloud cover being observed and mesoscale circulations predicted. More recently, Wang et al. (2009) further analyzed satellite cloud maps combined with local sounding measurements and confirmed that the heterogeneous "fish-bone" pattern has left detectable marks in the Amazon cloud climatology. They found that shallow clouds are prone to appear over deforested surfaces, whereas less-frequent deep clouds favor forested surfaces due to greater humidity and resultant larger values of convective available potential energy (CAPE). Based on low-level flight measurements, Segal (1989) showed that the atmospheric boundary layer is shallower, cooler, moister, and less turbulent over irrigated cropland than over adjacent bare soil surfaces. Similarly, an observational study conducted by Rabin et al. (1990) showed that convective clouds are first formed over a harvested wheat field surrounded by growing vegetation and are suppressed immediately downstream of lakes and forests. In addition, Koster et al. (1986) indicated that the observed growing season precipitation peak may be due, in part, to the local recycling of water.

Since the 1980s, many land surface models have been developed and incorporated into various mesoscale and global-scale atmospheric models to account for the observed effects of land surface processes on weather and climate. These land surface models, which are referred to as Simple Vegetation–Atmosphere Transfer Scheme (SVATS), include the Biosphere–Atmosphere Transfer Scheme (BATS) of Dickinson et al. (1986, 1993), the Simple Biosphere Scheme of Sellers et al. (1986), the Simple SiB (Xue et al., 1991), the Bare Essentials of Surface Transfer scheme (Pitman, 1991), the Interaction Soil–Biosphere–Atmosphere (Noilhan and Planton, 1989), the Land Ecosystem–Atmosphere Feedback model (Lee, 1992; Walko et al., 2000), Land–Air Parameterization Scheme (Mihailovic, 1996), the Land Surface Model of Bonan (1996), and the Community Land Model (Levis et al., 2004; Oleson et al., 2004; Dickinson et al., 2006). Reviews of SVATS and their evaluations are reported in Henderson-Sellers et al. (1993, 1995), Avissar (1995), Dickinson (1995), Chen et al. (2000), and Pitman (2003). An extensive review of modeling and observational studies of the importance of landscape processes on weather and climate can be found in Pielke and Avissar (1990), Pielke et al. (1998, 2002), and Pielke and Niyogi (2010), where it is concluded that land surface processes play a significant role in defining local, regional, and global climate.

The seasonal cycle and interannual variation of vegetation exert a significant control on surface–atmosphere interactions. Dirmeyer (1994), for instance, showed that the inclusion of dormant vegetation during the spring and early summer in a Global Climate Model (GCM) run greatly reduces surface moisture fluxes by eliminating transpiration from leaves, and prevents further depletion of moisture in the root zone of soil, leading to soil moisture recovery during the subsequent summer. A study conducted by Xue et al. (1996) found that the erroneous prescription of crop vegetation phenology in the surface model contributed greatly to the temperature biases of summer simulation in the United States. Bounoua et al. (2000) examined the sensitivity of a coupled atmosphere–biosphere GCM to changes in vegetation density and found that increasing vegetation density globally caused both evapotranspiration and precipitation to increase. Lu and Shuttleworth (2002) showed that Regional Atmospheric Modeling System (RAMS) is sensitive to remotely sensed vegetation product, and the introduced spatial heterogeneity in leaf area index produces a wetter and cooler climate over central United States.

Incorporating dynamic vegetation into climate models is a fairly new endeavor, but research in this area has already provided important insights in to Earth system modeling. Foley et al. (1998) directly coupled the GENESIS (version 2) GCM and IBIS (version 1) Dynamic Global Vegetation Model, and found that the atmospheric portion of the model correctly simulates the basic zonal distribution of temperature and precipitation with several important regional biases, and the biogeographic vegetation model was able to roughly capture the general placement of forests and grasslands. An interactive canopy model (Dickinson et al., 1998) was derived and added to the BATS (Dickinson et al., 1986, 1993) to describe the seasonal growth of leaf area as needed in an atmospheric model, and to provide carbon fluxes and net primary productivity; this scheme differs from other studies by focusing on short-timescale leaf dynamics. Tsvetsinskaya (2001) introduced daily plant growth and development functions into BATS and coupled it to the National Center for Atmospheric Research's Regional Climate Model to simulate the effect of seasonal plant development and growth on the atmosphere–land surface heat, moisture, and momentum exchange. She found that the coupled model is in better agreement with observations compared to the noninteractive mode. Using the coupled atmosphere and ecosystem model, RAMS–CENTURY, Lu et al. (2001) showed that the variation in vegetation phenology and its associated land surface heterogeneity play a sizeable role in surface energy partitions and influence predictions of surface temperature and precipitation over United States for a climatologically average year.

Advances in satellite remote sensing technology have provided unprecedented opportunities for earth science study. One of the most commonly used remote sensing products is the land-cover map. Figure 18.1 shows the vegetation distribution at three resolutions. Combined with land surface parameterization, these vegetation spatial distributions can have significant impacts on surface energy and moisture fluxes, as well as the related atmospheric boundary layer processes. Normalized Difference Vegetation Index (NDVI) derived from optical satellite sensors are mainly dependent on the green leaf material of the vegetation cover that is directly related to the

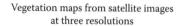

Vegetation maps from satellite images
at three resolutions

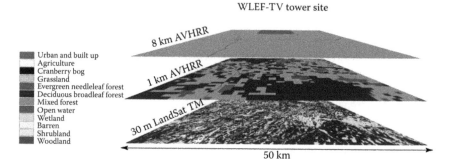

FIGURE 18.1
Vegetation maps at three resolutions. These spatial distributions can have a significant impact
on surface energy and moisture fluxes, and the related atmospheric boundary layer processes
(Data from http://biocycle.atmos.colostate.edu).

photosynthetic capacity (i.e., the live green material of the vegetation), and
hence energy absorption of plant canopies (Myneni and Ganapol, 1992;
Sellers et al., 1992). Numerous vegetation properties have been derived from
NDVI. Most relevant examples include Leaf Area Index (LAI) (Turner et al.,
1999; Haboudane et al., 2004), Photosynthetically Active Radiation (PAR)
(Myneni et al., 1997; Cohen et al., 2003), above-ground biomass (Todd et al.,
1998; Labus et al., 2002; Foody et al., 2003), and fractional vegetation cover
(Purevdorj et al., 1998; Gitelson et al., 2002). Calculation of the NDVI is sensitive to a number of perturbing factors that include atmospheric composition
(water vapor and aerosols), clouds (deep, thin, shadow), soil effects (moisture
state, color), snow cover, anisotropic effects (geometry of the target), and spectral effects (different instruments), which introduce uncertainty in quantitative assessments. NDVI observations have significantly improved our
understanding of the characteristics and variability of vegetation cover at
pixel, local, regional, and global scales, and are now being used to evaluate
ecosystem model and improve the Four-Dimensional Data Assimilation
(4DDA) weather forecasts and climate modeling.
    The majority of previous studies have focused on appraising impacts of
assimilating observed vegetation density into GCMs or stand-alone land
surface models. In the current study, we integrate previous researches, further understand the results, and explore the best way to incorporate realistic
vegetation descriptions into regional climate modeling. First, we introduce
Colorado State University's RAMS, build the assimilation RAMS (ASSM-
RAMS) to ingest NDVI-derived LAI, and examine the sensitivity of simulated regional climate to multiyear satellite observations of vegetation
changes (NDVI). Second, recognizing the large sensitivity of RAMS to seasonal variation in vegetation phenology, we adopt the CENTURY ecosystem

model to simulate vegetation growth, and link it with RAMS to form a regional climate modeling system (RAMS–CENTURY) that includes the two-way feedback between the atmosphere and biosphere. Third, we compare the results from both ASSM-RAMS and the coupled RAMS and CENTURY modeling system to comprehensively understand the impact of heterogeneous vegetation distribution on the seasonal climate prediction.

In its current form, RAMS land surface hydrological processes (e.g., evaporation and transpiration), energy exchanges (e.g., latent heat and sensible heat fluxes), momentum exchanges (e.g., roughness length), and biophysical parameters (e.g., vegetation albedo, transmissivity, and stomatal conductance) are parameterized to have a strong dependence on the value of LAI. Consequently, inadequate and unrealistic description of the vegetation distribution and its evolution in the current RAMS land surface models is considered a major deficiency. Both using NDVI datasets to derive LAI and using CENTURY model to simulate LAI can provide a more realistic vegetation distribution, which has the potential to improve the regional climate model simulations. The goal of the present study was to gain insight into the likely impact of heterogeneous LAI specification on the near-surface climate variables modeled by RAMS at regional spatial scales and seasonal timescales.

18.2 Regional Atmospheric Modeling System

18.2.1 Model Description

RAMS is a three-dimensional, nonhydrostatic, general-purpose atmospheric simulation modeling system consisting of equations of motion, heat, moisture, and mass continuity in a terrain-following coordinate system (Pielke et al., 1992; Cotton et al., 2003). RAMS was developed at the Colorado State University primarily to facilitate research into mesoscale and regional, cloud and land surface–atmospheric phenomena and interactions (Pielke, 1974; Tripoli and Cotton, 1982; Tremback et al., 1985; Pielke et al., 1992; Nicholls et al., 1995; Walko et al., 1995a; Lu et al., 2001, 2005; Lu and Shuttleworth, 2002). The model is three dimensional, nonhydrostatic (Tripoli and Cotton, 1980); includes telescoping, interactive nested grid capabilities (Clark and Farley, 1984; Walko et al., 1995b); supports various turbulence closures (Deardorff, 1980; McNider and Pielke, 1981; Tripoli and Cotton, 1986), shortwave and longwave radiation (Mahrer and Pielke, 1977; Chen and Cotton, 1983, 1987; Harrington, 1997), initialization (Tremback, 1990), and boundary condition schemes (Pielke et al., 1992); includes a land surface energy balance submodel that accounts for vegetation, open water, and snow-related surface fluxes (Mahrer and Pielke, 1977; McCumber and Pielke, 1981; Avissar et al., 1985; Tremback and Kessler, 1985; Avissar and Mahrer, 1988; Lee, 1992); and includes explicit cloud microphysical submodels describing liquid and ice processes related to clouds and precipitation (Meyers et al., 1992; Meyers, 1995; Walko et al., 1995a). A modified Kuo (1974) scheme is used for convection-produced precipitation. The RAMS horizontal grid uses an oblique (or rotated) polar-stereo-graphic projection, where the projection pole is near the center of the simulation domain. The vertical grid uses a σ_z terrain-following coordinate system (Gal-Chen and Somerville, 1975; Clark, 1977; Tripoli and Cotton, 1982), where the top of the model is flat and the bottom follows the terrain. An Arakawa-C-grid configuration is used in the model, where the velocity components u, v, and w are defined at locations staggered one-half a grid length in the x, y, and z directions, respectively, from the thermodynamic, moisture, and pressure variables (Arakawa and Lamb, 1977).

The soil submodel used in this version of RAMS provides prognostic temperature and moisture for both soil and vegetation. For bare soil, RAMS uses a multilayer soil model described by Tremback and Kessler (1985). The moisture diffusivity, hydrologic conductivity, and moisture potential is given by Clapp and Hornberger (1978). The thermal properties of the soil are a function of the soil moisture. The boundary condition for moisture at the deepest soil level is held constant in time and equal to the initial value. The temperature of the bottom soil layer varies following the deep soil temperature model of Deardorff (1978). For the vegetated surface, RAMS uses the "big leaf" approach where there is a layer of vegetation overlying a shaded soil (Avissar et al., 1985; Avissar and Mahrer, 1988; Lee, 1992; Walko et al., 2000). The moisture taken from soil by transpiration is accomplished by defining a vertical root profile (Dickinson et al., 1986) and extracting the water masses from the soil depending on the fraction of roots in each soil layer. The surface layer fluxes of heat, momentum, and water vapor are computed using the method of Louis (1979) and Louis et al. (1982).

The model's setup and parameterizations are summarized in Table 18.1.

18.2.2 Study Area and Experiment Design

The model domain used in this study is shown in Figure 18.2. It comprises a coarse grid covering the entire conterminous United States at 200-km grid spacing and a finer, nested grid covering Kansas, Nebraska, South Dakota, Wyoming, and Colorado at 50-km grid spacing. The finer grid covers an area of 1500 km in the east–west direction and 1300 km in the north–south direction, which includes Rocky Mountain National Park, Yellowstone National Park, Arapaho and Roosevelt National Forests, Pawnee National Grassland, etc. that are designated as protected natural areas. The pole point for the oblique polar stereographic projection used to define the grid is 40°N–100°W. There are 20 vertical levels in the modeled atmosphere, with a layer thickness of 119 m at the surface, stretching to 2000 m at the (23 km) top of the domain. The model is driven by 6-hourly lateral boundary conditions derived from National Centers for Environmental Prediction (NCEP) atmospheric reanalysis

TABLE 18.1

Model Options

Category	Options Selected	References
Basic equations	Nonhydrostatic; compressible	Tripoli and Cotton (1980)
Vertical coordinates	Terrain-following sigma z	Clark (1977); Tripoli and Cotton (1982)
Horizontal coordinates	Oblique polar-stereographic projection	
Grid stagger and structure	Arakawa C grid, multiple nested grid (fixed)	Arakawa and Lamb (1977)
Time differencing	Hybrid	
Large-scale precipitation	Dump-bucket	Cotton et al. (1995); Rhea (1978)
Convective parameterization	Modified—Kuo for Grid #1 and #2	Tremback (1990)
Radiation	Mahrer/Pielke	Mahrer and Pielke (1977)
Cloud	Thompson	Thompson (1993)
Surface layer	Louis	Louis (1979); Louis et al. (1982)
	Prognostic soil model	Tremback and Kessler (1985)
	Vegetation parameterization	McCumber and Pielke (1981); Avissar and Mahrer (1988); Lee (1992); Walko (2000)

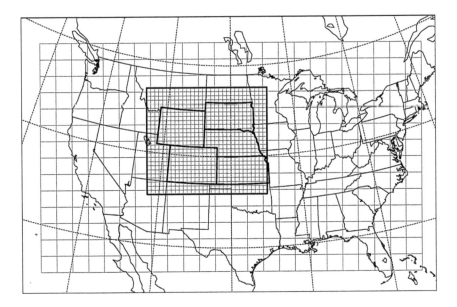

FIGURE 18.2
The simulation domain and grid configuration used for RAMS in this study. The grid intervals used in the coarse- and fine-grid domains were 200 and 50 km, respectively.

products (Kalnay et al., 1996). Lateral boundary condition nudging, which includes horizontal wind speed, relative humidity, air temperature, and geopotential height on pressure levels, is performed on the two outer-boundary grid cells of the coarse grid. The initial atmospheric fields are also provided from the NCEP reanalysis. The time step for the atmospheric model integrations is 2 min.

This domain has rather complex topographic features because it covers parts of the Great Plains and the Rocky Mountains. Heterogeneous soil types, based on the U.S. Department of Agriculture (USDA) State Soil Geographic Database (STATSGO), were used within the model domain (Miller and White, 1998). The model has 10 soil layers with boundaries at 2.0, 1.65, 1.3, 0.95, 0.65, 0.45, 0.3, 0.2, 0.125, and 0.05 m below the surface. The initial distribution of soil moisture is generated by first defining spatially constant soil moisture content (i.e., 40% of the total water capacity) over the domain, and then running the model for 1 year. The modeled soil moisture distribution on the last day of that 1-year simulation was then used as the initial condition for the next year's simulation.

The model vegetation distribution is defined using the International Geosphere–Biosphere Programme (IGBP) land-cover classification. Associated with this complex topography and soils, the domain also includes rather diverse vegetation classes, including C3 and C4 grassland, various agricultural croplands, evergreen needle-leaf trees, shrub land, and tundra.

18.3 Assimilating NDVI-Derived LAI into RAMS

18.3.1 NDVI Dataset

The Pathfinder AVHRR 10-day composite NDVI dataset for North America was used in this study. These data are available at 8-km resolution for the period August 1981 to present. The high spatial variation within these NDVI data emphasizes the significant differences in vegetation phenology across North America. For the purposes of the current study, these Pathfinder NDVI data were aggregated from their original 8 km × 8 km pixel scale to the 50 km × 50 km fine-grid scale (Figure 18.2).

Meanwhile, we obtained 3800 first-order Summary of the Day (SOD) meteorological station observations data from the National Climatic Data Center (NCDC), which include observations of the daily precipitation, maximum and minimum screen-height air temperatures (T_{max} and T_{min}). Analyzing the monthly mean maximum and minimum screen-height air temperatures and precipitation, the year 1989 was identified as most close to near-average year, and was therefore chosen for control simulation.

To illustrate the correlation between SOD and NDVI, the monthly, domain-averaged T_{max}, T_{min}, precipitation, and NDVI values were averaged over June–July–August (JJA), that is, the peak growing season in Northern Hemisphere midlatitudes. The resulting time series are shown in Figure 18.3. NDVI and rainfall in the central United States are positively correlated, reflecting the fact that vegetation growth depends strongly on soil moisture availability, which in turn depends on rainfall amount and frequency. In part, this is also why the NDVI of grassland exhibits greater

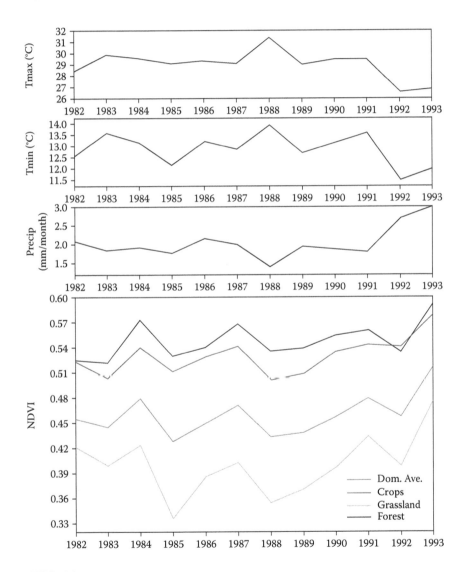

FIGURE 18.3
Time series of climate variables and NDVI averaged over the domain of the fine grid (Figure 18.2) for the period 1982–1993. JJA average values, selected to represent the values during the peak growing season, are shown.

interannual variability than trees and crops: the deeper and more extensive rooting of trees and the irrigation of many crops may well allow them to have more consistent access to soil moisture than grasslands. The fact that NDVI and precipitation are significantly correlated argues for the use of realistic vegetation growth models in climate simulations. The bottom panel in Figure 18.3 shows that natural grasslands have the smallest value of NDVI, trees have the largest, and crops lie between the two.

The NDVI-to-LAI conversion algorithm follows Sellers et al. (1996), and the detail is provided by Lu and Shuttleworth (2002).

18.3.2 Results from Assimilation Runs

Three experiments were carried out to evaluate the impact of directly assimilating NDVI-derived estimates of LAI into RAMS. In the first run, here called the "assimilation" (ASSM) run, the LAI is derived from the NDVI observations. In the second run, here called the "default" (DEF) run, LAI was prescribed to follow the standard used in RAMS (which follows that of the equivalent BATS classes). The third run was a sensitivity test (SEN), in which the LAI has the same spatial distribution as for the DEF run, but the domain-averaged value of LAI (and each contributing LAI) was reduced to agree with the LAI derived from the NDVI observations. The spatial and temporal LAI distributions used in these three experiments are shown in Figure 18.4.

18.3.2.1 Assimilation Run (ASSM) Compared to Observations

The simulated climate given by the RAMS run in which NDVI was assimilated (ASSM) is summarized in Figure 18.5, where the model's ability to reproduce the observed domain-averaged daily maximum (T_{max}) and minimum (T_{min}) screen-height air temperature and the daily precipitation is presented. It is reassuring that the model simulation adequately captures the synoptic signal and evolution in seasonal temperature. Over the year, the model-simulated T_{max} and T_{min} are, on average, 0.75°C and 2.78°C lower than observations, respectively, while the modeled daily precipitation is, on average, 0.58 mm/day too high with respect to observations. The overestimation of precipitation occurs mainly in JJA, that is, during the growing season; consequently, the simulated Northern Hemisphere summer is colder and wetter than observations in this domain.

18.3.2.2 Assimilation Run (ASSM) Compared to Default Run (DEF)

Figure 18.6a shows that ASSM run is generally colder and rains more during the growing season when compared with DEF run. In the DEF run, seasonal evolution of LAI is defined by the BATS classification and has a sinusoidal

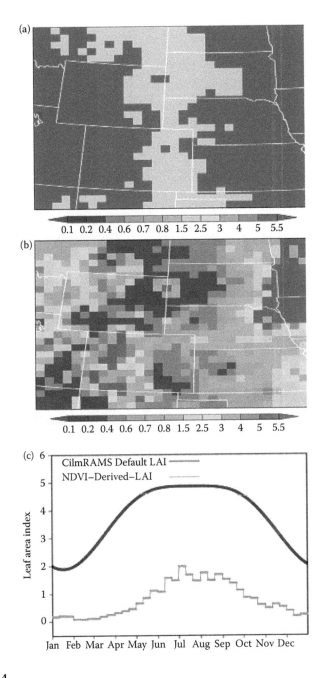

FIGURE 18.4
(a) The spatial distribution of LAI across the fine grid, averaged May through August 1989 from RAMS default LAI specification; (b) the same as (a), but aggregated from 8-km satellite NDVI product to 50-km grid-spacing LAI; and (c) time series of both default LAI and NDVI-derived-LAI, averaged over the entire domain of the fine grid (Figure 18.2).

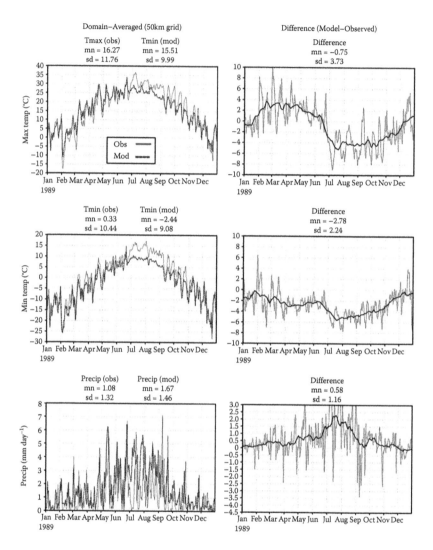

FIGURE 18.5

Observed, domain-averaged daily maximum and minimum screen-height air temperature and daily precipitation compared with the equivalent values calculated by RAMS with assimilation of LAI derived from NDVI. The variables have all been averaged over the (50 km) fine grid. Also shown are the difference between the model and observations and the 30-day running mean of these differences. The mean (mn) and standard deviation (sd) for each panel and variable are included.

variation with the day of the year. Consequently, the spatial pattern and magnitude of the LAI stay much the same from June through October. On the other hand, the ASSM run uses the 10-day composite NDVI data to derive the specification of LAI, which results in much greater heterogeneities across the modeled domain, in terms of both the spatial distribution and the

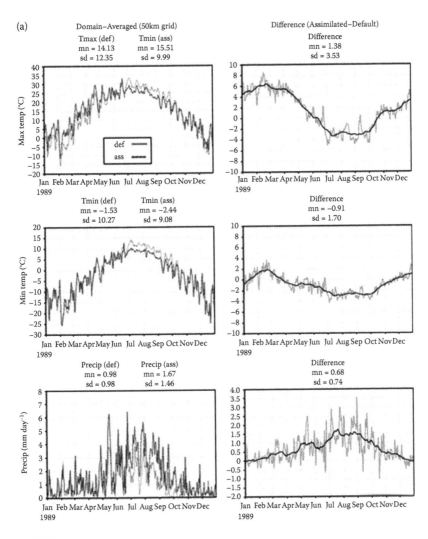

FIGURE 18.6
Domain-averaged daily maximum and minimum screen-height air temperature and daily precipitation calculated by RAMS for (a) the DEF and ASSM runs, and (b) the SEN and DEF runs. In each case, the variables have all been averaged over the (50 km) fine grid. Also shown are the difference between the model and observations and the 30-day running mean of these differences. The mean (mn) and standard deviation (sd) for each panel and variable are included.

seasonal evolution of the LAI (Figure 18.4). It is well known that such land surface heterogeneity can induce mesoscale circulations in the atmosphere that not only influence the surface layer immediately above the vegetation, but which can also trigger moist convection and precipitation in preferred areas.

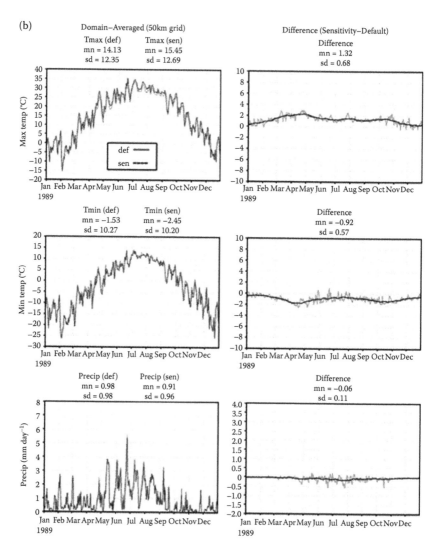

FIGURE 18.6 Continued.

18.3.2.3 Sensitivity Run (SEN) Compared to Default Run (DEF)

However, there are two striking differences between the specification of LAI used in the ASSM and DEF runs. First, the magnitude of the domain-averaged LAI during the summer season is approximately a factor of 2 less for the ASSM run than for the DEF run. Second, the LAI spatial distribution of the LAI used in the ASSM run is much more heterogeneous than that used in the DEF run. The question arises: Which factor leads to the extra precipitation produced during the ASSM run? To clarify this issue, a third experiment (SEN) was carried out in which the LAI was specified to have the same domain-averaged value as the ASSM run but the spatial distribution of the DEF run (Figure 18.4). Specifically, the LAI for each default vegetation class was scaled down by a factor equal to the ratio of the domain-averaged LAI for the ASSM run divided by the domain-averaged LAI for the DEF run, this latter ratio being a function of the day of the year. Comparison between the simulated climates given in the SEN and ASSM runs relative to that given by the DEF run should define whether it is the overall magnitude of the NDVI-derived LAI or greater LAI heterogeneity that is most important.

Figure 18.6b shows that the climate simulated by the SEN run follows much more closely that simulated by the DEF run. On average, the climate simulated in the SEN run, in fact, has a drier and warmer summer than the DEF run, the opposite result to that for the ASSM run. Consequently, reducing the magnitude of LAI alone does not contribute to the extra precipitation simulated in the ASSM run. On the contrary, the simulated precipitation decreased (as might be expected if the overall magnitude of the LAI decreases). We conclude that it is the introduction of spatial heterogeneity into NDVI-derived fields that is the primary cause of the generally wetter and colder summer climate produced in the ASSM simulation.

18.4 The Coupled RAMS and CENTURY Modeling System

Section 18.3 demonstrates that modeled climate is indeed sensitive to the LAI specifications, and that the changes in LAI have first-order effect on the model-simulated weather and climate. Next, we attempt to use an ecosystem model to account for the temporal and spatial heterogeneity introduced by vegetation growth. A coupled RAMS and CENTURY modeling system is developed to study the regional-scale two-way interactions between the atmosphere and biosphere (Figure 18.7). Both atmospheric forcings and ecological parameters are prognostic variables in the linked system. The atmospheric and ecosystem models exchange information on a weekly time step. CENTURY receives as input air temperature, precipitation, radiation, wind speed, and relative humidity simulated by RAMS. From CENTURY-produced outputs, leaf area index, and vegetation transmissivity are computed and returned to RAMS. In this way, vegetation responses to weekly and seasonal atmospheric changes are simulated and fed back to the atmospheric–land surface hydrology model.

FIGURE 18.7
Coupled RAMS and CENTURY modeling system.

18.4.1 CENTURY Ecosystem Model

CENTURY is a biogeochemistry model that was originally designed to simulate long-term dynamics of carbon (C), nitrogen (N), phosphorus (P), and sulfur (S) for different plant–soil systems (Figure 18.8). Since the mid-1980s, the CENTURY model has been developed, modified, and applied to simulate various ecosystem dynamics over a wide range of spatial and temporal scales (Parton et al., 1987, 1988, 1993, 1994a,b, 1995, 1996; Ojima et al., 1993, 1994; Parton and Rasmussen, 1994; Parton, 1996). The grassland, agriculture crop, forest, and savanna ecosystems have different plant production submodels that are linked to a common soil organic matter submodel (SOM). The SOM simulates the flow of C, N, P, and S through plant litter and the different inorganic and organic pools in the soil. The model includes three soil organic matter pools (active, slow, and passive) with different potential decomposition rates, above- and below-ground litter pools, and a surface microbial pool that is associated with decomposing surface litter. The plant production models assume that plant production is controlled by moisture and temperature, and that plant production rates are decreased if nutrient supplies are insufficient. The fraction of the mineralized pools that are available for plant growth is a function of the root biomass increases.

The versions of the CENTURY model used in our current study are CENTURY version 4 (Parton, 1996) and daily time step CENTURY (DayCENT). The input variables are available for most natural and agricultural ecosystems and can generally be estimated from the existing literature (Parton et al., 1987). DayCENT (Parton et al., 1998; Kelly et al., 2000) uses a daily time step for the water and nutrient cycles, and the above- and below-ground

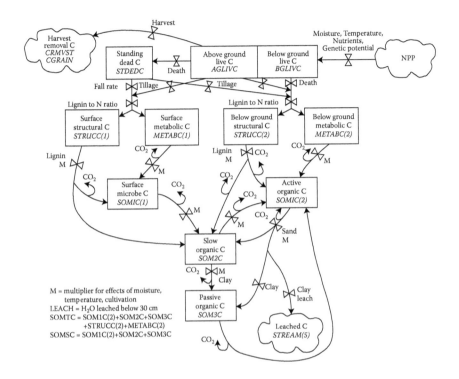

FIGURE 18.8
Flow diagram of carbon pools and flows of the CENTURY ecosystem model (CENTURY Technical Documentation, 1993).

biomass are updated weekly. When an event or management practice was scheduled for a given month, it occurs weekly (irrigation), in the first week of the month (organic matter addition, fertilization, and cultivation), or in the last week of the month (grazing, fire, tree removal, harvest).

18.4.2 Vegetation Growth Simulated by DayCENT

DayCENT is driven by observed daily atmospheric forcings derived from the U.S. SOD dataset for each grid cell of the fine-grid domain (Figure 18.2). The above-ground live carbon (aglivc) produced by DayCENT was then converted to LAI following the algorithm presented by Lu et al. (2001). Figure 18.9 shows monthly normalized difference vegetation index (NDVI)-derived LAI and DayCENT-simulated LAI. The LAI for the fine grid was then averaged over the domain that includes grasslands, trees, and crops. The agreement between the two grassland time series is good, with both the seasonal cycle and interannual variation well captured by the model. The simulated LAI maxima for trees, however, are generally 25% higher than observed. The simulated minima are around 4 LAI units, while the observed minima are around 0.5 LAI units. A possible explanation for the large difference in the minima is that the NDVI data may

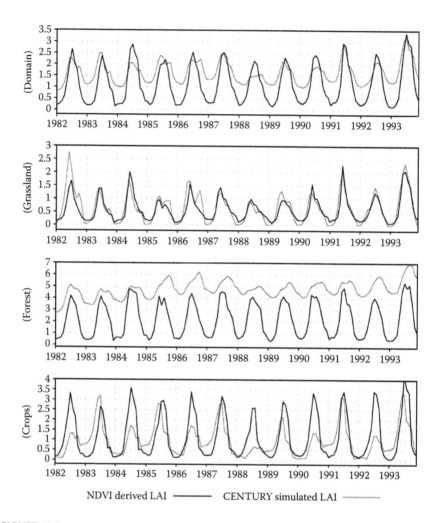

FIGURE 18.9
Monthly NDVI-derived LAI and DayCENT-simulated LAI over the fine grid for the period 1982–1993, averaged over the domain, grasslands, trees, and crops.

have been contaminated by snow cover during the winter months so that the evergreen forest cannot be detected by the satellite sensors. The large seasonal variation in the forest NDVI-derived LAI data may not be realistic for this reason. Thus, we expect that the seasonal cycle of DayCENT-produced forest LAI is actually more realistic than those derived from the NDVI data. The model-simulated crop LAI clearly shows a 2-year alternating "low–high" pattern, introduced by DayCENT's crop management practices; field fallow has been scheduled every other year for the winter wheat, which makes up 50% of the croplands. Comparison of the modeled and observed LAI shows that DayCENT is capable of representing seasonal and interannual biomass variabilities (Figure 18.10).

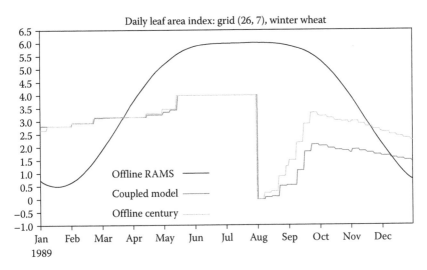

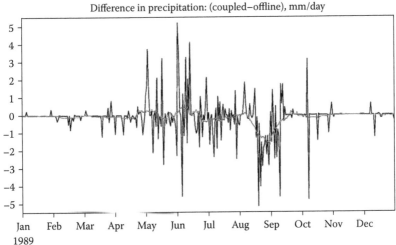

FIGURE 18.10
Daily LAI prescribed by RAMS (offline RAMS), simulated by DayCENT (Offline DayCENT), and simulated by the coupled model, for a single grid cell near Salina, Kansas. Also shown are the precipitation difference between the coupled and offline simulation and the 30-day running mean. The land-use type for this grid cell is winter wheat.

18.4.3 Results from the Coupled RAMS–CENTURY Modeling System

18.4.3.1 *Simulated Climate*

The comprehensive evaluation of the coupled RAMS–CENTURY modeling can be found in Lu et al. (2001). When compared with observations, the coupled model's temperature and precipitation simulation captures the synoptic signals as well as the seasonal evolution. Also, the coupled model produced more precipitation during the summer, which led to a colder summer compared to the offline RAMS (DEF) simulations, as shown in Figure 18.11a. The results show that introducing dynamic vegetation descriptions in regional climate modeling can cause significant changes in atmospheric conditions that, in turn, influence vegetation growth and evolution.

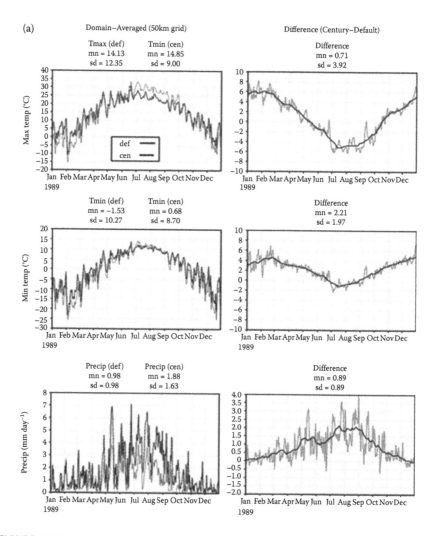

FIGURE 18.11
Domain-averaged daily maximum and minimum screen-height air temperature and daily precipitation calculated by RAMS for (a) the coupled RAMS–CENTURY run and the default run, and (b) the NDVI-derived-LAI assimilation run and the default runs. In each case, the variables have all been averaged over the (50 km) fine grid. Also shown are the difference between the model and observations and the 30-day running mean of these differences. The mean (mn) and standard deviation (sd) for each panel and variable are included.

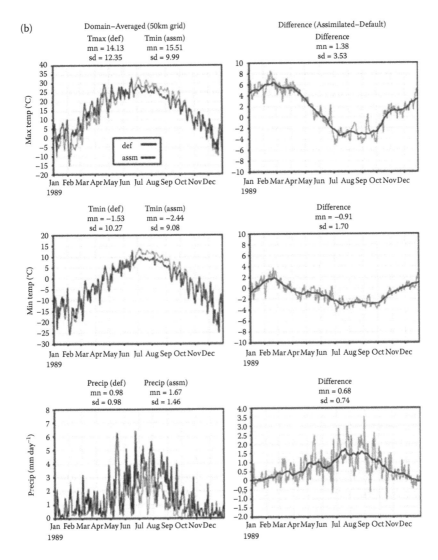

FIGURE 18.11 Continued.

18.4.3.2 Simulated Atmosphere and Vegetation Feedbacks

The coupled model allows us to look into the atmosphere and vegetation feedback dynamics in detail. Figure 18.10 shows the daily LAI prescribed by RAMS, generated by DayCENT, and simulated by the coupled model, for a single grid cell having a winter wheat land-use type. Also shown are the precipitation difference between the coupled and offline simulation and the 30-day running mean. At the beginning of August, the winter wheat harvest occurred that brought the LAI value down to near zero. Corresponding to the harvest event, the coupled model has a dramatic precipitation decrease from August to mid-September due to the sudden shutdown of vegetation transpiration that greatly reduced the local moisture availability. Accordingly, the coupled model starts another year's winter wheat growth with less precipitation and drier soil, resulting in less biomass growth compared to the offline DayCENT that is driven by the offline RAMS climate.

18.5 Summary, Discussions, and Conclusions

This study first focused on implementing the assimilation of NDVI-derived LAI into RAMS and evaluating the impact of doing so on the climate simulated by the model. Then, CENTURY ecosystem model was introduced and linked to RAMS to prognose vegetation growth. Collectively, the regional climate calculated by both ASSM-RAMS and RAMS–CENTURY modeling system remains reasonable relative to observations without any attempt to tune the model from its default state. We found the modeled climate to be sensitive to the LAI specification in RAMS, and that changes in LAI have a first-order effect on the model-simulated weather and climate. The seasonal vegetation phenological variation strongly influences regional climate patterns through its control over land surface water and energy exchange. The coupled model captures the key aspects of weekly, seasonal, and annual feedbacks between the atmospheric and ecological systems.

Putting it together, both satellite-derived and model-calculated LAI have significant impact on the modeled seasonal climate simulated by RAMS. In both cases, as shown in Figure 18.11, the simulated climate is cooler and has more precipitation relative to using RAMS default LAI distribution. Further sensitivity experiment highlights the fact that the effect of spatial heterogeneity in LAI appears to dominate over the effect of a difference in LAI magnitude. Our primary conclusion is, therefore, that including realistic description of heterogeneous vegetation phenology does influence the prediction of seasonal climate.

Four-Dimensional Data Assimilation is widely used as the basis of initiating real-time weather forecasts, and the required real-time global network for atmospheric observation is well established. The recent advent of land surface information derived from satellite observations opens up the possibility of assimilating new variables, such as leaf area index (LAI), surface albedo, and soil moisture, into numerical models. This study suggests that there may be potential benefit in assimilating at least indirectly measured LAI in real time into regional models.

While conducting our sensitivity experiments with a coupled modeling system, we realize that regional atmospheric models are necessarily highly constrained by their prescribed lateral boundary conditions, and some of the effect of vegetation feedback may be lost. Land surface properties can act as a mechanism to provide "triggering effects" in regional climate models, with the overall water and energy budget still largely prescribed by boundary conditions, and the effect of land surface characteristics being mainly a redistribution of water and energy within the simulation domain. In the present study, this was mitigated by two-way nesting of the high-resolution domain of interest within a much larger domain, but some remnant influences of boundary forcing may still be present in our results.

Overall, our study makes a case for the real-time use of satellite-derived LAI in fine-resolution climate models, preferably in fine-resolution climate models operating at a global scale, but at least in fine-resolution regional climate models that are two way nested within general circulation models. Associated with this is the continuing need for improved NDVI data retrieval processes and the NDVI-to-LAI conversion algorithms. Fortunately, this need is now receiving increased attention with the advent of relevant satellite systems within the Earth Observing System. There is also a continuing need to advance high-resolution regional- and global-scale modeling capability. Hopefully, these new modeling systems will be able to resolve mesoscale circulations generated by land surface heterogeneities—an important feature in the present study—and at the same time represent the possible long-term feedbacks from the vegetation and soil moisture. Thus, on the basis of this study, we argue that assimilating the observed vegetation distribution into models is worthwhile, and we anticipate that including remotely sensed observations of vegetation cover will soon become standard in 4DDA systems.

Acknowledgments

The funding of writing this chapter is provided by NOAA CPPA program under NA17RJ1228, NSF AGS-1011975, and MOST Program under 2010CB428502.

References

Arakawa, A. and R. V. Lamb, 1977: Computational design of the basic dynamical processes of the UCLA general circulation model. In: J. Chang (Ed.), *Methods in Computational Physics*, Vol. 17, Academic Press, California, USA, pp. 174–265.

Avissar, R., 1995: Recent advances in the representation of land–atmosphere interactions in general circulation models. *Rev. Geo. Phys.*, 33 (Suppl.), 1005–1010.

Avissar, R., P. Avissar, Y. Mahrer, and B. A. Bravdo, 1985: A model to simulate response of plant stomata to environmental conditions. *Agric. For. Meteor.*, 34, 21–29.

Avissar, R. and Y. Mahrer, 1988: Mapping frost-sensitive areas with a three-dimensional local-scale numerical model. Part I: Physical and numerical aspects. *J. Appl. Meteor.*, 27, 400–413.

Bonan, G. B., 1996: A land surface model (LSM Version 1.0) for ecological, hydrological, and atmospheric studies: Technical description and user's guide. NCAR Technical Note 417, 150 pp.

Bounoua, L., G. J. Gollatz, S. O. Los, P. J. Sellers, D. A. Dazlich, C. J. Tucker, and D. A. Randall, 2000: Sensitivity of climate to changes in NDVI. *J. Climate*, 13, 2277–2292.

Chen, C. and W. R. Cotton, 1983: A one-dimensional simulation of the stratocumulus-capped mixed layer. *Bound. Layer Meteor.*, 25, 289–321.

Chen, C. and W. R. Cotton, 1987: The physics of the marine stratocumulus-capped mixed layer. *J. Atmos. Sci.*, 44, 2951–2977.

Clapp, R. and G. Hornberger, 1978: Empirical equations for some soil hydraulic properties. *Water Resour. Res.*, 14, 601–604.

Clark, T. L., 1977: A small-scale dynamic model using a terrain-following coordinate transformation. *J. Comput. Phys.*, 24, 186–215.

Clark, T. L. and R. D. Farley, 1984: Severe downslope windstorm calculations in two and three spatial dimensions using anelastic interactive grid nesting: A possible mechanism or gustiness. *J. Atmos. Sci.*, 41, 329–350.

Cohen, W. B., T. K. Maierpserger, S. T. Gower, and D. P. Turner, 2003: An improved strategy for regression of biophysical variables and Landsat ETM + data. *Remote Sens. Environ.*, 84, 561–571.

Cotton, W. R., R. A. Pielke, Sr., R. L. Walko, G. E. Liston, C. J. Tremback, H. Jiang, R. L McAnelly et al. 2003: RAMS 2001: Current status and future directions. *Meteorol. Atmos. Phys.*, 82, 5–29.

Deardorff, J. W., 1978: Efficient prediction of ground surface temperature and moisture, with inclusion of layer of vegetation. *J. Geophys. Res.*, 83, 1889–1903.

Deardorff, J. W., 1980: Stratocumulus-capped mixed layers derived from a three-dimensional model. *Bound. Layer Meteor.*, 18, 495–527.

Dickinson, R. E., 1995: Land–atmosphere interaction. *Rev. Geophys.*, 33 (Suppl.), 917–922.

Dickinson, R. E., A. Henderson-Sellers, and P. J. Kennedy, 1993: Biosphere–Atmosphere Transfer Scheme (BATS) version 1E as coupled to the NCAR community climate model. Technical Note NCAR/TN-387 + STR, 72 pp.

Dickinson, R. E., A. Henderson-Sellers, P. J. Kennedy, and M. F. Wilson, 1986: Biosphere–Atmosphere Transfer Schemes (BATS) for the NCAR community climate model. NCAR/TN-275, STR, 69 pp.

Dickinson, R. E., K. W. Oleson, G. Bonan, F. Hoffman, P. Thornton, M. Vertenstein, Z.-L. Yang, and X. Zeng, 2006: The community land model and its climate statistics as a component of the community climate system model. *J. Climate*, 19, 2302–2324. doi: 10.1175/JCLI3742.1.

Dickinson, R. E., M. Shaikh, R. Bryant, and L. Graumlich, 1998: Interactive canopies for a climate model. *J. Climate*, 11, 2823–2836.

Dirmeyer, P. A., 1994: Vegetation stress as a feedback mechanism in midlatitude drought. *J. Climate*, 7, 1463–1483.

Foley, J. A., S. Levis, I. C. Prentice, D. Pollard, and S. L. Thompson, 1998: Coupling dynamic models of climate and vegetation. *Global Change Biol.*, 4, 561–579.

Foody, G. M., D. S. Boyd, and M. E. J. Cutler, 2003: Predictive relations of tropical forest biomass from Landsat TM data their transferability between regions. *Remote Sens. Environ.*, 85, 463–474.

Gal-Chen, T. and R. C. J. Somerville, 1975: On the use of coordinate transformation for the solution of the Navier–Stokes equations. *J. Comput. Phys.*, 17, 209–228.

Gash, J. H. C. and C. A. Nobre, 1997: Climatic effects of Amazonian deforestation: Some results from ABRACOS. *Bull. Am. Meteor. Soc.*, 78, 823–830.

Gitelson, A. A., Y. J. Kaufman, R. Stark, and D. Rundquist, 2002: Novel algorithms for remote estimation of vegetation fraction. *Remote Sens. Environ.*, 80, 76–87.

Haboudane, D., J. R. Miller, E. Pattey, P. J. Zarco-Tejada, and I. Strachan, 2004: Hyperspectral vegetation indices and novel algorithms for predicting green LAI of crop canopies: Modeling and validation in the context of precision agriculture. *Remote Sens. Environ.*, 90, 337–352.

Harrington, J. Y., 1997: The effects of radiative and microphysical processes on simulated warm and transition season Arctic stratus. PhD dissertation, Atmospheric Science Paper 637, Department of Atmospheric Science, Colorado State University, 289 pp.

Henderson-Sellers, A., A. J. Pitman, P. K. Love, P. Irannejad, and T. H. Chen, 1995: The Project for Intercomparison of Land Surface Parameterization Schemes (PILPS) Phases 2 and 3. *Bull. Am. Meteor. Soc.*, 76, 489–503.

Henderson-Sellers, A., Z. L. Yang, and R. E. Dickinson, 1993: The project for intercomparison of land-surface parameterization schemes. *Bull. Am. Meteor. Soc.*, 74, 1335–1349.

Kalnay, E., M., Kanamitsu, R., Kistler, W., Collins, D., Deaven, L., Gandin, M., Iredell et al. 1996: The NCEP/NCAR 40-Year Reanalysis Project. *Bull. Am. Meteor. Soc.*, 77, 437–471.

Kelly, R. H., W. J. Parton, M. D. Hartman, L. K. Stretch, D. S. Ojima, and D. S. Schimel, 2000: Intra-annual and interannual variability of ecosystem processes in shortgrass steppe. *J. Geophys. Res.*, 105, 20 093–20 100.

Koster, R., J. Jouawl, R. Suozzo, and G. Russel, 1986: Global sources of local precipitation as determined by the NASA/GISS GCM. *Geophys. Res. Lett.*, 13, 121–124.

Kuo, H. L., 1974: Further studies of the parameterization of the influence of cumulus convection on large-scale flow. *J. Atmos. Sci.*, 31, 1232–1240.

Labus, M. P., G. A. Nielsen, R. L. Lawrence, R. Engel, and D. S. Long, 2002: Wheat yield estimates using multitemporal NDVI satellite imagery. *Int. J. Remote Sens.*, 23, 4169–4180.

Lee, T. J., 1992: The impact of vegetation on the atmospheric boundary layer and convective storms. PhD dissertation, Department of Atmospheric Science Paper 509, Colorado State University, 137 pp.

Levis, S., G. B., Bonan, M. Vertenstein, and K. W. Oleson, 2004: The Community Land Model's Dynamic Global Vegetation Model (CLM-DGVM). Technical Description and User's Guide NCAR Technical Note, NCAR/TN-459 + IA.

Louis, J. F., 1979: A parametric model of vertical eddy fluxes in the atmosphere. *Bound. Layer Meteor.*, 17, 187–202.

Louis, J. F., M. Tiedtke, and J. F. Geleyn, 1982: A short history of the PBL parameterization at ECMWF. *Proceedings of the 1981 Workshop on PBL Parameterization*, Shinfield Park, Reading, United Kingdom, ECMWF, pp. 59–71.

Lu, L., A. S. Denning, M. A. da Silva Dias, P. Silva-Dias, M. Longo, S. R. Freitas, and S. Saatchi, 2005: Mesoscale circulation and atmospheric CO_2 variation in the Tapajos Region, Para, Brazil. *J. Geophys. Res.*, 110, D21102. doi:10.1029/2004JD005757.

Lu, L., R. A. Pielke, G. E. Liston, W. J. Parton, D. Ojima, and M. Hartman, 2001: Implementation of a two-way interactive atmospheric and ecological model and its application to the central United States. *J. Climate*, 14, 900–919.

Lu, L. and W. J. Shuttleworth, 2002: Incorporating NDVI-derived LAI into the climate version of RAMS and its impact on regional climate. *J. Hydrometeorology*, 3, 347–362.

Mahrer, Y. and R. A. Pielke, 1977: A numerical study of the airflow over irregular terrain. *Beitr. Phys. Atmos.*, 50, 98–113.

McCumber, M. C. and R. A. Pielke, 1981: Simulation of the effects of surface fluxes of heat and moisture in a mesoscale numerical model, Part I: Soil layer. *J. Geophys. Res.*, 86, 9929–9938.

McNider, R. T. and R. A. Pielke, 1981: Diurnal boundary-layer development over sloping terrain. *J. Atmos. Sci.*, 38, 2198–2212.

Meyers, M. P., 1995: The impact of a two-moment cloud model on the microphysical structure of two precipitation events. PhD dissertation, Atmospheric Science Paper 575, Department of Atmospheric Science, Colorado State University, 165 pp.

Meyers, M. P., P. J. DeMott, and W. R. Cotton, 1992: New primary ice nucleation parameterizations in an explicit cloud model. *J. Appl. Meteor.*, 31, 708–721.

Mihailovic, T., 1996: Description of a land-air parameterization scheme (LAPS). *Global Planet. Change*, 13(1–4), 207–215.

Miller, D. A. and R. A. White, 1998: A conterminous United States multi-layer soil characteristics data set for regional climate and hydrology modeling. *Earth Interactions*, 2.

Myneni, R. B. and B. D. Ganapol, 1992: Remote sensing of vegetation canopy photosynthetic and stomatal conductance efficiencies. *Remote Sens. Environ.*, 42, 217–238.

Myneni, R. B., R. R. Nemani, and S. W. Running, 1997: Estimation of global leaf area index and absorbed PAR using radiative transfer models. *IEEE Trans. Geosci. Remote Sens.*, 35, 1380–1393.

Nicholls, M. E., R. A. Pielke, J. L. Eastman, C. A. Finley, W. A. Lyons, C. J. Tremback, R. L. Walko, and W. R. Cotton, 1995: Applications of the RAMS numerical model to dispersion over urban areas. In: J. E. Cermak et al. (Ed.), *Wind Climate in Cities*. Kluwer Academic Publishers, Amsterdam, pp. 703–732.

Noilhan, J. and S. Planton, 1989: A simple parameterization of land surface processes for meteorological models. *Mon. Wea. Rev.*, 117, 536–549.

Ojima, D. S., W. J. Paron, D. S. Schimel, J. M. O. Scurlock, and T. G. F. Kittel, 1993: Modeling the effect of climatic and CO_2 changes on grassland storage of soil C. *Water Air Soil Pollut.*, 70, 643–657.

Ojima, D. S., D. S. Schimel, W. J. Parton, and C. Owensby, 1994: Short- and long-term effects of fire on N cycling in tallgrass prairie. *Biogeochemistry*, 24, 67–84.

Oleson, K. W., Y. Dai, G. Bonan, M. Bosilovich, R. Dickinson, P. Dirmeyer, F. Hoffman et al. 2004: Technical description of the Community Land Model (CLM). NCAR Technical Note, NCAR/TN-461 + STR.

Parton, W. J., 1996: The CENTURY model. In: D. S. Powlson, P. Smith, and J. U. Smith (Eds.), *Evaluation of Soil Organic Matter Models*, NATO ASI Series, Vol. I3, Springer-Verlag, Berlin, Germany, pp. 283–291.

Parton, W. J., M. Hartman, D. Ojima, and D. Schimel, 1998: DAYCENT and its land-surface submodel: Description and testing. *Global Planet. Change*, 19, 35–48.

Parton, W. J., D. S. Ojima, and D. S. Schimel, 1996: Models to evaluate soil organic matter storage and dynamics. In: M. R. Carter (Ed.), *Structure and Organic Matter Storage in Agricultural Soils*. CRC Press, Inc., Boca Raton, Florida, USA, pp. 421–448.

Parton, W. J. and P. E. Rasmussen, 1994: Long-term effects of crop management in wheat-fallow: II. CENTURY model simulations. *Soil Sci. Soc. Am. J.*, 58, 530–536.

Parton, W. J., D. S. Schimel, C. V. Cole, and D. S. Ojima, 1987: Analysis of factors controlling soil organic matter in Great Plains Grassland. *Soil Sci. Soc. Am. J.*, 51, 1173–1179.

Parton, W. J., J. W. B. Steward, and C. V. Cole, 1988: Dynamics of C, N, P, and S in grassland soils: A model. *Biogeochemistry*, 5, 109–131.

Parton, W. J. and coauthors, 1993: Observations and modeling of biomass and soil organic matter dynamics for the grassland biome world-wide. *Global Biogeochem. Cycles*, 7, 785–809.

Parton, W. J., D. S. Schimel, D. S. Ojima, and C. V. Cole, 1994a: A general model for soil organic matter dynamics: Sensitivity to litter chemistry, texture and management. *Quantitative Modeling of Soil Forming Processes*, SSSA Special Publication 39, R. B. Bryant and R. W. Arnold, (Eds.), ASA, CSSA, and SSA, Madison, Wisconsin, USA, pp. 137–167.

Parton, W. J., J. M. O. Scurlock, D. S. Ojima, and K. Paustian, 1994b: Modelling soil biology and biochemical processes for sustainable agricultural research. In: C. E. Pankhurst et al. (Eds.) *Soil Biota: Management in Sustainable Farming Systems*, CSIRO, Information Services, Melbourne, Australia, pp. 182–193.

Parton, W. J. and coauthors, 1995: Impact of climate change on grassland production and soil carbon worldwide. *Global Change Biol.*, 1, 13–22.

Pielke, R. A., 1974: A three-dimensional numerical model of the sea breezes over south Florida. *Mon. Wea. Rev.*, 102, 115–139.

Pielke, R. A. and Avissar, 1990: Influence of landscape structure on local and regional climate. *Landscape Ecol.*, 4, 133–155.

Pielke, R. A., R. A. Avissar, M. Raupach, H. Dolman, X. Zeng, and S. Denning, 1998: Interactions between the atmospheric and terrestrial ecosystems: Influence on weather and climate. *Global Change Biol.*, 4, 101–115.

Pielke, R. A. and coauthors, 1992: A comprehensive meteorological modeling system—RAMS. *Meteor. Atmos. Phys.*, 49, 69–91.

Pielke, R. A., Sr, G. Marland, R. A. Betts, T. N. Chase, J. L. Eastman, J. O. Nile, D. S. Niyogi, and S. W. Running, 2002: The influence of land-use change and landscape dynamics on the climate system: Relevance to climate-change policy beyond the radiative effect of greenhouse gases. *Phil. Trans. R. Soc. Lond. A*, 360, 1705–1719. doi: 10.1098/rsta.2002.1027.

Pielke, R. A. and D. Niyogi, 2010: The role of landscape processes within the climate system. *Landform-Struct. Evol. Process Control Lect. Notes Earth Sci.*, 115, 67–85. doi: 10.1007/978-3-540-75761-0_5.

Pitman, A. J., 1991: A simple parameterization of sub-grid scale open water for climate models. *Climate Dyn.*, 6, 99–112.

Pitman, A. J., 2003: Review—The evolution of, and revolution in, land surface schemes designed for climate models. *Int. J. Climatol.*, 23, 479–510.

Purevdorj, T., Tateishi, R., Ishiyama, T., and Honda, Y., 1998: Relationships between percent vegetation cover and vegetation indices. *Int. J. Remote Sen.*, 19, 3519–3535.

Rabin, R. M., S. Stadler, P. J. Wetzel, D. J. Stensrud, and M. Gregory, 1990: Observed effects of landscape variability on convective clouds. *Bull. Am. Meteor. Soc.*, 71, 272–280.

Segal, M., 1989: Impact of crop areas in northeast Colorado on mid- summer meso-scale thermal circulations. *Mon. Wea. Rev.*, 117, 809–825.

Sellers, P. J., J. A. Berry, G. J. Gollatz, C. B. Field, and F. G. Hall, 1992: Canopy reflectance, photosynthesis and transpiration, III, A reanalysis using improved leaf models and a new canopy integration scheme. *Remote Sens. Environ.*, 42, 187–216.

Sellers, P. J., S. O. Los, C. J. Tucker, C. O. Justice, D. A. Dazlich, G. J. Collatz, and D. A. Randall, 1996: A revised land surface parameterization (SiB2) for atmospheric GCMs. Part II: The generation of global fields of terrestrial biophysical parameters from satellite data. *J. Climate*, 9, 706–737.

Sellers, P. J., Y. Mints, R. C. Sud, and A. Delcher, 1986: A simple biosphere model (SiB) for use within general circulation models. *J. Atmos. Sci.*, 43, 505–531.

Todd, S. W., R. M. Hoffer, and D. G. Milchunas, 1998: Biomass estimation on grazed and ungrazed rangelands using spectral indices. *Int. J. Remote Sens.*, 19(3), 427–438.

Tremback, C. J., 1990: Numerical simulation of a mesoscale convective complex: Model development and numerical results. PhD dissertation, Atmospheric Science Paper 465, Department of Atmospheric Science, Colorado State University, 247 pp.

Tremback, C. J. and R. Kessler, 1985: A surface temperature and moisture parameterization for use in mesoscale numerical models. Preprints, *Seventh Conference on Numerical Weather Prediction*, Montreal, PQ, Canada, American Meteorological Society, pp. 355–358.

Tripoli, G. J. and W. R. Cotton, 1980: A numerical investigation of several factors contributing to the observed variable intensity of deep convection over South Florida. *J. Appl. Meteor.*, 19, 1037–1063.

Tripoli, G. J. and W. R. Cotton, 1982: The Colorado State University three-dimensional cloud/mesoscale model—1982. Part I: General theoretical framework and sensitivity experiments. *J. Rech. Atmos.*, 16, 185–220.

Tripoli, G. J. and W. R. Cotton, 1986: An intense, quasi-steady thunderstorm over mountainous terrain. Part IV: Three-dimensional numerical simulation. *J. Atmos. Sci.*, 43, 896–914.

Tsvetsinskaya, E. A., L. O. Mearns, and W. E. Easterling, 2001: Investigating the effect of seasonal plant growth and development in three-dimensional atmospheric simulations. Part II: Atmospheric response to crop growth and development. *J. Climate*, 14, 711–729.

Turner, D. P., W. B. Cohen, R. E. Kennedy, K. S. Fassnacht, and J .M. Briggs, 1999: Relationships between leaf area index and Landsat TM spectral vegetation indices across three temperate zone sites. *Remote Sens. Environ.*, 70, 52–68.

Wang, J. F., F. J. F. Chagnon, E. R. Williams, A. K. Betts, N. O. Renno, L. A. T. Machado, G. Bisht, R. Knox, and R. L. Brase, 2009: Impact of deforestation in the Amazon basin on cloud climatology. *PNAS*, 106, 10, 3670–3674. doi: 10.1073.

Walko, R. L., W. R. Cotton, M. P. Meyers, and J. Y. Harrington, 1995a: New RAMS cloud microphysics parameterization. Part I: The single-moment scheme. *Atmos. Res.*, 38, 29–62.

Walko, R. L., C. J. Tremback, R. A. Pielke, and W. R. Cotton, 1995b: An interactive nesting algorithm for stretched grids and variable nesting ratios. *J. Appl. Meteor.*, 34, 994–999.

Walko, R. L. and coauthors, 2000: Coupled atmosphere–biophysics–hydrology models for environmental modeling. *J. Appl. Meteor.*, 39, 931–944.

Xue, Y., P. J. Sellers, J. L. Kinter, and J. Shukla, 1991: A simplified biosphere model for global climate studies. *J. Climate*, 4, 345–364.

Xue, Y., M. J. Fennessy, and P. J. Sellers, 1996: Impact of vegetation properties on U.S. summer weather prediction. *J. Geophys. Res.*, 101, 7419–7430.

19

Monitoring Natural Hazards in Protected Lands Using Interferometric Synthetic Aperture Radar

Zhong Lu, Daniel Dzurisin, and Hyung-Sup Jung

CONTENTS

19.1 Introduction

Natural hazards in protected lands frequently result in damage to or loss of lives, livelihoods, structures, or ecosystem productivity. Often, natural hazards occur in remote areas of protected lands or impact on such large areas that *in situ* measurements are not feasible, accurate, or timely enough for monitoring or early warning. Satellite interferometric synthetic aperture radar (InSAR) provides an all-weather imaging capability for measuring ground-surface deformation with centimeter-to-subcentimeter precision and inferring changes in landscape characteristics over a large region. With its global coverage and all-weather imaging capability, InSAR is an important remote sensing technique for measuring ground-surface deformation of various natural hazards and the associated landscape changes. The spatial distribution of surface deformation data, derived from InSAR imagery, enables the construction of detailed numerical models to enhance the study of physical processes of natural hazards. This chapter (a) introduces the basics of InSAR for deformation mapping and landscape change detection, (b) discusses state-of-the-art technical issues in InSAR processing and interpretation, and (c) showcases the application of InSAR to the study of volcano, earthquake, landslide, and glacier movement and the mapping of high-resolution digital elevation model and fire progression with InSAR imagery over protected lands.

19.2 Principles of Interferometric Synthetic Aperture Radar

19.2.1 Imaging Radar

Radar is an acronym for radio detection and ranging, which hints at some of the technique's uses and capabilities (Levanon, 1988). Radar operates by broadcasting a pulse of electromagnetic energy into space. If the pulse encounters an object, some of the energy is redirected to the radar. The same antenna can be used to transmit the initial pulse and receive the return signal. Precise timing of the delay between the initial and return signals allows determination of the distance from radar to object, and the Doppler frequency shift between the two signals is a measure of the object's velocity relative to the radar. Thus, radar can be used to detect and measure the velocity of things such as aircraft or highway vehicles—two common uses of Doppler radar. But radar's capabilities extend far beyond air traffic control and law enforcement.

Imaging radar systems operate on the same principles but have additional capability to distinguish among return signals from individual resolution elements within a target footprint. This enables processing of resulting data into an image of the target area, which contains information about topography and radar-reflective properties of the surface. The resolution of real-aperture imaging radar systems depends on, among other factors, the size of the antenna (bigger is better), which, for practical reasons, is limited to a few meters or decimeters. This limitation can be overcome, however, by creating a much larger "synthetic" radar antenna.

19.2.2 Synthetic Aperture Radar

Synthetic aperture radar (SAR) is an advanced radar system that utilizes image processing techniques to synthesize a large virtual antenna, which provides much higher spatial resolution than is practical using a real-aperture radar. Because SAR actively transmits and receives signals backscattered from the target area, and because radar wavelengths are mostly unaffected by weather clouds, SAR can operate effectively during day and night under most weather conditions. Using a sophisticated image processing technique, called SAR processing (Bamler and Hartl, 1998; Curlander and McDonough, 1991; Henderson and Lewis, 1998), both intensity and phase of the reflected (or backscattered) signal of each ground resolution element (a few meters to tens of meters) can be calculated in the form of a complex-valued SAR image representing the reflectivity of the ground surface. The intensity of the SAR image (Figure 19.1a) is controlled primarily by terrain slope, surface roughness, and dielectric constants, whereas the phase of the SAR image (Figure 19.1b) is controlled primarily by the distance from the satellite antenna to ground targets, the atmospheric delays, and the interaction of electromagnetic waves with the ground surface.

19.2.3 Basics of Interferometric SAR

InSAR is formed by interfering signals from two spatially or temporally separated antennas. The term "interferometry" draws its meaning from two root words: interfere and measure. The interaction of electromagnetic waves, referred to as interference, is used to measure precise distances and angles. Interference of electromagnetic waves that are transmitted and received by SAR, an advanced imaging radar instrument, is called InSAR. Very simply, InSAR involves the use of two or more SAR images of the same area to extract the land surface topography and its deformation patterns.

For InSAR purposes, the spatial separation between two SAR antennas, or between two vantage points of the same SAR antenna, is called the baseline. The two antennas may be mounted on a single platform for simultaneous interferometry, the usual implementation for aircraft and spaceborne systems such as Topographic SAR (TOPSAR) and Shuttle Radar Topography Mission (SRTM) systems (Farr et al., 2007; Zebker et al., 1992). Alternatively, InSAR images can be formed by using a single antenna on an airborne or spaceborne platform in nearly identical repeating flight lines or orbits for repeat-pass

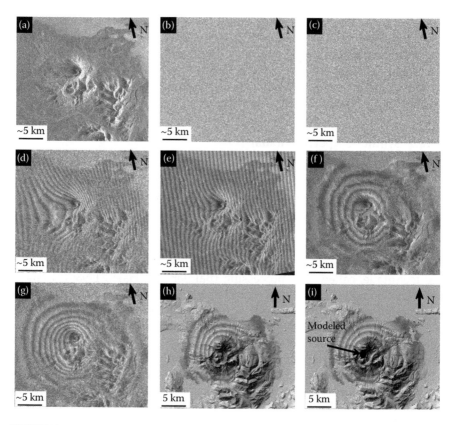

FIGURE 19.1

(See color insert.) (a) The amplitude component of an SLC SAR image acquired on October 4, 1995, by ERS-1 satellite over Peulik Volcano, Alaska. (b) The phase component of the SAR image acquired on October 4, 1995, corresponding to the amplitude image in (a). (c) The phase of an SLC SAR image acquired on October 9, 1997, by ERS-2 satellite over Peulik Volcano, Alaska. The amplitude image is similar to that in (a) and therefore is not shown. (d) An original interferogram formed by differencing the phase values of two coregistered SAR images (b and c). The resulting interferogram contains fringes produced by the differing viewing geometries, topography, any atmospheric delays, surface deformation, and noise. (e) An interferogram simulated to represent the topographic contribution in the original interferogram (d). The perpendicular component of the InSAR baseline is 35 m. (f) A topography-removed interferogram produced by subtracting the interferogram in (e) from the original interferogram in (d). The resulting interferogram contains fringes produced by surface deformation, any atmospheric delays, and noise. (g) A flattened interferogram that was produced by removing the effect of an ellipsoidal earth surface from the original interferogram (d). The resulting interferogram contains fringes produced by topography, surface deformation, any atmospheric delays, and noise. (h) A georeferenced topography-removed interferogram (f) overlaid on a shaded relief image produced from a DEM. The concentric pattern indicates ~17 cm of uplift centered on the southwest flank of Peulik Volcano, Alaska, which occurred during an aseismic inflation episode (Lu et al., 2002). (i) A modeled interferogram produced using a best-fit inflationary point source at ~6.5- km depth with a volume change of ~0.043 km³ on the observed deformation image in (g). Each interferometric fringe (full-color cycle) represents 360° of phase change (or 2.83 cm of range change between the ground and the satellite). Areas of loss of radar coherence are uncolored in (h) and (i).

interferometry (Gray and Farris-Manning, 1993; Massonnet and Feigl, 1998). In this case, even though successive observations of the target area are separated in time, the observations will be highly correlated if the backscattering properties of the surface have not changed in the interim. In this way, InSAR is capable of measuring ground-surface deformation with subcentimeter precision for X-band and C-band sensors (wavelength λ = 2–8 cm), or few-centimeter precision for L-band sensors (λ = 15–30 cm), in both cases at a spatial resolution of tens of meters over an image swath (width) of a few tens of hundreds of kilometers. This is the typical implementation for spaceborne sensors, including European Space Agency (ESA) European Remote-sensing Satellite 1 (ERS-1) (operated 1991–2000, C-band, λ = 5.66 cm), Japan Aerospace Exploration Agency (JAXA) Japanese Earth Resources Satellite 1 (JERS-1) (1992–1998, L-band, λ = 23.5 cm), ESA European Remote-sensing Satellite 2 (ERS-2) (1995–July 2011 C-band, λ = 5.66 cm), Canadian Space Agency (CSA) Canadian Radar Satellite 1 (RADARSAT-1) (1995–present, C-band, λ = 5.66 cm), ESA European Environmental Satellite (Envisat) (2002–present, C-band, λ = 5.63 cm), JAXA Japanese Advanced Land Observing Satellite (ALOS) (January 2006–May 2011, L-band, λ = 23.6 cm), CSA RADARSAT-2 (2007–present, C-band, λ = 5.55 cm), German Aerospace Agency (DLR) TerraSAR-X (2007–present, X-band, λ = 3.1 cm), Italian COSMO-SkyMed satellite constellation (2007–present, X-band, λ = 3.1 cm), and German Aerospace Agency (DLR) TerraSAR add-on for Digital Elevation Measurements (TanDEM-X) (2010–present, X-band, λ = 3.1 cm).

The generation of an interferogram requires two single-look-complex (SLC) SAR images. Neglecting phase shifts induced by the transmitting/receiving antenna and SAR processing algorithms, the phase value of a pixel in an SLC SAR image (Figure 19.1b) can be represented as

$$\phi_1 = -\frac{4\pi}{\lambda} r_1 + \varepsilon_1 \qquad (19.1)$$

where r_1 (a deterministic variable) is the apparent range distance (including possible atmospheric delay) from the antenna to the ground target, λ is the wavelength of radar, and ε_1 is the sum of phase shift due to the interaction between the incident radar wave and scatterers within the resolution cell. Because the backscattering phase (ε_1) is a stochastic (randomly distributed, unknown) variable, the phase value (ϕ_1) in a single SAR image cannot be used to calculate the range (r_1) and is of no practical use. However, a second SLC SAR image (with the phase image shown in Figure 19.1c) could be obtained over the same area at a different time with a phase value of

$$\phi_2 = -\frac{4\pi}{\lambda} r_2 + \varepsilon_2 \qquad (19.2)$$

Note that phase values in the second SAR image (Figure 19.1c) cannot provide range information (r_2) either, due to the stochastic nature of the back-scattering phase ε_2.

An interferogram (Figure 19.1d) is created by coregistering two SAR images and differencing the corresponding phase values of the two SAR images (Figure 19.1b and c) on a pixel-by-pixel basis. The phase value of the resulting interferogram (Figure 19.1d) is

$$\phi = \phi_1 - \phi_2 = -\frac{4\pi(r_1 - r_2)}{\lambda} + (\varepsilon_1 - \varepsilon_2) \tag{19.3}$$

The fundamental assumption in repeat-pass InSAR is that the scattering characteristics of the ground surface remain undisturbed. The degree of changes in backscattering characteristics can be quantified by the interferometric coherence, which is discussed further in a later section. Assuming that the interactions between the radar waves and scatterers remain the same when the two SAR images were acquired (i.e., $\varepsilon_1 = \varepsilon_2$), the interferometric phase value can be expressed as

$$\phi = -\frac{4\pi(r_1 - r_2)}{\lambda} \tag{19.4}$$

Nominal values for the range difference ($r_1 - r_2$) extend from a few meters to several hundred meters. The SAR wavelength (λ) is of the order of several centimeters. As the measured interferometric phase value (ϕ) is modulated by 2π, ranging from $-\pi$ to π, there is an ambiguity of many cycles (i.e., numerous 2π values) in the interferometric phase value. Therefore, the phase value of a single pixel in an interferogram is of no practical use. However, the change in range difference, $\delta(r_1 - r_2)$, between two neighboring pixels that are a few meters apart is normally much smaller than the SAR wavelength. So the phase difference between two nearby pixels, $\delta\phi$, can be used to infer the range distance difference ($r_1 - r_2$) to a subwavelength precision. This explains how InSAR uses the phase difference to infer the change in range distance to an accuracy of centimeters or millimeters.

The phase (or range distance difference) in the original interferogram represented by Equation 19.4 and exemplified by Figure 19.1d contains contributions from both the topography and any possible ground-surface deformation. In order to derive a deformation map, the topographic contribution needs to be removed from the original interferogram (Figure 19.1d). The most common procedure is to use an existing digital elevation model (DEM) and the InSAR imaging geometry to produce a synthetic interferogram and subtract it from the interferogram to be studied (e.g., Massonnet and Feigl, 1998; Rosen et al., 2000). This is the so-called two-pass InSAR. Alternatively, the synthetic interferogram that represents the topographic contribution can come from a different interferogram of the same area. The procedures are then called three-pass or four-pass InSAR (Zebker et al., 1994). As the two-pass InSAR method is commonly used in deformation mapping, a brief explanation of how to simulate the topographic effect based on an existing DEM follows.

Two steps are required to simulate a topography-only interferogram based on a DEM. First, the DEM needs to be resampled to project heights from a map coordinate into the appropriate radar geometry via geometric simulation of the imaging process. The InSAR imaging geometry is shown in Figure 19.2. The InSAR system acquires two images of the same scene with SAR platforms located at A_1 and A_2. The baseline, defined as the vector from A_1 to A_2, has a length B and is tilted with respect to the horizontal by an angle α. The slant range r from the SAR to a ground target T with an elevation value h is linearly related to the measured phase values in the SAR

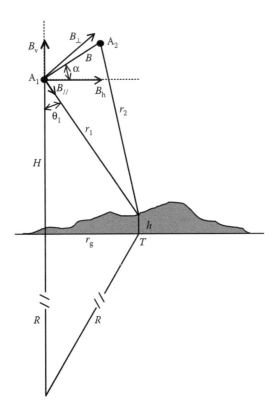

FIGURE 19.2
InSAR imaging geometry. The InSAR system acquires two images of the same scene with SARs located at A_1 and A_2. The spatial distance between A_1 and A_2 is called baseline, which has a length B and is tilted with respect to the horizontal by an angle α. The baseline B can be expressed by a pair of horizontal (B_h) and vertical (B_v) components, or a pair of parallel ($B_{//}$) and perpendicular ($B_\perp$) components. The range distances from the SARs to a ground target T with elevation h are r_1 and r_2, respectively. The look angle from A_1 to the ground point T is θ_1.

images by Equations 19.1 and 19.2. The look angle from A_1 to the ground point T is θ_1. For each ground resolution cell at ground range r_g with elevation h, the slant range value (r_1) should satisfy

$$r_1 = \sqrt{(H + R)^2 + (R + h)^2 - 2(H + R)(R + h)\cos\left(\frac{r_g}{R}\right)} \qquad (19.5)$$

where H is the satellite altitude above a reference Earth surface, which is assumed to be a sphere with radius R. The radar slant range and azimuth coordinates are calculated for each point in the DEM. This set of coordinates forms a nonuniformly sampled grid in the SAR coordinate space. The DEM height data are then resampled into a uniform grid in the radar coordinates using the values over the nonuniform grid.

Second, the precise look angle from A_1 to ground target T at the ground range r_g (slant range r_1) and elevation h is calculated by

$$\theta_1 = \arccos\left[\frac{(H + R)^2 + r_1^2 - (R + h)^2}{2(H + R)r_1}\right] \qquad (19.6)$$

By knowing θ_1, the interferometric phase value due to the topographic effect at target T can be calculated by

$$\phi_{dem} = -\frac{4\pi}{\lambda}(r_1 - r_2) = \frac{4\pi}{\lambda}\left(\sqrt{r_1^2 - 2(B_h \sin\theta_1 - B_v \cos\theta_1)r_1 + B^2} - r_1\right) \qquad (19.7)$$

where B_h and B_v are horizontal and vertical components of the baseline B (Figure 19.1).

Figure 19.1e shows the simulated topographic effect in the interferogram in Figure 19.1d, using an existing DEM and the InSAR imaging geometry for the interferometric pair (Figure 19.2). Removing the topographic effects (Figure 19.1e) from the original interferogram (Figure 19.1d) results in an interferogram containing the ground-surface deformation during the time duration and the measurement noise (Figure 19.1f) with the phase value given as

$$\phi_{def} = \phi - \phi_{dem} \qquad (19.8)$$

In practice, an ellipsoidal Earth surface, characterized by its major axis, e_{maj}, and minor axis, e_{min}, is used to replace the spherical Earth. The radius of the Earth over the imaged area is then

$$R = \sqrt{(e_{min} \sin\beta)^2 + (e_{maj} \cos\beta)^2} \qquad (19.9)$$

where β is the latitude of the center of the imaged region.

If h is taken as zero, the procedure outlined in Equations 19.5 through 19.9 will remove the effect of an ellipsoidal Earth surface on the interferogram. This results in a flattened interferogram, where its phase value can be mathematically approximated as

$$\phi_{flat} = -\frac{4\pi}{\lambda}\frac{B\cos(\theta_1 - \alpha)}{r_1\sin\theta_1}h + \phi_{def} = -\frac{4\pi}{\lambda}\frac{B_\perp}{H\tan\theta_1}h + \phi_{def} \qquad (19.10)$$

where $B_\perp$ is the perpendicular component of the baseline with respect to the incidence angle θ_1 (Figure 19.2). Removing the effect of an ellipsoidal Earth surface on the original interferogram (Figure 19.1d) will result in a flattened interferogram (Figure 19.1g).

If ϕ_{def} is negligible in Equation 19.10, the phase value in Equation 19.10 can be used to calculate height h. This explains how InSAR can be used to produce an accurate, high-resolution DEM over a large region such as the SRTM DEM (Farr et al., 2007). For the ERS-1/-2 satellites, H is about 800 km, θ_1 is about $23° \pm 3°$, λ is 5.66 cm, and $B_\perp$ should be less than 1100 m for a coherent interferogram. Therefore, Equation 19.10 can be approximated as

$$\phi_{flat} \approx -\frac{2\pi}{9600}B_\perp h + \phi_{def} \qquad (19.11)$$

For an interferogram with $B_\perp$ of 100 m, 1 m of topographic relief produces a phase value of about 4°. However, producing the same phase value requires only 0.3 mm of surface deformation. Therefore, it is evident that the interferogram phase value can be much more sensitive to changes in topography (i.e., the surface deformation ϕ_{def}) than to the topography itself (i.e., h). That explains why repeat-pass InSAR is capable of detecting surface deformation at a theoretical accuracy of subcentimeters.

In the two-pass InSAR deformation mapping, errors in the DEM can be mapped into deformation measurement. This is characterized by a term called the "altitude of ambiguity," which is the amount of topographic error required to generate one interferometric fringe in a topography-removed interferogram (Massonnet and Feigl, 1998). As the altitude of ambiguity is inversely proportional to the baseline $B_\perp$, interferometric pairs with smaller baselines are better suited for deformation analysis.

The final procedure in two-pass InSAR is to rectify the SAR images and interferograms into a map coordinate, which is a backward transformation of Equation 19.5. The geo-referenced interferogram (Figure 19.1h) and derived products can be readily overlaid with other data layers to enhance the utility of the interferograms and facilitate data interpretation. Figure 19.1h shows six concentric fringes that represent about 17 cm of range decrease (mostly uplift) centered on the southwest flank of Mount Peulik, Alaska. The volcano inflated aseismically from October 1996 to September 1998, a period that included an intense earthquake swarm that started in May 1998 over 30 km northwest of Peulik Volcano (Lu et al., 2002).

In-depth description of InSAR processing can be found in Zebker et al. (1994), Bamler and Hartl (1998), Henderson and Lewis (1998), Massonnet and Feigl (1998), Rosen et al. (2000), Hensley et al. (2001), and Hanssen (2001). Interested readers should consult these references.

19.2.4 InSAR Image Interpretation and Modeling

To understand geophysical processes that cause the observed surface deformation, numerical models are often employed to infer physical parameters of deformation sources based on the observed deformation. The spatial resolution of surface displacement data provided by InSAR makes it possible to constrain different deformation models. For earthquake studies, the dislocation source (Okada, 1985) is commonly used. For volcano studies, a variety of source geometries, such as the spherical point pressure source (Mogi source) (Mogi, 1958), the dislocation source (sill or dike source) (Okada, 1985), the ellipsoid source (Davis, 1986; Yang et al., 1988), the penny-crack source (Fialko et al., 2001), and so on can be used to model ground-surface deformation due to magmatic activity. The most widely used source in volcano deformation modeling is the spherical point pressure source (also called Mogi source) embedded in an elastic homogeneous half-space (Mogi, 1958). The predicted displacement (u) at the free surface of an elastic homogeneous half-space due to a change in volume (ΔV) or pressure (ΔP) of a sphere (i.e., a presumed magma reservoir) is

$$u_i(x_1 - x_1', x_2 - x_2', -x_3) = C\frac{x_i - x_i'}{\left|R^3\right|} \tag{19.12}$$

where x_1', x_2', and x_3' are the horizontal locations and depth of the center of the sphere, R is the distance between the sphere and the location of observation (x_1, x_2, and 0), and C is a combination of material properties and source strength

$$C = \Delta P(1 - v)\frac{r_s^3}{G} = \Delta V\frac{(1 - v)}{\pi} \tag{19.13}$$

where ΔP and ΔV are the pressure and volume changes of the magma chamber, respectively, v is Poisson's ratio of the host rock (typical value of 0.25), r_s is the radius of the reservoir sphere and G is the shear modulus (Delaney and McTigue, 1994; Johnson, 1987).

A nonlinear least-squares inversion approach is often used to optimize the source parameters in Equations 19.12 and 19.13 (Cervelli et al., 2001; Press et al., 1992). Modeling the observed interferogram in Figure 19.1h using a Mogi source results in a best-fit source located at a depth of 6.5 ± 0.2 km. The calculated volume change is 0.043 ± 0.002 km^3. Figure 19.1i shows the modeled interferogram based on the best-fit source parameters. It is obvious that the Mogi source fits the observed deformation in Figure 19.1h very well.

19.2.5 InSAR Products

Typical InSAR processing includes precise registration of an interferometric SAR image pair, interferogram generation, removal of curved Earth phase trend, adaptive filtering, phase unwrapping, precise estimation of interfero-metric baseline, generation of a surface deformation image (or DEM map), estimation of interferometric correlation, and rectification of interferometric products. Using a single pair of SAR images as input, a typical InSAR pro-cessing chain outputs two SAR intensity images, a deformation or DEM map, and an interferometric correlation image.

19.2.5.1 SAR Intensity Image

SAR intensity images are sensitive to terrain slope, surface roughness, and target dielectric constant. Surface roughness refers to the SAR wavelength-scale variation in the surface relief, and the radar dielectric constant is an electric property of material that influences radar return strength. Therefore, SAR intensity images alone can be used to map hazard-related landscape changes, whether natural or manmade. Multiple temporal SAR intensity images can be used to monitor the progression of landscape changes due to hazards such as flooding, wildfire, volcanic eruption, earthquake shaking, or landslide. In cloud-prone areas, all-weather SAR intensity imagery can be the most useful data source available to track the course of hazardous events.

19.2.5.2 InSAR Coherence Image

An InSAR coherence image is a cross-correlation product derived from two coregistered complex-valued (both intensity and phase components) SAR images (Lu and Freymueller, 1998; Zebker and Villasenor, 1992). It depicts changes in backscattering characteristics on the scale of radar wavelength. Loss of InSAR coherence is often referred to as decorrelation. Decorrelation can be caused by the effects of (1) thermal decorrelation caused by uncorre-lated noise sources in radar instruments, (2) geometric decorrelation result-ing from imaging a target from different look angles, (3) volume decorrelation caused by volume backscattering effects, and (4) temporal decorrelation due to environmental changes over time (Lu and Kwoun, 2008; Zebker and Villasenor, 1992). Decorrelation renders an InSAR image useless for measur-ing ground-surface deformation. On the one hand, geometric and temporal decorrelation can be mitigated by choosing an image pair with short base-line and brief temporal separation, respectively; so choosing such a pair is recommended when the goal is to measure surface deformation, for exam-ple. On the other, the pattern of decorrelation within an image can indicate surface modifications caused by flooding, wildfire, volcanic eruption, or earthquake shaking. In this way, time-sequential InSAR coherence maps can be used to map the extent and progression of hazardous events.

19.2.5.3 InSAR Deformation Image

Unlike an SAR intensity image, an InSAR deformation image is derived from phase components of two overlapping SAR images. SAR is a side-looking sensor; so an InSAR deformation image depicts ground-surface displacements in the SAR line-of-sight (LOS) direction, which include both vertical and horizontal motion. Typical look angles for satellite-borne SARs are less than 45° from vertical, and so LOS displacements in InSAR deformation images are more sensitive to vertical motion (uplift/subsidence) than horizontal motion. Here and henceforth we conform to common usage by sometimes using the terms "displacement" and "deformation" interchangeably. Readers should keep in mind that, strictly speaking, displacement refers to a change in position (e.g., LOS displacement of a given resolution element or group of elements in an InSAR image), whereas deformation refers to differential motion among several elements or groups (i.e., strain). The spatial distribution of surface deformation data from InSAR images can be used to constrain numerical models of subsurface deformation sources. By comparing the deformation patterns predicted by such idealized sources to the actual patterns observed with InSAR, one can identify a best-fitting source. InSAR deformation images have an advantage for modeling purposes over point measurements made with GPS, for example, because InSAR images provide more complete spatial coverage than is possible with even a dense network of GPS stations. On the other hand, continuous GPS stations provide better precision and much better temporal resolution than is possible with InSAR images, which is constrained by the orbital repeat times of SAR satellites. For hazards monitoring, a combination of periodic InSAR observations and continuous data streams from networks of *in situ* instruments is ideal.

19.2.5.4 Digital Elevation Model

As described earlier, the ideal SAR configuration for DEM production is a single-pass (simultaneous) two-antenna system (e.g., SRTM). However, repeat-pass single-antenna InSAR also can be used to produce useful DEMs. Either technique is advantageous in areas where the photogrammetric approach to DEM generation is hindered by persistent clouds or other factors (Lu et al., 2003a). There are many sources of error in DEM construction from repeat-pass SAR images, for example, inaccurate determination of the InSAR baseline, atmospheric delay anomalies, possible surface deformation due to tectonic, volcanic, or other sources during the time interval spanned by the images, and so on. To generate a high-quality DEM, these errors must be identified and corrected using a multi-interferogram approach (Lu et al.,

2003a). A data fusion technique such as the wavelet method can be used to combine DEMs from several interferograms with different spatial resolution, coherence, and vertical accuracy to generate the final DEM product (Baek et al., 2005; Ferretti et al., 1999).

19.3 Issues in InSAR Processing

19.3.1 Phase Anomalies due to SAR Processor

An SAR processor is required to transform a scene of raw SAR data into an SLC image through the matched filtering of raw SAR data in both range and azimuth directions with corresponding reference functions (e.g., Curlander and McDonough, 1991). However, imperfect geometric calculations could result in a ramping phase in an InSAR image. Figure 19.3 shows two InSAR images processed with two different SAR processors. The ramping phase in Figure 19.3a is likely due to SAR processing error. Note that this ramping phase can be easily confused with that caused by baseline error (see Section 19.3.3).

19.3.2 InSAR Coherence Improvement

Interferometric coherence is a qualitative assessment of the correlation of SAR images acquired at different times. It determines the amount of phase error and thus the accuracy of deformation estimates or DEM products. Constructing a coherent interferogram requires that SAR images must correlate with each other; that is, the backscattering spectrum must be substantially similar over the observation period. Physically, this translates into a requirement that the ground scattering surface be relatively undisturbed at the radar wavelength scale between measurements (Li and Goldstein, 1990; Zebker and Villasenor, 1992). The comparison of L-band and C-band interferometric coherence suggests that L-band is far superior to C-band for surfaces covered with thick vegetation or loose material (Lu, 2007; Lu et al., 2005a, 2005b). Therefore, chances for producing coherent interferograms are assured by using C-band images separated in time by a few months to a few years over sparsely vegetated terrains and by using L-band imagery over surfaces with thick vegetation or loose material (Figure 19.4).

Another factor that can enhance InSAR coherence is image coregistration. One stringent prerequisite in InSAR processing is the precise registration of reference and slave SAR images and resampling the slave image to the geometry of the reference image. For conventional InSAR processing, coregistration is done by cross-correlating the reference and slave images at a dense grid of pixel locations and using the results to construct range and azimuth offset polynomials for the entire image. The range and azimuth offset polynomials are expressed as functions of range and azimuth pixel position.

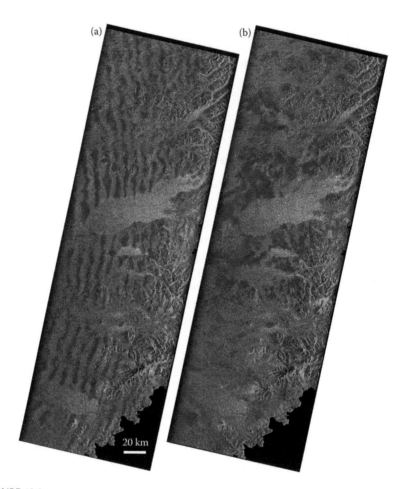

FIGURE 19.3
Interferograms produced from two different SAR processors based on the same pair of SAR images. The ramping fringes in Figure 19.2a are likely due to the systematic phase error in SAR processing.

A problem arises when, for an interferogram with a large perpendicular baseline, topographic variations introduce additional localized offsets between the reference and slave images (Lu and Dzurisin, 2010). The range offset due to topographic relief is linearly dependent on topography and can be approximated as

$$\Delta r_{\text{off}} = -\frac{B_\perp}{H \tan\theta} \Delta h \tag{19.14}$$

where Δr_{off} is the range offset due to height difference Δh, $B_\perp$ is the perpendicular baseline of the interferogram, H is the altitude of the satellite above Earth, and θ is the SAR look angle. For the ERS and Envisat SAR sensor,

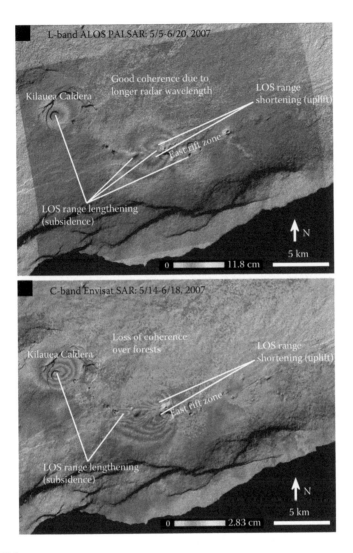

FIGURE 19.4
(See color insert.) (a) L-band ALOS and (b) C-band Envisat InSAR images capturing ground-surface deformation associated with the June 2007 eruption at Kilauea volcano. Each fringe (full color cycle) represents a line-of-sight range change of 11.8 and 2.83 cm for ALOS and Envisat interferograms, respectively. InSAR deformation values are draped over the shaded relief map. Areas of loss of coherence are not colored. Note that C-band InSAR image loses coherence over areas of dense vegetation.

normal values for H and θ are about 790 km and 23° (beam mode IS2 for Envisat), respectively. A 1 km difference in topography can induce about ~1.5 m range offset for an Envisat interferogram with a perpendicular baseline of 500 m. This offset is about 8% of the range pixel size. For the ALOS PALSAR sensor, normal values for H and θ are about 700 km and 34°, respectively. The topography-induced range offset for a fine-beam PALSAR interferogram with a perpendicular baseline of 1 km can be as large as ~2.1 m, or about 23% of the range pixel size. In other words, range offsets due to topographic relief over rugged terrains can be large enough to degrade InSAR coherence if the offsets are not taken into account during image coregistration. Therefore, we recommend using a DEM and the SAR imaging geometry to compute direct functions that map the position of each pixel in the reference image to a corresponding pixel location in the slave image. This can result in significant improvement in coherence for interferograms with relatively large baselines (Lu and Dzurisin, 2010).

19.3.3 InSAR Baseline

A significant error source in InSAR deformation mapping is baseline uncertainty due to inaccurate determination of SAR antenna positions. For ERS-1, ERS-2, Envisat, ALOS, and TerraSAR-X satellites, the refined precision orbit data should be used for InSAR processing. The accuracy of the satellite position vectors provided in RADARSAT-1 and JERS-1 metadata is much poorer than that for ERS-1, ERS-2, Envisat, ALOS, and TerraSAR-X. Therefore, baseline refinement is particularly required for RADARSAT-1 or JERS-1 interferogram processing. Even for ERS-1, ERS-2, and Envisat satellites where precise restitute vectors are available, baseline errors in interferograms can often be present. Figure 19.2a is an interferogram of Okmok Volcano from a pair of ERS-2 images acquired on August 18, 2000, and July 19, 2002. The precision position vectors are used for the InSAR processing. The apparent range changes due to baseline errors are obvious, and the volcanic deformation over the island can be easily confused as there are more than three fringes outside the 10-km-wide caldera. Therefore, interferogram baselines should be refined for InSAR deformation mapping. A commonly used method is to determine the baseline vector based on an existing DEM via a least-squares approach (Rosen et al., 1996). In this approach, areas of the interferogram that are used to refine the baseline should have negligible deformation or known deformation from an independent source. Assuming that the deformation away from the caldera is insignificant in the interferogram in Figure 19.5a, the baseline for this interferogram can be refined. Figure 19.5b shows the deformation interferogram produced with the refined baseline for the interferogram shown in Figure 19.5a. Volcanic deformation of more than three fringes (about 8–10 cm inflation) over the island can be observed from this interferogram, and it now becomes obvious that most of the deformation occurred within the caldera (Figure 19.5b).

19.3.4 Atmospheric Artifacts

A significant error source in repeat-pass InSAR deformation measurement is due to atmospheric delay anomalies caused by small variations in the index

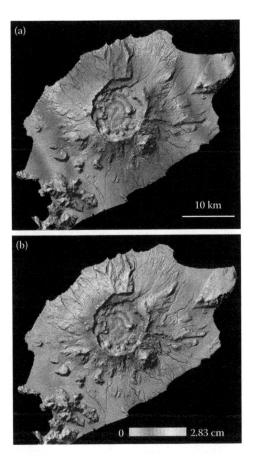

FIGURE 19.5
Topography-removed interferograms of Okmok Volcano, Alaska (a) before and (b) after base-line refinement. Each interferometric fringe (full color cycle) represents a 2.83-cm range change between the ground and the satellite.

of refraction along the line of propagation. Spaceborne SAR sensors such as ERS-1/-2, JERS-1, RADARSAT-1, Envisat, ALOS, and TerraSAR-X satellites orbit at altitudes of about 600–800 km. The electromagnetic wave from these sensors must propagate through the ionosphere, the stratosphere, and the troposphere. Therefore, the radar pulses are subject to small variations in the index of refraction along the line of propagation. Changes in temperature, pressure, and water vapor content of the atmosphere during the two observation instances will result in variations of phase of signals. These variations introduce errors in the observed interferogram. Zebker et al. (1997) indicated that variations of atmospheric water vapor contributed the most to atmospheric anomaly delays. Spatial and temporal changes of 20% in relative humidity could lead to 10-cm errors in repeat-pass interferometric deformation maps.

Over cloud-prone and rainy regions, the range change caused by atmospheric delays can be significant. Figure 19.6a shows a topography-removed interferogram, covering the southeastern part of Okmok Volcano, Alaska. This topography-removed interferogram is constructed using a pair of SAR images acquired in May and July 1997, respectively. Range change up to about 5 cm is observable. To confirm that the range changes in Figure 19.6a were caused by a difference in atmospheric conditions rather than by volcanic activity, another two interferograms were generated for the same area: one interferogram (Figure 19.6b) was produced using the image acquired in July 1997 (which was used in Figure 19.6a) and an image acquired in September 1997; the other interferogram (Figure 19.6c) was produced using the May and September 1997 images. In Figure 19.6b, apparent fringes similar to that in the May–July interferogram (Figure 19.6a) were observed. As the change in color in Figure 19.6a is opposite that in Figure 19.6b, and because no fringe was observed in the May–September interferogram (Figure 19.6c), it is concluded that the fringes in Figure 19.6a and b were most

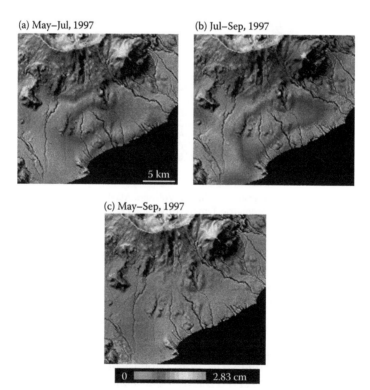

(a) May–Jul, 1997   (b) Jul–Sep, 1997

5 km

(c) May–Sep, 1997

0   2.83 cm

FIGURE 19.6
A portion of a topography-removed interferogram of Okmok Volcano, Alaska shows severe atmospheric anomalies associated with the July 1997 acquisition used for interferograms in (a) and (b).

likely caused by an atmospheric anomaly that occurred primarily on the July 1997 image. Therefore, multiple observations from independent interferograms for similar time intervals should be used to verify any apparent deformation (Lu et al., 2000; Zebker et al., 1997). As atmospheric artifacts do not correlate in time, multi-interferogram InSAR processing (see Section 19.4) can be used to model and reduce atmospheric noise and enhance the signal-to-noise ratio of the deformation signal.

There are three kinds of techniques proposed to estimate the water vapor content from external sources/images and remedy the atmospheric effect on deformation interferograms. The first method is to estimate water vapor concentrations in the target area at the times of SAR image acquisitions using short-term predictions from operational weather models (e.g., Foster et al., 2006). Predicted atmospheric delays from the weather model are used to generate a synthetic interferogram that is subtracted from the observed interferogram, thus reducing atmospheric delay artifacts and improving the ability to identify any remaining ground deformation signal. The problem with this approach is that the current weather models have much coarser resolution (a few kilometers) than InSAR measurements (tens of meters). This deficiency can be remedied to some extent by integrating weather models with high-resolution atmospheric measurements, but this approach requires intensive computation. The second method is to estimate water vapor concentration from continuous Global Positioning System (CGPS) observations in the target area. CGPS is capable of retrieving precipitable water vapor content along the satellite-to-ground LOS with an accuracy that corresponds to 1–2 mm of surface displacement (Bevis et al., 1992; Niell et al., 2001). The spatial resolution (i.e., station spacing) of local or regional CGPS networks is typically several kilometers to a few tens of kilometers, which is sparse relative to the decimeter-scale spatial resolution of SAR images. Therefore, spatial interpolations that take into account the covariance properties of CGPS zenith wet delay (ZWD) measurements and the effect of local topography are required. Jarlemark and Emardson (1998) applied a topography-independent, turbulence-based method to spatially interpolate ZWD values. In a follow-up study, Emardson et al. (2003) found that the spatio-temporal average variance of water vapor content depends not only on the distance between GPS observations, but also on the height difference between stations (i.e., topography). These and other studies led to a topography-dependent turbulence model for InSAR atmospheric correction using ZWD values from CGPS data (Li et al., 2005). The third approach to correcting atmospheric delay anomalies in InSAR observations is to utilize water vapor measurements from optical satellite sensors such as the Moderate Resolution Imaging Spectroradiometer (MODIS), Advanced Spaceborne Thermal Emission and Reflection Radiometer (ASTER), and European Medium Resolution Imaging Spectrometer (MERIS) (Li et al., 2003). The disadvantage of this method is the requirement of nearly simultaneous acquisitions of SAR and cloud-free optical images.

19.3.5 Ionospheric Artifacts

It has been demonstrated that fluctuations in ionospheric electron density can result in modulations in SAR and InSAR images (Gray et al., 2000; Mattar and Gray 2002; Meyer et al., 2006; Wegmuller et al., 2006). The ionospheric disturbance on InSAR images could produce azimuth pixel shift in SAR image correlation and can affect a region of a few kilometers in scale. Accordingly, interferometric phase values over the affected region are also biased. The ionospheric effect is more pronounced over regions of strong magnetic disturbance. The effects are more severe on long-wavelength SAR (e.g., L-band) than short-length SAR (e.g., C-band). Figure 19.7b shows the image offset in azimuth direction by correlating the two SAR images used to produce the interferogram (Figure 19.7a) over northern Alaska. Azimuth offsets range from –5 to 10 m. This is mostly likely caused by ionosphere

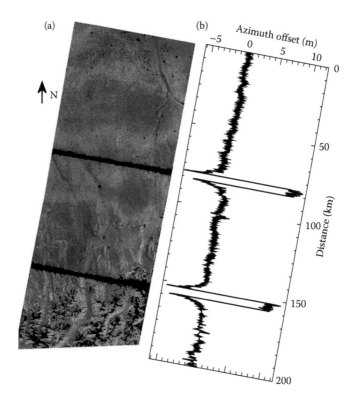

FIGURE 19.7
(a) An interferogram of northern Alaska from a pair of L-band JERS-1 SAR images of June 17 and July 31, 1996. The horizontal stripes of uncolored areas represent loss of InSAR coherence due to extreme azimuth offsets caused by ionospheric anomalies. (b) Azimuth offsets between two JERS-1 images used to generate the interferogram in (a). The azimuth offsets, which are most likely due to ionospheric disturbance, render loss of interferogram coherence over ionosphere-affected areas.

disturbance at the times when the SAR images were acquired. The iono-spheric anomalies affect not only the azimuth offset but also the interfero-gram phase by introducing phase delays as well as reducing the coherence of interferogram (Figure 19.7a). Even though the interferogram coherence can be improved by taking into account the localized offset anomalies (Wegmuller et al., 2006), the phase anomalies cannot be easily removed. Research on removing ionosphere effects in InSAR phase variations is an ongoing hot topic (Meyer et al., 2010).

19.4 Multi-Interferogram InSAR Processing

Multi-interferogram InSAR (Berardino et al., 2002; Ferretti et al., 1999, 2001; Hooper et al., 2007; Rocca 2007) is one of the most significant recent advances in InSAR image fusion to improve deformation measurement accuracy for improved hazards assessment and monitoring. "Multi-" in this context refers to a series of InSAR observations in time, thus affording the opportunity to recognize spurious effects. The objective is to fuse multiple interferogram measurements of the same area to characterize the spatial and temporal behaviors of the deformation signal and various artifacts and noise sources (atmospheric delay anomalies, orbit errors, DEM-induced artifacts), then to remove the artifacts and anomalies to retrieve time-series deformation mea-surements at the SAR pixel level.

Among several approaches to multi-interferogram analysis, persistent scatterer InSAR (PSInSAR) is one of the newest and most promising data fusion techniques. PSInSAR uses the distinctive backscattering characteris-tics of certain ground targets (PS) and unique characteristics of atmospheric delay anomalies to improve the accuracy of conventional InSAR deformation measurements (Ferretti et al., 1999, 2001). The SAR backscattering signal of a PS target has a broadband spectrum in the frequency domain, implying that the radar phase of this kind of scatterer correlates over much longer time intervals and over much longer baselines than other scatterers. As a result, if the backscatter signal from a given pixel is dominated by return from one or more PS(s), the pixel remains coherent over long time intervals. Therefore, at PS pixels, the limitation imposed by loss of coherence in conventional InSAR analysis can be overcome. Because InSAR coherence is maintained at PS pix-els, the atmospheric contribution to the backscattered signal, DEM error, and orbit error can be identified and removed from the data using a multi-inter-ferogram iterative approach. After these errors are removed, displacement histories at PS pixels can be resolved with millimeter accuracy (Figure 19.8) (Lee et al., 2010). If a sufficient number of PS pixels exist in a series of inter-ferograms, relative displacements among them can provide a relatively detailed picture of the surface deformation field.

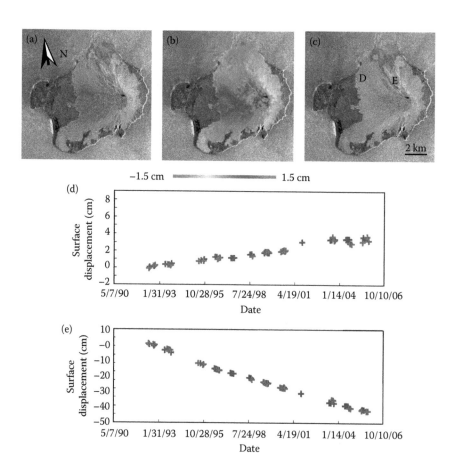

FIGURE 19.8
An example of multi-interferogram InSAR processing to improve deformation mapping. (a)
An observed interferogram of Augustine Volcano, Alaska, from two SAR images acquired on
June 5, 1992, and July 30, 1993. (b) A synthetic interferogram showing range changes due to
atmospheric effects. This image is produced using spatial low-pass and temporal high-pass
filtering during multi-interferogram InSAR processing. (c) Refined InSAR image after multi-
interferogram processing, which effectively removes artifacts due to baseline and atmospheric delay anomalies in (a). (d, e) Line-of-sight (LOS) time-series surface displacements at
points D and E (c) during 1992–2005. Red and purple symbols indicate time-series displacement histories that are produced by multi-interferogram InSAR processing of ERS-2 SAR
images from tracks 229 and 501, respectively. Ground surface at location D inflated at a rate of
3 mm/year presumably due to magma intrusion while ground surface at E subsided at a rate
of 3 cm/year due to thermal compaction of the pyroclastic flows from 1986 eruption. (Adapted
from Lee, C.W. et al. 2010. Surface deformation of Augustine Volcano (Alaska), 1992–2005,
from multiple-interferogram processing using a refined SBAS InSAR approach. USGS
Professional Paper, 1769:453–465.)

19.5 Application of InSAR Imagery to Monitor Natural Hazards in Protected Lands

19.5.1 DEM Generation

A precise digital elevation model can be a very important dataset for characterizing, monitoring, and mitigating various natural hazards in protected lands. For example, a DEM is needed to simulate potential mudflows (lahars) that are commonly associated with volcanic eruptions, large earthquakes, and flooding caused by heavy rainfall. As described earlier, the ideal SAR configuration for DEM production is a single-pass (simultaneous) two-antenna system (Farr et al., 2007). However, repeat-pass single-antenna InSAR also can be used to produce useful DEMs (Ruffino et al., 1998, Sansosti et al., 1999). Either technique is advantageous in areas where the photogrammetric approach to DEM generation is hindered by persistent clouds or other factors (Lu et al., 2003a). To generate a high-quality DEM, a multi-interferogram approach is needed to identify and correct various artifacts in repeat-pass InSAR imagery (Lu et al., 2003a). One example of the utility of precise InSAR-derived DEMs is illustrated in Figure 19.9, which shows the extent and thickness of a lava flow extruded during the 1997 Okmok eruption. The flow's three-dimensional (3-D) distribution was derived by differencing two DEMs that represent the surface topography before and after the eruption. Multiple repeat-pass interferograms were used to correct various error sources and generate the high-quality DEMs (Lu et al., 2003a).

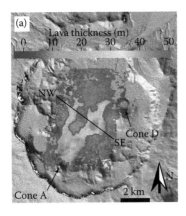
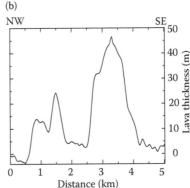

FIGURE 19.9
Thickness of lava flows emplaced during the April 1997 eruption at Okmok Volcano, Alaska. Flow thickness was derived from the height difference between preeruption and posteruption DEMs that were constructed from multiple interferograms. (b) Lava thickness along profile S–S', which reached nearly 50 m in the thickest part of the flow.

19.5.2 Mapping Volcanic Deformation

Volcanic deformation patterns can provide an important insight into the structure, plumbing, and state of restless volcanoes, and can be the first sign of increasing unrest that might include earthquake swarms or other precursors to an impending intrusion or eruption (Dzurisin, 2007). Here we show an example of how multitemporal InSAR images can be used to monitor volcanic activity and construct a magma plumbing system.

Okmok Volcano, a broad shield volcano in protected lands over the central Aleutian, produced blocky, basaltic flows during relatively large, effusive eruptions in 1945, 1958, and 1997 from Cone A, and a hydrovolcanic eruption in 2008 from near Cone D (Figure 19.10) (Lu et al., 1998, 2000, 2005a, 2010). InSAR images constructed from ERS-1, ERS-2, RADARSAT-1 and Envisat SAR data depict volcanic deformation before, during, and after the 1997 eruption and prior to the 2008 eruption (Figure 19.10). More than five fringes appear inside the caldera in the 1992–1993 interferogram (Figure 19.10a), but only two fringes appear in the 1993–1995 interferogram (Figure 19.10b). It can be inferred from these two interferograms that the center of the caldera rose more than 14 cm during 1992–1993 and about 6 cm during 1993–1995. The 1995–1996 interferogram (Figure 19.10c) indicates that the caldera subsided 1 ~ 2 cm between 1.5 and 0.5 years before the 1997 eruption. Therefore, the preeruption inflation rate decreased with time during 1992–1995, and inflation stopped sometime during 1995–1996. More than 140 cm of surface deflation associated with the 1997 eruption can be inferred from the ERS interferogram (Figure 19.10d). The deflation presumably is due to the withdrawal of magma during the 1997 eruption (Lu et al., 1998, 2000, 2005a, 2010). Progressive posteruptive inflation rates generally decreased with time during 1997–2001 (Figure 19.10e through h): from ~10 cm/year during 1997–1998 to ~8 cm/year during 1998–2000 and ~4 cm/year during 2000–2001. Then the inflation rate increased to ~12 cm/year during 2001–2002 (Figure 19.10i)and reached a maximum of ~17–20 cm/year during 2002–2004 (Figure 19.10j and k). The inflation trend was interrupted during 2004–2005, when ~3–5 cm of subsidence occurred (Figure 19.10l). A similar amount of uplift occurred during 2005–2006 (Figure 19.10m), followed by nearly no volcano-wide deformation during 2006–2007 (Figure 19.10n). About 15 cm of uplift occurred from summer 2007 to July 10, 2008, shortly before the July 12, 2008 eruption (Figure 19.10o). Based on the shape and radial pattern of the displacement field, Lu et al. (2005a, 2010) assumed that deformation was caused by a volume change in a spherical magma reservoir, and modeled the surface displacement field using a point source within a homogenous isotropic elastic half-space (Mogi, 1958). Point-source models indicate that a magma reservoir at a depth of 3.2 km below sea level, located beneath the center of the caldera, is responsible for the observed volcano-wide deformation. Magma filled this reservoir at a rate that varied both before and after the eruption, causing volcano-wide inflation. When the magma pressure within the reservoir reached a certain

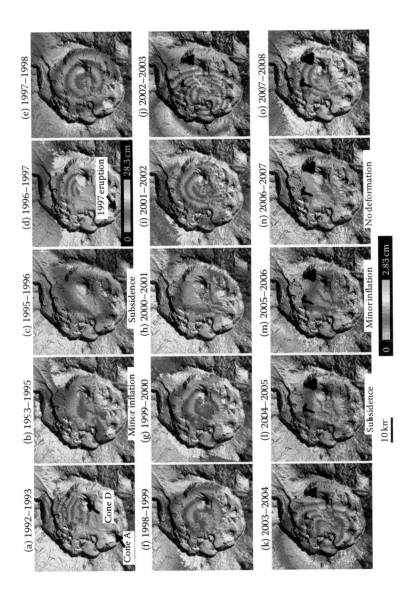

FIGURE 19.10

(**See color insert.**) Deformation interferograms of Okmok Volcano for the periods (a–c) before, (d) during, and (e–o) after the 1997 eruption. Areas of loss of radar coherence are uncolored. Unless otherwise noted, each interferometric fringe (full-color cycle) represents a 2.83-cm range change between the ground and the satellite.

threshold, an eruption ensued. Withdrawal of magma via an eruption depressurized the reservoir, causing volcano-wide deflation, and fed surface lava flows. Magma started to accumulate in the reservoir soon after the eruption stopped, initiating a new intereruption strain cycle.

19.5.3 Mapping Earthquake Displacement

Frequent earthquakes shake the protected lands. Using a pair of SAR images, one acquired before the earthquake and the other after the earthquake, InSAR can be used to map the co-seismic deformation field, and in turn to estimate earthquake location, fault geometry, and rupture dynamics (Figure 19.11). Multiple-temporal InSAR images can be used to estimate interseismic strain accumulation, which is crucial to understanding continental deformation, the earthquake cycle, and seismic hazards (Biggs et al., 2007). InSAR can map ground-surface deformation immediately after an earthquake (i.e., postseismic deformation), which yields important information for inferring properties of the Earth's crust and upper mantle (e.g., Biggs et al., 2009). InSAR is playing an increasingly important role in mapping triggered slip, which occurs during an earthquake on faults not involved in the main shock and therefore is extremely difficult to capture with conventional geodetic techniques (e.g., Fialko et al., 2002). In addition, InSAR can identify blind faults (i.e., buried faults that do not intersect the ground surface) from

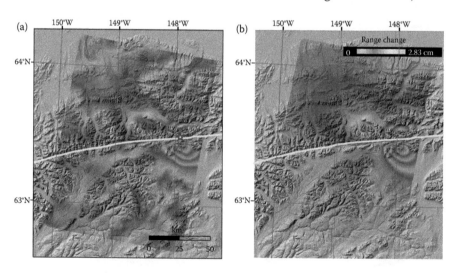

FIGURE 19.11

(a) RADARSAT-1 InSAR image (August 16–October 27, 2002) showing coseismic ground-surface deformation associated with the October 23, 2002, M 6.7 Nenana Mountain earthquake along the Denali Fault (bright line), Alaska. (Adapted from Lu, Z., T. Wright, and C. Wicks. 2003b. Deformation of the 2002 Denali Fault earthquakes, Alaska, mapped by RADARSAT-1 interferometry. *EOS Trans.*, 84:425–431.) (b) Modeled InSAR image using fault parameters that best fit the observed interferogram shown in (a).

surface deformation patterns. Furthermore, loss of InSAR coherence images can be used to infer ground damages (e.g., Fielding et al., 2005). Combined with seismology and other geophysical and geodetic techniques, InSAR can be expected to foster many breakthroughs in understanding the physics of the earthquake cycle (Wright, 2002).

19.5.4 Landslide Monitoring

Landslides in protected lands constitute a major natural hazard. InSAR can be used to map movement of landslides (Figure 19.12). InSAR's remote sensing capability provides a nonintrusive way to monitor landslide hazards in protected lands. Ground-surface deformation associated with landslide movement, when combined with other observation and analysis, can render a better understanding of landslide behavior and contribute more effective procedures to reduce landslide hazards.

19.5.5 Glacier Monitoring

Monitoring changes in glaciers and ice sheets can provide an improved understanding on their role in global warming and sea-level rise. InSAR has been used to record the movement of glaciers and ice fields, and has significantly advanced the study of glacier, ice flow, and ice-sheet mass balance (Figure 19.13). InSAR can not only determine ice velocity and discharge by ice streams and glaciers and quantify their contributions to sea-level rise, but

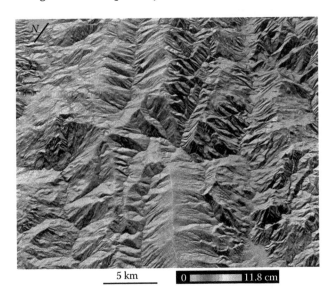

FIGURE 19.12
An L-band PALSAR InSAR image showing landslide movement over Six Rivers National Forest in northern California during April 29 and July 30, 2007.

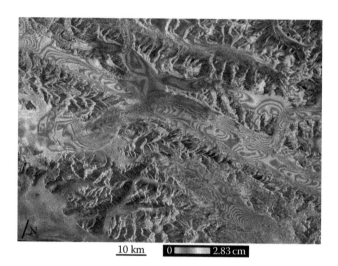

FIGURE 19.13
Interferogram showing movement of Wrangell glaciers over southern Alaska in 24 h between March 31 and April 1, 1996. The interferometric phase image is draped over the radar intensity image. Each fringe (full-color cycle) represents a 2.8-cm change in range distance.

also characterize the temporal variability in ice flow well enough to separate short-term fluctuations from long-term change (Rignot and Thomas, 2002).

19.5.6 Fire Scar Mapping

Fires occur frequently on protected lands. InSAR images and their associated products (e.g., SAR intensity images and InSAR coherence images) have proven useful for mapping landscape changes (Kwoun and Lu, 2009; Ramsey et al., 2006). Wildfires modify vegetation structure as well as moisture conditions that induce changes in SAR backscattering signal. Therefore, multiple SAR images can be used to map the progression of fire and to estimate fire severity (Figure 19.14). InSAR products that characterize changes in SAR backscattering return (both intensity and coherence images) are indispensable for precise mapping of fire scar extents and severities (Rykhus and Lu, 2011). Similarly, SAR and InSAR coherence images provide all-weather imaging capability for flood monitoring over protected lands (Rykhus and Lu, 2007).

19.6 Conclusions

The satellite InSAR technique has proven to be a powerful all-weather remote sensing tool for monitoring and studying a variety of geological hazards in protected lands by analyzing surface deformation patterns and inferring landscape changes based on InSAR coherence and SAR intensity imagery. Timely

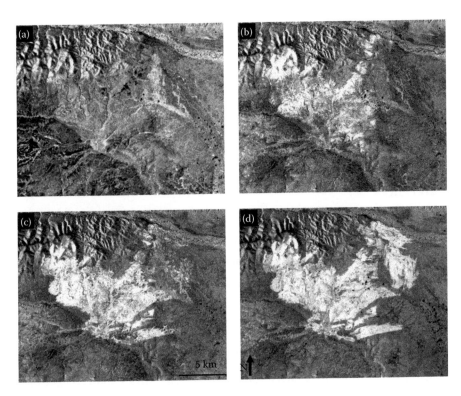

FIGURE 19.14
RADARSAT-1 SAR intensity images of the Yukon River Basin, Alaska, acquired on (a) July 24, 2003, (b) August 17, 2003, (c) September 10, 2003, and (d) October 4, 2003. These SAR images show the progression of a fire that started in July 2003 (successively larger bright areas).

observations of precise land surface topography and time-transient surface changes will accelerate development of predictive models that can anticipate the effects of natural processes such as volcanic eruptions, earthquakes, landslides, and wildfires. With more and more operational SAR sensors available for timely data acquisitions, InSAR—coupled with state-of-the-art information technologies such as data-mining and grid computation—will continue to address and provide solutions to many scientific questions related to natural hazards monitoring and characterization in protected lands and elsewhere.

Acknowledgments

ERS-1/-2 and Envisat, RADARSAT-1, and JERS-1 SAR images are copyrighted © European Space Agency (ESA), Canadian Space Agency (CSA), and Japan Aerospace Exploration Agency (JAXA), respectively. ALOS PALSAR data are copyrighted @ JAXA and Japan Ministry of Economy, Trade and Industry

(METI). SAR data were provided by the Alaska Satellite Facility (ASF), JAXA, and ESA. The research summarized in this chapter was supported by funding from the NASA Earth Science & Interior Program and USGS Volcano Hazards Program. We thank many colleagues for their unselfish contributions to this research, and ASF, JAXA, and ESA staff members for their special efforts in making the SAR data available to us on a timely basis.

References

Baek, S., O. Kwoun, A. Braun, Z. Lu, and C.K. Shum. 2005. Digital elevation model of King Edward VII Peninsula, West Antarctica, from SAR interferometry and ICESat laser altimetry. *IEEE Geosci. Remote Sens. Lett.*, 2:413–417.

Bamler, R. and P. Hartl. 1998. Synthetic aperture radar interferometry. *Inverse Probl.*, 14:R1–54.

Berardino, P., G. Fornaro, R. Lanari, and E. Sansosti. 2002. A new algorithm for surface deformation monitoring based on small baseline differential SAR interferograms. *IEEE Trans. Geosci. Remote Sens.*, 40:2375–2383.

Bevis, M., S. Businger, T. Herring, C. Rocken, R. Anthes, and R. Ware. 1992. GPS meteorology: Remote sensing of the atmospheric water vapor using the Global Positioning System. *J. Geophys. Res.*, 97:15787–15801.

Biggs, J., R. Burgmann, J. Freymueller, Z. Lu, I. Ryder, B. Parsons, and T. Wright. 2009. The postseismic response to the 2002 M7.9 Denali Fault Earthquake: Constraints from InSAR. *Geophys. J. Int.*, 176:353–367.

Biggs, J., T. Wright, Z. Lu, and B. Parsons. 2007. Multi-interferogram method for measuring interseismic deformation: Denali Fault, Alaska. *Geophys. J. Int.*, 173:1165–1179.

Cervelli, P., M.H. Murray, P. Segall, Y. Aoki, and T. Kato. 2001. Estimating source parameters from deformation data, with an application to the March 1997 earthquake swarm of the Izu Peninsula, Japan. *J. Geophys. Res.*, 106:11,217–11,237.

Curlander, J. and R. McDonough. 1991. *Synthetic Aperture Radar Systems and Signal Processing* (New York: John Wiley & Sons).

Davis, P.M. 1986. Surface deformation due to inflation of an arbitrarily oriented triaxial ellipsoidal cavity in an elastic half-space, with reference to Kilauea Volcano, Hawaii. *J. Geophys. Res.*, 91:7429–7438.

Delaney, P.T. and D.F. McTigue. 1994. Volume of magma accumulation or withdrawal estimated from surface uplift or subsidence, with application to the 1960 collapse of Kilauea Volcano. *Bull. Volcanol.*, 56:417–424.

Dzurisin, D. 2007. *Volcano Deformation—Geodetic Monitoring Techniques* (Chichester, UK: Springer-Praxis Publishing Ltd).

Emardson, T.R., M. Simons, and F.H. Webb. 2003. Neutral atmospheric delay in interferometric synthetic aperture radar applications: Statistical description and mitigation. *J. Geophys. Res.*, 108:2231, doi:10.1029/2002JB001781.

Farr, T.G. et al. 2007. The shuttle radar topography mission. *Rev. Geophys.*, 45:RG2004, doi:10.1029/2005RG000183.

Ferretti, A., C. Prati, and F. Rocca. 1999. Multibaseline InSAR DEM Reconstruction: The wavelet approach. *IEEE Trans. Geosci. Remote Sens.*, 37:705–715.

Ferretti, A., C. Prati, and F. Rocca. 2001. Permanent scatterers in SAR interferometry. *IEEE Trans. Geosci. Remote Sens.*, 39:8–20.

Fialko, Y., Y. Khazan, and M. Simons. 2001. Deformation due to a pressurized horizontal circular crack in an elastic half-space, with applications to volcano geodesy. *Geophys. J. Int.*, 146:181–190.

Fialko, Y., D. Sandwell, D. Agnew, M. Simons, P. Shearer, and B. Minster. 2002. Deformation on nearby faults induced by the 1999 Hector Mine earthquake. *Science*, 297:1858–1862.

Fielding, E.J., M. Talebian, P.A. Rosen, H. Nazari, J.A. Jackson, M. Ghorashi, and R. Walker. 2005. Surface ruptures and building damage of the 2003 Bam, Iran, earthquake mapped by satellite synthetic aperture radar interferometric correlation. *J. Geophys. Res.*, 110:B03302, doi:10.1029/2004JB003299.

Foster, J. et al. 2006. Mitigating atmospheric noise for InSAR using a high-resolution weather model. *Geophy. Res. Lett.*, 33:L16304, doi:10.1029/2006GL026781.

Gray, A.L., K.E. Mattar, and G. Sofko. 2000. Influence of ionospheric electron density fluctuations on satellite radar interferometry. *Geophys. Res. Lett.*, 27: 1451–1454.

Gray, L. and P.J. Farris-Manning. 1993. Repeat-pass interferometry with airborne synthetic aperture radar. *IEEE Trans. Geosci. Remote Sens.*, 31:180–191.

Hanssen, R. 2001. *Radar Interferometry: Data Interpretation and Error Analysis* (Netherlands: Kluwer Academic Publishers).

Henderson F. and A. Lewis. 1998. *Principals and Applications of Imaging Radar—Manual of Remote Sensing* (New York: John Wiley & Sons Inc.).

Hensley S., R. Munjy, and P. Rosen. 2001. Interferometric synthetic aperture radar (IFSAR). In *Digital Elevation Model Technologies and Applications: The DEM Users Manual*, D.F. Maune (Ed.) (Bethesda, MD: American Society for Photogrammetry and Remote Sensing).

Hooper, A., P. Segall, and H. Zebker. 2007. Persistent scatterer interferometric synthetic aperture radar for crustal deformation analysis, with application to Volcán Alcedo, Galápagos. *J. Geophys. Res.*, 112:B07407, doi:10.1029/2006JB004763.

Jarlemark, P.O.J. and G. Elgered. 1998. Characterization of temporal variations in atmospheric water vapor. *IEEE Trans. Geosci. Remote Sens.*, 36:319–321.

Johnson, D. 1987. Elastic and inelastic magma storage at Kilauea volcano, Volcanism. In Hawaii, R. Decker, T. Wright, and P. Stauffer (Eds.). *USGS Professional Paper*, 1350:1297–1306.

Kwoun, O. and Z. Lu. 2009. Multi-temporal RADARSAT-1 and ERS backscattering signatures of coastal wetlands at southeastern Louisiana. *Photogramm. Eng. Remote Sens.*, 75:607–617.

Lee, C.W., Z. Lu, H.S. Jung, J.S. Won, and D. Dzurisin. 2010. Surface deformation of Augustine Volcano (Alaska), 1992–2005, from multiple-interferogram processing using a refined SBAS InSAR approach. USGS Professional Paper, 1769:453–465.

Levanon, N. 1988. *Radar Principles* (New York: John Wiley & Sons).

Li, F.K. and R.M. Goldstein. 1990. Studies of multibaseline spaceborne interferometric synthetic aperture radars. *IEEE Trans. Geosci. Remote Sens.*, 28:88–96.

Li, Z., J.-P. Muller, and P. Cross. 2003. Comparison of precipitable water vapor derived from radiosonde, GPS, and Moderate-Resolution Imaging Spectroradiometer measurements. *J. Geophys. Res.*, 108:4651, doi:10.1029/2003JD003372.

Li, Z., E. Fielding, P. Cross, and J.-P. Muller. 2005. InSAR atmospheric correction: GPS Topography-dependent Turbulence Model (GTTM). *J. Geophys. Res.* 110:B02404, doi:10.1029/2005JB003711.

Lu, Z. and J. Freymueller. 1998. Synthetic aperture radar interferometry coherence analysis over Katmai volcano group, Alaska. *J. Geophys. Res.*, 103:29887–29894.

Lu, Z., D. Mann, and J. Freymueller. 1998. Satellite radar interferometry measures deformation at Okmok volcano. *EOS Trans.*, 79:461–468.

Lu, Z., D. Mann, J. Freymueller, and D. Meyer. 2000. Synthetic aperture radar interferometry of Okmok volcano, Alaska: Radar observations. *J. Geophys. Res.*, 105: 10791–10806.

Lu, Z., E. Fielding, M. Patrick, and C. Trautwein. 2003a. Estimating lava volume by precision combination of multiple baseline spaceborne and airborne interferometric synthetic aperture radar: The 1997 eruption of Okmok Volcano, Alaska. *IEEE Trans. Geosci. Remote Sens.*, 41:1428–1436.

Lu, Z., T. Wright, and C. Wicks. 2003b. Deformation of the 2002 Denali Fault earthquakes, Alaska, mapped by RADARSAT-1 interferometry. *EOS Trans.*, 84: 425–431.

Lu, Z., T. Masterlark, and D. Dzurisin. 2005a. Interferometric synthetic aperture radar (InSAR) study of Okmok Volcano, Alaska, 1992–2003: Magma supply dynamics and post-emplacement lava flow deformation. *J. Geophys. Res.*, 110:B02403, doi:10.1029/2004JB003148.

Lu, Z., C. Wicks, O. Kwoun, J. Power, and D. Dzurisin. 2005b. Surface deformation associated with the March 1996 earthquake swarm at Akutan Island, Alaska, revealed by C-band ERS and L-band JERS radar interferometry. *Can. J. Remote Sens.*, 31(1):7–20.

Lu, Z. 2007. ALOS PALSAR InSAR. *NASA Alaska Satellite Facility News Notes*, 4(4):1–2.

Lu, Z. and O. Kwoun. 2008. RADARSAT-1 and ERS interferometric analysis over southeastern coastal Louisiana: Implication for mapping water-level changes beneath swamp forests. *IEEE Trans. Geosci. Remote Sens.*, 46:2167–2184.

Lu, Z. and D. Dzurisin. 2010. Ground surface deformation patterns, magma supply, and magma storage at Okmok volcano, Alaska, inferred from InSAR analysis: II. Co-eruptive deflation, July-August 2008. *J. Geophys. Res.*, 115:B00B02, doi:10.1029/2009JB006970.

Lu, Z., Dzurisin, D., Biggs, J., Wicks, C. Jr., and S. McNutt. 2010. Ground surface deformation patterns, magma supply, and magma storage at Okmok volcano, Alaska, from InSAR analysis: I. Inter-eruption deformation, 1997–2008. *J. Geophys. Res.*, 115:B00B03, doi:10.1029/2009JB006969.

Lu, Z., C. Wicks, D. Dzurisin, J. Power, S. Moran, and W. Thatcher. 2002. Magmatic inflation at a dormant stratovolcano: 1996–98 activity at mount Peulik Volcano, Alaska, revealed by satellite radar interferometry. *J. Geophys. Res.*, 107(B7):2134, doi:10.1029/2001JB000471.

Massonnet, D. and K. Feigl. 1998. Radar interferometry and its application to changes in the Earth's surface. *Rev. Geophys.* 36:441–500.

Meyer, F., R. Bamler, N. Jakowski, and T. Fritz. 2006. The potential of low-frequency SAR systems for mapping ionospheric TEC distributions. *IEEE Geosci. Remote Sens. Lett.*, 3:560–564.

Meyer, F. et al. 2010. A review of ionospheric effects in low-frequency SAR data (abstract), *2010 International Geoscience and Remote Sensing Symposium*, Honolulu, HI.

Mogi, K. 1958. Relations between the eruptions of various volcanoes and the deformations of the ground surface around them. *Bull. Earthquake Res. Inst. Univ. Tokyo*, 36:99–134.

Niell, A.E., A.J. Coster, F.S. Solheim, V.B. Mendes, P.C. Toor, R.B. Langley, and C.A. Upham. 2001. Comparison of measurements of Atmospheric Wet Delay by

Radiosonde, Water Vapor Radiometer, GPS, and VLBI. *J. Atmos. Oceanic Technol.*, 18:830–850.

Okada, Y. 1985. Surface deformation due to shear and tensile faults in a half-space. *Bull. Seismol. Soc. Am.*, 75:1135–1154.

Press, W., S. Teukolsky, W. Vetterling, and B. Flannery. 1992. *Numerical Recipes in C, the Art of Scientific Computing*, (Cambridge: Cambridge University Press).

Ramsey, E. III, Z. Lu, A. Rangoonwala, and R. Rykhus. 2006. Multiple baseline radar interferometry applied to coastal landscape classification and changes. *GISci. Remote Sens.*, 43:283–309.

Rignot, E. and R. Thomas. 2002. Mass balance of polar ice sheets. *Science*, 297:1502–1506.

Rocca, F. 2007. Modeling interferogram stacks. *IEEE Trans. Geosci. Remote Sens.*, 45:3289–3299.

Rosen, P., S. Hensley, H. Zebker, F.H. Webb, and E.J. Fielding. 1996. Surface deformation and coherence measurements of Kilauea volcano, Hawaii, from SIR-C radar interferometry. *J. Geophys. Res.*, 101:23109–23125.

Rosen, P., S. Hensley, I.R. Joughin, F.K. Li, S.N. Madsen, E. Rodriguez, and R.M. Goldstein. 2000. Synthetic aperture radar interferometry. *Proc. IEEE*, 88:333–380.

Ruffino, G., A. Moccia, and S. Esposito. 1998. DEM generation by means of ERS tandem data. *IEEE Trans. Geosci. Remote Sens.*, 36:1905–1912.

Rykhus, R. and Z. Lu. 2011, Monitoring a 2003, Yukon Flats, Alaska wildfire using multi-temporal Radarsat-1 intensity and interferometric coherence images. *Geomatics Nat. Hazards Risk*, 2:15–32.

Rykhus, R. and Z. Lu. 2007. Hurricane Katrina flooding and possible oil slicks mapped with satellite imagery. USGS Circular 1306: Science and the Storms—The USGS Response to the Hurricanes of 2005, pp. 50–53.

Sansosti, E. et al. 1999. Digital elevation model generation using ascending and descending ERS-1/ERS-2 tandem data. *Int. J. Remote Sens.*, 20:1527–1547.

Wegmuller, U., C. Werner, T. Strozzi, and A. Wiesmann. 2006. Ionospheric electron concentration effects on SAR and INSAR. In *International Geoscience and Remote Sensing Symposium*, Denver, Colorado, USA.

Wright, T. 2002. Remote monitoring of the earthquake cycle using satellite radar interferometry. *Phil. Trans. R. Soc. Lond.*, 360:2873–2888.

Yang, X.M., P. Davis, and J. Dieterich. 1988. Deformation from inflation of a dipping finite prolate spheroid in an elastic half-space as a model for volcanic stressing. *J. Geophys. Res.*, 93:4249–4257.

Zebker H. et al. 1992. The TOPSAR interferometric radar topographic mapping instrument. *IEEE Trans. Geosci. Remote Sens.*, 30:933–940.

Zebker, H.A. and J. Villasenor. 1992. Decorrelation in interferometric radar echoes. *IEEE Trans. Geosci. Remote Sens.*, 30:950–959.

Zebker, H.A., P.A. Rosen, R.M. Goldstein, A. Gabriel, and C.L. Werner. 1994. On the derivation of coseismic displacement fields using differential radar interferometry: The Landers earthquake. *J. Geophys. Res.*, 99:19617–19634.

Zebker, H., P. Rosen, and S. Hensley. 1997. Atmospheric effects in interferometric synthetic aperture radar surface deformation and topographic maps. *J. Geophys. Res.*, 102:7547–7563.

20

Characterizing Biophysical Properties in Protected Tropical Forests with Synergistic Use of Optical and SAR Imagery

Cuizhen Wang and Jiaguo Qi

CONTENTS

20.1 Introduction

Tropical forests account for 40% of terrestrial biomass and play vital roles in regulating carbon, water, and energy exchange in the atmosphere (Foody et al., 1997). Over the past two decades, tropical forests in Southeast Asia have experienced more rapid decline than those in Central Africa and Latin America (Mayaus et al., 2005). Under economic and political pressures in this region, vast landscape of forests is converted into agricultural lands or commercial plantation (Jepson et al., 2001).

Protected lands in many Southeast Asian countries have been established to leverage deforestation and forest degradation, and the associated environmental and ecological problems in this region. For example, half of the forests in Thailand had been lost in the past 50 years (Inoguchi et al., 2005). In 1989, the Thai government introduced a complete ban on logging and established a network of protected areas that covered over 20% of the nation's land. Despite those logging bans and conservation activities, however, deforestation and forest degradation continue to be serious problems. Particularly, encroachment of indigenous people and illegal logging of economically valuable trees (e.g., teak) result in changes in forest structures and reduced canopy cover (Mayaus et al., 2005). Different from reported status of the Amazonian forests, the changing forests are not well documented in Southeast Asia. Therefore, there is a need in quantifying biophysical properties to better understand the current conditions of protected forests in Southeast Asia.

Satellite remote sensing has proven useful in mapping forest extents and identifying forest changes under various natural and human-induced stresses (Skole and Tucker, 1993; Tucker and Townshend, 2000; Fuller, 2006). With the Landsat imagery over the National Parks and the Appalachian Trails in Northeast United States, Wang et al. (2009) found similar patterns of forest degradation inside the protected forests and adjacent buffer zones due to increased human activities. Optical signals are mostly sensitive to green leaf properties and therefore, optical remote sensing is widely applied to extract green properties such as fractional cover and green leaf area index (LAI) (Jasinski and Eagleson, 1990; Qi et al., 2000). LAI is often used to represent green biomass (Wang and Qi, 2008a,b) and is widely used to parameterize and validate models of ecosystem functioning, net primary production and environmental processes (Sellers and Schimel, 1993).

Microwave remote sensing, with its unique cloud-free imaging capabilities, provides important alternatives in monitoring tropical forests. In the global moist forest monitoring project (Podest and Saatchi, 2002), image mosaics acquired by the Synthetic Aperture Radar (SAR) onboard the Japan Earth Resources Satellite (JERS-1) developed the 100-m spatial resolution vegetation maps with acceptable accuracies in tropical forests in Southeast Asia, Central Africa, and the Amazon (Sgrenzaroli et al., 2002). Radar signals at lower frequencies could penetrate green canopies and record information about woody structures and aboveground woody biomass. Le Toan et al. (1992) found that there was a good correlation between forest biomass and SAR intensity P- and L-bands. C-band data primarily interact with the crown layer (Saatchi and McDonald, 1997) and rapid saturation of C-band backscatter occurs in dense forests (Dobson et al., 1992). Cross-polarized backscatter (HV) has high correlation with forest biomass. HH backscatter comes from both trunk and crown layers, whereas VV and particularly HV returns are from crown layers (Beaudoin et al., 1994).

Luckman et al. (1997) found that L-band SAR imagery could discriminate different levels of forest biomass up to 110 ton/ha. With topography effects, the biomass threshold, above which there appears to have no further increase in L-band radar backscatter, was around 60 ton/ha (Luckman, 1998). Other studies (Dobson et al., 1992; Le Toan et al., 1992) using P- and L-band data showed that biomass threshold could reach 200 ton/ha in temperate forests. Several canopy scattering models have been developed to simulate radar backscatter when woody structures of forests are taken into account (Ulaby et al., 1990; Sun et al., 1991; Karam et al., 1992). These models are a theoretical approximation of forest scattering based on a 1st- or 2nd-order resolution of radiative transfer functions (RTF). Both empirical regression and RTF model approaches of biomass retrieval only work on forests with sparse to medium biomass. In dense tropical forests, because of strong attenuation from green canopies, radar backscatter is less sensitive to woody biomass. As green canopies could be quantitatively described with optical imagery, a synergistic use of optical and SAR remote sensing could quantify leaf contribution in forest backscatter and therefore, improve the accuracy of woody biomass estimation.

    In this study, a forest canopy scattering model was modified to simulate forest backscatter in a tropical region. While linking forest green biomass to vegetation index (*VI*) retrieved from the Landsat ETM+ image, the model was synergistically applied to estimate aboveground woody biomass from the JERS-1 SAR image in the national parks in northern Thailand.

20.2 Method Development

20.2.1 Site Description and Data Collection

The Mae Chaem Watershed is located at 98°00′–99°00′E and 18°00′–19°00′N, covering an area of 6692 km² in northern Thailand. The Chiang Mai city, the second largest city in Thailand, is at the downhill flatlands in the east of the watershed. In this well-known touring city, the Thai government established a set of national parks in its vast area of mountainous landscapes. Five national parks (NP) are located in the Mae Chaem Watershed including the Mae Tho, the Op Luang, the Doi Inthanon, the Mae Wang, and the Op Khan (Figure 20.1). Part of the Lhun Khan NP in the north is also covered in the watershed. Historically large-scale forests in the watershed had been cleared and converted into agricultural land use (Figure 20.2a). After the logging ban, human disturbances in these parks primarily include shift cultivation that burns trees and converts into dryland crops in a short rotation (Figure 20.2b), cutting in secondary forests for firewood (Figure 20.2c), and illegal logging of valuable trees such as teak (Figure 20.2d).

    The watershed has a typical tropical monsoon climate in which wet season is during May through November and dry season during November through February. Topography in the watershed varies greatly in a range from 250 to

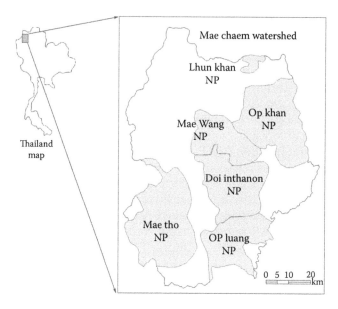

FIGURE 20.1
The Great Mae Chaem Watershed and the six national parks (NP) in the watershed. (Modified from data sources at http://www.thaiforestbooking.com/nationalpark-eng.htm.)

FIGURE 20.2
Human disturbances in national parks in the watershed: historical clear cutting (a), shift culti-vation (b), firewood cutting (c), and illegal logging for timber (d).

2500 m above the mean sea level (Figure 20.3a). The Mount Doi Inthanon (2565 m) is located in the Doi Inthanon NP at the center of the watershed and is the highest mount in Thailand. Except for the relative flat areas in the southeast (to Chiang Mai city) and in the middle (the Mae Chaem town), the watershed has an elevation higher than 500 m. Tropical forests cover 80% of the watershed (Thailand LUCC Case Study, 1997). Clear transition of forest types could be observed along the elevation profile (Figure 20.3b): dry dipterocarps, mixed deciduous, pine transition, and tropical evergreen (dry evergreen and moist evergreen). The dipterocarps are deciduous trees grown at lower elevation and are mostly secondary forests, frequently disturbed by shift cultivation and firewood cutting by local residence. At higher elevation, both deciduous and evergreen species compose the mixed deciduous forests, which also suffer from logging disturbance. However, due to topographical

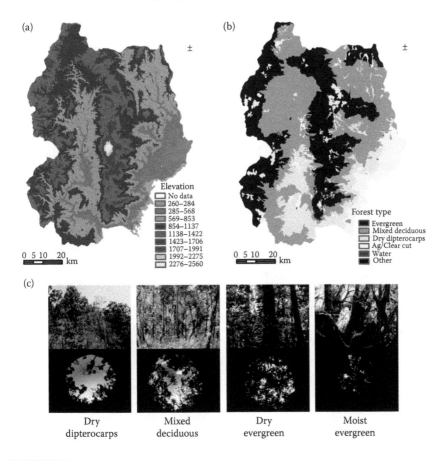

FIGURE 20.3
(See color insert.) The DEM map with hillshade effect (a), forest-type map (b), and example field and fisheye photos on the ground (c). The white area in the DEM map is the Doi Inthanon Mount, the highest point in Thailand. Photos are taken during the field trip in January, 2002.

difficulty in accessing these areas, the disturbance is less than dry diptero-carps. Tropical evergreen forests occur at higher elevations. Pine transition is a narrow zone distributed between the mixed deciduous and tropical ever-green forests (not categorized in Figure 20.2a). Tropical dry evergreen forests grow above the pine transition zone. Tropical moist evergreen forests grow on top of mountains, where forests are often covered by cloud and are moist all year round. Due to the extreme difficulty in accessing these areas, tropical evergreen forests are generally dense natural forests, almost undisturbed by human activities. As shown in the fisheye photographs and ground pictures in Figure 20.3c, dry dipterocarps have the lowest canopy cover, followed by mixed deciduous and evergreen forests. In moist evergreen forests on the very top of the Doi Inthanon Mount, tree canopies are almost fully covered.

Both optical and SAR images were acquired in the watershed. One Landsat ETM+ scene was acquired on March 5, 2000 (Figure 20.4a). Dry dipterocarps are senescent in the dry season and have a bluish tone in the false-color image composite. The image was atmospherically corrected with the MODTRAN 4.0 to calculate surface reflectance. Using ground control points collected dur-ing field trips, the image was geometrically corrected with a 2nd-order poly-nomial model and reached a root-mean-square error of 23.6 m, comparative to pixel size. Two JERS-1 SAR scenes (path/row of 132/269 and 132/270) were acquired on February 27, 1998 and were mosaiced to cover most area of the watershed (Figure 20.4b). The JERS-1 SAR images were in L-band HH polar-ization at a pixel size of 18 m. The SAR image was co-registered to the ETM+ image and topographic correction was performed with DEM data.

FIGURE 20.4
The ETM+ image in the false-color composition (a) acquired on 03/05/2000 and the JERS-1 SAR image (b) acquired on February 27, 1998 in the watershed.

TABLE 20.1

Average Ground Measures of Biophysical Parameters at the Ground Sites

	Tree H (m)	Stem H (m)	DBH (cm)	Density (#/ha)	Canopy H (m)	Biomass (ton/ha)
Dry dipterocarps	10.624	7.736	17.242	711.871	5.775	47.756
Mixed deciduous	13.298	9.640	22.899	544.754	7.316	92.713
Pine transition	14.265	10.212	24.564	754.859	8.105	95.786
Dry evergreen	14.489	10.262	23.338	581.271	8.454	200.409
Moist evergreen	22.025	17.121	35.340	777.622	9.809	426.682

Two field trips were carried out to conduct field surveys and to collect ground data in the watershed. The first trip was in the wet season between August 10 and 18, 2001, and the second one was in the dry season between January 20 and 27, 2002. Due to logistical difficulties in these mountainous areas, only 32 ground sites were visited along the Doi Inthanon 1009 Royal Road in the Doi Inthanon National Park. Among those sites included nine dry dipterocarps, nine mixed deciduous, five pine transition, six dry evergreen, and three moist evergreen forests. At each ground site, 10 sample points were evenly selected along a 300 m transect perpendicular to the road. At each sample point, a hemispherical photograph (fisheye photo) was taken and later processed in an GLA software (Frazer et al., 1999) to extract canopy fractional cover and LAI. The point-quadrant method was used to measure tree height, stem height, diameter at breast height (DBH), and stand density at the sample point. Values at the 10 sample points were averaged to represent biophysical measures at each ground site. Aboveground woody biomass was calculated with commonly applied allometric equations of tropical forests (Luckman et al., 1997). Table 20.1 listed the average measures of these biophysical measures in each forest type.

20.2.2 Estimating Green Biophysical Properties: A Linear Unmixing Model

Forests in the watershed were composed of green canopies and open gaps. In this study, we applied a linear unmixing model to estimate forest fractional cover. In the dry season, open areas were mostly bare ground and senescent grasses that had similar spectral responses all over the watershed. Consequently, each ETM+ pixel was assumed a mixed reflectance of green tree canopy and open area.

In a linear unmixing model, reflectance of the two components was independent of each other. At certain spectral band, total reflectance (R) of a mixed pixel was a linear combination of reflectance values of tree canopy (R_{canopy}) and open area (R_{open}), weighted by the percentage cover of these two components:

$$R = R_{canopy} fc + R_{open}(1 - fc) + \varepsilon, \tag{20.1}$$

where fc is the forest fractional cover in a pixel unit. An error term ε is introduced to account for insignificant remaining components within the pixel.

Surface reflectance of vegetation changed greatly at different bands. The values of R_{canopy} and R_{open} were also highly influenced by various factors such as leaf wetness and structure, soil moisture and texture, and sun-target-sensor geometry (Jasinski, 1990; Townshend, 1999). To optimally select the independent variables in Equation 20.1, some studies have expanded the reflectance-based unmixing model to a VI domain (Myneni et al., 1992; Wittich, 1997; Wang et al., 2005). As a mathematical combination of multiple spectral bands, VI can suppress the above-mentioned external effects (Gutman and Ignatov, 1998; Qi et al., 2000). Equation 20.1 could thus be written as

$$VI = VI_{canopy}\,fc + VI_{open}(1 - fc) + \varepsilon, \tag{20.2}$$

where VI_{canopy} is the vegetation index of green canopies in a pixel, and VI_{open} is the vegetation index of open area. In dry-season tropical forests, these two components are dominant and the error term (ε) could be ignored. The fc can then be calculated as

$$fc = \frac{VI - VI_{open}}{VI_{canopy} - VI_{open}}. \tag{20.3}$$

Wang et al. (2005) explored the sensitivities of a set of commonly applied vegetation indices such as the normalized difference vegetation index (NDVI) (Tucker, 1979), soil-adjusted vegetation index (SAVI) (Huete et al., 1988), modified soil-adjusted vegetation index (MSAVI) (Qi et al., 1994), and enhanced vegetation index (EVI) (Huete et al., 1999) in tropical forests. The MSAVI was found most linearly sensitive to green properties and was adopted in the linear unmixing model in this study. The two end members, VI_{canopy} and VI_{open}, were extracted from the ETM+ image. A subset of moist evergreen forest on the top of Mount Doi Inthanon was selected to calculate VI of full-cover canopy ($VI_{canopy} = 0.71$). Several small pieces of open areas in the watershed were used to calculate the average VI of open area ($VI_{open} = 0.16$).

LAI is logarithmically related to fractional cover in light radiation models (Welles, 1990; Welles and Cohen, 1996; Frazer et al., 1999). With fisheye photos at the 32 ground sites, the relationship between LAI and fc at a scale of 0–100% was empirically determined:

$$LAI = 0.211 + 0.032 \times e^{0.048 \times fc}. \tag{20.4}$$

Both 4-ring and 5-ring LAI values were displayed in Figure 20.5a demonstrating the relationship. The 4-ring LAI was the effective LAI integrated over zenith angles 0–60°, while the 5-ring LAI was integrated over zenith

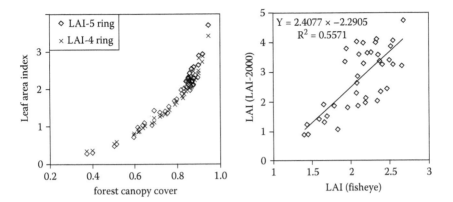

FIGURE 20.5

Relationships between *LAI* and fractional cover extracted from fisheye photos (a). The fisheye *LAI* measures are linearly related to LAI-2000 measures (b).

TABLE 20.2

List of Forests in Northern Michigan That Were Visited to Simultaneously Collect LAI-2000 Measures and Fisheye Photos

Broadleaf Forests	GPS Location (UTM, WGS84)	Date	Ground Sites
Muskegon River Watershed	(552741, 4789298)	10/1/2002	5
Burchfield Park	(697576, 4720093)	10/2/2002	5
Rose Lake Wildlife Conservation	(716891, 4742766)	10/5/2002	16
		11/12/2002	10
MSU baker woodlot	(706798,4732483)	11/12/2002	6

angles 0–75°. It was noticed in the GLA software (Frazer et al. 1999) that the *LAI* extracted from fisheye photos were significantly underestimated in dense forests. To compensate for the problem, we conducted an independent experiment in Northern Michigan in October and November 2002 to collect ground measures with both LAI-2000 and fisheye photos in four broadleaf forests (Table 20.2). The LAI-2000 was commonly accepted to collect LAI as "ground truth." Although *LAI* values from fisheye photos were relatively lower than LAI-2000 measures, the two measures were linearly correlated (Figure 20.5b). In this study, the *fc*-retrieved *LAI* was linearly adjusted with the equation in Figure 20.5b to better represent LAI in the watershed.

20.2.3 Estimating Woody Biomass: A Microwave/Optical Synergistic Model

Linking optical remote sensing variables to microwave scattering model provides quantitative information of green and woody biomass. A forest can be simplified as a multilayer composition of green canopy layer and woody structure (branches and trunks) layers atop of a rough soil surface. In a 1st-order solution of radiative transfer equations, total backscatter of a forest consists of (1) volume scattering from layers of leaves, branches, and trunks, (2) surface scattering from soil ground, (3) a double bounce between each layer and soil surface (Karam et al. 1992). It was found in Wang and Qi (2008b) that soil surface scattering in tropical forests was far less than other components.

Leaf contribution in the model could be quantified after green biomass (represented by *LAI*) was extracted with the ETM+ image. The amount of green leaves determined volume scattering from green canopies and leaf attenuation (τ) to other scattering components. The total backscatter could thus be simplified (in power unit):

$$\sigma_{total} = \sigma_{leaf} + \tau^2 \sigma_{woody}, \tag{20.5}$$

where σ_{leaf} represents combined contribution of leaf volume scattering and double bounce between leaves and soil surface; σ_{woody} represents combined contribution of volume scattering and double bounce of branches and trunks in the forest (Karam et al. 1992; Wang and Qi 2008b). Soil surface scattering is not considered in this study.

Assuming that the height of green canopy was H, and a and b were the half-length and half-width of an elliptic leaf, respectively, leaf attenuation factor in the forward or backward direction can be quantified when *LAI* was known:

$$\tau_t = e^{-k_t H / \cos \eta}$$

$$k_t = \frac{4\pi N}{k_0} \text{Im} < F_{tt} > \tag{20.6}$$

$$N = \frac{LAI}{\pi a b H}$$

where $t = H$ or V polarization. The extinction coefficient (k_t) is the total extinction cross section of all leaves statistically averaged over a spherical probability distribution. F_{tt} is the copolarization component of the scattering matrix, or scattering amplitude tensor, of a single leaf. N is the leaf number density ($\#/m^3$) of the green canopy.

With *LAI* estimated from the ETM+ image, both σ_{leaf} and τ can be quantified in Equation 20.6. Backscatter of woody structures could thus be extracted from the JERS-1 SAR image:

$$\sigma_{woody} = \frac{\sigma_{SAR} - \sigma_{leaf}}{\tau^2}, \tag{20.7}$$

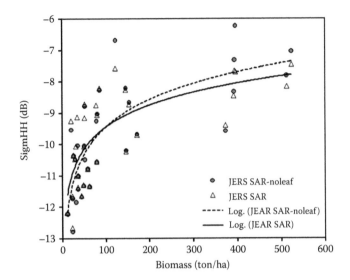

FIGURE 20.6
Log-linear relationships between SAR backscatter and woody biomass. Both raw JERS-1 SAR backscatter and woody backscatter (with leaf contribution removed) are displayed in the figure.

where σ_{SAR}, σ_{woody}, and σ_{leaf} are backscatter intensity (in power unit) of JERS-1 SAR, woody structures, and leaf contribution, respectively.

Woody backscatter σ_{woody} was directly related to aboveground woody biomass. Based on biomass measures at the 32 ground sites, both σ_{SAR}^0 and σ_{woody}^0 (in dB unit) displayed a log-linear relationship with biomass (Figure 20.6). Both backscatter coefficients turned to saturate in dense forests with high biomass. After leaf backscatter was removed and attenuation to woody backscatter was compensated, σ_{woody}^0 was slightly lower than σ_{SAR}^0 in sparse forests and was higher when biomass became high. Although leaf contribution was less dominant in dry season in the watershed, the σ_{woody}^0 demonstrated a higher sensitivity with forest biomass than σ_{SAR}^0. It increased more rapidly at low biomass and had slower rate of saturation when biomass was higher. The log-linear relationship between woody backscatter (dB) and biomass (ton/ha) was determined in a curve-fitting process:

$$\text{Biomass} = 8.314 + e^{(\sigma_{woody}^0 + 18.732)/1.714}. \tag{20.8}$$

At degree of freedom = 29 (32 study sites − 3 model coefficients), the χ^2 of curve fitting was 30.002. It was smaller than the critical $\chi_{0.05,29}^2$ (46.19) at confidence level of 95%, indicating that the regression model in Equation 20.8 was valid. In this way, biomass distribution of the watershed was quantified with synergistic use of optical and SAR images.

20.3 Results and Discussion

20.3.1 Green Biophysical Properties: Forest Fractional Cover and LAI

Figure 20.7a was the forest fractional cover (fc) map estimated from the ETM+ image, binned into 10 categories at an interval of 10%. Nonforest land covers (in orange) were masked out from the forest-type map. Forests with fc less than 20% were displayed in dark brown and orange. Forests with fc higher than 20% followed a graduate color scheme from grayish green to dark green. The fc map showed a similar pattern with the forest-type map. Along with the elevation profile, forest types changed in a sequence of dry diptero-carps, mixed deciduous, dry evergreen, and moist evergreen. In accordance, the fc distribution changed from low values in dry dipterocarps to saturation (100%) in moist evergreen forests.

Forest fractional cover in the watershed was affected by various natural and human disturbances. Since the ETM+ image was acquired during the dry season, dry dipterocarps at lower elevations were mostly senescent and leaf-off. Their VI values were much lower than other forests and therefore, their fractional cover was low. Some dry dipterocarps had fractional cover lower than 20%. Aside from seasonal effects, dry dipterocarps suffered from fre-quent human disturbance of burning for shift agriculture, cutting for fire-wood, and illegal logging for valuable trees such as teak. As a result, most of the dry dipterocarps forests were young, secondary regrowth with fractional cover less than 40%. The fc values were lower in forests close to villages and

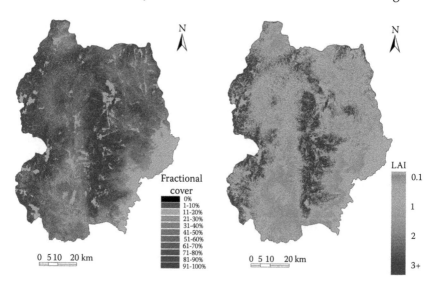

FIGURE 20.7

Forest fractional cover (a) and LAI (b) estimated from the ETM+ image. Nonforest land covers are masked out.

higher in remote mountainous areas. Mixed deciduous forests were found at higher elevations and were less affected by seasonal variation and human activities. Small-area shift cultivation was primary types of deforestation. The fractional cover of mixed deciduous forests ranged from 40% to 80%. Evergreen forests grew at high elevations in the watershed and human disturbance was much lower than forests at lower elevations. The fractional cover of evergreen forests ranges from 70% to 100%. Fractional cover values of moist evergreen forests on the top of the mountains turned to be saturated.

The *LAI* distribution (Figure 20.7b) had similar pattern as the fractional cover. In dry season, deciduous trees became senescent and were partially leaf-off and, therefore, most of the dry dipterocarps and mixed deciduous forests had low *LAI* values (less than or around 1.0). Tropical evergreen forests could reach *LAI* values of 3.0 and higher. In moist evergreen forests, the *LAI* values saturate at around 10.9. Similarly, the *LAI* dynamics in each forest type revealed the extent of human disturbance and recovery in the watershed.

20.3.2 Aboveground Woody Biomass

In the forests of the study area, biomass in a range of 10–200 ton/ha was shown in the aboveground woody biomass map (Figure 20.8). Biomass higher than 200 ton/ha was not further examined because it was found in

FIGURE 20.8
Aboveground woody biomass estimated with synergistic use of JERS-1 SAR and Landsat ETM+ images. Nonforest land covers are masked out.

past studies that SAR signals saturated at this threshold or lower (Dobson et al., 1992; Le Toan et al., 1992; Luckman et al., 1997). Woody biomass estimation in the watershed was intrinsically affected by speckle noises of the SAR image. For this reason, a 5×5 low-pass filter was applied to the biomass map to reduce its salt-and-pepper effects.

In contrast to the fractional cover and LAI distributions which were primarily determined by green properties in the dry season, biomass distribution in the watershed did not have clear correlation with elevation and seasonal variation. However, it demonstrated similar characteristics that areas close to nonforest land uses often had lower biomass of less than 50 ton/ha. Forests in the west of the watershed were closer to the borders between Thailand and Myanmar and had least human disturbance. High biomass was observed in this area. In general, biomass increased from the east of the watershed closer to Chiang Mai city to the west.

Biomass estimation in the highly mountainous watershed was inevitably affected by topographic distortions that cannot be effectively removed. For example, as the highest point in Thailand, the Doi Inthanon Mount in the middle of the watershed was covered with moist evergreen forests at high densities and biomass levels. However, the estimated biomass was very low (around 10 ton/ha) along the north–south ridge. Controlled by flight orbit of the satellite, the track direction of JERS-1 SAR signals was from east to west. Correspondingly, low biomass was commonly observed on north–south distributed slopes all over the watershed, a common problem of topographic effects at steep slopes, for example, layover and radar shadow, facing east–west directions.

20.3.3 Uncertainty Analysis

The ETM-estimated fc values at ground sites were compared with those extracted from the fisheye photos that were assumed "ground truth" (Figure 20.9a). Forest types were also specified in the figure with different marks. The ETM-estimated fc was correlated with ground fisheye measurements ($R^2 = 0.757$) although the correlation varied in different forest types. Both the ETM-estimated and ground-measured fc values increased in a sequence of dry dipterocarps, mixed deciduous, pine transition, and evergreen forests. In dry dipterocarps where the fc was low, the ETM-estimation was lower than ground measurements. On the contrary the ETM estimation was higher than ground measurement in dense evergreen forests. The values matched well in mixed deciduous and pine transition forests with fc in a range of 70–85%. In very dense forests, the ETM-estimated fc was saturated, whereas the maximal fisheye measure was only 95%.

The discrepancy in Figure 20.9a came from different processing mechanisms. The ETM-estimated fc was calculated in a linear unmixing model in *VI* domain. Only green leaves in forests contributed to the calculation. Therefore, the ETM-estimated fc was actually the "green" fractional cover.

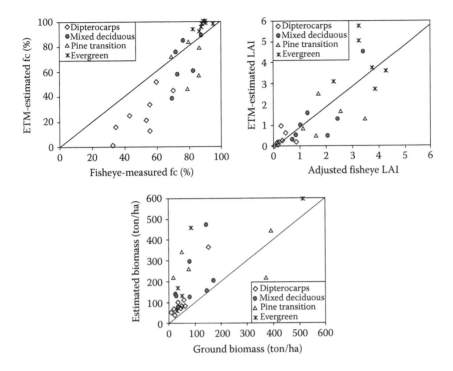

FIGURE 20.9
Scatterplots of image-estimated and ground-measured biophysical attributes: forest fractional cover (a), LAI (b), and aboveground woody biomass (c).

The ground-measured *fc*, however, was calculated from a binary classification of a hemispherical photograph taken on the ground. All nonsky elements including green leaves, senescent leaves, stems, branches and trunks, contributed to the value. Therefore, the ground-measured *fc* was the total cover of the projected area of the forest. In the dry season, the ETM-estimated *fc* of dry dipterocarps was low. The ground-measured *fc*, however, was much higher because of significant contribution from woody components. In mixed deciduous and pine transition forests, some species were partially leaf-off and others remained evergreen. Consequently, both the ETM-estimated and ground-measured *fc* values were higher and matched better. In evergreen forests, both the ETM-estimated and ground-measured *fc* values were high (>90%). In dense evergreen forests, the in-tree gaps became dominant when calculating *fc* in circular hemispherical photographs. However, these small-size gaps were unrecognizable in the ETM+ image at 30-m resolution. As a result, the ETM-estimated *fc* was higher than ground measurements. In very dense, moist evergreen forests, the ETM-estimated *fc* reached saturation (100%) while ground measurement was less than 95%.

Owing to logistical difficulties of accessing the mountainous forests, it was impractical to collect "ground truth" of LAI in the watershed. Here we used the adjusted *LAI* from fisheye photos as "ground truth" to be compared with the ETM-estimated *LAI*. The overall correlation between ETM-estimated and fisheye-adjusted *LAI* values reached R^2 of 0.67 when all ground sites were considered. In Figure 20.9b, the estimated *LAI* fitted well with adjusted fisheye *LAI* when it was less than 3. It indicated that the *LAI* estimation from ETM+ image was feasible in dry dipterocarps, mixed deciduous, pine transition, and some evergreen forests. These areas covered most of the watershed. Higher discrepancy was observed in dense forests, in which the adjusted fisheye *LAI* turned to be underestimated and became less valid to serve as ground truth. In moist evergreen forests, the ETM-estimated *LAI* values were saturated at 10.9 while the adjusted fisheye *LAI* values were still less than 5.

The optical/SAR synergistically estimated woody biomass was less consistent with ground measurements (Figure 20.9c). Ground measured biomass values at most of the study sites were less than 200 ton/ha, while the estimated biomass turned to be higher (above the 1:1 line in the scatterplot). The total root-mean-square error (RMSE) was 121 ton/ha when all ground sites were considered. The discrepancy also varied with forest types. Both estimated and ground measured biomass values were low in dry dipterocarps but reached relatively good agreement. Most values in mixed deciduous forests matched well. More discrepancy was observed in pine transition and evergreen forests. It may come from topographic effects because these forests grew in high mountains with steep slopes. The estimation was less reliable in these dense tropical forests.

20.3.4 Monitoring Biophysical Properties in Protected Areas

More than half of the watershed was managed as national parks. We can see in Figure 20.7 that green properties in these national parks were mainly determined by forest types that have distinct seasonal variations. It was still clear, however, green biophysical properties in each forest type were affected by historic and current human disturbances. Lower forest fractional cover and LAI were observed in areas closer to human settlements and indigenous highland uses across the parks. Seasonal variation was less dominant in the woody biomass map in Figure 20.8. Rather, human disturbances became major effects of deforestation and forest degradation. Woody biomass in the national parks turned to be lower than un-touched forests close to country borders in the west.

Estimation of biophysical characteristics of national parks in tropical mountains is a challenging task. Ground measurements are time/labor intensive and are often limited because of logistical difficulties in accessing the areas. Moreover, ground measurements are often not actual ground "truth." For example, ground-measured fractional cover and LAI with fisheye photos are actually projected cover and foliage area index. They are a combined contribution from green leaves, branches, trunks, and any other nonopen components. Theoretically, it is different from remotely sensed fractional cover and LAI that are based on green foliage. Ground measurements of woody structures are also problematic. The fixed-radium plot method often provides an accurate measure of these parameters in an area, but is too time/labor intensive and thus not realistic in tropical forests. The point-quadrant plot method is more efficient, but often results in high bias of ground measurements. In late succession of forests, many young trees grow surrounding old parent trees. Therefore, there is high possibility of picking up young trees in each quadrant. Assuming these data as ground "truth," biophysical parameters from remotely sensed data are often overestimated. Validation of remote sensing estimated biophysical properties with ground measurements is thus biased.

Topographic distortion on radar backscatter was the largest obstacle in SAR remote sensing in mountainous tropical forests. The quality of topographic correction depends on both characteristics of SAR systems and quality of DEM data. High relief and steep slopes in tropical national parks often introduce mis-correction and data loss of SAR imagery, which results in high errors in biophysical estimation in these protected areas.

Acquisition of paired optical and SAR imagery is critical in synergistic approaches explored in this study. Although rich optical imagery, for example, Landsat TM/ETM+, are available, spaceborne SAR systems are limited. Most of the SAR systems that are currently operative, such as the ERS-1/2, Radarsat 1/2, and ENVISAT ASAR, are in C-band and their application in dense tropical forests could be limited due to strong attenuation and saturation of canopy scattering. As one of the very few L-band SAR sensors that are currently available, the Phrase-Array-type L-band Synthetic Aperture Radar (PALSAR) onboard the Advanced Land Observing Satellite (ALOS) launched in 2006 brings more opportunities in effectively monitoring protected areas in tropical mountains.

20.4 Conclusion

This study applied both optical and SAR imagery to estimate biophysical properties of tropical forests. Forest fractional cover was estimated from optical imagery with a linear unmixing model, while LAI was calculated based on its log-linear relationship with fractional cover. As leaf contribution on radar backscatter was quantified using the estimated LAI, aboveground woody biomass was extracted by developing an empirical regression model between woody backscatter and ground biomass. Linking radar backscatter with optical imagery reduced the effects of leaf attenuation on woody structures and therefore, improved the sensitivity of radar backscatter to woody biomass. The SAR/optical synergistic approach developed in this study has potential to improve the accuracy of woody biomass retrieval, which could provide supportive information in managing protected forests in tropical mountains.

Acknowledgments

This study was partially funded by NASA GOFC grant (NAG 5-9286) and NASA Land Use Land Cover Program grant (NNG05GD49G) at the Center for Global Change and Earth Observations (CEODE), Michigan State University. The authors would like to thank Dr. C. Navanujraha and S. Lawavirojwong at Mahidol University, Thailand, and Drs. N. Wiangwang and P. Vamakovida at Michigan State University and local foresters in Chiang Mai for their tremendous help in field trips and data collection.

References

Beaudoin, A., T. Le Toan, S. Goze, E. Nezry, A. Lopes, and E. Mougin. 1994. Retrieval of forest biomass from SAR data. *International Journal of Remote Sensing* 15:2777–2796.

Dobson, M. C., F. T. Ulaby, L. E. Pierce, T. L. Sharick, K. M. Bergen, J. Kellndorfer, J. R. Kendra et al. 1992. Estimation of forest biophysical characteristics in Northern Michigan with SIR-C/X-SAR. *IEEE Transactions on Geoscience and Remote Sensing* 33:877–895.

Foody, G. M., R. M. Lucas, P. C. Curran, and M. Honzak. 1997. Mapping tropical forest fractional cover from coarse-resolution remote sensing imagery. *Plant Ecology* 131:143–154.

Frazer, G. W., C. D. Canham, and K. P. Lertzman. 1999. Gap Light Analyzer (GLA), Version 2.0: Imaging software to extract canopy structure and gap light transmission indices from true-color fisheye photographs, users manual and program documentation. Simon Fraser University, Burnaby, British Columbia, Canada, and the Institute of Ecosystem Studies, Millbrook, New York, USA.

Fuller, D. O. 2006. Tropical forest monitoring and remote sensing: A new era of transparency in forest governance? *Singapore Journal of Tropical Geography* 27:15–29.

Gutman, G. and A. Ignatov. 1998. The derivation of the green vegetation fraction from NOAA/AVHRR data for use in numerical weather prediction models. *International Journal of Remote Sensing* 19:1533–1543.

Huete, A. R. 1988. A soil-adjusted vegetation index (SAVI). *Remote Sensing of Environment* 25:295–309.

Huete, A. R., C. Justice, and W. Van Leeuwen. 1999. MODIS vegetation index. MODIS algorithm theoretical basis document, NASA Goddard Space Flight Center, Greenbelt.

Inoguchi, A., R. Soriaga, and P. Walpole. 2005. Approaches to controlling illegal forests activities: Considerations from Southeast Asia. Working Paper Series No. 7, Asia Forest Network, Bohol, Philippines.

Jasinski, M. and P. S. Eagleson. 1990. Estimation of subpixel vegetation cover using red-infrared scattergrams. *IEEE Transactions on Geoscience and Remote Sensing.* 28:253–267.

Jepson, P., J. K. Jarvie, K. MacKinnon, and K. A. Monk. 2001. The end for Indonesia's lowland forests? *Science* 292:859–861.

Karam, M. A., F. Amar, A. K. Fung, E. Mougin, A. Lopes, D. M. Le Vine, and A. Beaudoin. 1992. A microwave scattering model for layered vegetation. *IEEE Transactions on Geoscience and Remote Sensing* 30:767–784.

Le Toan, T., A. Beaudoin, and D. Guyon. 1992. Relating forest biomass to SAR data. *IEEE Transactions on Geoscience and Remote Sensing* 30:403–411.

Luckman A. J. 1998. The effects of topography on mechanisms of radar backscatter from a coniferous forest and upland pasture. *IEEE Transactions on Geoscience and Remote Sensing* 36:1830–1834.

Luckman, A., J. Baker, T. M. Kuplich, C. C. F. Yanasse, and A. C. Frery. 1997. A study of the relationship between radar backscatter and regenerating tropical forest biomass for spaceborne SAR instruments. *Remote Sensing of Environment* 60:1–13.

Mayaus, P., P. Holmgren, F. Achard, H. Eva, H. Stibig, and A. Branthomme. 2005. Tropical forest cover change in the 1990s and options for future monitoring. *Philosophical Transactions of the Royal Society B-Biological Sciences* 360:373–384.

Myneni, R. B., G. Asrar, D. Tanre, and B. J. Choudhury. 1992. Remote sensing of solar radiation absorbed and reflected by vegetated land surfaces. *IEEE Transactions on Geoscience and Remote Sensing* 30:303–314.

Podest, E. and S. Saatchi. 2002. Application of multiscale texture in classifying JERS-1 radar data over tropical vegetation. *International Journal of Remote Sensing* 23:1487–1506.

Qi, J., A. Chehbouni, A. R. Huete, and Y. Kerr. 1994. A modified soil adjusted vegetation index (MSAVI). *Remote Sensing of Environment* 48:119–126.

Qi, J., Y. H. Kerr, M. S. Moran, M. Weltz, A. R. Huete, S. Sorooshian, and R. Bryant. 2000. Leaf area index estimates using remotely sensed data and BRDF models in a semiarid region. *Remote Sensing of Environment* 73:18–30.

Saatchi, S. S. and K. C. McDonald. 1997. Coherent effects on microwave backscattering models for forest canopies. *IEEE Transactions on Geoscience and Remote Sensing* 35:1032–1044.

Sellers, P. J. and D. Schimel. 1993. Remote sensing of the land biosphere and biochemistry in the EOS era: Science priorities, methods and implementation—EOS land biosphere and biogeochemical panels. *Global and Planetary Change* 7:279–297.

Sgrenzaroli, M., G. F. de Grandi, H. Eva, and F. Achard. 2002. Tropical forest cover monitoring: Estimates from the GRFM JERS-1 radar mosaics using wavelet zooming techniques and validation. *International Journal of Remote Sensing* 7:1329–1355.

Skole, D. L. and C. J. Tucker. 1993. Tropical deforestation and habitat fragmentation in the Amazon: Satellite data from 1978 to 1988. *Science* 260:1905–1910.

Sun, G., D. Simonett, and A. Strahler. 1991. A radar backscatter model for discontinuous coniferous forests. *IEEE Transactions on Geoscience and Remote Sensing* 29:639–650.

Thailand Land Use and Land Cover Change Case Study. 1997. Southeast Asia, IHDP, IGBP, WCRP program.

Townshend, J. R. G. 1999. MODIS enhanced land cover and land cover change product, Algorithm Theoretical Basis Documents (ATBD), Version 2.0, NASA, GSFC, Greenbelt, MD, 94pp. Available on-line at http://modis.gsfc.nasa.gov/data/atbd/atbd_mod29.pdf.

Tucker, J. J. 1979. Red and photographic infrared linear combinations for monitoring vegetation. *Remote Sensing of Environment* 8:127–150.

Tucker, C. J. and J. R. G. Townshend. 2000. Strategies for monitoring tropical deforestation using satellite date. *International Journal of Remote Sensing* 21:1461–1471.

Ulaby, F. T., K. Sarabandi, K. Mcdonald, M. Whitt, and M. C. Dobson. 1990. Michigan microwave canopy scattering model. *International Journal of Remote Sensing* 11:1223–1253.

Wang, C., J. Qi, and M. Cochrane. 2005. Assessment of Tropical Forest Degradation with Canopy Fractional Cover from Landsat ETM+ and IKONOS Imagery. *Earth Interactions* 9:1–18.

Wang, C. and J. Qi. 2008a. Biophysical estimation in tropical forests using JERS-1 VNIR and SAR imagery: I—leaf area index. *International Journal of Remote Sensing* 29:6811–6826.

Wang, C. and J. Qi. 2008b. Biophysical estimation in tropical forests using JERS-1 SAR and VNIR Imagery: II—aboveground woody biomass. *International Journal of Remote Sensing* 29:6827–6849.

Wang, Y., B. R. Mitchell, J. Nugranad-Marzilli, G. Bonynge, Y. Zhou, and G. Shriver. 2009. Remote sensing of land-cover change and landscape context of the National Parks: A case study of the Northeast Temperate Network. *Remote Sensing of Environment* 113:1453–1461.

Welles, J. M. 1990. Some indirect methods of estimating canopy structure. *Remote Sensing Reviews* 5:31–43.

Welles, J. M. and S. Cohen. 1996. Canopy structure measurement by gap fraction analysis using commercial instrumentation. *Journal of Experimental Botany* 47:1335–1342.

Wittich, K. P. 1997. Some simple relationships between land-surface emissivity, greenness and the plant cover fraction for use in satellite remote sensing. *International Journal of Biometeorology* 41:58–64.

Section IV

Remote Sensing in Decision Support for Management of Protected Lands

21

Monitoring and Modeling Environmental Change in Protected Areas: Integration of Focal Species Populations and Remote Sensing[*]

Robert L. Crabtree and Jennifer W. Sheldon

CONTENTS

21.1 Introduction

Humans have created protected areas (PAs) over the past millennia for a multitude of reasons. The establishment of Yellowstone National Park in 1872 by the United States Congress ushered in the modern era of governmental

---

[*] J. Sheldon and R. Crabtree contributed equally to this work as co-senior authors.

protection of natural areas that catalyzed a global movement (IUCN, 2008; Heinen, 2007). Today, approximately 13% of terrestrial and 1% of marine environments are designated as protected and a tremendous variety of monitoring programs and conservation planning efforts are underway (e.g., UNEP the United Nations Environment Programme http://www.unep.org/; see also http://www.wdpa.org/). Despite these conservation achievements, species' population declines, biodiversity loss, extinctions, system degradation, pathogen spread, and state change events are occurring at unprecedented rates (Hoffmann et al., 2010; Pereira et al., 2010). These effects are augmented by continued changes in land-use, alongside the direct, indirect, and interactive effects of climate change and disruption. It is clear that designation and monitoring of PAs are necessary, but insufficient, in providing the required levels of support to effectively maintain the attributes of ecosystem function and species preservation into the indefinite future (Sinclair et al., 1995).

Monitoring, when coupled with *modeling*, can provide a powerful basis for guiding management actions, while simultaneously advancing science through hypothesis testing and predictions. Modeling of ecosystem indicators informed by monitoring information such as that provided from remote sensing (RS) is essential for efficient, transparent, repeatable, and defensible decision-making in ecological systems. These decisions should include outcomes-based activities driving toward goals such as population recovery, critical habitat restoration, biodiversity increase, and improved ecosystem services. Models—whether narrative, conceptual, visual, ecological, or statistical—provide a common language for scientists and practitioners, permitting hypothesis testing about the mechanisms or drivers underlying the observed variation in data. Of particular importance is the idea that statistical models can be improved by careful attention to ecological concepts and mechanisms (Austin, 2002) with the goal of clarifying the "sometimes tenuous link between observed pattern and significant ecological process" (Gross et al., 2009).

Many PAs are centered on an explicit or implicit valuation of the species they support, and many monitoring and management programs are focused on species persistence, as both an ecological goal and as part of a set of legislated mandates. In this context, long-term species data sets are of unique value because they provide insight into the mechanisms that support the desired attributes of the PA systems. What would be the value of Yellowstone National Park without its iconic charismatic megafauna? Leaving the difficulties of shifting baselines aside (Baum and Myers, 2004), the intrinsic value of PAs is in part set by the unique and rare species and processes they encompass. For these reasons, we focus in this chapter on the role of long-term, legacy species data sets ("focal species" in the parlance of the U.S. Fish and Wildlife Service, hereafter FWS). What is their monitoring value, in a PA setting, in helping to elucidate the metrics of resilient, intact, and sustainable ecosystems? What are their limitations?

Practitioners such as agency biologists and managers commonly fund and conduct monitoring programs, while scientists seek access to the data that have been collected. This provides a common ground for collaboration and bridging of the gap between scientists and practitioners (Marris, 2007). Science and decision making should go hand in hand, because they both measure success by their ability to predict the consequences of actions (Pielke, 2003). Although monitoring of focal species, as well as other ecological indicators or vital signs, are the critical first steps needed for science-based decisions, it is often treated as an end unto itself. How do we explain their variation across space and over time? To what is it attributable? What are the anticipated long- and short-term outcomes from our management activities? At its best, ecological modeling provides answers to these questions by (1) exploratory and synthetic analysis of possible causes using explanatory variables, (2) investigating possible causes while testing hypotheses (diagnostic models), and (3) predicting consequences, for example, under future scenarios (prognostic models). Models serve the essential function of connecting cause and effect in otherwise intractably complex natural systems (Paola, 2011).

Arguably, ecological models are as powerful as the quality and relevance of the causal and explanatory variables ("covariates") included. In a sense, lack of explanatory covariates in predictive modeling is equivalent to conducting science without alternate hypotheses. Ecosystem indicators, whether process based (e.g., productivity), pattern based (e.g., land-use activities), or component based (species populations) vary in space and time, yet a major limiting factor in comprehensive ecological models is lack of explanatory geospatial data. Although these geospatial data may exist, impediments to access by ecologists in PA contexts may include issues as varied as data formats, technical barriers to data integration (e.g., CPU time), validation issues, documentation issues, uniform standards and protocols accompanying RS data, increasing specialization of disciplines, cost, lack of requisite technical expertise, and time for dealing with all of the above complexities. These issues conspire against the ready, standardized integration of RS into ecological research for PA management.

Nonetheless, RS science is a universal tool for managers and researchers across many domains (Kennedy et al., 2009). The lack of standardized protocols, workflow architecture, guidelines, training, and software tools has led us into a baffling jungle of complexity. Our goal in this chapter is to take steps toward standardized yet flexible workflow architecture alongside a set of decision support (software) tools that can be modified as needed, all for integration of RS data into ecological applications for PA systems, in order to support an enhanced role for science at the decision-making table. In this chapter, we propose that RS data/data products coupled with user-friendly data exploration, data management, analyses and modeling tools, in an accessible common platform, can assist scientists and practitioners toward a better understanding of how environmental impacts affect species populations and the ecosystem services that sustain them. It is our intent to describe a science-based approach that makes the most out of existing data (i.e., legacy data) from monitoring programs in and adjacent to PAs by analysis and modeling of focal species populations.

21.2 PAs and the Focal Species Concept

Focal species (listed, sensitive, endangered, or otherwise noteworthy species) are often of substantial socioeconomic and ecological importance. Data from long-term monitoring programs related to focal species and their habitats (Figure 21.1) have the potential to provide invaluable insight into both stable and changing ecosystem function and, if addressed carefully, provide insight into the cause and consequence of dynamic environmental impacts. In order to work with a focal species for monitoring and modeling, variation in space and time discloses environmental impacts at landscape scales and at both short and long timescales. Building empirically based diagnostic models and prognostic models (ecological forecasting) using focal species data (time-series, legacy data sets) and making use of the concept of benchmark ecosystems (see Section 21.4) may be the best approach to craft successful adaptation

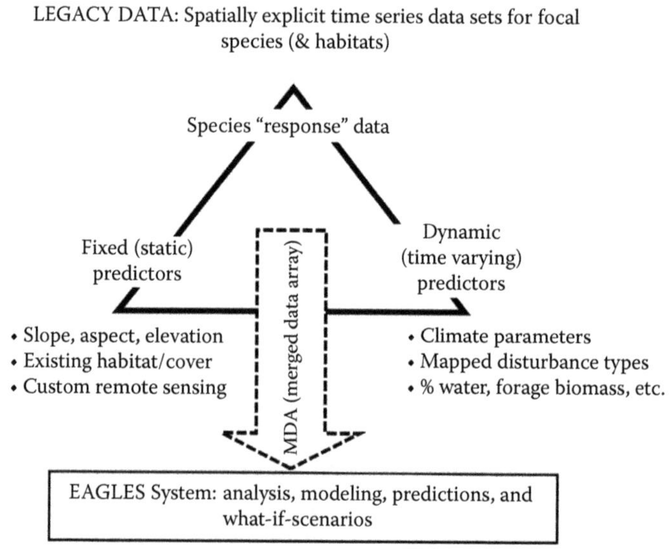

LEGACY DATA: Spatially explicit time series data sets for focal species (& habitats)

Species "response" data

Fixed (static) predictors

Dynamic (time varying) predictors

MDA (merged data array)

- Slope, aspect, elevation
- Existing habitat/cover
- Custom remote sensing

- Climate parameters
- Mapped disturbance types
- % water, forage biomass, etc.

EAGLES System: analysis, modeling, predictions, and what-if-scenarios

FIGURE 21.1
Schematic representation of the process of matching (1) fixed and (2) temporally dynamic geospatial covariates with spatio-temporal response data from legacy data sets to create a merged data array (MDA) for analysis and modeling in EAGLES (Ecosystem Assessment, Geospatial Analysis and Landscape Evaluation).

strategies that protect biodiversity and ecosystem services for human value and those of nature in its own right and intrinsic value (Wiens et al., 2007).

Any number of species, vital rates, vital signs, or ecosystem indicators could have been chosen for our monitoring and modeling approach, for example, using annual NPP (net primary production) as in Crabtree et al. (2009). Here, we focus on species populations (legacy data) as ecosystem indicators to understand complex environmental impacts for three major reasons. First, "species" constitute the longest standing indicator within the ecological/wildlife/biological management professions. For example, the 55-year waterfowl monitoring program (Figure 21.2, for mallards) is believed to be the most extensive, comprehensive, long-term wildlife survey effort in the world. The concept of species as indicators, keystones, and umbrellas has been a dominant theme in ecology and conservation worldwide (Landres et al., 1988; Mills and Soule, 1993) and is still the focus of land management agencies in the United States (e.g., FWS). Second, and most importantly, the legacy of species as ecosystem indicators has created long-term, time-series data sets. These legacy data serve as continuously running experimental units responding to both natural and policy experiments. Because conducting randomized, replicated experiments at ecosystem and regional scales is rarely feasible, we can use these "quasi-experimental" approaches to study impacts on ecosystem structure and function (Hargrove and Pickering, 1992). Analysis of variability in legacy data sets may be our best means to understand the complex interactions of climate disruption with existing

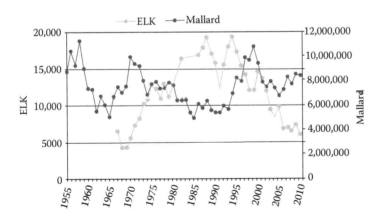

FIGURE 21.2
Legacy data on focal species populations are regularly collected by state and federal agency monitoring programs. Explaining their variation across space and time requires models that include geospatial explanatory variables that change over time. A multitude of these spatio-temporal variables have been recently provided by recent advances in RS technologies. Elk (*Cervus elaphus*) data are derived from winter aerial surveys across Yellowstone National Park's northern winter range. Mallard (*Anas platyrhynchos*) data are from the joint FWS/Canadian Wildlife Service annual aerial survey of breeding waterfowl.

environmental impacts such as changing land-use and invasive spread. With modeling, we can detangle multicausal impacts and build the foundation of predictive models (ecological forecasting). These data have the potential to explain response to climate (multidecadal) and thus allow us to craft adaptation strategies to climate, both direct and indirect (e.g., disturbance), and their interaction with land-use activities.

The third reason for using focal species as indictors of ecosystem impacts is due to their link to human management systems and their socioeconomic importance. Focal species are often protected and managed by state and federal agencies and can bridge gaps between science and practitioners (biologists, conservationists, managers) on-the-ground (Anonymous, 2007). This becomes the key charge of those concerned with successful, long-term, adaptive management strategies for PAs—many of which are undergoing rapid environmental changes.

Figure 21.1 presents the essential function of our approach: matching variation in species legacy data to covariates (candidate hypotheses) across space and time to create a candidate model. Ideally, legacy data can be modeled with known, suspected, and causal spatially explicit and time-varying covariates. However, difficulties in obtaining the needed geospatial covariates to explain observed variation in response remain a prominent challenge for scientists and decision-makers. Forward-looking models (scenarios, forecasts, prognostic models, or projections) present even greater challenges due to lack of stationarity, threshold phenomenon, interaction effects, and discontinuities. We submit that RS data, data products, and output from models that assimilate RS data constitute our best chance at creating the needed covariates required to construct accurate and realistic models of species populations that can be generalized across ecosystems.

21.3 Remote Sensing: Data Integration for Monitoring and Modeling

On-the-ground monitoring of PA ecosystems is expensive primarily due to the size and logistical constraints of national parks, designated wilderness, wildlife refuges, and other large PA ecosystems. However, recent advances in technology—primarily RS—provide new avenues for monitoring and modeling of large PA ecosystems at multiple spatial and temporal scales. Unlike traditional field plots and surveys, RS provides wall-to-wall coverage of geospatial products at unprecedented scales across landscapes, ecosystems, ecoregions, and the globe, revealing continuous empirical patterns of environmental variables in space and time. These patterns are crucial to uncovering cause and consequence in analyses and modeling of a wide variety of ecosystem indicators and vital signs. Of course, this view requires a belief that these observed patterns indicate a comprehensible system of control underlying the material under observation, and that these controls may be disclosed by rule-based (e.g., logic-based) approaches.

In the past, RS data seldom provided the information needed to fulfill the requirements (e.g., accuracy, scale) for the specific objectives of an investigation. Today, integration of data from multiple sources allows creation of previously unavailable geospatial data for monitoring and modeling. Remote sensors on satellites, airborne platforms, and a wide variety of ground-based instruments such as weather stations, sensor networks, and wildlife GPS collars are producing rich streams of environmental data about ecosystems and the species that inhabit them. Even satellite data can now be readily acquired at varying spatial (1 m to 1 km) and temporal (daily to annual) resolutions globally to track many environmental impacts such as land use, disturbance, and climate change. For example, recent deployment of MODIS (Moderate Resolution Imaging Spectroradiometer, http://modis.gsfc.nasa.gov/) sensors have systematically generated multiple ecological indicators available at many resolutions across the entire globe at no or low cost (Justice et al., 1998).

Based on the existing literature of optimal trade-space between spectral, spatial, and temporal resolutions, RS data can provide direct and accurate measurement of important environmental variables with both single and multiple sensors (i.e., fusion). And when direct and/or accurate measurement is not feasible, data integration and data assimilation models can provide direct, relative or by-proxy measurement of previously unavailable parameters of interest that can be validated. The TOPS (Terrestrial Observation and Prediction System) program (Nemani et al., 2009), for example, is a data integration system that uses data from multiple sources (airborne, spaceborne, and ground-based) to produce continuous, gridded variables for monitoring and modeling (see Nemani et al., this volume). Concurrent with this exponential growth of geospatial information is the parallel development of statistical and mathematical techniques, as well as the CPU (computer muscle) capacity to handle these enormous analytical tasks. It is now feasible to create analyses of multiple responses to multiple factors at even global scales (see Zhao and Running, 2010).

21.4 PAs and the Concept of Benchmark Ecosystems

Following the concept of a benchmark ("a standard for evaluation or measurement, a standard of reference"), we propose the concept of a *benchmark ecosystem*. A benchmark ecosystem is one in which all entities (*species, biotic components, abiotic/geophysical components, hydrological processes,* and other *ecological processes,* e.g., migration and predation) function within an integrated and potentially self-sustaining unit. Given the encroachment of threats on even the most remote ecological systems, the idea of a benchmark system represents an idealization. Yet even as an abstraction, the construct serves a heuristic and pragmatic role, forming the basis for evaluation and comparison (e.g., with other systems under restoration). In an era dominated by shifting ecological baselines, the benchmark ecosystem concept provides a set of reference standards for restoration, conservation, and related management activities. Without a specified suite of standards, of metrics, the entities and components of PAs will be increasingly vulnerable to the systemic degradation characteristic of expedience, as well as economic and societal pressures. The benchmark state is to some degree embodied in all PAs, and as such, justifies their creation, perpetuation, and legitimacy. An analogy, from human epidemiology, is the "well-normal" set of criteria for system values. Just as human physiological parameters fall within a defined specified *a priori* range of values for a healthy human body system, so too can the well-normal ranges of ecological systems be specified ("bench-marked"). Precedents already exist for many ecological parameters, such as minimum viable population standards for breeding vertebrate populations, and air- and water-quality standards. We are still short on the overall metrics conferring resilience and stability for the long term.

Benchmark ecosystems can then be used (1) to derive the well-normal attributes of a functioning ecosystem, (2) as a reference standard for restoration efforts, and (3) to provide an empirical testing ground for ecological research into the mechanisms driving sustainable systems and the focal species they contain. The relationship of this concept to PAs becomes one of quantitative and qualitative measurement and justification of the validity of the protection and restoration/conservation efforts. Is the wildlife refuge large enough to sustain its population and the processes that sustain them, such as staging for migration? As research moves forward hand-in-hand with management activities, the outcomes from proposed and forecast activities can be validated or rejected on the basis of the benchmarked goals in a quasi-experimental PA setting (Hargrove and Pickering, 1992).

The concept of a benchmark system can help bridge the gap between scientific measurement and actionable management goals, by finding the common agreement on outcomes-driven, prespecified activities. For example, the ecological process of herbivory (amenable to time-series analysis of RS measurement), is well characterized in a functioning ecosystem, with published values for production, offtake, and patterns of occupancy by species of interest. The absence of this process then, is demonstrable as the process of species depletion occurs, and the restoration of this absent, yet well characterized process, can be set as a specified goal. General reluctance among ecologists and managers to set hard targets for the maintenance, in perpetuity, of species and habitats, may be overcome by acceptance of standardized metrics of ecological function based on empirical standards.

In this intentional design, standardized RS data and data products serve an absolutely critical role. Given the current incompleteness of our ability to characterize, in a deterministic manner, those system attributes required for effective preservation, our best hope may be to attempt functional RS representations, which characterize, with appropriate resolution, and temporal repeat rates, then to store these characterizations as attribute sets of PAs. The novel application of *dynamic attribute mapping* as benchmark metrics is perhaps the most interesting application of RS in PA ecosystems in need of adaptive management and research strategies into our uncertain future.

As ecological insight improves, we begin to understand the mechanisms driving the components that PAs were originally intended to preserve. As this characterization improves, we will be able to more formally set value ranges for functioning and sustainable ecosystem components, and, by extension, apply those to the restoration of other systems. These challenges are directly related to those metrics needed in the developing theory of long-term system health and resilience (Holling, 1973).

As PA monitoring techniques become increasingly powerful, standardizing the metrics of management outcomes in a systems framework becomes an increasingly important objective. PAs exist because they embody ideals of content and condition such as the charismatic panda bear (*Ailuropoda melanoleuca*), the historic legacy of bison (*Bison bison*), and ecological processes like migration and natural disturbance regimes that humans decide to perpetuate. PAs are created, set aside, and managed with the implicit justification of maintaining specific components and functions whose presence originally caused the PA set-aside in the first place, with recognition of their intrinsic worth. Yet managers and scientists still lack clear standards of how to measure success.

In a sense, then, benchmark ecosystems, the best of the remaining PAs globally, offer researchers a "Rosetta Stone" for decoding system attributes which we will need to actively protect. Even as we struggle with a still foggy comprehension of what constitutes an intact ecosystem, we can decode the architecture of attributes of those systems, in order to accomplish the mandates of the Endangered Species Act, as well as the more profound ethical obligations that underlie that Act.

21.5 Augmenting Decision Support Systems: Workflow Architecture and Tools as Unifying Elements

Recent efforts within the US Department of Interior (DOI) provide a case-study opportunity to narrow the gap between research and applications, by bridge building from the side of practitioners. Recently, the National Park Service (NPS) began a nationwide Inventory and Monitoring program (Fancy et al., 2009). The FWS has begun a similar Inventory and Monitoring Program for the National Refuge system. At the same time, the FWS and its federal partner, the U.S. Geological Survey (USGS) formed a National Ecological

Assessment Team to reevaluate how the FWS makes trust resource management decisions, encompassing fish, wildlife, and plants as well as the habitats necessary to sustain them. The team developed a Strategic Habitat Conservation (SHC) framework (NEAT, 2006; http://www.fws.gov/science/doc/SHCTechnicalHandbook.pdf), which was adopted formally in October 2006. This framework became the conceptual basis for the creation of Landscape Conservation Cooperatives or LCC program (http://www.fws.gov/science/shc/lcc.html) and is intended to provide funding to enact the SHC through partnerships that link science and conservation delivery. The LCC effort is being led by the FWS, an agency with a long and profound history of PA management and conservation of biodiversity. The FWS, whose responsibilities include protecting threatened and endangered species, maintaining migratory wildlife populations, and managing national wildlife refuges, has an urgent need to understand how environmental change affects species populations as well as the habitats and ecological functions that support them. In order to achieve these goals, an integrated systems approach or architecture of methods and tools will be required. It must be both consistent enough to be reproducible and defensible, and flexible enough to operate in a variety of contexts. From a modeling standpoint, the ability to census the attributes at landscape to regional scales, through the use of RS technologies, confers that single most necessary tool in unraveling cause and consequence.

21.6 EAGLES: Workflow Architecture and Tools for Scientists and Practitioners

Below, we describe an initial series of linked Decision Support Tools (DSTs) organized as an adaptable, unifying workflow architecture, collectively referred to as EAGLES (*Ecosystem Assessment, Geospatial analysis, and Landscape Evaluation System*). The goal of EAGLES is to lower the barrier of entry to allow scientists and practitioners the ability to understand the cause and consequence of environmental change using focal species (legacy) data as key ecosystem indicators. It stresses the use of *common* data sources (e.g., standardized libraries of RS data products), standardized protocols, and a set of transparent, robust, defensible analysis techniques across jurisdictional and ecological boundaries to provide site-specific, actionable outcomes.

Because species legacy data sets vary with regard to management objectives, spatial and temporal extent, data drop-out in space and time and sampling design, we focused the development of EAGLES to be flexible and user-friendly. We understand the tension between a workflow that is general enough to be applicable across a broad array of data types, yet specific enough to be useful, and emphasize the need for a clear set of approaches, providing a road map for practitioners arriving at management decisions that are well supported by science and common data standards. Thus, EAGLES is a prototype framework for modeling, using a variety of tools for integrating species legacy data (response) and geospatial covariates (explanatory variables), in a variety of forms, within a standardized and documentable workflow for decision-making.

EAGLES was developed through grants awarded from NASA (Ecological Forecasting Program—Award no. NNX08AO58G) and the FWS to the authors in order to enhance existing agency DSSs, in this case the SHC and subsequent LCC program within the Department of Interior (DOI). The project's overriding goal was to integrate RS data into species habitat and demography models. Its three main objectives follow the three sequential segments of the EAGLES acronym: (1) *assess* the ecological conditions of focal region or *ecosystem* where climate and other related environmental drivers are having significant impacts, (2) conduct *geospatial analysis* of focal species populations responding to these impacts, and (3) use these statistical (diagnostic) models to create forecasts (prognostic models) of land cover change and future climate, hence, *landscape evaluations*.

EAGLES's linked software tools (Table 21.1) operate in a user-friendly software environment, allowing user control of data processing, analysis, and

TABLE 21.1

List of EAGLES Tools

Tool	Platform	Code	Readiness
Remote Sensing (RS) data WIKI http://geospatialdatawiki.wikidot.com/	Internet	Web application	Online
www.ClimateScape.net	Internet	Web application	Online
Data integration and assimilation models for covariates (CASA, TOPS, NCEP, etc.); there are many that will be listed in RS WIKI	Various	Application specific	Beta
Remote Sensing Classification techniques, for example, percent surface water from MODIS	Various	In publications	Beta
COASTER: data extraction, visualization (www.COASTERdata.net)	Internet	C#	Online (beta)
Data Exploration Tools (DET)	ArcMap	R w/in Arc interface	Beta
Animal-habitat Models (RSPF)	ArcMap	R w/in Arc interface	Beta
GLM, MHRA, others for animal habitat	ArcMap	R w/documentation	Prototype
SWAP tool nonclimate scenarios	ArcMap	Various	Prototype
FcModelBuilder for Future Climate scenarios	Various	Python	Beta

predictive modeling capabilities. The work flow components of EAGLES within a generic DSS are presented in Figure 21.3. End-users access tools through multiple pathways of data processing, beginning with matching species data sets with RS data/data products (data products include modeled products such as forage quality). The end-to-end nature of the work flow— from data input to visualization, analysis and forecasting of species populations—is intended to provide a platform for flexible yet repeatable analytical pathways. For the remainder of this chapter, we highlight the major processing components and tools within the EAGLES architecture (Figure 21.3) and provide focal species and covariate examples from the legacy data sets accessed through our partnership with FWS and from NASA data/data products. We also provide some general guidelines to covariate selection criteria.

EAGLES tools were designed for a personal computer (PC) platform, including accessing and sharing large data sets on the internet to support a community of users involved in advanced data acquisition, management and manipulations, exploration, data integration, data mining, analysis and modeling, visualization, and other computing and information processing

EAGLES general work flow

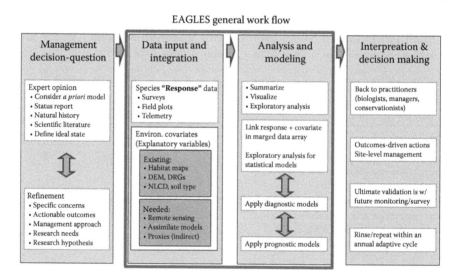

FIGURE 21.3

The EAGLES (Ecosystem Assessment, Geospatial analysis, and Landscape Evelation System) workflow schematic diagram. EAGLES is a workflow architecture that includes both tools (software based) and workflow to allow modeling of species legacy data sets to address management and conservation decision making. It is flexible and provides multiple workflow pathways based on the specifics of the species response data and management question(s). The general idea is to provide a systematic yet flexible architecture for integration of species data with geospatial covariates, most of which are derived from NASA data, data products, and ecosystem models that assimilate sensor data. As the degree of complexity in statistical analyses and RS data increases, the need for a set of standardized techniques and common data protocols becomes more essential, if we are to support repeatable, transparent methods for ecological modeling.

services. The ArcGIS (ESRI, 2009) software interface is chosen because this is the most widely employed and available platform across city, county, state, federal, and international levels. The Data Exploration Toolset (DET) and models were developed primarily in the open-source, statistical software "R" (R Development Core Team, 2009) by Zuur et al. (2010), subsequently translated into ArcGIS user interface, to lower the barrier for end-users (specifically the steep learning curve for command line interface). Both R and ArcGIS are in widespread use among government, NGOs, and in the private sector. The documented underlying code arrays are available to the expert user community, which may include ecological statisticians, academics, and scientists interested in continuing code development.

21.6.1 Outside and Inside EAGLES

We also designed EAGLES with the explicit intent of augmenting existing agency and organizational DSSs. Because EAGLES provides a variety of tools (Table 21.1) for decision-makers it is not a DSS, but is rather a framework populated with DSTs to enhance existing DSS structures. Thus, columns 1 and 4 in Figure 21.3 exist outside EAGLES but are essential activities in order to achieve decision-making. They depict how practitioners might interact with EAGLES given a species legacy data set. At the outset, a specific set of questions need to be defined in order to extract necessary information from a legacy data set that leads to a science-based management action. As is the goal of any science application, outcomes-driven questions must originate from decision-makers. They will eventually interpret the results of model analyses before proceeding to site-level actions (column 4, Figure 21.3). Thus, EAGLES provides the intermediate decision support tools needed in an iterative process starting with the management or decision concern and ending with site-level action (e.g., habitat restoration, harvest levels, land-use activities).

21.6.2 EAGLES Workflow

We envision that experts/managers convene either informally or formally, using, for example, new Structured Decision Making (Lyons et al., 2008) procedures in order to develop a basic understanding of the factors affecting their focal species and/or its habitat. Species recovery plans serve as excellent examples of extant information that can form the basis for *a priori* model structures. Another important objective of this process is to develop the required set of covariates. After a list of candidate covariates is assembled, the user will be able to integrate these into the modeled structure. Given the ideal situation, EAGLES end-users will also be faced with two data input and integration tasks: (1) bias correction and data drop-out issues in the legacy response data (which is beyond the scope of this chapter), and (2) accessing the needed covariates and matching them to the response data.

21.6.3 Covariate Selection Criteria

With rapid technological advances, another gap is being created between method and underlying theory. As RS technologies accelerate delivering new data products, the underlying theory of covariate selection is still catching up. Ecologists are waiting for theoretical and statistical frameworks that can help guide the complex and rapidly expanding world of RS data and RS-modeled products. Indeed, covariate selection for ecological modeling is both an art and a science (Wiens, 2002). Little guidance exists for ecologists charged with building diagnostic and prognostic models for focal species populations that requires a cadre of needed covariates that may or may not vary adequately in space and time.

In an ideal world, the covariates should reflect a minimum and sufficient set of known factors that affect species response. Whether the statistical theory underlying the modeling effort relies on information-theoretic approaches and *a priori* model selection, or involves other approaches, we are after a clear-minded, interpretable method to deconstruct cause and effect—regardless of the RS inputs, PA status, and inferential techniques. In our view, building an ecological model with RS data and data products may be greatly facilitated by access to standardized, documented, accessible, and cost-controlled geospatial layers (see Reichman et al., 2011 for a more formal exposition of these structures and their importance). As a modest beginning, we have started this process by building a variety of web-based tools to access these data sources (rows 1–3, Table 21.1) (see GeoSpatial data wiki at http://geospatialdatawiki.wikidot.com/, which lists RS data for ecological applications).

All ecological models are to some extent *incomplete*, and this incompleteness can be characterized by overall goodness-of-fit, expert knowledge, visually plotting the predictions on the landscape, hypothesis testing, and most importantly, independent validation. A *deficient* model, however, is one that proceeds without inclusion of *known and expected* causal covariates, or at least, interpretable proxies of these covariates. And this violation of scientific principle is in addition to the perennial problem of creating the needed covariates when they usually do not exist or are too costly to acquire. Below, we provide initial guidelines and suggested solutions to this conundrum as well as the overall problem of covariate selection. At this point in a hypothetical EAGLES workflow, we assume that we have identified the ideal situation, independent of availability or cost, that is, a full list of causal or mechanistic covariates based on the discovery process described above. However, given their sensitivity and fundamental role in the scientific method, careful consideration must be given to the selection of covariates prior to preliminary analysis and final modeling. There is surprisingly little attention provided in the published literature (but see Scott et al., 2002). Too often we gather what is easily available without further consideration of either creating the needed covariate or selection of a proxy that is interpretable.

21.6.3.1 Scale

Issues of scale present fundamental challenges for all ecological data. These challenges are further compounded by the interpretation and classification of the abstractions of rasters, pixels, and postings. Most spatial ecological data are inherently continuous in nature, but we discretize these in space, time, and, often, level of organization. Scale determination can be made based on prior research grounded in empirical biology, and is often determined by spatio-temporal resolution available in covariate sets as well. Based on research, natural history, field observation and expert opinion, scale of covariates can be selected from the perspective of the species of study. When in doubt, select several scales then use, for example, metrics like log-likelihood values in model selection approaches (Burnham and Anderson, 2002) to arrive at a reduced set.

Measurement error in the response data should also be used. For example, error polygons for radio telemetry locations set the lower limit on analytical resolution for species studied with telemetry. It is also important to keep in mind the inherent level of spatial and temporal heterogeneity of the covariate of interest, for example, 30 m snow-water-equivalent (SWE) measurements may vary little over a 1 km distance. Continuous observations are recommended over categorical classifications of the same covariate, for example, percent cover of sagebrush captures a more meaningful aspect of selection criteria for a given species compared to a multinomial category that may or may not include sagebrush presence. Finally, selection of temporal resolution is very challenging. Consider not only scale but "critical windows" or seasonality given the life-history strategy for your species of interest. Species response to specific events (e.g., previous disturbance) and cumulative lags are common, for example, peak phytomass production responding to precipitation in both the current spring and previous year (Potter et al., 2007).

21.6.3.2 Collinearity

Collinearity can be considered from both biological and statistical points of view. Rigorous and clear-minded approaches to using data exploration tools (DET) and diagnostics before, during, and after the modeling process (e.g., see Zuur et al., 2010) give the best outcomes for deconstructing the relevance of collinearity metrics in the data under investigations. There are accepted statistical criteria for collinearity, but ecological insights can be of equal importance. This is troublesome because two correlated covariates are often ecologically related to one another from the perspective of the species of interest. For example, assume prey availability, shrub cover, and southern aspects are all significantly correlated and ecologically related covariates. Is a predator, for example, selecting for one, two, or all three? Also, because many approaches offer only loose guidelines with respect to collinearity in model selection procedures, we strongly suggest using our data exploration toolset (DET) which is patterned after Zuur et al. (2010).

21.6.3.3 Proxies and Interpretation

Owing to the cost and/or unavailability of important covariates, we are often left with selection of proxy covariates that can be difficult to interpret. As proxy covariates exist as abstractions on a gradient ranging from gross approximations to specific causal agents, caution must be taken when interpreting model results. The strong relationship between level of proxy and ability to interpret model results, can lead to erroneous conclusions. For example, annual plant production may serve as an interpretable proxy for herbivores responding to forage biomass but not forage quality. Finally, any suspected proxy can be further examined with field validation and field observation efforts. In a sense, a proxy relationship can be thought of as a hypothetical relationship, and is therefore one that should be tested and validated.

21.7 Final Covariate Selection and Creation

Owing to the information needs of a particular focal species data set, it is very likely that some important covariates will have to be created or existing proxies interpreted. We provide just four examples (Figure 21.4a–d) of the many covariates one might need in order to build predictive models and test hypotheses. These examples underscore the vast potential that RS data holds in providing covariates, ranging from free or low-cost to high-cost solutions, for ecological modeling. In our experience, temporally dynamic covariates are most often those in critically short supply.

The most time-consuming step we have experienced in the analysis of focal species data is in data integration techniques (column 2, Figure 21.3) and in creating covariates for modeling and further analysis. This vast subject lies beyond the scope of this chapter. There are many early and developed utilities that are moving rapidly in the right direction to remedy this shortfall. TOPS (see chapter in this volume) serves as an excellent example of a data integration system useful for creation of many covariates. We also refer the reader to http://geospatialdatawiki.wikidot.com/ and www.climatescape.net as examples of both a source and portal to examine further links and techniques to discover and then access covariates or proxy measures. Table 21.2 gives examples of recently created covariates using data integration techniques, in particular, using data from space-borne, airborne, and ground sensor networks to create previously unavailable geospatial covariate layers at various temporal resolutions.

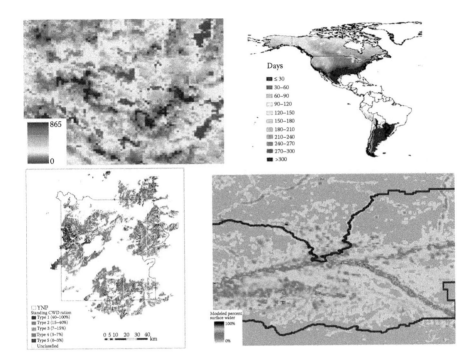

FIGURE 21.4
Four different geospatial covariates created to avoid "deficient" models which lead to errone-ous conclusions and poor inference. Percent surface water (PSW) at Yukon Flats National Wildlife Refuge, Alaska (lower right). Lesser scaup (*Aythya affinis*) breeding pairs were mea-sured in response to the intra- and interannual variability of PSW. The number of frost-free days (Kim et al. 2010) across North America were used to estimate changes in the timing and duration of the growing season for analysis of migratory waterfowl (upper right). These data were generated using merged SSMR and SSM/I data and can be retrieved at (http://freez-ethaw.ntsg.umt.edu). We created classification maps of coarse woody debris biomass in the areas burned in the great fires of 1988 in Yellowstone National Park (YNP) using AIRSAR and AVIRIS data (Huang et al., 2009). The proportion of standing vs. downed CWD was also esti-mated (lower left) using these procedures. Inclusion of the CWD data into the CASA ecosystem model resulted in large portions of YNP switching from a carbon sink to a carbon source. CASA_Express, an ArcGIS version of CASA was used to generate estimates of annual forage production (biomass available to ungulates) using MODIS EVI data (upper left). These data were then used in spatio-temporal models to predict seasonal movements of elk (*Cervus ela-phus*) and bison (*Bison bison*) in and out of YNP (Geremia et al., 2011).

21.8 Ecosystem Assessment and Diagnosis

Once the needed covariates have been identified and created they can be accessed and manipulated using a recently developed software tool called the Customized Online Aggregation & Summarization Tool for Environmental Rasters or COASTER (www.COASTERdata.com). This tool has two major

TABLE 21.2

A List of the Available Geospatial Data Products We Considered for Use in the Analysis of Focal Species Populations in the Northern Rockies Ecoregion

NASA Data Product (Variables and Covariates)	Frequency	Period	Resolution
Terrestrial Observation and Prediction System			
Evapotranspiration (TOPS ET)	Daily	1950–Present	1 km
Solar radiation (TOPS SRAD)	Daily	1950–Present	1 km
Snow water equivalent (TOPS SWE)	Daily	1950–Present	1 km
Temperature (min/max) (TOPS MIN, MAX	Daily	1950–Present	1 km
Precipitation (TOPS PRCP)	Daily	1950–Present	1 km
Snow temperature (TOPS SNWTMP)	Daily	1950–Present	1 km
Vapor Pressure Deficit (TOPS VPD)	Daily	1950–Present	1 km
Gross primary productivity (TOPS GPP)	8-Day	2000–Present	1 km
*Soil cover (% sand, %silt, %clay, hydro, root depth	Static	Recent	1 km
Snow Extent	16-day	2000–Present	1 km
Landcover (IGBP & UMD)	Static	Single Years	1 km
EVI and NDVI – Vegetation indices	8-Day	2000–Present	1 km
FPAR – Photosynthetic Active Radiation	8-Day	2000–Present	1 km
LAI – Leaf Area Index	8-Day	2000–Present	1 km
CASA_Express Wetlands version			
NPP, Net primary productivity	W,M,A	'84/'00–Present	250 m/30 m
***Soil Moisture—3 layers to root depth**	W,M,A	'84/'00–Present	250 m/30 m
***PET, Potential Evapotranspiration**	W,M,A	'84/'00–Present	250 m/30 m
***Herbaceous (Foliar) Biomass Production**	M,A	'84/'00–Present	250 m/30 m
***SWE, Snow Water Equivalent**	D,W,M	'84/'00–Present	250 m/30 m
***Snowmelt Rate**	W,M	'84/'00–Present	250 m/30 m
***Water Temperature in Rivers and Lakes**	W,M	'84/'00–Present	250 m/30 m
Growing Season Length (in days)	Annual	'84/'00–Present	250 m/30 m
Drought—User Specified and Probabilistic	Annual	'84/'00–Present	250 m/30 m
Disturbance Classes (MODIS derived)			
Urban Expansion	Annual	2001–Present	250 m
*Agriculture Expansion—New Irrigated Cropland	Annual	2001–Present	250 m
*Agriculture Expansion—CRP for two years or more	Annual	2001–Present	250 m
***Wetland Conversion to cropland**	Annual	2001–Present	250 m
***Wetland Loss (drained or dried out)**	Annual	2001–Present	250 m
***Wetland Expansion**	Annual	2001–Present	250 m
Fires (nonforest)	Annual	2001–Present	250 m
Fires (forested)	Annual	2001–Present	250 m
Insect kill (forested)	Annual	2001–Present	250 m
Logging (forested)	Annual	2001–Present	250 m

TABLE 21.2 (continued)

A List of the Available Geospatial Data Products We Considered for Use in the Analysis of Focal Species Populations in the Northern Rockies Ecoregion

NASA Data Product (Variables and Covariates)	Frequency	Period	Resolution
Other Remote Sensing Products			
*Percent Cover of Shrubs, Herbacious, Soil**	Static	Recent	30 m
*PSW, Percent Surface Water**	8-Day	2000–Present	1 km
Freeze–Thaw Parameters (Frozen/Thawed/ Trans.)	Daily	1979–Present	25 km

Note: Temporal resolution, temporal extent, and spatial resolution must all be carefully considered in the selection and interpretation steps of modeling efforts in EAGLES. D = Daily, W = Weekly, M = Monthly, and A = Annual. We laid great emphasis on those products related to water and wetlands (in bold) due to the strong relationship between water and biodiversity. Products marked with an asterisk had to be validated in select locations prior to their use in predictive modeling.

functions within EAGLES: (1) data discovery and (2) data access. COASTER allows end-users to identify and extract desired subsets of RS covariates from archived geospatial databases (~1 TB and larger). These typically reside on servers and supercomputers. The subsetted covariates are accessed via COASTER, then posted to an ftp site for input into the EAGLES environment for further analysis (column 3, Figure 21.3; see http://coasterdata.net/Examples. aspx for examples). This allows end-users to conduct ecosystem and vulnerability assessments through use of basic functions that summarize, threshold, and compare covariates across space and time. Additional functions allow the creation of visualization models to assess, for example, how key climate metrics are driving changes in plant productivity. Most simply, this can be done through the creation of trend maps and anomaly detection. Figure 21.5 provides a trend map of climate anomalies for the Northern Rockies region created in COASTER. The combination of functions within COASTER provides desktop capability for characterization of ecosystem condition, function, pattern, and process in a specific area of interest, for example, a refuge or a management region. It can also be used to create new covariates needed for further analysis of species legacy data. At a more streamlined level, end-users may arrive at a sufficient level of decision support directly within the COASTER suite of utilities (data visualization and change detection, trend, threshold, and anomaly metrics) and elect not to proceed further in the EAGLES framework.

The conversion of input data types into a standardized protocol is accomplished with the software tools that reside in the EAGLES workflow architecture. These functions are programmed to be compatible with, or are already included in, the ArcGIS environment. They can also be further modified by expert users. These include resampling, interpolation, simulation modeling, statistical correction of bias, and calibration/validation of the final mapped products. Two components occur outside the workflow, (1) datum and

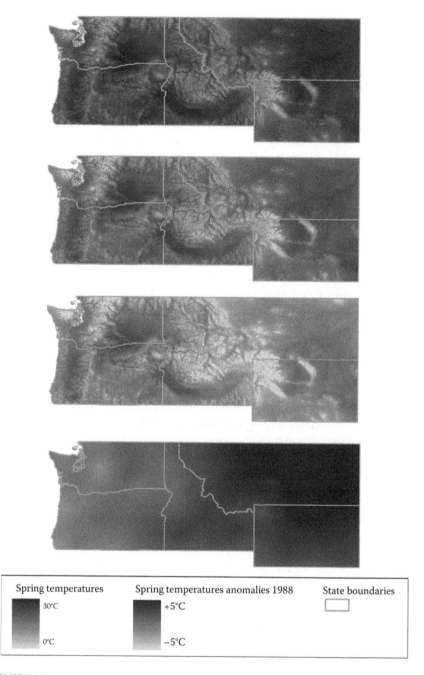

FIGURE 21.5
These maps illustrate some of the functionality available within the COASTER system. To create these maps, gridded, daily maximum temperature datasets from March 21st through June

scaling integration and (2) creation of unique or nonstandard RS products, for example, geospatial covariates specific to analysis and predictive modeling of a given focal species. These unique RS products may be essential for successful modeling efforts. For example, in order to support a waterfowl model from annual counts of breeding pairs, we created a fractional surface water product (see Figure 21.4b) from MODIS that produces estimates every eight days (Weiss and Crabtree, in press). This RS product met an identified critical need suggested by a federal agency practitioner working group focused on waterfowl and shorebird populations.

21.9 Statistical Analysis and Modeling

Figure 21.6 provides an example of our Data Exploration Toolset or DET (Zuur et al., 2010) as applied to analysis of pronghorn (*Antilocapra americana*) data in Yellowstone National Park, WY. The utility of data exploration before, during, and after modeling cannot be overstated. Data exploration serves a vital role in testing assumptions (e.g., about the distribution of the data, the relative contributions of extreme values, and the presence of underlying patterns that may require more thought). As the sophistication of RS data inputs increases, the investigation of unexpected patterning of relationships within and among covariates becomes increasingly important. Making use of a readily available and standardized toolbox for investigation the patterning of dependent and independent variables has proven to be of great utility.

After data exploration procedures, end-users are now ready for diagnostic analysis using various statistical models. For this, we created a conceptual and practical framework for analysis of species populations and their habitats called Risk-Reward Spatial Capacity (RRSC) models. These are a progressive series of spatially explicit species population models that can be diagnostic and/or predictive. Historically, species–environment models (e.g., Scott et al., 2002) focus on desirable habitat conditions, leaving out important risk or hazard conditions and their impacts on vital rates (Wittmer et al.,

---

FIGURE 21.5 Continued.
21st were summarized for each year and for all years from 1955 to 2009. The top three maps show (from top to bottom) the minimum, mean, and maximum average springtime temperatures for the Pacific Northwest. These graphs are conceptually similar to the normal high, record high, and record low values typically provided within weather reports on the local news. However, instead of the conditions for a single day the maps show conditions summarized for all days within spring. The lower map, in contrast, shows how conditions in the spring of a single year (1988) differ from normal conditions. This map was created by subtracting the average springtime high temperature (i.e., the second map) from the springtime temperatures summarized for 1988, which was one of the warmest years on record in the United States.

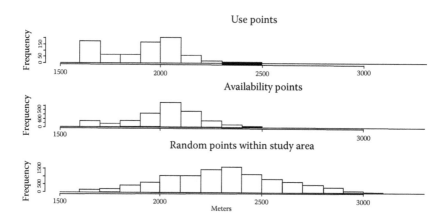

FIGURE 21.6

Example of the utilization function in the EAGLES Data Exploration Toolset (DET). The summer 2008 relocations presented in Figure 21.7 (purple spears) are summarized above as a three-panel frequency histogram for the covariate "elevation." The top displays the actual response data or "use points"; the middle displays a random sample of points within a 1 km buffer area around the response use points and; the bottom panel displays a random sample of points within the study area (user-defined region of interest for model inference) ($n = 10,000$ random points).

2007). It is clear that inclusion of spatial information on hazards such as predation risk can expose sublethal effects on species distributions and vital rates. Similarly, in the temporal domain, extreme weather and disturbance events, which can often be characterized as single-event hazards, are essential to integrate into species modeling efforts and can have long-lasting impacts on populations.

The RRSC framework provides a logical progression of legacy data set analysis starting with basic coarse-scale species distribution models (SDMs) to identify and delineate critical habitat components. Next, fine-scale habitat selection models that include temporally dynamic information, either implicitly (time for space substitution) or explicitly (time-varying model structure) can be constructed. Both SDMs and habitat selection models can then be used to identify and filter out important covariate structures prior to demographic analysis and modeling (vital rates and abundance as related, ultimately, to fitness). Species must survive environmental risks and then utilize resource (habitat and prey/forage) to successfully reproduce and increase fitness. Species abundance at a given time is essentially the result of survival (loss) and reproduction (gain).

As a first RRSC model within EAGLES, we developed an ArcGIS tool for analysis of species location data using a recent modification of a widely used technique called Resource Selection Functions or RSF (Manly et al., 2002). This falls under a more general category referred to as resource selection analysis, hence our RRSC-RSA tool. This type of RSF model is described in

Lele and Keim (2006) and Lele (2009), and provides a probabilistic approach to identifying regions and resources/risks that are used more/less than expected. These approaches are robust and are additionally appealing because they create empirical response plots and probabilistic maps (resource selection surfaces) across the study area of interest. These are particularly useful for threshold delineation, for example, for explicating the patterning of animal response data against each covariate (e.g., polar bear, *Ursus maritimus*, vs. sea ice density). Originally developed in R, we have translated this approach into an ArcGIS platform, and have developed numerous diagnostics and interactive decision-points for end-users, providing a transparent, repeatable process or "white-box" approach. We have also developed and documented a suite of RRSC modeling approaches (e.g., GLM, GLM with spatial autocorrelation, and GAMs) for habitat and demographic analysis in R and are in the process of translating them into the ArcGIS environment.

Figure 21.7 provides a visualization of how a merged data array (MDA) is created from the legacy and covariate data sets and then input into a habitat model for diagnostic analysis and final model output. These and related functions and tools are available as ArcGIS plug-ins. Again, the specifics of the data sets will dictate the specific processing chain or pathways through EAGLES. The original or modified MDA can also be run through other pathways and various statistical models to explore further relationships and test hypotheses. The output of the RSPF analysis is shown at the bottom of Figure 21.7 as a predicted probability surface within the region-of-interest or sampling universe (bottom of Figure 21.6). This probability surface was used to estimate a species distribution, delineate critical habitat, identify critical habitat components, and provided a probabilistic resource selection/avoidance surface with measures of uncertainty per pixel. Table 21.3 provides results of the pronghorn RSPF model. Many other diagnostics are provided as well as intermediate steps and decision points that guide the end-user through the statistical analysis.

Once a diagnostic model has been developed, we have developed other tools to create prognostic models and explore "what-if" scenarios. One tool ingests and manipulates downscaled future climate scenarios and another allows end-users to apply RRSC models to different areas or under different habitat scenarios such as habitat restoration, development, and disturbance events (fire, drought, snow-hardening events, and a variety of land-use activities). The "SWAP tool" is an ArcGIS utility that allows the end-user to modify a covariate value in an existing model structure, holding all else constant, and examining resultant changes in modeled output. For example, an existing diagnostic model structure (Figure 21.7) created the output in Table 21.3. The SWAP tool allows the end-user to simulate (or forecast) a new covariate (or a new range of values for an existing covariate, e.g., increase biomass, 50% increase in precipitation, new road as in Figure 21.8) by swapping out the previous covariate value with the new one. The predictive model structure and coefficients stay the same while only the new covariate values—new or

FIGURE 21.7
(See color insert.) Visualization of co-registered GIS layers generated using various data integration procedures in EAGLES, a set of decision-support tools integrated into ESRI's ArcGIS environment. Four geospatial data layers, generated for resource selection analysis (RSPF, Lele and Keim, 2006) of pronghorn summer habitat selection, are (from top to bottom): (a) forage biomass created by the CASA model assimilating MODIS EVI data (Potter, 2007), (b) coyote utilization created from kernel density smoothing of relocations of radio-marked adults, (c) elevation and aspect from a USGS 30 m DEM, and (d) a remotely sensed classification map of forest (dark green) and sagebrush (light green) using PALSAR and Landsat ETM data. The locations of radio-tracked pronghorn adults (response data used in the RSPF model) are indicated as coregistered purple "spears." The bottom layer is the resource selection probability surface generated by the EAGLES RSPF model tool where "warmer" temperature colors indicate higher probability of use (ranging from selected to avoided habitat areas) by pronghorn during the summer.

future scenarios—are inserted and the model rerun. The original diagnostic model was modified to create a "what-if" scenario by adding a road (Figure 21.8). It was then rerun to create prognostic model output that predicts how pronghorn would respond as well as how critical habitat may be functionally removed or improved.

The Future Climate Tool or FcModelBuilder (listed in Table 21.1) is a similar tool that ingests and manipulates downscaled future climate scenarios (IPCC, AR4, etc.) and another allows end-users to apply RRSC models to different areas or under different habitat scenarios such as habitat restoration, development, and disturbance events (fire, drought, snow-hardening events, and a variety of land-use activities). Through a disciplined, documented, and transparent diagnostic modeling processes, end-users can then utilize these tools to build prognostic models to effectively support scenario construction guiding site-level actionable outcomes. Such ecological forecasting (Clark et al., 2002) is an emerging subdiscipline in ecology that has many

TABLE 21.3

Model Results for the Pronghorn Example Depicted in Figure 21.7

Parameter	*t*-Value	*p*-Value	v-i-f
(Intercept)	−2.76	5.92E−03	NA
Distance to road (+)	3.59	3.52E−04	1.8
Distance to road ^2 (-)	−6.98	6.51E−12	1.3
Coyote (+)	4.64	4.11E−06	1.4
Elevation (−)	−5.91	5.19E−09	2.6
Forage (+)	2.51	1.23E−02	2.8
June NPP (−)	−2.15	3.19E−02	2.3
% Sage (−)	−3.1	2.01E−03	1.5
% Forest (−)	−3.45	5.92E−04	3.5
% Herbaceous (+)	2.45	1.45E−02	2.4
Slope (−)	−4.68	3.40E−06	1.4
Wolf (−)	−8.35	3.29E−16	1.1

Note: The VIF, variance inflation factor, is a measure of multicollinearity among the independent variables, that is, it gives a measure of correlation.

unsolved issues, not the least of which is how one independently validates such future forecasts for conservation action. Making sense of the chaotic and disarticulated array of old and emerging issues involved in future predictions and decision-making (issues of scale, components, instrumentation, algorithm function, information about uncertainty and assumptions, and statistical issues) (see Fulton, 2010) can be helped by beginning with a standardized, highly documented, and inter-comparable common set of inputs and methods (e.g., Hijmans et al., 2005) as we have outlined in this chapter.

21.10 Conclusions

Maintaining resilient plant and animal communities within managed PA ecosystems is a daunting challenge. Yet, PAs often have internal monitoring programs in place with existing legacy data sets as well as national and global monitoring efforts producing vast streams of environmental data (ground-, air-, and space-borne sensors). We contend that the analysis and modeling of these "merged" data sets provide our best hope for crafting successful adaptation strategies to environmental change agents, especially those being complicated by climate disruption. The generalized modeling approach—from narrative, conceptual models to visual, spatially explicit population models—provided by EAGLES allows a solution to these challenges as well as a bridge between science and decision-makers. Conservation of species and the

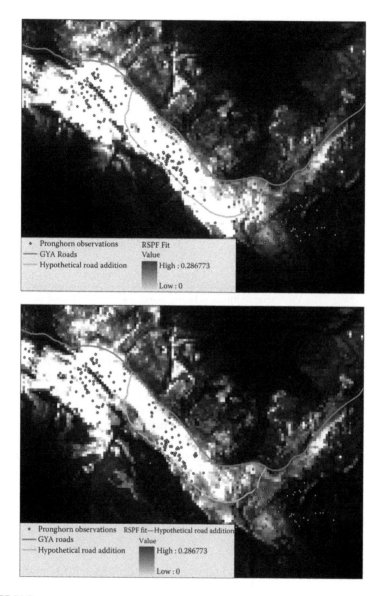

FIGURE 21.8
A portion of the original RSPF model output (bottom layer of Figure 21.7) indicating the resource selection function for pronghorn in Yellowstone National Park (top). The SWAP was used to create an additional road shown in orange. The new prognostic RSPF model output for pronghorn (bottom) indicates that pronghorn are excluded from portions of their original selected habitats. These types of What-if-Scenario (WIS) will provide practitioners with important decision support to guide site-level action plans, restoration efforts, and understand the environmental impacts from climate disruptions, invasive species, changing land-use, and disturbance regimes.

ecosystem processes that support them will require a more effective set of programmatic linkages (e.g., DSTs), narrowing the gap between scientists (researchers and academics), who are forging ahead with new methodologies, and the end-user practitioners (biologists, managers, conservationists), who require straightforward, cost-effective tools in order to make informed and defensible decisions based on diagnostic and predictive modeling. Casting and comparing these activities against the concepts and metrics of "bench-mark ecosystems" will provide yet another set of powerful adaptive management techniques leading to resilient species and ecosystems.

The modeling approaches we describe are not new. But lowering the barrier of entry for users as they approach an increasingly complex analytical environment seems useful. Among various factors limiting the development and use of predictive landscape models, the lack of access to known, causal covariates may be the most problematic. This leads to "deficient" models the scientific principles: predictive models that exclude explanatory variables are equivalent to conducting science without explanatory alternate hypotheses. It is unfortunate that many needed explanatory covariates simply do not exist or are too expensive to acquire. However, with the advent of many new RS sensors, data products (e.g., free MODIS data products), and assimilation models that ingest RS data, we can now create direct or proxy measures of those needed covariates. For example, Geremia et al. (2011, accepted) derived annual estimates of above-ground forage biomass and weekly snow-water-equivalents from sensor-assimilation models. Those two factors were crucial in predicting winter bison (*Bison bison*) movements outside of Yellowstone National Park, a famous benchmark ecosystem.

While new information and technology are emerging at unprecedented rates, environmental impacts on species populations are in a state of flux due to the cumulative impacts of human activities on landscapes. Invasive spread, pathogen outbreaks, land-use activities, and especially climate disruption and its associated impacts—severe drought, reduced stream flow, increased wildfire frequency, extended growing season, and extreme weather events—are increasing, and in some cases accelerating. These changes are outpacing management and conservation actions. In particular, the increase in frequency and severity of climate-mediated impacts (e.g., record drought in the Northern Rockies) are now occurring at larger, landscape to regional scales. This combined with unpredictability and unexpected interactions are rendering traditional management strategies ineffective at sustaining ecological function, resiliency, and viable species populations.

Facing this uncertain future, we have described the EAGLES framework and provided an example of DSTs, important geospatial covariates, ecosystem assessment tools, and initial RRSC species population models for analysis of focal species. This will allow agencies like the FWS (and the new LCC program) the ability to examine what-if scenarios guiding outcomes-driven, on-the-ground actions to recover or maintain populations. Restoration in response to disturbance, acquisition of additional habitat, and regulation of risk factors such as hunting and predation are examples of such outcome-driven actions. These necessitate the need for models that are spatially explicit because decision-makers regulate land-use activities or conduct habitat management at the site level (e.g., forest stand, allotment, and refuge scales). At the same time, multiple impacts are occurring at unpredictable, multiple temporal scales from periodic oscillations (e.g., ENSO), planned disturbance events, and daily extreme weather events. Thus, spatially explicit models also need to incorporate such temporally dynamic information. In this highly uncertain spatiotemporal frame it seems imperative to adhere to standardized, transparent, and defensible processes in arriving at science-based management decisions, so that we are able to systematically track, measure, and evaluate outcomes, such as species persistence, as clearly as possible.

References

Anonymous. 2007. The great divide: The gap between theory and practice remains surprisingly wide in conservation biology (editorial). *Nature* 450: 136.

Austin, M. P. 2002. Spatial prediction of species distribution: An interface between ecological theory and statistical modelling. *Ecological Modelling* 157 (2–3): 101–118.

Baum, J. K. and R. A. Myers. 2004. Shifting baselines and the decline of pelagic sharks in the Gulf of Mexico. *Ecology Letters* 7 (2): 135–145.

Burnham, K. P. and D. R. Anderson. 2002. *Model Selection and Multimodel Inference: A Practical Information-Theoretic Approach*, Second Edition. New York: Springer-Verlag.

Clark, J. S., S. R. Carpenter, M. Barber, S. Collins, A. Dobson, J. A. Foley, D. M. Lodge et al. 2002. Ecological forecasts: An emerging imperative. *Science* 293: 657–660.

Crabtree, R., C. Potter, R. Mullen et al. 2009. A modeling and spatio-temporal analysis framework for monitoring environmental change using NPP as an ecosystem indicator. *Remote Sensing of Environment* 113 (7): 1486–1496.

ESRI. 2009. *ESRI ArcMap 9.3*. Redlands, CA: Environmental Systems Research Institute

Fancy, S. G., Gross, J. E., and S. L. Carter. 2009. Monitoring the condition of natural resources in US national parks. *Environmental Monitoring and Assessment* 151: 161–174.

Fancy, S. G., J. E. Gross., and S. L. Carter. 2009. Monitoring the condition of natural resources in US national parks. *Environmental Monitoring and Assessment* 151 (1–4): 161–174.

Ferrier, S., G. V. N. Powell, K. S. Richardson, G. Manion, J. M. Overton, T. F. Allnutt, S. E. Cameron et al. 2004. Mapping more of terrestrial biodiversity for global conservation assessment. *Bioscience* 54 (12): 1101–1109.

Fulton, E. A. 2010. Approaches to end-to-end ecosystem models. *Journal of Marine Systems* 81: 171–183.

Geremia, C., P. J. White, R. L. Wallen, F. G. R. Watson, J. J. Treanor, J. Borkowski, C. S. Potter, and R. L. Crabtree. 2011. Predicting bison migration out of Yellowstone National Park using Bayesian models. *PLoSOne.* 6(2): e16848, doi:10.1371/journal.pone.0016848.

Gross, J. E., S. J. Goetz., and J. Cihlar. 2009. Application of remote sensing to parks and protected area monitoring: Introduction to the special issue. *Remote Sensing of Environment* 113 (7): 1343–1345.

Hargrove, W. W. and J. Pickering. 1992. Pseudoreplication: A *sine qua non* for regional ecology. *Landscape Ecology* 6: 251–258.

Heinen, J. 2007. *Protected Natural Areas. The Encyclopedia of Earth.* http://www.eoearth.org/article/Protected_natural_areas

Hijmans R. J., S. E. Cameron, J. L. Parra, P. G. Jones., and A. Jarvis. 2005. Very high resolution interpolated climate surfaces for global land areas. *International Journal Climatology* 25: 1965–1978.

Hoffmann, M., C. Hilton-Taylor, A. Angulo et al. 2010. The impact of conservation on the status of the world's vertebrates. *Science* 330 (6010): 1503–1509.

Holling, C. S. 1973 Resilience and stability of ecological systems *Annual Review of Ecology and Systematics* 4: 1–23.

IUCN. 2008. Shaping a sustainable future. In *Gland.* IUCN: The IUCN Programme 2009–2012.

Justice, C., E. Vermote, J. Townshend et al. 1998. The Moderate Resolution Imaging Spectroradiometer (MODIS): Land remote sensing for global change research. *IEEE Transactions on Geoscience and Remote Sensing* 36: 1228 – 1249.

Kennedy, R. E., P. A. Townsend, J. E. Gross et al. 2009. Remote sensing change detection tools for natural resource managers: Understanding concepts and tradeoffs in the design of landscape monitoring projects. *Remote Sensing of Environment* 113 (7): 1382–1396.

Landres, P. B., J. Verner, and J. W. Thomas. 1988. Ecological uses of vertebrate indicator species—A critique. *Conservation Biology* 2: 316–328.

Lele, S. R. and J. L. Keim. 2006. Weighted distributions and estimation of resource selection probability functions. *Ecology* 87: 3021–3028.

Lele, S. R. 2009. A new method for estimation of resource selection probability function. *Journal of Wildlife Management* 73 (1): 122–127.

Lyons, J. E., M. C. Runge, H. P. Laskowski, and W. L. Kendall. 2008. Monitoring in the context of structured decision-making and adaptive management. *Journal of Wildlife Management* 72: 1683–1692.

Manly, B. F., L. L. McDonald, D. L. Thomas, T. L. McDonald, and W. P. Erickson. 2002. *Resource Selection by Animals: Statistical Design and Analysis for Field Studies,* Second Edition. Dordrecht: Kluwer Academic Publishers.

Marris, E. 2007. What to let go. *Nature* 450: 152 – 155.

Mills, L. S. and M. E. Soule. 1993. The keystone-species concept in ecology and conservation. *Bioscience* 43: 219.

NEAT. 2006. Strategic habitat conservation: a report from the National Ecological Assessment Team—June 29, 2006. Washington, DC: U.S. Fish and Wildlife Service; Reston, VA: U.S. Geological Survey, 2006.

Nemani, R., H. Hashimoto, P. Votava et al. 2009. Monitoring and forecasting ecosystem dynamics using the Terrestrial Observation and Prediction System (TOPS). *Remote Sensing of Environment* 113 (7): 1497–1509.

Paola, C. 2011. Simplicity versus complexity: In modelling simplicity isn't simple. *Nature* 469: 38–39.

Pielke, R. A. Jr., C. D. Canham, J. J. Cole., and W. K. Lauenroth (Ed.) 2003. The role of models in prediction for decision. *Models in Ecosystem Science*, pp. 111–135. Princeton, NJ: Princeton University Press.

Pereira, H. M., P. W. Leadley, V. Proença et al. 2010. Scenarios for global biodiversity in the 21st century. *Science* 330 (6010): 1496–1501.

Potter, C. S., Klooster, S., Huete, A., and V. Genovese. 2007. Terrestrial carbon sinks for the United States predicted from MODIS satellite data and ecosystem modeling. *Earth Interactions* 11: 1 – 21.

Reichman, O. J., M. B. Jones, and M. P. Schildhauer. 2011. Challenges and opportunities of open data in ecology. *Science* 331 (6018): 703–705.

Scott, J. M., P. J. Heglund, M. L. Morrison, J. B. Haufler, M. G. Raphael, W. A. Wall., and F. B. Santon (eds.) 2002. *Predicting Species Occurrences*.Washington, DC: Island Press.

Sinclair, A. R. E., D. S. Hik, O. J. Schmitz, G. G. E. Scudder, D. H. Turpin., and N. C. Larter. 1995. Biodiversity and the need for habitat renewal. *Ecological Applications* 5: 579–587.

Weiss, D. J. and R. L. Crabtree. 2011 (in press). Subpixel percent surface water estimation from MODIS BRDF 16-day image composites. *Remote Sensing of Environment*.

Wiens, J. A. 2002. Predicting species occurrrences: Progress, problems and prospects. In J. M. Scott, P. J. Heglund, F. Samson, J. Haufler, M. Morrison, M. Raphael, and B. Wall (eds.). *Predicting Species Occurrences: Issues of Accuracy and Scale*, pp. 739–749. Covelo, CA: Island Press.

Wiens, J. A., M. R. Moss, D. J. Mladenoff, and M. G. Turner (eds.). 2007. *Foundation Papers in Landscape Ecology*. New Yo rk: Columbia University Press, 582pp.

Wittmer, H. U., B. N. McLellan, R. Serrouya, and C. D. Apps. 2007. Changes in landscape composition influence the decline of a threatened woodland caribou population. *Journal of Animal Ecology* 76 (3): 568–579.

Zhao, M. S. and S. W. Running. 2010. Drought-induced reduction in global terrestrial net primary production from 2000 through 2009. *Science* 329 (5994): 940–943.

Zuur, A. F., E. N. Ieno., and C. S. Elphick. 2010. A protocol for data exploration to avoid common statistical problems. *Methods in Ecology and Evolution* 1: 3–14.

22

Monitoring and Forecasting Climate Impacts on Ecosystem Dynamics in Protected Areas Using the Terrestrial Observation and Prediction System

Hirofumi Hashimoto, Samuel H. Hiatt, Cristina Milesi,
Forrest S. Melton, Andrew R. Michaelis, Petr Votava,
Weile Wang, and Ramakrishna R. Nemani

CONTENTS

22.1 Introduction

Assessing the impacts of climate change on protected areas such as national parks is crucial for decision makers to develop adaptive management strategies to help protect natural resources for the benefit of the society. The National Park Service (NPS) recently developed a Climate Change Response Strategy (CCRS) featuring four integrated components: science, adaptation, mitigation, and communication (NPS, 2010). Scientists can contribute to this strategy by providing data and information about projected climate change impacts on ecosystems within and surrounding national parks and other protected areas. As outlined in the NPS CCRS, climate change can affect parks and protected areas in a number of ways. Changes in temperature and precipitation are leading to reductions in winter snow pack and shifts in the timing and amount of streamflow (Stewart et al., 2005). These changes, in turn, can lead to shifts in the distribution of species (e.g., Moritz et al., 2008) and alterations to the fire cycle, including increasing frequency of wildfires (Westerling et al., 2006). In addition, warmer temperatures are leading to conditions which are favorable to outbreaks of insects and pathogens (Bentz et al., 2010) which are impacting on forested ecosystems throughout the United States, and increasing tree mortality rates have been reported for forests in the western United States (van Mantgem et al., 2009). As impacts from climate change to species and ecosystems begin to emerge, there is an increasing need to develop a standard approach to produce long-term forecasts of key ecological parameters at spatial scales that are relevant to park managers.

The procedure for climate impact assessments has been well implemented in global carbon and water cycle studies, where it contributes to answering general questions about the effect of climate change on energy, water, and nutrient fluxes at regional to continental scales. We can take advantage of this heritage with respect to the data and methodologies adopted in global carbon and water cycle studies, and apply it to protected lands to understand climate change impacts at finer spatial scales. The assessment process includes the monitoring of present vegetation status, simulating the influence of recent climate variability on vegetation, and projecting the vegetation response to future climate scenarios. Among the various ecophysiological variables, Net Primary Production (NPP) and Leaf Area Index (LAI) are indicators of ecosystem state and canopy structure that can be modeled using biogeochemical cycle models, and which describe vegetation status at the landscape scale. Monitoring of these parameters helps identify where the greatest impacts of recent climate changes took place, allows one to identify climate drivers of changes in vegetation, and what would happen to the ecosystems under future climate scenarios. The information derived using this approach can be used by decision makers to craft future mitigation or adaptation plans for protected lands and the vicinity areas.

The US Global Change Research Program (USGCRP, 2009) summarized past and future climates in the United States as follows. In the last century, the Southwest showed a small temperature increase and a significant increase in precipitation of 30%, especially during the fall season. Meanwhile, from 1970 onward, an increase in temperature of more than 1°C and a decrease in precipitation accompanied by drought were observed in the Southeast. In the Northeast, for the last 50 years temperature has been significantly increasing along with the intensity of heavy precipitation events. Overall in the United States, the temperature will keep increasing in the twenty-first century though the rate of increase is dependent on future CO_2 emission scenarios. An increase in precipitation is projected for the northern conterminous United States, while the southern region is expected to experience a decrease in precipitation. While these summaries capture the broad patterns of expected changes in climate over the United States, they cannot be simply extended to protected areas, as they need to be adjusted to the particular local conditions associated with each park or network of parks and protected areas, such as higher elevation and rugged terrain.

To analyze climate impacts on ecosystems of protected lands narrowly distributed across multiple ecoregions, it is reasonable to downscale low-resolution climate fields with statistical or physical methods. The Terrestrial Observation and Prediction System (TOPS, Nemani et al., 2009), developed at NASA Ames Research Center, is suitable for this kind of processing. In this chapter, we describe the TOPS framework structure and components and demonstrate examples of the use of TOPS for analyzing current vegetation status, assessing the impact of recent climate change on vegetation, and projecting future vegetation conditions using climate scenarios downscaled for the protected lands and ecosystems along the Appalachian Trail (A.T.). The ultimate goal of this A.T. case study is to provide seamless data and derived information for the development of a decision support system for monitoring, reporting, and forecasting ecological conditions of the A.T. region (Wang et al., 2010).

22.2 About TOPS

22.2.1 Overview of TOPS

TOPS is a data and modeling software system that streamlines the integration of data from ground, aircraft, and satellite sensors with weather, climate, and application models, with the goal of efficiently producing operational nowcasts and forecasts of ecological conditions (Figure 22.1, Nemani et al., 2009). Through TOPS, the automation of the retrieval, preprocessing, and reprojection of time series of Earth-observing satellite images,

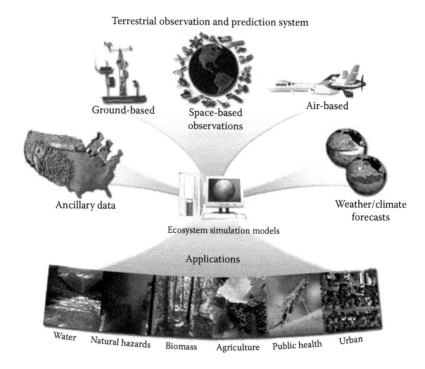

FIGURE 22.1
Schematic diagram of the Terrestrial Observation and Prediction System (TOPS).

climate data, and other geographical information (i.e., land cover, soils, and topography) provides a continuous stream of inputs to ecosystem models, which can therefore be run in near real time. With the replacement of climate observations with climate projections, TOPS serves as an ecological forecasting tool that can be used to predict the effects of climate change scenarios on ecosystems over timescales from decades to centuries.

22.2.2 Architecture Framework of TOPS in Data Integration and Modeling

The data enter into the TOPS system through a pool of automated data-acquisition agents that are tasked with data retrieval from a number of different locations. The data are then preprocessed, which often includes conversion into a common format, projection and geographic extent, and then archived in the TOPS database system. To streamline the data integration process, TOPS deploys a flexible plug-in framework that provides a unified interface to data formatting, processing, analysis, and visualization. The TOPS framework also facilitates access to a number of satellite, model, and climate data sets by interfacing with TOPS database system. Data are internally stored in self-describing formats (HDF4, HDF5, netCDF) so that they

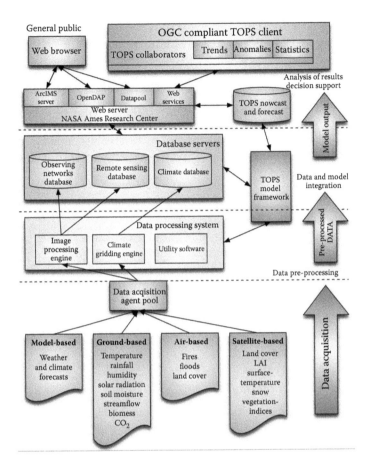

FIGURE 22.2
Architecture framework of the TOPS.

can be easily integrated with the processing components. In addition, TOPS includes a number of data transformation utilities capable of moving data between different formats depending on the requirements of different ecosystem models and analysis tools. This enables TOPS to run any number of land surface models without extensive manual data integration as well as to present model outputs in a format that adheres to TOPS clients' specifications (Figure 22.2).

22.2.2.1 Climate Gridding

Gridded climate surfaces are a key input to many component models within TOPS, and TOPS includes functionality for deriving gridded climate surfaces from networks of meteorological stations. To create a gridded climate surface, the automated TOPS data-acquisition system first fetches the necessary weather data from weather stations in the region of interest. Next, the data are preprocessed by checking for consistency against historical averages, filling missing values from additional sources, flagging missing values, and finally converting these observation records into data structures suitable for processing by the Surface Observation and Gridding System (SOGS, Jolly et al., 2005), a component within TOPS. SOGS is a climate gridding system that uses maximum, minimum, and dewpoint temperatures, in addition to precipitation, to create spatially continuous surfaces for air temperatures, precipitation, vapor pressure deficits, and incident radiation.

22.2.2.2 Remote Sensing Data

In addition to the periodic retrieval of standard satellite products, TOPS has the ability to ingest and process near-real-time global MODIS and AMSR-E data from the NASA LANCE system as well as global Landsat data from the USGS Landsat processing facility. The data are acquired on an hourly basis using the TOPS data-acquisition system and once they are staged locally, the TOPS framework initiates a set of actions including the creation of mosaics, interpolation of missing values, regridding, and reprojection to common grids used within TOPS.

22.2.2.3 Ecosystem Modeling

The ecosystem modeling component integrated into TOPS consists in the incorporation of a number of ecosystem process model routines including Biome-BGC (Thornton et al., 2002), and publicly available versions of CASA (Potter et al., 1993), LPJ (Sitch et al., 2003), VSIM (Pierce et al., 2006), SWAT (Gassman et al., 2007), and WRF (Skamarock et al., 2005). TOPS uses these models to simulate states and fluxes of ecosystem carbon, nitrogen, and water.

TOPS model outputs are routinely compared against observed data to assess spatio-temporal biases and general model performance. In the case of snow pack dynamics, for example, we performed a three-way comparison among model-, observation-, and satellite-derived fields of snow cover expansion and contraction over the Columbia River Basin (Ichii et al., 2008). Variables related to water and carbon fluxes, such as evapotranspiration, gross primary production (GPP), and net primary production (NPP) (Yang et al., 2007), are tested against FLUXNET-derived data at select locations representing a variety of land-cover/climate combinations. Similarly, the Soil Climate Analysis Network (SCAN) of soil moisture measurements, USGS National Streamflow Information Program (NSIP) streamflow measurements, and the United States Department of Agriculture SNOw TELemetry snow data provide valuable information for verifying the hydrology predictions from TOPS.

22.3 Application of TOPS for Protected Lands

TOPS is being used in a variety of applications, from local irrigation scheduling in vineyards (Pierce et al., 2006) to global monitoring of net primary productivity (Nemani et al., 2003). TOPS has previously been applied to develop measures of ecosystem change at different temporal scales within protected areas, and has been used for monitoring and forecasting of landscape-level indicators of ecosystem conditions including changes in snow cover, vegetation phenology, and productivity (Nemani et al., 2009; Gross et al., 2011).

22.3.1 Appalachian Trail Study Area

In this chapter, we focus on a case study using TOPS to assess recent and future changes within the protected lands along the A.T. The A.T. traverses most of the high-elevation ridges of the eastern United States, extending 3676 km across 14 states, from Springer Mountain in Northern Georgia to Mount Katahdin in central Maine. The A.T.'s gradients in elevation, latitude, and moisture sustain a rich biological assemblage of temperate zone forest species. The A.T. and its surrounding protected lands harbor forests with some of the greatest biological diversity in the United States, including rare, threatened, and endangered flora and fauna. The study area for this analysis was selected to correspond to the region falling within the Hydrologic Unit Code (HUC) watersheds encompassing the A.T. The boundaries of the watersheds studied were extended approximately 10–50 km away from the central corridor of the A.T. A detailed description of the study site can be found in Chapter 1 of this book (Wang, 2011).

22.3.2 Meteorological Surfaces

For this study, we used meteorological observations from the Global Surface of Summary Data (GSSD) version 7 and from the U.S. Cooperative Summary of Day data (TD3200) for the period 1982–2006. The daily data were gridded using SOGS to produce spatially continuous 8-km daily meteorological surfaces, including maximum and minimum temperature, precipitation, and shortwave radiation. These spatially continuous meteorological surfaces were used to define the climatology for the study area, as well as inputs to ecosystem models used to assess changes in vegetation productivity and runoff.

22.3.3 Satellite Observations

For the purpose of assessing vegetation conditions over the A.T., TOPS ingested and preprocessed time series from the Moderate Resolution Imaging Spectroradiometer (MODIS) and from the Advanced Very High

Resolution Radiometer (AVHRR). AVHRR data are useful for analyzing interannual variation in vegetation because of their long record, starting from 1981, and global-scale coverage. In particular, we used the Global Inventory Modeling and Mapping Studies (GIMMS) (Tucker et al., 2005), which is a data set derived from AVHRR that has been used extensively for global-scale vegetation monitoring and detection of trends in vegetation condition (e.g., Goetz et al., 2005; Nemani et al., 2009). The GIMMS data sets provide a global time series of Normalized Differential Vegetation Index (NDVI) measurements at a spatial resolution of 8 km × 8 km. NDVI is calculated from the red and near-infrared spectral bands, and provides a measure of vegetation greenness on the land surface. We used monthly composites of GIMMS data for the A.T. from 1982 to 2006.

The MODIS sensor onboard the Terra (launched in 1999) and Aqua (launched in 2002) satellites is specifically designed to monitor the Earth's environment, including aspects of the atmosphere, land, and oceans. MODIS has 36 spectral bands with a spatial resolution of 250 m to 1 km, and daily temporal resolution. The standard MODIS land data products include land cover, snow cover, vegetation indices, LAI, and Gross Primary Production (GPP). MODIS data are useful in analyses of current vegetation status. For this study, TOPS ingested MODIS LAI (MOD15A2) and NPP (MOD17EN) data from 2001 to 2006, the time period for which MOD17EN data are currently available.

22.4 Current Patterns in Climate and Vegetation

Figure 22.3 shows the land-cover map of the A.T. using the MODIS land-cover product (MOD12Q1) with the IGBP land-cover classification scheme. The dominant land cover along the trail is deciduous broadleaf forest. The area north of latitude 42°N includes mixed forests, which make up almost all of the vegetation cover of the trail crossing the State of Maine. The central portion of the study area, between latitude 38°N and 41°N, is partially covered by cropland and urban areas. The southern portion of the trail extending to 37°N is covered by mixed forests and deciduous broadleaf forests, similar to the tracts north of 41°N.

Patterns along the latitudinal gradient of the study site from 2001 to 2006 are summarized in Figure 22.4, including temperature, precipitation, LAI, and NPP. Mean temperature derived from SOGS increases from north to south, with a difference of over 10°C between the northern and southern terminus of the trail. During the 2001–2006, precipitation derived from SOGS decreases steeply from 35°N to 37°N by 500 mm/year, increased from 37°N to 41°N, and decreases again north of latitude 41°N. Most of the study area receives more than 900 mm/year of precipitation. Annual peak MODIS LAI (MOD15A2) shows a large decline from 39°N to 41°N because of the amount of cropland and urban area in the landscape. MODIS NPP (MOD17EN)

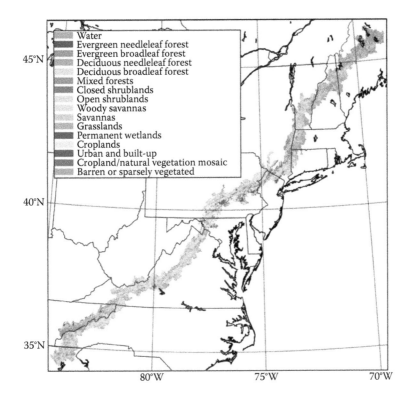

FIGURE 22.3
MOD12Q1 land-cover map with IGBP land-cover scheme for the study area. The study area was based on the HUC watershed area along the A.T.

decreases from South to North, and the decrease can be explained by the temperature gradient, with the exception of a decline induced by the cropland and urban area around 40°N. A comparison between climate variables and NPP shows that the latitudinal gradient of NPP is mostly controlled by temperature through its effect on modulating growing season length (Jenkins et al., 2002).

22.5 Recent Climate Patterns and Vegetation Response

We produced time series of climate variables for the A.T. from 1982 to 2006 using the SOGS component of TOPS (Figure 22.5). Both annual mean temperature and annual precipitation show no clear trend over the study period. This result does not contradict the studies reporting a trend in the eastern United States (USGCRP, 2009), because our study period is shorter than the other studies and also average of narrow-shaped study region. However, GIMMS NDVI

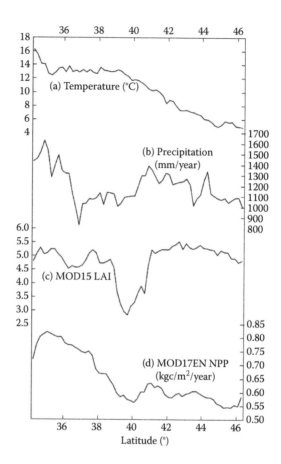

FIGURE 22.4
Latitudinal climatology (2001–2006) along the A.T. of mean temperature (a), precipitation (b), MOD15A2 Leaf Area Index (LAI) (c), and MOD17EN Net Primary Production (NPP) (d).

data show a significant decrease over the same period. Given that neither temperature nor precipitation shows a correlation with NDVI, we hypothesize that the observed decrease in NDVI is driven by the combination of urban expansion and recent outbreaks of hemlock wooly adelgids (*Adelges tsugae*) (Eschtruth et al., 2006; Lovett et al., 2006; Orwig et al., 2008) and other insects.

Consistent with the patterns observed in the AVHRR GIMMS data along the A.T., analyses of Landsat data have shown a decrease in forest cover in the eastern United States mainly due to the growth of urban areas since the 1970s (Potere et al., 2007; Wang et al., 2009; Drummond and Loveland, 2010; Hansen et al., 2010). Meanwhile, forest inventory studies have found that forest regrowth caused the aboveground biomass in the eastern United States to increase (Caspersen et al., 2000; Smith et al., 2009). Model studies also simulated an increase in NPP driven by climate change (Hicke et al., 2002). Thus, we suggest that while photosynthetic activity increased at the forest stand

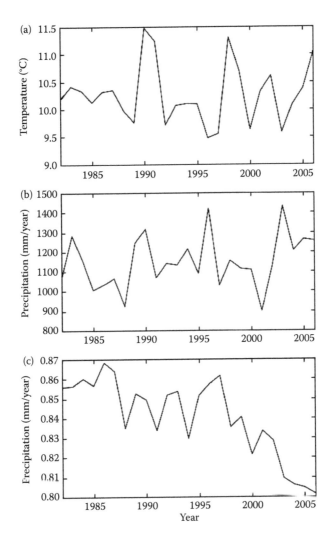

FIGURE 22.5

Time series of temperature (a), precipitation (b), and Global Inventory Modeling and Mapping Studies (GIMMS) Normalized Differential Vegetation Index (NDVI) (c) averaged over the entire study area for the period 1982–2006.

level, an overall loss of forest cover due to urban development, outbreaks of herbivorous insects, and other disturbances caused a decrease in peak NDVI since the 1970s. Although the signal of the forest cover loss is prominent in the satellite data, the net carbon exchange of forests (flux from forests to the atmosphere) in the eastern United States is still expected to become sink over time according to modeling studies and airborne CO_2 composition analysis (Peters et al., 2007).

22.6 Predicting Future Changes Using Ecosystem Modeling and Downscaled Climate Scenarios

We used TOPS to project the regional impacts of climate change along the A.T. to the end of the twenty-first century by downscaling general circulation model (GCM) scenarios, and using the scenarios to drive dynamic ecosystem models to assess the vegetation response to the projected climate scenarios. We used climate scenarios derived from the World Climate Research Program (WCRP) Coupled Model Intercomparison Project (CMIP3) multimodel data sets. The data sets are based on the climate scenarios produced for the Fourth Assessment Report (AR4) of the Intergovernmental Panel on Climate Change (IPCC) (IPCC, 2007).

For the purpose of this study, we used the outputs from 11 models for the Special Report on Emission Scenarios A1B (SRES A1B) scenario, which assumes a future with high economic growth, a well-balanced energy sources resource portfolio, and new technology development, with atmospheric CO_2 concentration stabilizing at 720 ppm. TOPS downscaled the SRES A1B outputs from the 11 GCMs onto an 8-km grid using the bias-correction algorithm described by Wood et al. (2002). The variability of the twenty-first century outputs deviated from the mean of the twentieth century experiments (20C3M) was adjusted to that of TD3200 station data, and the anomaly was calculated at each station and interpolated into spatially continuous meteorological surfaces. The spatially interpolated anomaly was then added to SOGS gridded data to produce the downscaled projection. All the models project a steady temperature increase, ranging from 2°C to 6°C by the end of the twenty-first century. The ensemble mean temperature increased from 11°C to 14.5°C, while precipitation did not show any significant trend or decadal variation.

To evaluate the regional impacts of these climate scenarios on ecosystems and protected areas along the A.T., we ran the dynamic ecosystem model LPJ (Sitch et al., 2003; Gerten et al., 2004) to simulate the ecosystem response to changes in climate from 1980 to the end of the twenty-first century. LPJ is carbon and water cycle biophysical model with a dynamic vegetation simulation component, which includes representations for photosynthesis, respiration, allocation, fire, establishment, evapotranspiration, and outflow. LPJ requires monthly temperature, precipitation, radiation, and CO_2 concentration data as inputs. The simulated outputs should be considered as a potential vegetation response, and do not account for human activities, such as urbanization and agricultural expansion. We spun up the model with SOGS climate data assuming that the soil carbon is in equilibrium during the last three decades from 1980 to 2009, and then conducted the model simulation experiment from 2009 to 2099 using the downscaled GCM data.

The simulated NPP for the region is projected to increase monotonously until the end of the twenty-first century. The ensemble mean NPP increases from 60 to 80 g C/m²/year. The increase can be attributed to the effect of increasing atmospheric CO_2 concentrations and CO_2 fertilization of vegetation in the region. However, NEE is predicted to be constant at a rate of approximately -10 g $C/m^2/year$ through the twenty-first century (negative NEE indicates a flux of carbon from terrestrial ecosystems to the atmosphere). This means that the predicted increase in respiration due to rising temperature exceeds the predicted increase in NPP. No trend is detected in the predicted watershed outflow due to the small variation in projected

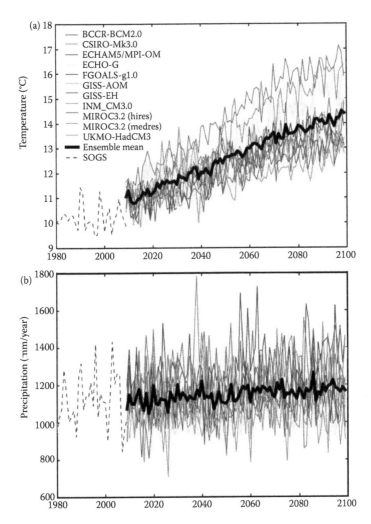

FIGURE 22.6
(See color insert.) Projection of mean temperature (a) and precipitation (b) until 2099 downscaled from Coupled Model Intercomparison Project (CMIP3) multimodel data set of SRES A1B scenario. The colored lines correspond to 11 General Circulation Model (GCM) data. The black thick line is the ensemble mean of the 11 GCM data. The dashed line is past time-series data derived from TOPS Surface Observation and Gridding System (SOGS) data.

precipitation. These results suggest that the current carbon sink in the eastern United States could turn into a carbon source in the future under the SRES A1B, which assumes a mild future growth in greenhouse gas emissions. If we were to use the projections that follow the trajectory of the highest-emission scenarios, the forests along the A.T. would be expected to start releasing even more carbon and eventually decline in growth.

Changes in the seasonal distribution of runoff have a great impact on downstream water availability and potentially affect water allocation planning. Although the projected annual runoff does not show a clear trend (Figure 22.6c), the peak in runoff is projected to take place earlier in the year, advancing from April to March by the end of the twenty-first century (Figure 22.7). Also, the cumulative winter runoff is projected to increase, while peak runoff will decrease. This projected change in runoff can be attributed to increased winter snowmelt caused by higher temperatures (Figure 22.8).

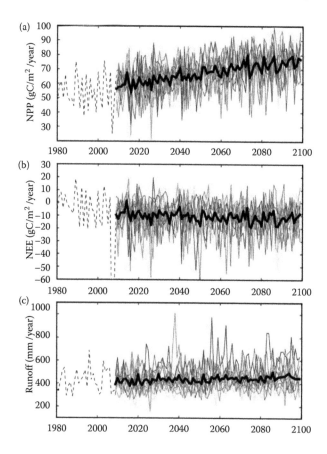

FIGURE 22.7
(**See color insert.**) Same as Figure 21.6 except NPP (a), Net Ecosystem Exchange (NEE) (b), and runoff (c).

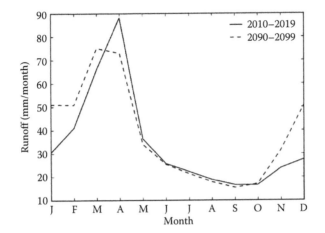

FIGURE 22.8
Monthly distribution of runoff for the study area. The straight line is the mean from 2010 to 2019. The dashed line is the mean from 2090 to 2099.

22.7 Summary

This chapter summarizes the TOPS design and functions and provides a case study in which TOPS is used to analyze recent trends in vegetation condition and predict the response of vegetation to climate change. Climate processing, satellite data ingestion, and ecosystem modeling are the main capabilities that are integrated into TOPS. We used TOPS to produce the climatology of recent temperature, precipitation, and vegetation patterns, and found a North–South temperature-dependent gradient of NPP along the A.T. corridor. Time series of climate and NDVI derived from TOPS reveal a forest cover decline during the last three decades. Finally, TOPS was used to downscale future climate change projections to a spatial resolution of 8 km, and run a biophysical ecosystem model to understand the impacts on the A.T. ecosystems. This analysis illustrated a projected increase in both NPP and carbon release, with the overall result that ecosystems along the A.T. are projected to serve as a net source of CO_2.

Acknowledgments

The A.T. case study was part of the efforts from a project titled "A Decision Support System for Monitoring, Reporting and Forecasting Ecological Conditions of the Appalachian National Scenic Trail," which is funded by

NASA Applied Sciences Program under Decision Support through Earth Science Research Results (Grant NNX09AV82G).

References

Bentz, B. J., J. Régnière, C. J. Fettig, E. M. Hansen, J. L. Hayes, J. A. Hicke, R. G. Kelsey, J. F. Negrón, and S. J. Seybold. 2010. Climate change and bark beetles of the Western United States and Canada: Direct and indirect effects. *BioScience*, 60(8): 602.

Caspersen, J. P., S. W. Pacala, J. C. Jenkins, G. C. Hurtt, P. R. Moorcroft, and R. A. Birdsey. 2000. Contributions of land-use history to carbon accumulation in U.S. forests. *Science* 290(5494): 1148–1151.

Drummond, M. A. and T. R. Loveland. 2010. Land-use pressure and a transition to forest-cover loss in the Eastern United States. *BioScience* 60(4): 286–298.

Eschtruth, A. K., N. L. Cleavitt, J. J. Battles, R. A. Evans, and T. J. Fahey, 2006. Vegetation dynamics in declining eastern hemlock stands: 9 years of forest response to hemlock woolly adelgid infestation. *Canadian Journal of Forest Research* 36(6): 1435–1450.

Gassman, P. W., M. Reyes, C. H. Green, and J. G. Arnold. 2007. The soil and water assessment tool: Historical development, applications, and future directions. *Transactions of the ASABE*, 50(4): 1211 – 1250.

Gerten, D., S. Schaphoff, U. Haberlandt, W. Lucht, and S. Sitch. 2004. Terrestrial vegetation and water balance—Hydrological evaluation of a dynamic global vegetation model. *Journal of Hydrology* 286(1–4): 249–270.

Gross, J. E., A. J. Hansen, S. J. Goetz, D. M. Theobald, F. M. Melton, N. B. Piekielek, and R. R. Nemani. 2011. Remote sensing for inventory and monitoring of the U.S. National Parks. In: Wang, Y. (Ed.), *Remote Sensing of Protected Lands*. Boca Raton, FL: CRC Press (this book).

Hansen, M. C., S. V. Stehman, and P. V. Potapov. 2010. Quantification of global gross forest cover loss. *Proceedings of the National Academy of Sciences* 107(19): 8650–8655.

Hicke, J. A., G. P. Asner, J. T. Randerson, C. Tucker, S. Los, R. Birdsey, J. C. Jenkins, and C. Field. 2002. Trends in North American net primary productivity derived from satellite observations, 1982–1998. *Global Biogeochemical Cycles* 16(2): 1018.

IPCC. 2007. The physical science basis. In: S. Solomon, D. Qin, M. Manning, Z. Chen, M. Marquis, K. B. Averyt, M. Tignor, and H. L. Miller (Eds.), *Contribution of Working Group I to the Fourth Assessment Report of the Intergovernmental Panel on Climate Change.* Cambridge: Cambridge University Press.

Jenkins, J. P., B. H. Braswell, S. E. Frolking, and J. D. Aber. 2002 Detecting and predicting spatial and interannual patterns of temperate forest springtime phenology in the eastern U.S. *Geophysical Research Letters* 29(24): 54.

Jolly, W. M., J. S. Graham, A. Michaelis, R. R. Nemani, and S. W. Running. 2005. A flexible, integrated system for generating meteorological surfaces derived from point sources across multiple geographic scales. *Environmental Modelling & Software* 20(7): 873–882.

Lovett, G. M., C. D. Canhan, M. A. Arthur, K. C. Weathers, and R. D. Fitzhugh. 2006. Forest ecosystem response to exotic pests and pathogens in eastern North America. *Bioscience* 56: 395–405.

Moritz, C., J. L. Patton, C. J. Conroy, J. L. Parra, G. C. White, and S. R. Beissinger. 2008. Impact of a century of climate change on small-mammal communities in Yosemite National Park, USA. *Science* 322: 261–264.

National Park Service (NPS). 2010. *National Park Service Climate Change Response Strategy.* Fort Collins, Colorado: National Park Service Climate Change Response Program.

Nemani, R., H. Hashimoto, P. Votava, F. Melton, W. Wang, A. Michaelis, L. Mutch, C. Milesi, S. Hiatt, and M. White. 2009. Monitoring and forecasting ecosystem dynamics using the Terrestrial Observation and Prediction System (TOPS). *Remote Sensing of Environment* 113(7), 1497–1509.

Nemani, R. R., C. D. Keeling, H. Hashimoto, W. M. Jolly, S. C. Piper, C. J. Tucker, R. B. Myneni, and S. W. Running. 2003. Climate-driven increases in global terrestrial net primary production from 1982 to 1999. *Science* 300(5625): 1560–1563.

Orwig, D. A., R. C. Cobb, A. W. D'Amato, M. L. Kizlinski, and R. D. Foster. 2008. Multi-year ecosystem response to hemlock woolly adelgid infestation in southern New England Forests. *Canadian Journal of Forest Research* 38: 834–843.

Peters, W., A. R. Jacobson, C. Sweeney, A. E Andrews, T. J Conway, K. Masarie, J. B. Miller et al. 2007. An atmospheric perspective on North American carbon dioxide exchange: CarbonTracker. *Proceedings of the National Academy of Sciences of the United States of America* 104(48): 18925–18930.

Pierce, L., R. Nemani, and L. Johnson. 2006. Vineyard Soil Irrigation Model (VSIM). User guide (http://geo.arc.nasa.gov/sge/vintage/vsim_030106_guide.pdf).

Potere, D., C. E. Woodcock, A. Schneider, M. Ozdogan, and A. Baccini. 2007. Patterns in forest clearing along the Appalachian Trail corridor. *Photogrammetric Engineering and Remote Sensing* 73(7): 783–791.

Potter, C. S., J. T. Randerson, C. B. Field, P. A. Matson, P. M. Vitousek, H. A. Mooney, and S. A. Klooster. 1993. Terrestrial ecosystem production: A process model based on global satellite and surface data. *Global Biogeochemical Cycles* 7(4): 811–841.

Skamarock, W. C., J. B. Klemp, J. Dudhia, D. O. Gill, D. M. Barker, W. Wang, and J. G. Powers. 2005. A description of the Advanced Research WRF Version 2, NCAR Technical Note, NCAR-TN-468 + STR 88pp.

Sitch, S., B. Smith, I. C. Prentice, A. Arneth, A. Bondeau, W. Cramer, J. O. Kaplan et al. 2003. Evaluation of ecosystem dynamics, plant geography and terrestrial carbon cycling in the LPJ dynamic global vegetation model. *Global Change Biology* 9(2): 161–185.

Smith, W. B., P. D. Miles, C. H. Perry, and S. A. Pugh. 2009. *Forest Resources of the United States, 2007.* Washington, DC, July.

Stewart, I. T., D. R. Cayan, and M. D. Dettinger. 2005. Changes toward earlier streamflow timing across Western North America. *Journal of Climate*, 18: 1136–1155.

Thornton, P., B. Law, H. Gholz, K. Clark, E. Falge, D. Ellsworth, A. Goldstein, R. Monson, D. Hollinger, and M. Falk. 2002. Modeling and measuring the effects of disturbance history and climate on carbon and water budgets in evergreen needleleaf forests. *Agricultural and Forest Meteorology* 113(1–4): 185–222.

Tucker, C., J. Pinzon, M. Brown, D. Slayback, E. Pak, R. Mahoney, E. Vermote, and N. El Saleous. 2005. An extended AVHRR 8-km NDVI dataset compatible with MODIS and SPOT vegetation NDVI data. *International Journal of Remote Sensing* 26(20): 4485–4498.

USGCRP. 2009. In: T. R. Karl, J. M. Melillo, and T. C. Peterson (Eds.), *Global Climate Change Impacts in the United States.* Cambridge: Cambridge University Press.

Van Mantgem, P. J., N. L. Stephenson, J. C. Byrne, L. D. Daniels, J. F. Franklin, P. Z. Fule, M. E. Harmon et al. 2009. Widespread increase of tree mortality rates in the western United States. *Science* 323(5913): 521–524.

Wang, Y. 2011. Remote sensing of protected lands: An overview. In: Wang, Y. (Ed.), *Remote Sensing of Protected Lands*. Boca Raton, FL: CRC Press (Chapter 1 this book).

Wang, Y., B. R. Mitchell, J. Nugranad-Marzilli, G. Bonynge, Y. Zhou, and G. Shriver. 2009. Remote sensing of land-cover change and landscape context of the National Parks: A case study of the Northeast Temperate Network. *Remote Sensing of Environment* 113(7): 1453–1461.

Wang, Y., R. Nemani, F. Dieffenbach, K. Stolte, G. Holcomb, M. Robinson, C. C. Reese et al. 2010. Development of a decision support system for monitoring, reporting and forecasting ecological conditions of the Appalachian Trail. In *Proceedings of the 2010 IEEE International Geoscience and Remote Sensing Symposium, IEEE Xplore* entry: 978–1–4244–9566–5, pp. 2095–2098.

Westerling, A. L., H. G. Hidalgo, D. R. Cayan, and T. W. Swetnam. 2006. Warmgin and earlier spring increase western U.S. forest wildfire activity. *Science* 313: 940–943.

Wood, A. W., E. P. Maurer, A. Kumar, and D. P. Lettenmaier. 2002. Long-range experimental hydrologic forecasting for the eastern United States. *Journal of Geophysical Research* 107(D20): 4429.

23

Geospatial Decision Models for Management of Protected Wetlands

Wei "Wayne" Ji

CONTENTS

23.1 Introduction

To manage protected wetlands effectively and efficiently, a variety of technical tools have been developed to address the needs of decision-making in change monitoring, impact assessment, restoration planning, and regulatory evaluation of wetland resources and related driving factors. Among these tools, the geospatial decision model is a remarkable methodology, which is consistently evolving and increasingly being applied to address various assessment issues in wetland management and research. There was a lack of systematic review on this methodology as a whole. This chapter intends to fill this information gap by analyzing some common characteristics of geospatial decision models, reviewing history of the methodology development and several widely used models, and describing a recent case project to demonstrate technical details in model development.

For the purpose of this chapter and for developers and users of decision-making tools, it is beneficial to identify some characteristics of geospatial decision models first. The term, geospatial decision model, refers to the assessment tools that usually generate ranked indices as results of a particular assessment task that typically involves geospatial data and across-scale analysis. They are also known as index models. As related to wetland applications, geospatial decision models are commonly developed and applied for following assessments:

- Quantifying the relationships between wildlife and habitat
- Evaluating anthropogenic and/or natural events that impose potential impacts on wetlands under regulatory management
- Assessing wetland conditions in inventory efforts or scientific studies
- Facilitating wetland restoration planning at regional or local scales
- Serving as baseline information for monitoring and assessment of wetland change

Geospatial decision models can be applied in a regulatory context, such as Clean Water Act Section 404 permit review, or for other management needs of various agencies. Thus, many geospatial decision models have been developed under well-defined management goals by corresponding agencies. Others could be originally developed as part of research efforts by researchers. Technically, geospatial decision models are characterized by their extensive use of geospatial data and spatial analysis, including utilizing remotely sensed imagery and GIS data, for environmental data acquisition, landscape characterization and analysis, or change detection and modeling. Geospatial decision models can be embedded in a commercial GIS package through software customization for user-friendly interfaces of model data handling, model implementation, and result visualization. This usually results in a GIS-based decision support system. Geospatial decision models also can be the decision-making output components of ecological process models or simulation models. It would be difficult, although possible for some cases, to verify a decision model's output, ranked indices, directly using field observations. This is because the decision rules or criteria used for evaluating and ranking the indices with a geospatial decision model are determined usually based on professional judgments of domain experts, which may be largely based on specific management goals and somewhat subjective.

23.2 Review of Geospatial Wetland Decision Models

In the past few decades, numerous wetland assessment techniques were developed in the United States. The Ecosystem Management and Restoration Information System of US Army Corps of Engineers (US ACE) has compiled the descriptions of over 40 wetland assessment decision support tools (http:// el.erdc.usace.army.mil/emrrp/emris/) which were developed mainly by federal or state governmental agencies. Clearly, there would be numerous geospatial wetland decision model tools that were created by individual researchers of academic or scientific institutions, which did not make the compilation of US ACE, but could be introduced in various academic publications as distinct projects or as part of research outcomes. Despite varied purposes of model applications, many of these models either fall in the category or contain a component of what we have defined "geospatial wetland decision model". The next section provides a review of selected geospatial wetland decision models which have been used at a regional or national level in the United States and have had lasting influence on developing other decision models or applications.

23.2.1 Habitat Suitability Index Model

Habitat Suitability Index (HSI) Model could be considered as a pioneering effort for the index-based decision model development in the United States. The HSI model is the key component of the Habitat Evaluation Procedures (HEP), a habitat-based assessment methodology of US Fish & Wildlife Service, whose initial development dates back to 1974[1–5]. Numerous HSI models were developed to assess quality of habitats, largely wetlands, of a variety of species. As a quantitative tool, an HSI model outputs an index, ranging from 0.0 to 1.0, to indicate the quality level ("suitability") of a habitat for a particular species. The value 1.0 represents a maximal condition of suitability. This index calculation technique has been adopted or adapted in developing other geospatial decision models for wetland-related assessment, for example, the Wetland Value Assessment (WVA) Model used for coastal Louisiana wetland restoration planning [6]. Initially, HSI model input parameters were mainly based on field data collection, thus the models were basically "non-geospatial" and used at a local scale. There have been efforts to make HSI models spatially explicit. According to Schamberger and Krohn [2], for example, using aerial photographs in lieu of field data collection to obtain environmental variables for model calculation was proposed in the early stage of the model development. Lai et al. [7] used GIS to generate parameters of HSI models, especially the spatial habitat parameters that are of explicit importance for the models, to study the species that occurred in Missouri Ozark Forest Ecosystem. Dijak et al. [8] developed a Microsoft Windows-based program, Landscape HSI models that can include landscape attributes, like edge effects, patch area, distance to resources, and habitat composition, to evaluate habitats for certain species This approach allows the evaluation of spatial relationships of habitat components at a landscape level and the creation of the model input and output with GIS such as ArcGIS (ESRI, Inc.).

23.2.2 Wetland Value Assessment Model

Wetland Value Assessment (WVA) models were developed specifically to evaluate and rank the proposed wetland restoration projects submitted for funding under the Coastal Wetlands Planning, Protection, and Restoration Act of 1990 [6]. The WVA models were modified HEP for a community-level application, specifically for coastal Louisiana wetland types, including fresh/intermediate marsh, brackish marsh, saline marsh, bottomland hardwoods, and fresh swamp. Similarly as HEP, the WVA method employs HSI models that calculate relevant suitability indices, each ranging from 0.0 to 1.0, to evaluate environmental variables identified for the wetland community under assessment. With the HSI model index output, for the restoration planning purpose, WVA models calculate habitat units (HUs = the index multiplies the habitat area) and then average annual habitat units (AAHUs) that would be created during the project duration by the proposed restoration project. Further, the cost of the proposed project is incorporated in to the modeling to determine the cost for AAHUs of a particular project, which is compared with other proposed restoration projects in order to rank the projects for funding in a given funding year. WVA method was revised several times. Also, a prototype decision support system was developed to adapt WVA models for operation in GIS environment [9], which aimed mainly to obtain environmental variable values from GIS data and to display and analyze project areas.

23.2.3 Hydrogeomorphic Approach

The hydrogeomorphic approach (HGM) was initially developed in the early 1990s by the efforts of US Army Corps of Engineers for assessing wetland functions under the Clean Water Act Section 404 Regulatory Program as well as for other regulatory and management applications [10–13]. Many HGM models have been developed to assess various wetland types in different regions in the United States. Generally, the HGM method includes two phases the development and the application. The former develops a guidebook for assessing wetland functions of a regional subclass with the models.

The latter applies the models to assess wetland functions through the following steps: (1) identify individual model variables, (2) list measures of the variables, (3) assign a score index (in 0.0–1.0 scale) for the condition of each variable, and (4) calculate functional capacity index (FCI) of a wetland under assessment. An FCI indicates how (rate or magnitude) a wetland performs a particular function. An FCI ranges from 0.0 to 1.0, with 1.0 representing the highest functional capability and 0.0 for no performance of the function. In determination of FCI, reference wetlands are needed within the wetland subclass. Based on FCI, functional capability units (FCUs) can be calculated by multiplying the FCI value with the size of the wetland under assessment. Geospatial data and tools like GIS could help many aspects of HGM model development and application, such as selecting reference wetlands and generating the measures of HGM model variables.

23.2.4 Index of Biotic Integrity

The Index of Biotic Integrity (IBI) method was first developed by Karr (1981) to help evaluate the condition of small warm water streams for fish species [14]. As this technique became popular, many IBI models in modified forms were developed to assess the integrity of various habitat types, including wetlands, using samples of living organisms in a particular region [15–20]. In the IBI model development, the user needs to determine a set of metrics that serve as reliable indicators of condition of the assessed habitat for particular biotic species. The range of metric values are changed to converted score of, for example, 1, 3, or 5, with 5 points indicating the impact of human on the site minimally, 3 moderately, and 1 severely, or of other ranges like 0–10 [21]. An index of biotic integrity (IBI) is then calculated for each site by summing the scores of multiple metric values. A reference site(s), which has been less disturbed by humans, is usually needed to determine the scores of the metrics. The IBI can be used as a tool for wetland management in, for instance, evaluating site-specific impacts, setting management goals, and assessing restoration efforts. GIS techniques can be used in developing IBI models in various ways, for example, GIS data visualization for reference site selection [21].

23.2.5 Landscape Development Intensity Index

The landscape development intensity (LDI) index method was first published by Brown and Vivas [22]. Its early development may trace back to 1980 [23]. This method is based on the consideration that the intensity of human-dominated land uses can affect adjacent ecological systems through various types of impact. The LDI index is a quantitative measure of land development intensity, which is derived from energy use per unit area and suitable to assess potential human disturbances on wetlands at various scales. Calculating the LDI index mainly involves the following processes [22]: (a) delineating area contributing to a particular landscape unit (e.g., a stream) under evaluation; (b) identifying land-use types in the area of influence; (c) quantifying human development intensity for the land-use types based on *emergy*—the amount of energy used in the past directly and indirectly to make a product or service [24], resulting in the LDI coefficient for each land-use type; (d) calculating an overall LDI by summing area-weighted LDI coefficient for each land-use type in the area of influence. While the method was originally developed based on the land uses in Florida [22–26], it can be adapted to evaluate rivers, streams, lakes, or wetlands in other regions [27–30]. Based on land-use and land-cover information, the LDI modeling could use various geospatial data, in addition to field surveys, including GIS-based land-use/land-cover data, aerial photographs, and high-resolution satellite images.

23.3 Case Studies

This section introduces a case study on geospatial decision model development for wetland vulnerability assessment. A related effort employed different indicators and metric evaluation methods in vulnerability assessment modeling were published earlier [31].

Rapidly identifying vulnerable wetlands, whose functions and values are more likely to be degraded, is critically needed in decision-making processes for wetland management and studies. For this purpose, we developed a decision model for wetland vulnerability assessment that is completely based on geospatial data and analysis using a customized GIS. The model generates ranked WVI that can be used to evaluate potential human impacts on protected wetlands at a scale of streams, lakes, and subwatersheds. This model tool can be used for Clean Water Act Section 404 permit review, wetland condition assessment, or wetland change monitoring as a few examples. Developing the wetland vulnerability assessment model basically follows the below procedures.

23.3.1 Defining Wetland Vulnerability for Geospatial Modeling

Wetland vulnerability is defined as the degradation likelihood of a wetland's functions and values under potential anthropogenic pressures [31]. The vulnerability of a wetland is largely determined by its spatial association or interaction with the source locations of human influence. For example, certain characteristics of a wetland (e.g., its size or recreational usage) and the spatial occurrence of certain concerned entities (e.g., endangered species or a historical site) relating to a particular wetland may determine the vulnerability of this wetland. Thus, wetland vulnerability can be assessed by well-designed geospatial analysis, which serves as a basis of developing geospatial models for wetland vulnerability assessment.

23.3.2 Determining Wetland Functions, Indicators, and Geospatial Metrics

The model development is focused on the assessment of major functions of wetlands in biological, hydrological, chemical, physical, and socioeconomic aspects. To assess the vulnerability of each of these functions using geospatial data, we have determined certain indicators, their metrics, and typical geospatial data that can be used to evaluate the metrics in modeling (Table 23.1).

The considerations of indicator and metrics selection are described below.

23.3.2.1 Biological Vulnerability

Species abundance: The population of fish and wildlife species is an important indicator for measuring the habitat-supporting function of a wetland. Considering that a species may move across habitats, Kernel density of selected species data is utilized in this study to create a density surface for the study watershed using the occurrences of species. Then, the average species density within a selected wetland's boundary can be calculated to represent the species abundance of the wetland under assessment. The weight scores for endangered, threatened, or at-risk species are assigned as 3.0, 2.0, and 1.0, respectively, which can be applied to the average species density to generate weighted density.

Habitat disturbance: Human activities can disturb species in wetland habitats. Among them, road-constructing projects are considered as a major disturbing factor regarding habitat modification. As a measure, the minimum edge-to-edge distance from a wetland to nearby roads is used, which can be obtained using a GIS to implement GIS road data.

Wetland connectivity: The spatial connectivity of wetlands may have direct impacts on the number and diversity of wetland species. Considering that coherent wetland habitats perform better in habitat supporting than dispersed ones, CONTAGION index is used as a metric of the dispersion and interspersion among wetlands. This index is calculated using FRAGSTATS, public domain software, on classified satellite data.

23.3.2.2 Hydrological Vulnerability

Flooding potential: Wetlands can influence regional water flow by intercepting storm runoffs and changing sharp runoffs during flooding events. The closer is a wetland to flooding plains, the higher is the possibility the wetland would be exposed to flooding, thus providing functions in flooding attenuation. In this study, the proximity (minimum edge-to-edge distance) between a wetland and FEMA 100-year flooding plains is calculated as the metric value of potential flooding control function of a wetland under assessment. (*Note*: FEMA—Federal Emergency Management Agency.)

TABLE 23.1

Vulnerability Indicators, Metrics, and Geospatial Data Sources

Indicators	Metrics	Geospatial Data Used
Biological vulnerability (Habitat support)		
Species abundance	Average species density with their conservation status as weights	Species data
Habitat disturbance	Proximity (minimum edge-to-edge distance) to the roads nearby	Road data
Habitat connectivity	Average *CONTAGION* index of wetlands in the landscape	Classified ETM+ image
Hydrological vulnerability (Flooding control)		
Flooding potential	Proximity (minimum edge-to-edge distance) to 100-year flooding plain	Flooding insurance map
Surface runoff	The size of a wetland's upstream contributing drainage multiplies the percentage of impervious surface in the drainage	Classified ETM+ & DEM
Flooding retention potential	Average depth multiplies wetland size	DEM
Chemical vulnerability (Nutrient sequestration)		
Pollution removal potential	The stream order of wetland's immediate downstream stream	NHD
Nonpoint source pollution	Simulated nonpoint source pollutants based on the land use/land cover in the upstream contributing drainage	Classified ETM+ & DEM
Pollution retention potential	Average flatness of a wetland multiplies its size and also its organic matters	DEM
Physical vulnerability (Sediment Retention)		
Vegetation density	Average NDVI in a wetland regime	ETM+
Accumulated sediment density	Simulated sediment density in surface inflow to a wetland	STATGO & DEM
Water–vegetation interaction	Average *COHESION* index of open water and vegetation areas in the landscape	Classified ETM+ & DEM
Socioeconomic vulnerability (Socioeconomic value)		
Proximity to recreation spots	Proximity (minimum edge-to-edge distance) to open waters	Classified ETM+, Park data
Proximity to urban clusters	Proximity (minimum edge-to-edge distance) to urban clusters	Classified ETM+
Proximity to agricultural land	Proximity (minimum edge-to-edge distance) to agricultural lands	Classified ETM+

Surface flow: Contributing drainage area is the total area that contributes runoffs to a wetland of interest. The surface runoff to a wetland is largely determined by the size of its contributing drainage area and the land use and land cover within it, especially indicated by impervious surface that is used as the metrics of this indicator. A programming script is written to automatically delineate the contributing drainage area and calculate the area of impervious surface within the drainage area.

Flooding retention capability: In this study, flooding retention capability of a wetland is characterized by its approximate volume (its size multiplies its depth). The depth of a wetland is derived from the Digital Elevation Model (DEM) data within the wetland's boundary.

23.3.2.3 Chemical Vulnerability

Pollution removal potential: Headwater wetlands are usually found in the highest reaches of a watershed at the head of and in association with headwater streams, small and often intermittent channels with a stream order of 1 or 2. These wetlands tend to be wider and flatter, therefore, contribute significantly to removing, reducing, or transforming pollutants. In this study, the stream order of a wetland's immediate downstream stream is utilized as the metric value of its nutrient removal potential, which can be determined from the national hydrological dataset (NHD).

Nonpoint source pollution: A wetland's nutrients and other types of nonpoint source pollutants are mainly received from surface runoffs, which run through a variety of land use and land-cover types before flowing into a wetland. P8 Urban Catchment Model (Walker, 2007, http://www.wwwalker.net/p8/index.htm) is utilized to simulate nonpoint source pollutants in surface runoffs to a wetland based on the land-use and land-cover types in its contributing drainage area. The metric value is represented by the annual average of simulated total amount of nonpoint source pollutants to a wetland.

Pollution retention potential: A flat wetland is more capable in pollution retention because the water can stay for a long time, and thus nutrient and other contaminants have a high probability to be sequestrated. According to this consideration, the slope of a wetland is used to indicate its intrinsic hydrogeomorphic characteristics for nutrient and sediment retention potential. Also, a bigger wetland obviously has more potential in pollutant sequestration than a smaller one. In addition, organic matters in a wetland's underlying soil can largely accelerate the sequestration of nutrients and other pollutants. The multiplication of a wetland's size, slope, and percent of organic matters that can be obtained from US general soil map—STATSGO is used as the metric value of its pollution retention potential in this study.

23.3.2.4 Physical Vulnerability

Vegetation density: Vegetation within a wetland, especially grow-up trees with deep roots, offers frictional resistance to wetland's bank erosion. In this study, vegetation density is indicated by NDVI (Normalized Difference Vegetation Index) of a wetland. The NDVI can be calculated from satellite image. The average NDVI value within a wetland's boundary is generated as the metric value of a wetland's function for its bank stabilization.

Accumulated sediment density: The sediments in a wetland come from surface runoffs in its contributing drainage area. In this study, a wetland's accumulated sediment density in inflow to a wetland is utilized as the metric value of this adverse anthropogenic pressure. It is represented by the maximum flow accumulation weighted by sediment erosion *K*-factor obtained from STATSGO.

Water–vegetation interaction: Vegetation that is highly interspersed within a wetland could offer frictional resistance to inflowing water, thus facilitating settlement of sediment. The COHESION index between wetlands and vegetated area is calculated with FRAGSTATS; then the average index within a wetland's boundary is used as the metric value of a wetland's water–vegetation interaction.

23.3.2.5 Socioeconomic Vulnerability

Proximity to recreation spots: People extensively count on open waters for a variety of recreational activities. Connecting to open water directly or indirectly, wetlands have a stewardship value in water-related recreational activities. In this study, the minimum edge-to-edge distance from a wetland to open waters is calculated as the metric value of this recreational value of a wetland.

Proximity to urban clusters: Wetlands close to urban clusters would obviously have a high possibility of being disturbed by human activities. Thus, the minimum edge-to-edge distance from a wetland to urban clusters is determined as the metric value of socioeconomic pressures on a wetland. The urban cluster information can be obtained from land-use/land-cover maps at a watershed level.

Proximity to agricultural lands: Wetlands close to agricultural lands have a great socioeconomic value because they could transform and remove nutrients and sediments from surface runoffs. In this study, the proximity of a wetland to its surrounding agricultural lands is determined as the metric value of its socioeconomic value.

23.3.3 Calculating WVI

After each indicator metric is evaluated as described above, a natural break algorithm is applied to break them into three levels, representing "highly vulnerable," "vulnerable," or "less vulnerable" status. A score of 1.0, 0.5, or 0.1 is then assigned to each status accordingly. A vulnerability index for each of the five wetland functions can be calculated by averaging the summed scores of all three indicators of a particular function. Then, adding the vulnerability index values for all functions will generate an overall WVI.

23.3.4 GIS-Based Decision Support System

A GIS-based decision support system is developed to implement the modeling process. The system is programmed using ArcObject in Visual Basic environment and delivered as an extension in ArcGIS (ESRI, Inc.). Fully utilizing advanced spatial analysis capabilities of a GIS, the decision support system provides user-friendly interfaces specifically designed to facilitate all evaluation procedures, such as manipulating geospatial data, converting indicator metric values into vulnerable scores, calculating wetland vulnerability indices, and visualizing assessed wetlands in a watershed landscape.

23.3.5 The Model Application

The developed geospatial model was applied to a pilot study area (Figure 23.1). Little Blue River watershed located at the southeast corner of greater Kansas City area in Missouri, USA, covering a total area of approximately 556 km². The Kansas City metropolitan has been expanding eastward rapidly through the watershed, largely enhancing the threats to the existing wetland habitats. Perennial and intermittent streams in Little Blue River watershed have also experienced substantial human alternations. To verify the model-generated wetland vulnerability, a field survey method was designed for this study, which was adapted from some published procedures

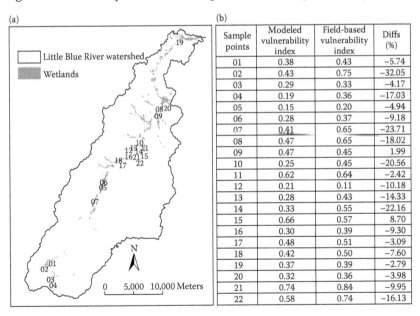

(a)

(b)

Sample points	Modeled vulnerability index	Field-based vulnerability index	Diffs (%)
01	0.38	0.43	−5.74
02	0.43	0.75	−32.05
03	0.29	0.33	−4.17
04	0.19	0.36	−17.03
05	0.15	0.20	−4.94
06	0.28	0.37	−9.18
07	0.41	0.65	−23.71
08	0.47	0.65	−18.02
09	0.47	0.45	1.99
10	0.25	0.45	−20.56
11	0.62	0.64	−2.42
12	0.21	0.11	−10.18
13	0.28	0.43	−14.33
14	0.33	0.55	−22.16
15	0.66	0.57	8.70
16	0.30	0.39	−9.30
17	0.48	0.51	−3.09
18	0.42	0.50	−7.60
19	0.37	0.39	−2.79
20	0.32	0.36	−3.98
21	0.74	0.84	−9.95
22	0.58	0.74	−16.13

FIGURE 23.1
(a) The location of sampling sites; (b) The model-based wetland vulnerability indices and field survey-generated vulnerability indices at 22 sites; the two sets of indices are correlated, with a correlation coefficient of 0.81, which is significant at a confidence level of 95%.

for stream, riparian, and wetland habitat assessments [32–34]. Based on the field survey results, the evaluation scores assigned to all relevant indicators were summed up to generate an overall WVI for a given site. For 22 survey sites selected throughout the wetlands in Little Blue River watershed, the correlation coefficient between the field survey-generated wetland vulnerability results and the model-generated results is as high as 0.81 (Figure 23.1b). This statistics validates the geospatial decision model for its indicators and evaluation criteria used. The model assessment results reveal certain spatial patterns of wetland vulnerability in relation to adjacent streams and landscape configurations.

23.4 Conclusion

The review and case study demonstrate that geospatial decision models have become an effective methodology widely used to address various assessment needs in wetland management and studies. The decision models are characterized by using ranked index, which makes the modeling outputs ready and efficient for being used in wetland-related decision-making. Utilization of GIS and remote sensing data as well as GIS-bsed spatial analysis and visualization techniques provides the following benefits: (1) allowing the use of huge volumes of environmental geospatial data in the development and application of decision models, (2) making rapid wetland assessments possible, and (3) facilitating landscape or watershed approaches for wetland assessment.

Acknowledgment

The case study described in the chapter was supported by U.S. Environmental Protection Agency through a funding award (CD 987009010) to the author. Jia Ma, a then doctoral student supervised by the author, involved in the study as part of her dissertation research at University of Missouri—Kansas City.

References

1. Daniel, C. and R. Lamaire. 1974. Evaluating effects of water resource developments on wildlife habitat. *Wildlife Society Bulletin* 2(3):114–118.
2. Schamberger, M. and W. Krohn. 1982. Status of the habitat evaluation procedures. In *Transactions of the Forty-Seventh North American Wildlife and Natural Resources Conference*, K. Sabol (ed.), University of Nebraska-Lincoln, Washington, DC, USA.

3. U.S. Fish and Wildlife Service. 1980. *Habitat as a Basis of Environmental Assessment.* U.S. Department of Interior Fish and Wildlife Service, Division of Ecological Services Manual 101, Washington, DC, USA.

4. U.S. Fish and Wildlife Service. 1980. *Habitat Evaluation Procedures (HEP).* U.S. Department of Interior Fish and Wildlife Service, Division of Ecological Services Manual 102, Washington, DC, USA.

5. U.S. Fish and Wildlife Service. 1981. *Standards for the Development of Habitat Suitability Index Models for Use in the Habitat Evaluation Procedure.* U.S. Department of Interior Fish and Wildlife Service, Division of Ecological Services Manual 103, Washington, DC, USA.

6. Environmental Work Group. 1998. *Wetland Value Assessment Methodology and Community Models.* Environmental Work Group, Coastal Wetland Planning, Protection, and Restoration Act (CWPPRA). 13pp. plus appendices.

7. Lai, Y. C., W. L. Mills, and C. Cheng. 2000. Implementation of a Geographic Information System (GIS) to Determine Wildlife Habitat Quality Using Habitat Suitability Index. In *Proceedings of the 21st Asian Conference on Remote Sensing,* Taipei, Taiwan, December 4–8.

8. Dijak, W. D., C. D. Rittenhouse, M. A. Larson, F. R. Thompson, and J. J. Millspaugh. 2007. Landscape habitat suitability index software. *Journal of Wildlife Management* 71(2):668–670.

9. Ji, W. and L. Mitchell, 1995. Analytical model-based decision support GIS for wetland resource management. In *Wetland and Environmental Applications of GIS,* J. Lyon and J. McCarthy (eds.), Chapter 4, Boca Raton, FL: CRC/Lewis Publishers.

10. Brinson, M. M. 1993. *A Hydrogeomorphic Classification for Wetlands.* Wetlands Research Program Technical Report WRP-DE-4. U.S. Army Corps of Engineers, Waterways Experiment Station, Vicksburg, Mississippi, USA. 79pp.

11. Smith, R. D., A. Ammann, C. Bartoldus, and M. M. Brinson. 1995. *An Approach for Assessing Wetland Functions Using Hydrogeomorphic Classification, Reference Wetlands, and Functional Indices.* Wetlands Research Program Technical Report WRP-DE-9. U.S. Army Corps of Engineers, Waterways Experiment Station, Vicksburg, Mississippi, USA. 88pp.

12. Brinson, M. M., R. Hauer, L. C. Lee, W. L. Nutter, R. Rheinhardt, R. D. Smith, and D. F. Whigham. 1996. *Guidebook for Application of Hydrogeomorphic Assessment to Riverine Wetlands* (Operational Draft). Wetlands Research Program Technical Report WRP-DE-11. U.S. Army Corps of Engineers, Waterways Experiment Station, Vicksburg, Mississippi, USA.

13. U.S. Army Corps of Engineers. 1997. National action plan to implement the hydrogeomorphic approach to assessing wetland functions. *Federal Register* 62(119):33607–33620.

14. Karr, J. R. 1981. Assessment of biotic integrity using fish communities. *Fisheries* 6(6):21–27.

15. Karr, J. R., P. R. Yant, and K. D. Furst. 1987. Spatial and temporal variability of the index of biotic integrity in three midwestern streams. *Transactions of the American Fisheries Society* 116(1):1–11.

16. Kerans, B. L. and J. R. Karr. 1994. A benthic index of biotic integrity (B-IBI) for rivers of the Tennessee Valley. *Ecological Applications* 4(4):768–785.

17. Hlass, Karr, J. R. 1997. Measuring biological integrity. In *Principles of Conservation Biology,* G. K. Meffe, C. R. Carroll, and Contributors (eds.), 2nd edn., pp. 483–485. Sunderland, MA: Sinauer.

18. Karr, J. R. and E. W. Chu. 1997. *Biological Monitoring and Assessment: Using Multimetric Indexes Effectively*. EPA 235-R97–001. University of Washington, Seattle. 149pp.
19. Fisher, L. J. W. L. and D. J. Turton. 1998. Use of the index of biotic integrity to assess water quality in forested streams of the Quachita Mountains Ecoregion, Arkansas. *Journal of Freshwater Ecology* 13:181–192.
20. Teels, B. M. and T. Danielson. 2001. *Using a Regional IBI to Characterize Condition of Northern Virginia Streams, with Emphasis on the Occoquan Watershed*. USDA Natural Resource Conservation Service, Technical Note 190-13-1. December 2001.
21. Rehn, A. C., J. T. May, and PR. Ode. 2008. *An Index of Biotic Integrity (IBI) for Perennial Streams in California's Central Valley*. Technical Report, Surface Water Ambient Monitoring Program, Water Boards, California (www.waterboards. ca.gov.swamp), December 2008.
22. Brown, M. T. and M. B. Vivas. 2005. Landscape development intensity index. *Environmental Monitoring and Assessment* 101:289–309.
23. Brown, M. T. 1980. *Energy Basis for Hierarchies in Urban and Regional Landscapes*. PhD Dissertation, Department of Environmental Engineering Sciences, University of Florida, Gainesville, 359pp.
24. Odum, H. T. 1996. *Environmental Accounting: Emergy and Environmental Decision Making*. New York: John Wiley and Sons, 370pp.
25. Reiss, K. C. and M. T. Brown. 2007. Evaluation of Florida Palustrine Wetlands: Application of USEPA Levels 1, 2, and 3 assessment methods. *EcoHealth* 4(2):206–218.
26. Lane, C. R. and M. T. Brown. 2007. Diatoms as indicators of isolated Herbaceous Wetland condition in Florida, USA. *Ecological Indicators* 7:521–540.
27. Mack, J. J. 2006. Landscape as a predictor of Wetland Condition an evaluation of the Landscape Development Index (LDI) with a Large Reference Wetland Dataset from Ohio. *Environmental Monitoring and Assessment* 120:221–241.
28. Mack, J. J. 2007. Developing a Wetland IBI with statewide application after multiple testing iterations. *Ecological Indicators* 7:864–881.
29. Mack, J. J., N. H. Avdis, E. C. Braig, IV, and D. L. Johnson. 2008. Application of a vegetation-based index of biotic integrity for Lake Erie coastal marshes in Ohio. *Aquatic Ecosystem Health and Management* 11:91–104.
30. Nestlerode, J. A., V. D. Engle, P. Bourgeois, P. T. Heitmuller, J. M. Macauley, and Y. C. Allen. 2009. An integrated approach to assess broad-scale condition of coastal Wetlands—The Gulf of Mexico Coastal Wetlands Pilot survey. *Environmental Monitoring and Assessment* 150:21–29.
31. Ji, W. and J. Ma. 2007. Geospatial decision models for assessing the vulnerability of Wetlands to potential human impacts. In W. Ji (ed.), *Wetland and Water Resource Modeling and Assessment: A Watershed Perspective*, pp. 280, CRC/Taylor & Francis Group, Boca Raton, Florida, USA.
32. Magee, D. W. 1998. *A Rapid Procedure for Assessing Wetland Functional Capacity*. Normandeau Associates, Bedford, NH, USA. (Available from the Association of State Wetland Managers, Berne, NY).190pp.
33. Mammoliti, C. S. and S. Schulte. Stream and riparian assessment system. In *Conference Proceedings of Self-Sustaining Solutions for Streams, Wetlands, and Watersheds*, pp. 12–15 September 2004, St. Paul, Minnesota, USA.
34. Brinson, M. M. 1996. Assessing Wetland functions using HGM. *National Wetlands Newsletter* 18(1):10–16.

Index